BOTANICAL RESEARCH AND PRACTICES

ELEMENTARY BOTANY

Botanical Research and Practices

Additional books and e-books in this series can be found
on Nova's website under the Series tab.

BOTANICAL RESEARCH AND PRACTICES

ELEMENTARY BOTANY

GEORGE FRANCIS ATKINSON

snova
New York

Library of Congress Cataloging-in-Publication Data

ISBN: 978-1-53619-448-7

Published by Nova Science Publishers, Inc. † New York

CONTENTS

PREFACE*

The present book is the result of a revision and elaboration of the author's "Elementary Botany," New York, 1898. The general plan of the parts on physiology and general morphology remains unchanged. A number of the chapters in the physiological part are practically untouched, while others are thoroughly revised and considerable new matter is added, especially on the subjects of nutrition and digestion. The principal chapters on general morphology are unchanged or only slightly modified, the greatest change being in a revision of the subject of the morphology of fertilization in the gymnosperms and angiosperms in order to bring this subject abreast of the discoveries of the past few years. One of the greatest modifications has been in the addition of chapters on the classification of the algæ and fungi with studies of additional examples for the benefit of those schools where the time allowed for the first year's course makes desirable the examination of a broader range of representative plants. The classification is also carried out with greater definiteness, so that the regular sequence of classes, orders, and families is given at the close of each of the subkingdoms. Thus all the classes, all the orders (except a few in the algæ), and many of the families, are given for the algæ, fungi, mosses, liverworts, pteridophytes, gymnosperms, and angiosperms.

But by far the greatest improvement has been in the complete reorganization, rewriting, and elaboration of the part dealing with ecology, which has been made possible by studies of the past few years, so that the subject can be presented in a more logical and coherent form. As a result the subject-matter of the book falls naturally into three parts, which may be passed in review as follows:

Part I. Physiology. This deals with the life processes of plants, as absorption, transpiration, conduction, photosynthesis, nutrition, assimilation, digestion, respiration, growth, and irritability. Since protoplasm is fundamental to all the life work of the plant, this subject is dealt with first, and the student is led through the study of, and experimentation with, the simpler as well as some of the higher plants, to a general

* This is an edited, augmented, and reformatted version of "Elementary Botany" by George Francis Atkinson, originally published by New York: Henry Holt and Company, dated 1905.

understanding of protoplasm and the special way in which it enables the plant to carry on its work and to adjust itself to the conditions of its existence. This study also serves the purpose of familiarizing the pupil with some of the lower and unfamiliar plants.

Some teachers will prefer to begin the study with general morphology and classification, thus studying first the representatives of the great groups of plants, and others will prefer to dwell first on the ecological aspects of vegetation. This can be done in the use of this book by beginning with Part II or with Part III.

But the author believes that morphology can best be comprehended after a general study of life processes and functions of the different parts of plants, including in this study some of the lower forms of plant life where some of these processes can more readily be observed. The pupil is then prepared for a more intelligent consideration of general and comparative morphology and relationships. Even more important is a first study of physiology before taking up the subject of ecology. The great value to be derived from a study of plants in their relation to environment lies in the ability to interpret the different states, conditions, behavior, and associations of the plant, and for this physiology is indispensable. It is true that a considerable measure of success can be obtained by a good teacher in beginning with either subject, but the writer believes that measure of success would be greater if the subjects were taken up in the order presented here.

Part II. Morphology and life history of representative plants. This includes a rather careful study of representative examples among the algæ, fungi, liverworts, mosses, ferns and their allies, gymnosperms and angiosperms, with especial emphasis on the form of plant parts, and a comparison of them in the different groups, with a comparative study of development, reproduction, and fertilization, rounding out the work with a study of life histories and noting progression and retrogression of certain organs and phases in proceeding from the lower to the higher plants. Thus, in the algæ a first critical study is made of four examples which illustrate in a marked way progressive stages of the plant body, sexual organs, and reproduction. Additional examples are then studied for the purpose of acquiring a knowledge of variations from these types and to give a broader basis for the brief consideration of general relationships and classification.

A similar plan is followed in the other great groups. The processes of fertilization and reproduction can be most easily observed in the lower plants like the algæ and fungi, and this is an additional argument in favor of giving emphasis to these forms of plant life as well as the advantage of proceeding logically from simpler to more complex forms. Having also learned some of these plants in our study of physiology, we are following another recognized rule of pedagogy, i.e., proceeding from known objects to unknown structures and processes. Through the study of the organs of reproduction of the lower plants and by general comparative morphology we have come to an understanding of the morphology of the parts of the flower, and of the true sexual organs of the seed plants, and no student can hope to properly interpret the significance of the flower, or the sexual organs of the seed plants who neglects a careful study of the general morphology of the lower plants.

Part III. Plant members in relation to environment. This part deals with the organization of the plant body as a whole in its relation to environment, the organization of plant tissues with a discussion of the principal tissues and a descriptive synopsis of the same. This is followed by a complete study from a biological standpoint of the different members of the plant, their special function and their special relations to environment. The stem, root, leaf, flower, etc., are carefully examined and their ecological relations pointed out. This together with the study of physiology and representatives in the groups of plants forms a thorough basis for pure plant ecology, or the special study of vegetation in its relation to environment.

There is a study of the factors of environment or ecological factors, which in general are grouped under the physical, climatic, and biotic factors. This is followed by an analysis of vegetation forms and structures, plant formations and societies. Then in order are treated briefly forest societies, prairie societies, desert societies, arctic and alpine societies, aquatic societies, and the special societies of sandy, rocky, and marshy places.

Acknowledgments. The author wishes to express his gratefulness to all those who have given aid in the preparation of this work, or of the earlier editions of Elementary Botany; to his associates, Dr. E. J. Durand, Dr. K. M. Wiegand, and Professor W. W. Rowlee, of the botanical department, and to Professor B. M. Duggar of the University of Missouri, Professor J. C. Arthur of Purdue University, and Professor W. F. Ganong of Smith College, for reading one or more portions of the text; as well as to all those who have contributed illustrations.

Illustrations. The large majority of the illustrations are new (or are the same as those used in earlier editions of the author's Elementary Botany) and were made with special reference to the method of treatment followed in the text. Many of the photographs were made by the author. Others were contributed by Professor Rowlee of Cornell University; Mr. John Gifford of New Jersey; Professor B. M. Duggar, University of Missouri; Professor C. E. Bessey, University of Nebraska; Dr. M. B. Howe, New York Botanical Garden; Mr. Gifford Pinchot, Chief of the Bureau of Forestry; Mr. B. T. Galloway, Chief of the Bureau of Plant Industry; Professor Tuomey of Yale University; and Mr. E. H. Harriman, who through Dr. C. H. Merriam of the National Museum allowed the use of several of his copyrighted photographs from Alaska. To those who have contributed drawings the author is indebted as follows: to Professor Margaret C. Ferguson, Wellesley College; Professor Bertha Stoneman of Huguenot College, South Africa; Mr. H. Hasselbring of Chicago; Dr. K. Miyake, formerly of Cornell University and now of Doshisha College, Japan; and Professors Ikeno and Hirase of the Tokio Imperial University. The author is also indebted to Ginn & Co., Boston, for the privilege to use from his "First Studies of Plant Life" the following Figures: 28, 29, 46, 48, 49, 56, 62, 66, 67, 87, 102, 103, 422-426, 429, 430, 438-440, 443, 444, 448, 449, 452, 472-475. A few others are acknowledged in the text.

Cornell University, April, 1905.

Cycas revoluta.

Part I. Physiology

Chapter 1

PROTOPLASM[1]

In the study of plant life and growth, it will be found convenient first to inquire into the nature of the substance which we call the living material of plants. For plant growth, as well as some of the other processes of plant life, are at bottom dependent on this living matter. This living matter is called in general protoplasm.

In most cases protoplasm cannot be seen without the help of a microscope, and it will be necessary for us here to employ one if we wish to see protoplasm, and to satisfy ourselves by examination that the substance we are dealing with is protoplasm.

We shall find it convenient first to examine protoplasm in some of the simpler plants; plants which from their minute size and simple structure are so transparent that when examined with the microscope the interior can be seen.

For our first study let us take a plant known as spirogyra, though there are a number of others which would serve the purpose quite as well, and may quite as easily be obtained for study.

PROTOPLASM IN SPIROGYRA

The Plant Spirogyra

This plant is found in the water of pools, ditches, ponds, or in streams of slow-running water. It is green in color, and occurs in loose mats, usually floating near the surface. The name "pond-scum" is sometimes given to this plant, along with others which are more or less closely related. It is an alga, and belongs to a group of plants known as algæ. If we lift a portion of it from the water, we see that the mat is made up of a great tangle of green

[1] For apparatus, reagents, collection and preservation of material, etc., see Appendix.

silky threads. Each one of these threads is a plant, so that the number contained in one of these floating mats is very great.

Let us place a bit of this thread tangle on a glass slip, and examine with the microscope and we will see certain things about the plant which are peculiar to it, and which enable us to distinguish it from other minute green water plants. We shall also wish to learn what these peculiar parts of the plant are, in order to demonstrate the protoplasm in the plant.[2]

Chlorophyll Bands in Spirogyra

We first observe the presence of bands; green in color, the edges of which are usually very irregularly notched. These bands course along in a spiral manner near the surface of the thread.

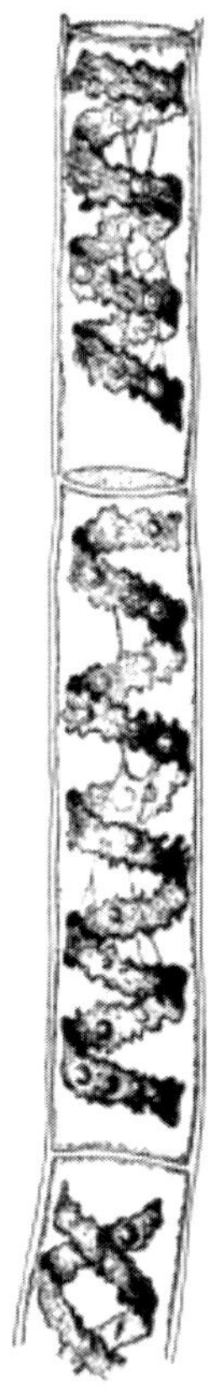

Figure 1. Thread of spirogyra, showing long cells, chlorophyll band, nucleus, strands of protoplasm, and the granular wall layer of protoplasm.

[2] If spirogyra is forming fruit some of the threads will be lying parallel in pairs, and connected with short tubes. In some of the cells there will be found rounded or oval bodies known as *zygospores*. These may be seen in Figure 86, and will be described in another part of the book.

There may be one or several of these spirals, according to the species which we happen to select for study. This green coloring matter of the band is chlorophyll, and this substance, which also occurs in the higher green plants, will be considered in a later chapter. At quite regular intervals in the chlorophyll band are small starch grains, grouped in a rounded mass enclosing a minute body, the pyrenoid, which is peculiar to many algæ. The spirogyra thread consists of cylindrical cells end to end.—Another thing which attracts our attention, as we examine a thread of spirogyra under the microscope, is that the thread is made up of cylindrical segments or compartments placed end to end. We can see a distinct separating line between the ends. Each one of these segments or compartments of the thread is a cell, and the boundary wall is in the form of a cylinder with closed ends.

Protoplasm

Having distinguished these parts of the plant we can look for the protoplasm. It occurs within the cells. It is colorless (i.e., hyaline) and consequently requires close observation. Near the center of the cell can be seen a rather dense granular body of an elliptical or irregular form, with its long diameter transverse to the axis of the cell in some species; or triangular, or quadrate in others. This is the nucleus. Around the nucleus is a granular layer from which delicate threads of a shiny granular substance radiate in a starlike manner, and terminate in the chlorophyll band at one of the pyrenoids. A granular layer of the same substance lines the inside of the cell wall, and can be seen through the microscope if it is properly focussed. This granular substance in the cell is protoplasm.

Cell-Sap in Spirogyra

The greater part of the interior space of the cell, that between the radiating strands of protoplasm, is occupied by a watery fluid, the "cell-sap."

Reaction of protoplasm to certain reagents.—We can employ certain tests to demonstrate that this granular substance which we have seen is protoplasm, for it has been found, by repeated experiments with a great many kinds of plants, that protoplasm gives a definite reaction in response to treatment with certain substances called reagents. Let us mount a few threads of the spirogyra in a drop of a solution of iodine, and observe the results with the aid of the microscope. The iodine gives a yellowish-brown color to the protoplasm, and it can be more distinctly seen. The nucleus is also much more prominent since it colors deeply, and we can perceive within the nucleus one small rounded body, sometimes more, the nucleolus. The iodine here kills and stains the protoplasm. The protoplasm, however, in a living condition will resist for a time some other reagents, as we shall see if we attempt to stain it with a one per cent aqueous solution of a dye known as eosin. Let us mount a few living threads in such a solution of eosin, and after a time wash off the stain. The protoplasm remains uncolored. Now let us place these threads for a short time, two or three minutes, in strong alcohol, which kills the protoplasm. Then mount them in the eosin solution. The protoplasm now takes the eosin stain. After the protoplasm has been killed we note that the nucleus is no longer elliptical or angular in outline, but is rounded. The strands of protoplasm are no longer in tension as they were when alive.

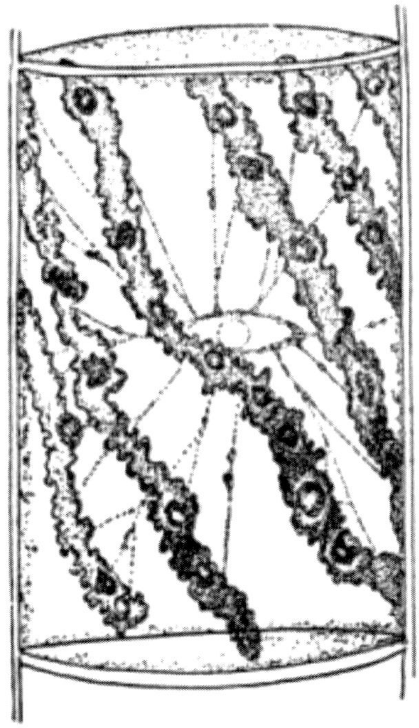

Figure 2. Cell of spirogyra before treatment with iodine.

Let us now take some fresh living threads and mount them in water. Place a small drop of dilute glycerine on the slip at one side of the cover glass, and with a bit of filter paper at the other side draw out the water. The glycerine will flow under the cover glass and come in contact with the spirogyra threads. Glycerine absorbs water promptly. Being in contact with the threads it draws water out of the cell cavity, thus causing the layer of protoplasm which lines the inside of the cell wall to collapse, and separate from the wall, drawing the chlorophyll band inward toward the center also. The wall layer of protoplasm can now be more distinctly seen and its granular character observed.

We have thus employed three tests to demonstrate that this substance with which we are dealing shows the reactions which we know by experience to be given by protoplasm. We therefore conclude that this colorless and partly granular, slimy substance in the spirogyra cell is protoplasm, and that when we have performed these experiments, and noted carefully the results, we have *seen* protoplasm.

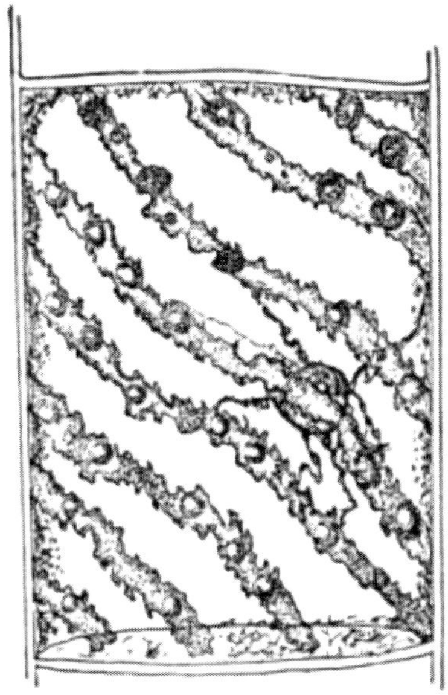

Figure 3. Cell of spirogyra after treatment with alcohol and iodine.

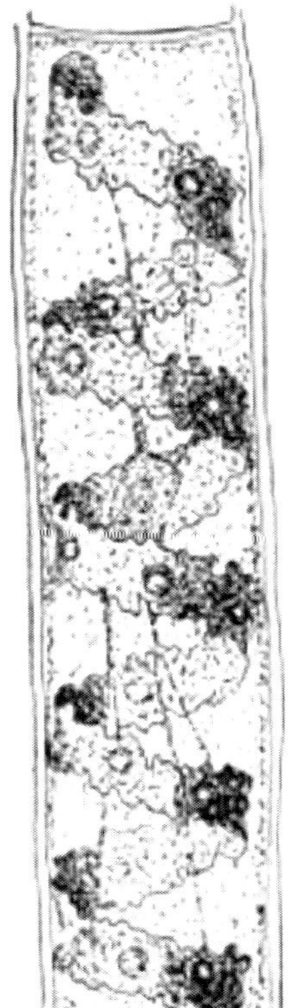

Figure 4. Cell of spirogyra before treatment with glycerine.

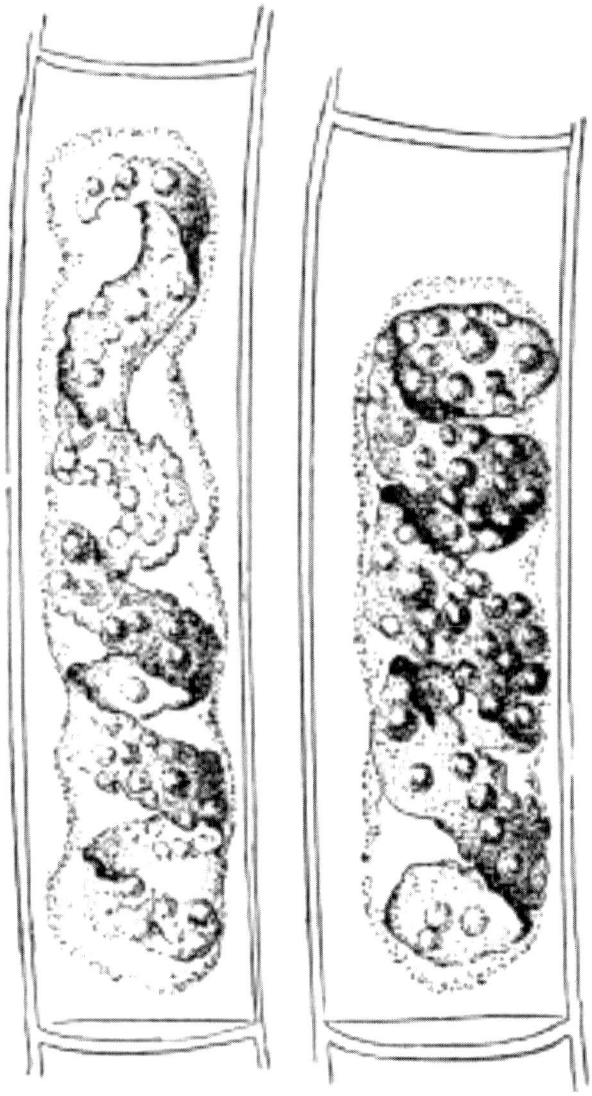

Figure 5. Cells of spirogyra after treatment with glycerine.

Earlier Use of the Term Protoplasm

Early students of the living matter in the cell considered it to be alike in substance, but differing in density; so the term protoplasm was applied to all of this living matter. The nucleus was looked upon as simply a denser portion of the protoplasm, and the nucleolus as a still denser portion. Now it is believed that the nucleus is a distinct substance, and a permanent organ of the cell. The remaining portion of the protoplasm is now usually spoken of as the *cytoplasm*.

In spirogyra then the cytoplasm in each cell consists of a layer which lines the inside of the cell wall, a nuclear layer, which surrounds the nucleus, and radiating strands which connect the nucleus and wall layers, thus suspending the nucleus near the center of the cell. But it seems best in this elementary study to use the term protoplasm in its general sense.

PROTOPLASM IN MUCOR

Let us now examine in a similar way another of the simple plants with the special object in view of demonstrating the protoplasm. For this purpose we may take one of the plants belonging to the group of *fungi*. These plants possess no chlorophyll. One of several species of *mucor*, a common mould, is readily obtainable, and very suitable for this study.[3]

Mycelium of Mucor

A few days after sowing in some gelatinous culture medium we find slender, hyaline threads, which are very much branched, and, radiating from a central point, form circular colonies, if the plant has not been too thickly sown, as shown in Figure 6. These threads of the fungus form the *mycelium*. From these characters of the plant, which we can readily see without the aid of a microscope, we note how different it is from spirogyra.

To examine for protoplasm let us lift carefully a thin block of gelatine containing the mucor threads, and mount it in water on a glass slip. Under the microscope we see only a small portion of the branched threads. In addition to the absence of chlorophyll, which we have already noted, we see that the mycelium is not divided at short intervals into cells, but appears like a delicate tube with branches, which become successively smaller toward the ends.

Appearance of the Protoplasm

Within the tube-like thread now note the protoplasm. It has the same general appearance as that which we noted in spirogyra. It is slimy, or semi-fluid, partly hyaline, and partly granular, the granules consisting of minute particles (the *microsomes*). While in mucor the protoplasm has the same general appearance as in spirogyra, its arrangement is very different. In the first place it is plainly continuous throughout the tube. We do not see the prominent radiations of strands around a large nucleus, but still the protoplasm does not fill the interior of the threads. Here and there are rounded clear spaces termed *vacuoles*,

[3] The most suitable preparations of mucor for study are made by growing the plant in a nutrient substance which largely consists of gelatine, or, better, agar-agar, a gelatinous preparation of certain seaweeds. This, after the plant is sown in it, should be poured into sterilised shallow glass plates, called Petrie dishes.

which are filled with the watery fluid, cell-sap. The nuclei in mucor are very minute, and cannot be seen except after careful treatment with special reagents.

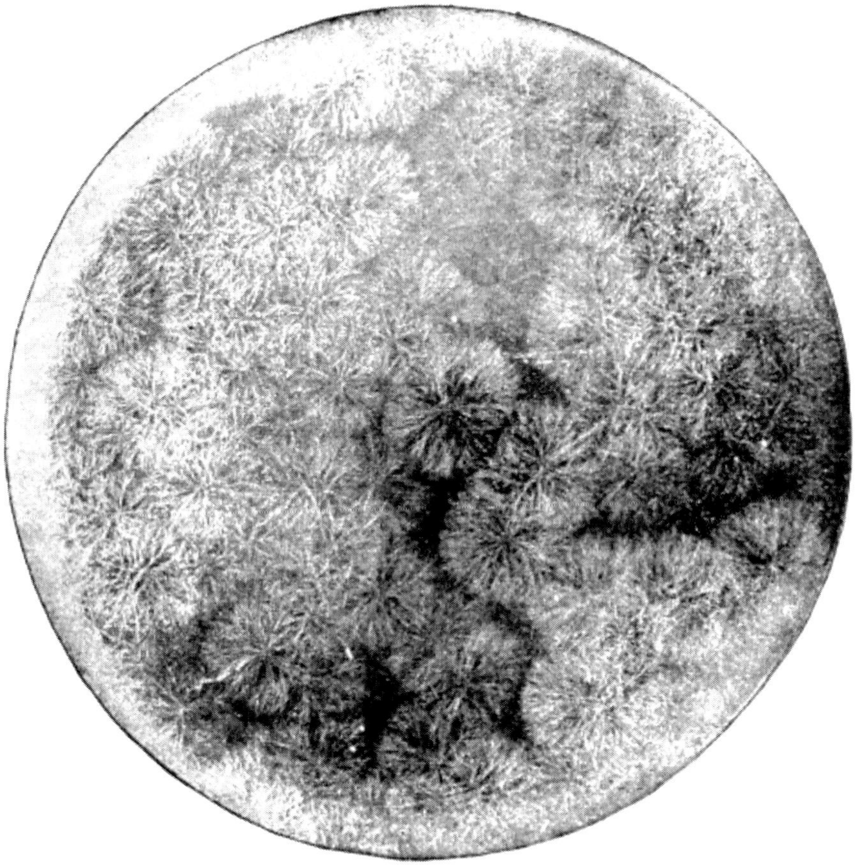

Figure 6. Colonies of mucor.

Movement of the Protoplasm in Mucor

While examining the protoplasm in mucor we are likely to note streaming movements. Often a current is seen flowing slowly down one side of the thread, and another flowing back on the other side, or it may all stream along in the same direction.

Test for Protoplasm

Now let us treat the threads with a solution of iodine. The yellowish-brown color appears which is characteristic of protoplasm when subject to this reagent. If we attempt to stain the living protoplasm with a one per cent aqueous solution of eosin it resists it for a time, but if we first kill the protoplasm with strong alcohol, it reacts quickly to the application of the eosin. If we treat the living threads with glycerine the protoplasm is contracted away from the wall, as we found to be the case with spirogyra. While the color, form and structure of the plant mucor is different from spirogyra, and the arrangement of the protoplasm within the plant is also quite different, the reactions when treated by certain reagents are the same. We are justified then in concluding that the two plants possess in common a substance which we call protoplasm.

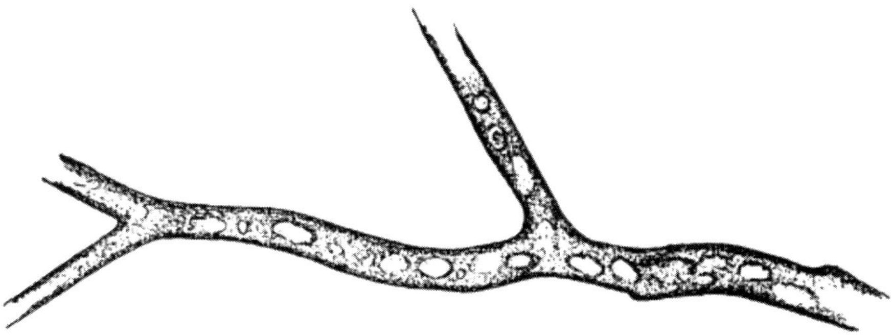

Figure 7. Thread of mucor, showing protoplasm and vacuoles.

Protoplasm in Nitella

One of the most interesting plants for the study of one remarkable peculiarity of protoplasm is *Nitella*. This plant belongs to a small group known as stoneworts. They possess chlorophyll, and, while they are still quite simple as compared with the higher plants, they are much higher in the scale than spirogyra or mucor.

Form of Nitella

A common species of nitella is *Nitella flexilis*. It grows in quiet pools of water. The plant consists of a main axis, in the form of a cylinder. At quite regular intervals are whorls of several smaller thread-like outgrowths, which, because of their position, are termed "leaves," though they are not true leaves. These are branched in a characteristic fashion at the tip. The main axis also branches, these branches arising in the axil of a whorl, usually singly. The portions of the axis where the whorls arise are the *nodes*. Each node is made up of a number of small cells definitely arranged. The portion of the axis between two adjacent whorls is an internode. These internodes are peculiar. They consist of but a single "cell," and are cylindrical, with closed ends. They are sometimes 5-10 cm. long.

Internode of Nitella

For the study of an internode of nitella, a small one, near the end, or the ends of one of the "leaves" is best suited, since it is more transparent. A small portion of the plant should be placed on the glass slip in water with the cover glass over a tuft of the branches near the growing end. Examined with the microscope the green chlorophyll bodies, which form oval or oblong discs, are seen to be very numerous. They lie quite closely side by side and form in perfect rows along the inner surface of the wall. One peculiar feature of the arrangement of the chlorophyll bodies is that there are two lines, extending from one end of the internode to the other on opposite sides, where the chlorophyll bodies are wanting. These are known as neutral lines. They run parallel with the axis of the internode, or in a more or less spiral manner as shown in Figure 9.

Figure 8. Portion of plant nitella.

Cyclosis in Nitella

The chlorophyll bodies are stationary on the inner surface of the wall, but if the microscope be properly focussed just beneath this layer we notice a rotary motion of particles in the protoplasm. There are small granules and quite large masses of granular matter which glide slowly along in one direction on a given side of the neutral line. If now we examine the protoplasm on the other side of the neutral line, we see that the movement is in the opposite direction. If we examine this movement at the end of an internode the particles are seen to glide around the end from one side of the neutral line to the other. So that when conditions are favorable, such as temperature, healthy state of the plant, etc., this gliding of the particles or apparent streaming of the protoplasm down one side of the "cell," and back upon the other, continues in an uninterrupted rotation, or cyclosis. There are many nuclei in an internode of nitella, and they move also.

Test for Protoplasm

If we treat the plant with a solution of iodine we get the same reaction as in the case of spirogyra and mucor. The protoplasm becomes yellowish-brown.

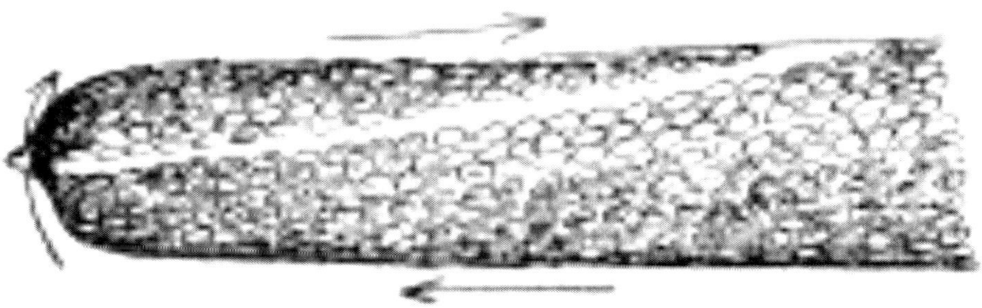

Figure 9. Cyclosis in nitella.

Protoplasm in One of the Higher Plants

We now wish to examine, and test for, protoplasm in one of the higher plants. Young or growing parts of any one of various plants—the petioles of young leaves, or young stems of growing plants—are suitable for study. Tissue from the pith of corn (Zea mays) in young shoots just back of the growing point or quite near the joints of older but growing corn stalks furnishes excellent material.

If we should place part of the stem of this plant under the microscope we should find it too opaque for observation of the interior of the cells. This is one striking difference which we note as we pass from the low and simple plants to the higher and more complex ones; not only in general is there an increase of size, but also in general an increase in thickness of the parts. The cells, instead of lying end to end or side by side, are massed together so that the parts are quite opaque. In order to study the interior of the plant we have selected it must be cut into such thin layers that the light will pass readily through them.

For this purpose we section the tissue selected by making with a razor, or other very sharp knife, very thin slices of it. These are mounted in water in the usual way for microscopic study. In this section we notice that the cells are polygonal in form. This is brought about by mutual pressure of all the cells. The granular protoplasm is seen to form a layer just inside the wall, which is connected with the nuclear layer by radiating strands of the same substance. The nucleus does not always lie at the middle of the cell, but often is near one side. If we now apply an alcohol solution of iodine the characteristic yellowish-brown color appears. So we conclude here also that this substance is identical with the living matter in the other very different plants which we have studied.

Movement of Protoplasm in the Higher Plants

Certain parts of the higher plants are suitable objects for the study of the so-called streaming movement of protoplasm, especially the delicate hairs, or thread-like outgrowths, such as the silk of corn, or the delicate staminal hairs of some plants, like those of the common spiderwort, tradescantia, or of thc tradescantias grown for ornament in greenhouses and plant conservatories.

Sometimes even in the living cells of the corn plant which we have just studied, slow streaming or gliding movements of the granules are seen along the strands of protoplasm where they radiate from the nucleus. See note at close of this chapter.

Movement of Protoplasm in Cells of the Staminal Hair of "Spiderwort"

A cell of one of these hairs from a stamen of a tradescantia grown in glass houses is shown in Figure 10. The nucleus is quite prominent, and its location in the cell varies considerably in different cells and at different times. There is a layer of protoplasm all around the nucleus, and from this the strands of protoplasm extend outward to the wall layer. The large spaces between the strands are, as we have found in other cases, filled with the cell-sap.

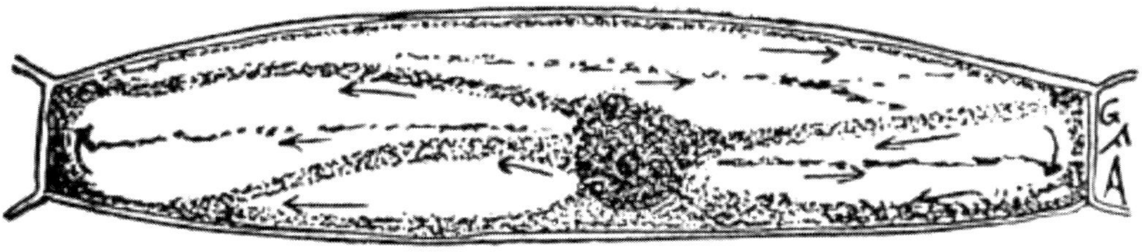

Figure 10. Cell from stamen hair of tradescantia showing movement of the protoplasm.

An entire stamen, or a portion of the stamen, having several hairs attached, should be carefully mounted in water. Care should be taken that the room be not cold, and if the weather is cold the water in which the preparation is mounted should be warm. With these precautions there should be little difficulty in observing the streaming movement.

The movement is detected by observing the gliding of the granules. These move down one of the strands from the nucleus along the wall layer, and in towards the nucleus in another strand. After a little the direction of the movement in any one portion may be reversed.

Cold Retards the Movement

While the protoplasm is moving, if we rest the glass slip on a block of ice, the movement will become slower, or will cease altogether. Then if we warm the slip gently, the movement becomes normal again. We may now apply here the usual tests for protoplasm. The result is the same as in the former cases.

Protoplasm Occurs in the Living Parts of All Plants

In these plants representing such widely different groups, we find a substance which is essentially alike in all. Though its arrangement in the cell or plant body may differ in the different plants or in different parts of the same plant, its general appearance is the same. Though in the different plants it presents, while alive, varying phenomena, as regards mobility, yet when killed and subjected to well known reagents the reaction is in general identical. Knowing by the experience of various investigators that protoplasm exhibits these reactions under given conditions, we have demonstrated to our satisfaction that we have seen protoplasm in the simple alga, spirogyra, in the common mould, mucor, in the more complex stonewort, nitella, and in the cells of tissues of the highest plants.

By this simple process of induction of these facts concerning this substance in these different plants, we have learned an important method in science study. Though these facts and deductions are well known, the repetition of the methods by which they are obtained on the part of each student helps to form habits of scientific carefulness and patience, and trains the mind to logical processes in the search for knowledge.

While we have by no means exhausted the study of protoplasm, we can, from this study, draw certain conclusions as to its occurrence and appearance in plants. Protoplasm is found in the living and growing parts of all plants. It is a semi-fluid, or slimy, granular, substance; in some plants, or parts of plants, the protoplasm exhibits a streaming or gliding movement of the granules. It is irritable. In the living condition it resists more or less for some time the absorption of certain coloring substances. The water may be withdrawn by glycerine. The protoplasm may be killed by alcohol. When treated with iodine it becomes a yellowish-brown color.

Note. In some plants, like elodea for example, it has been found that the streaming of the protoplasm is often induced by some injury or stimulus, while in the normal condition the protoplasm does not move.

Chapter 2

ABSORPTION, DIFFUSION, OSMOSE

We may next endeavor to learn how plants absorb water or nutrient substances in solution. There are several very instructive experiments, which can be easily performed, and here again some of the lower plants will be found useful.

Osmose in spirogyra.—Let us mount a few threads of this plant in water for microscopic examination, and then draw under the cover glass a five per cent solution of ordinary table salt (NaCl) with the aid of filter paper. We shall soon see that the result is similar to that which was obtained when glycerine was used to extract the water from the cell-sap, and to contract the protoplasmic membrane from the cell wall. But the process goes on evenly and the plant is not injured. The protoplasmic layer contracts slowly from the cell wall, and the movement of the membrane can be watched by looking through the microscope. The membrane contracts in such a way that all the contents of the cell are finally collected into a rounded or oval mass which occupies the center of the cell.

If we now add fresh water and draw off the salt solution, we can see the protoplasmic membrane expand again, or move out in all directions, and occupy its former position against the inner surface of the cell wall. This would indicate that there is some pressure from within while this process of absorption is going on, which causes the membrane to move out against the cell wall.

The salt solution draws water from the cell-sap. There is thus a tendency to form a vacuum in the cell, and the pressure on the outside of the protoplasmic membrane causes it to move toward the center of the cell. When the salt solution is removed and the thread of spirogyra is again bathed with water, the movement of the water is inward in the cell. This would suggest that there is some substance dissolved in the cell-sap which does not readily filter out through the membrane, but draws on the water outside. It is this which produces the pressure from within and crowds the membrane out against the cell wall.

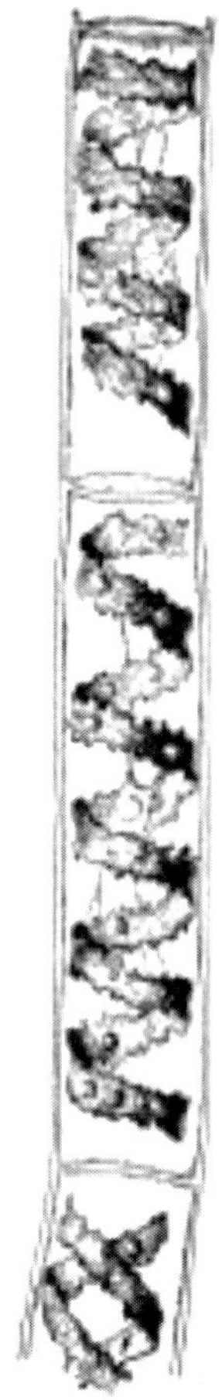

Figure 11. Spirogyra before placing in salt solution.

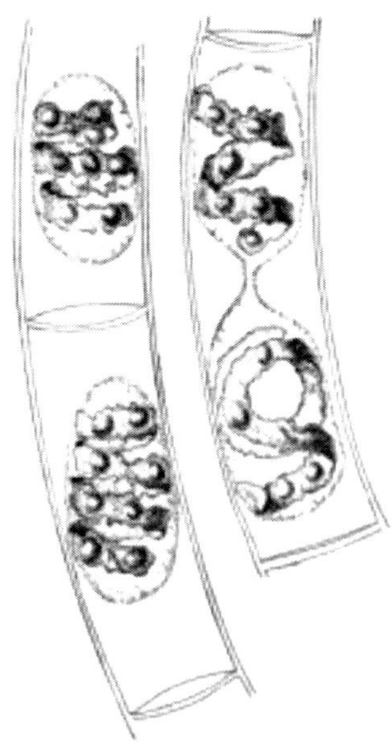

Figure 12. Spirogyra in 5% salt solution.

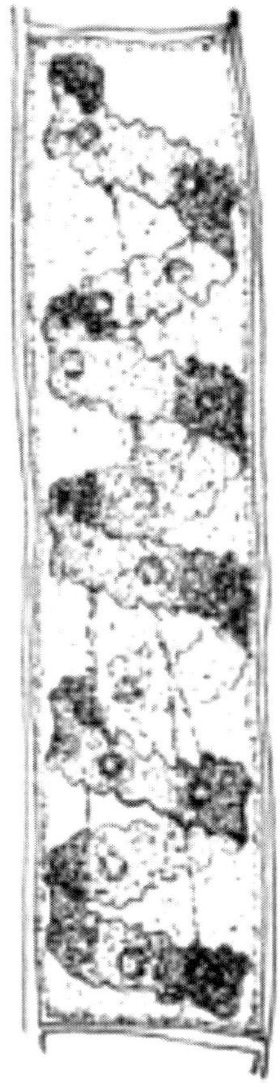

Figure 13. Spirogyra from salt solution into water.

TURGESCENCE

Were it not for the resistance which the cell wall offers to the pressure from within, the delicate protoplasmic membrane would stretch to such an extent that it would be ruptured, and the protoplasm therefore would be killed. If we examine the cells at the ends of the threads of spirogyra we shall see in most cases that the cell wall at the free end is arched outward. This is brought about by the pressure from within upon the protoplasmic membrane which itself presses against the cell wall, and causes it to arch outward. This is beautifully shown in the case of threads which are recently broken. The cell wall is therefore *elastic*; it yields to a certain extent to the pressure from within, but a point is soon reached beyond which it will not stretch, and an equilibrium then exists between the pressure from within on the protoplasmic membrane, and the pressure from without by the elastic cell wall. This state of equilibrium in a cell is *turgescence*, or such a cell is said to be *turgescent*, or *turgid*.

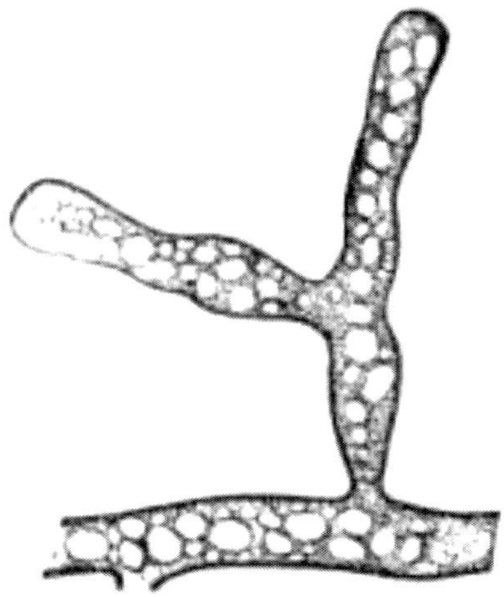

Figure 14. Before treatment with salt solution.

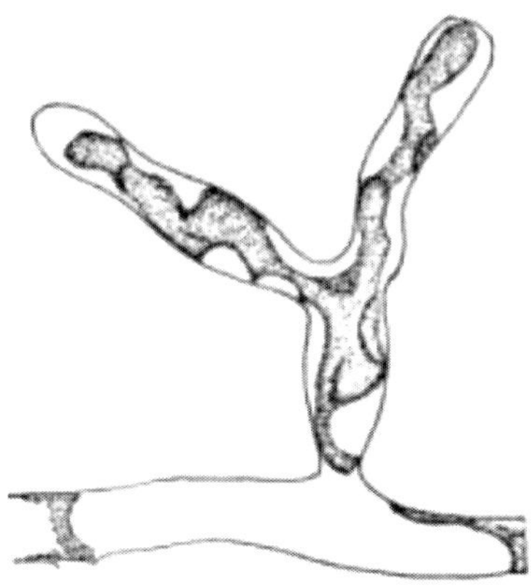

Figure 15. After treatment with salt solution.

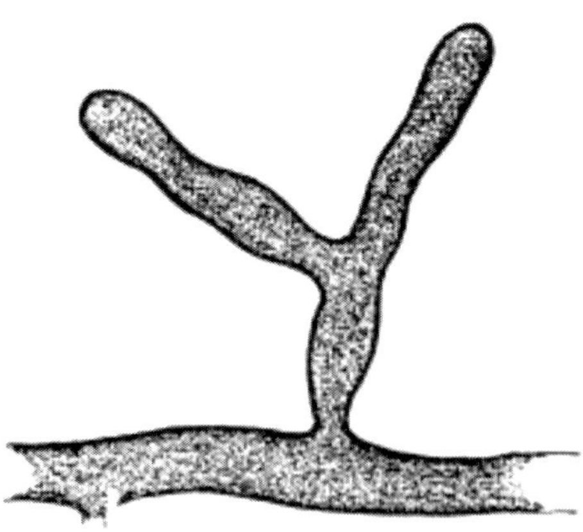

Figure 16. From salt solution placed in water.

Figures 14-16.—Osmosis in threads of mucor.

Experiment With Beet in Salt and Sugar Solutions

We may now test the effect of a five per cent salt solution on a portion of the tissues of a beet or carrot. Let us cut several slices of equal size and about 5mm in thickness. Immerse a few slices in water, a few in a five per cent salt solution and a few in a strong sugar solution. It should be first noted that all the slices are quite rigid when an attempt is made to bend them between the fingers. In the course of one or two hours or less, if we examine the slices we shall find that those in water remain, as at first, quite rigid, while those in the salt and sugar solutions are more or less flaccid or limp, and readily bend by pressure between the fingers, the specimens in the salt solution, perhaps, being more flaccid than those in the sugar solution. The salt solution, we judge after our experiment with spirogyra, withdraws some of the water from the cell-sap, the cells thus losing their turgidity and the tissues becoming limp or flaccid from the loss of water.

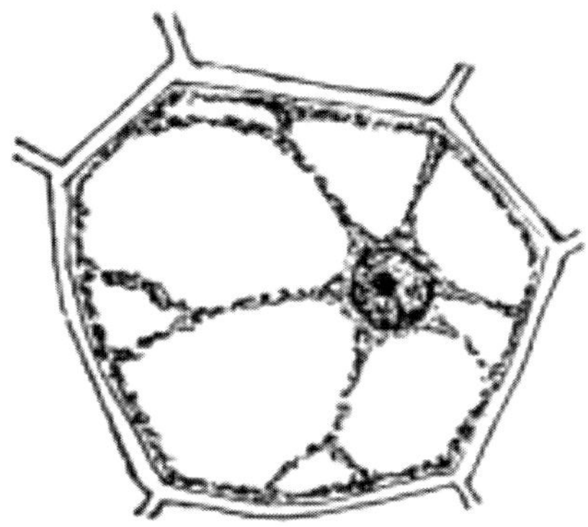

Figure 17. Before treatment with salt solution.

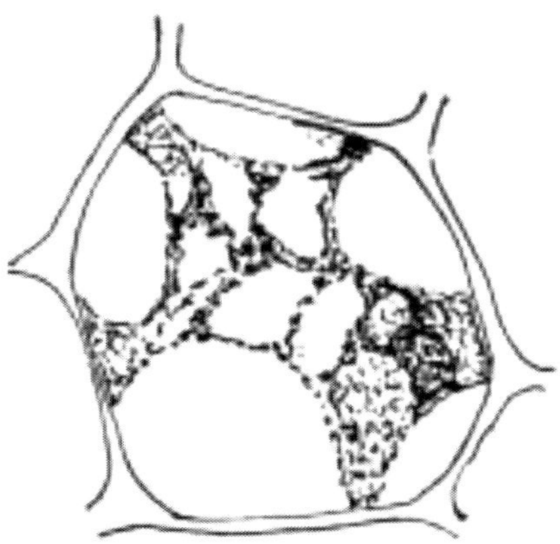

Figure 18. After treatment with salt solution.

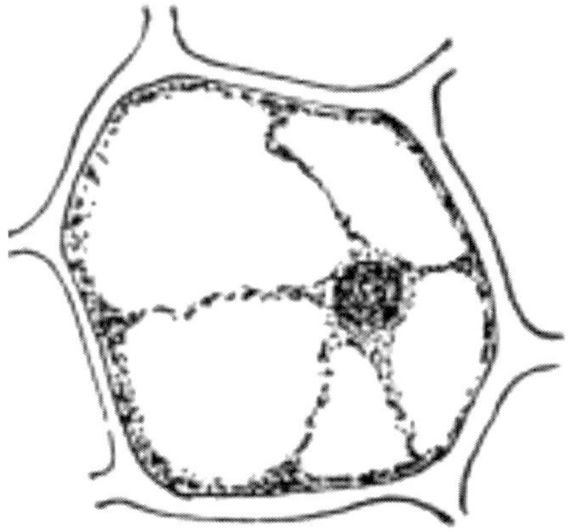

Figure 19. From salt solution intowater again.

Figures 17-19.—Osmosis in cells of Indian corn.

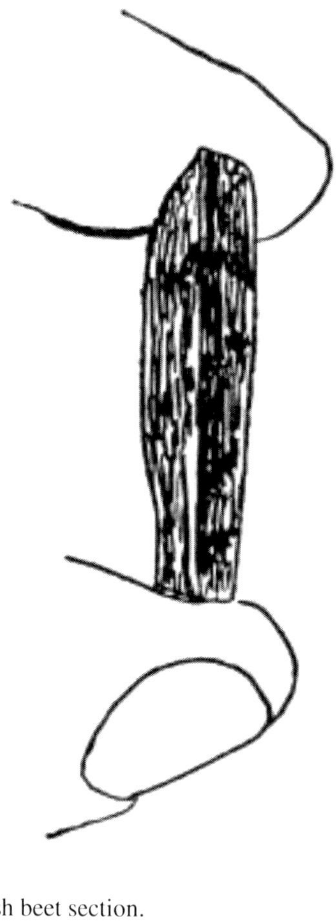

Figure 20. Rigid condition of fresh beet section.

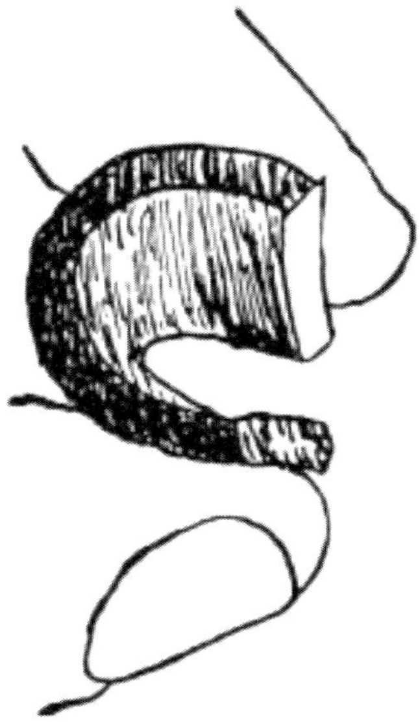

Figure 21. Limp condition after lying in salt solution.

Figure 22. Rigid again after lying again in water.

Figures 20-22.—Turgor and osmosis in slices of beet.

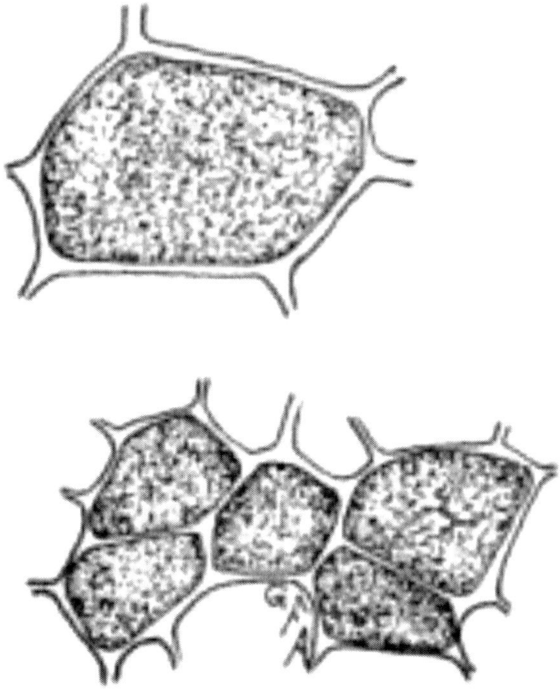

Figure 23. Before treatment with salt solution.

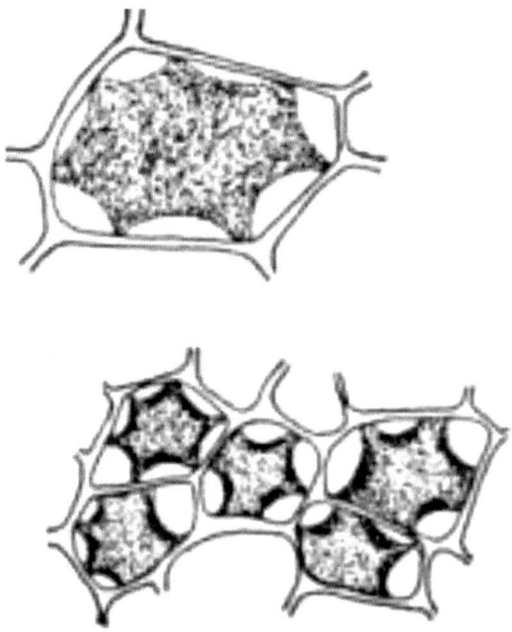

Figure 24. After treatment with salt solution.

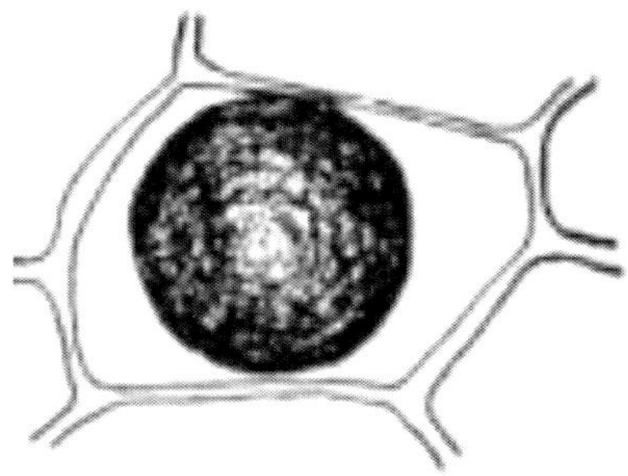

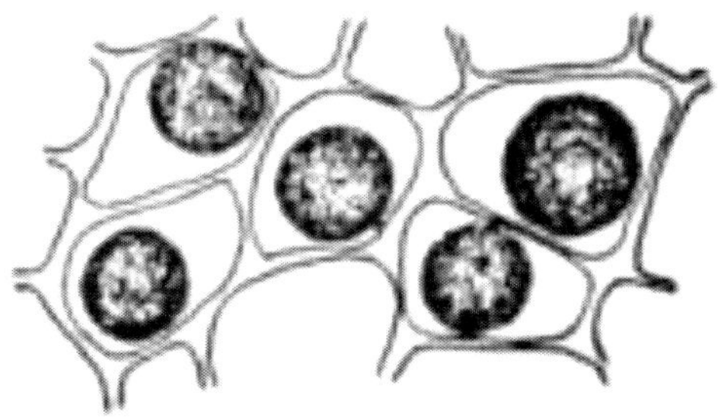

Figure 25. Later stage of the same.

Let us now remove some of the slices of the beet from the sugar and salt solutions, wash them with water and then immerse them in fresh water. In the course of thirty minutes to one hour, if we examine them again, we find that they have regained, partly or completely, their rigidity. Here again we infer from the former experiment with spirogyra that the substances in the cell-sap now draw water inward; that is, the diffusion current is inward through the cell walls and the protoplasmic membrane, and the tissue becomes turgid again.

Figures 23-25.—Cells from beet treated with salt solution to show osmosis and movement of the protoplasmic membrane.

Osmose in the Cells of the Beet

We should now make a section of the fresh tissue of a red colored beet for examination with the microscope, and treat this section with the salt solution. Here we can see that the effect of the salt solution is to draw water out of the cell, so that the protoplasmic membrane can be seen to move inward from the cell wall just as was observed in the case of spirogyra.[4] Now treating the section with water and removing the salt solution, the diffusion current is

[4] We should note that the coloring matter of the beet resides in the cell-sap. It is in these colored cells that we can best see the movement take place, since the red color serves to differentiate well the moving mass from the cell wall. The protoplasmic membrane at several points usually clings tenaciously so that at several places the membrane is arched strongly away from the cell wall as shown in Figure 24. While water is removed from the cell-sap, we note that the coloring matter does not escape through the protoplasmic membrane.

in the opposite direction, that is inward through the protoplasmic membrane, so that the latter is pressed outward until it comes in contact with the cell wall again, which by its elasticity soon resists the pressure and the cells again become turgid.

The Coloring Matter in the Cell-Sap Does Not Readily Escape from the Living Protoplasm of the Beet

The red coloring matter, as seen in the section under the microscope, does not escape from the cell-sap through the protoplasmic membrane. When the slices are placed in water, the water is not colored thereby. The same is true when the slices are placed in the salt or sugar solutions. Although water is withdrawn from the cell-sap, this coloring substance does not escape, or if it does it escapes slowly and after a considerable time.

The Coloring Matter Escapes from Dead Protoplasm

If, however, we heat the water containing a slice of beet up to a point which is sufficient to kill the protoplasm, the red coloring matter in the cell-sap filters out through the protoplasmic membrane and colors the water. If we heat a preparation made for study under the microscope up to the thermal death point we can see here that the red coloring matter escapes through the membrane into the water outside. This teaches that certain substances cannot readily filter through the living membrane of protoplasm, but that they can filter through when the protoplasm is dead. A very important condition, then, for the successful operation of some of the physical processes connected with absorption in plants is that the protoplasm should be in a living condition.

Osmose Experiments with Leaves

We may next take the leaves of certain plants like the geranium, coleus or other plant, and place them in shallow vessels containing water, salt, and sugar solutions respectively. The leaves should be immersed, but the petioles should project out of the water or solutions. Seedlings of corn or beans, especially the latter, may also be placed in these solutions, so that the leafy ends are immersed. After one or two hours an examination shows that the specimens in the water are still turgid. But if we lift a leaf or a bean plant from the salt or sugar solution, we find that it is flaccid and limp. The blade, or lamina, of the leaf droops as if wilted, though it is still wet. The bean seedling also is flaccid, the succulent stem bending nearly double as the lower part of the stem is held upright. This loss of turgidity is brought about by the loss of water from the tissues, and judging from the experiments on spirogyra and the beet, we conclude that the loss of turgidity is caused by the withdrawal of some of the water from the cell-sap by the strong salt solution.

Now if we wash carefully these leaves and seedlings, which have been in the salt and sugar solutions, with water, and then immerse them in fresh water for a few hours, they will regain their turgidity. Here again we are led to infer that the diffusion current is now inward through the protoplasmic membranes of all the living cells of the leaf, and that the resulting turgidity of the individual cells causes the turgidity of the leaf or stem.

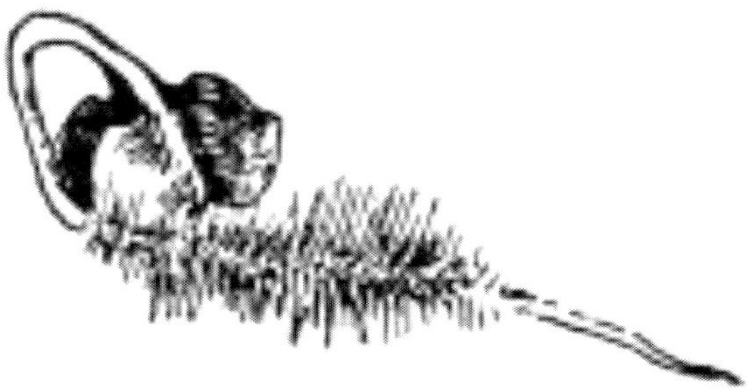

Figure 26. Seedling of radish, showing root hairs.

ABSORPTION BY ROOT HAIRS

If we examine seedlings, which have been grown in a germinator or in the folds of paper or cloths so that the roots will be free from particles of soil, we see near the growing point of the roots that the surface is covered with numerous slender, delicate, thread-like bodies, the root hairs. Let us place a portion of a small root containing some of these root hairs in water on a glass slip, and prepare it for examination with the microscope. We see that each thread, or root hair, is a continuous tube, or in other words it is a single cell which has become very much elongated. The protoplasmic membrane lines the wall, and strands of protoplasm extend across at irregular intervals, the interspaces being occupied by the cell-sap.

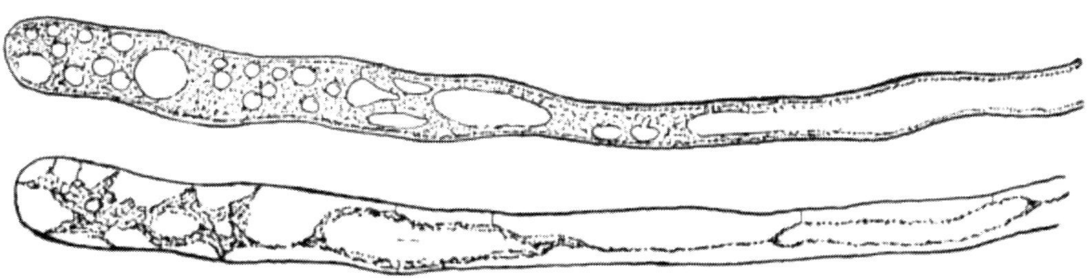

Figure 27. Root hair of corn before and after treatment with 5% salt solution.

We should now draw under the cover glass some of the five per cent salt solution. The protoplasmic membrane moves away from the cell wall at certain points, showing that *plasmolysis* is taking place, that is, the diffusion current is outward so that the cell-sap loses some of its water, and the pressure from the outside moves the membrane inward. We should not allow the salt solution to work on the root hairs long. It should be very soon removed by drawing in fresh water before the protoplasmic membrane has been broken at intervals, as is apt to be the case by the strong diffusion current and the consequent strong pressure from without. The membrane of protoplasm now moves outward as the diffusion current is inward, and soon regains its former position next the inner side of the cell wall. The root hairs then, like other parts of the plant which we have investigated, have the power of taking up water under pressure.

Cell-Sap a Solution of Certain Substances

From these experiments we are led to believe that certain substances reside in the cell-sap of plants, which behave very much like the salt solution when separated from water by the protoplasmic membrane. Let us attempt to interpret these phenomena by recourse to diffusion experiments, where an animal membrane separates two liquids of different concentration.

An Artificial Cell To Illustrate Turgor

Fill a small wide-mouthed vial with a *very strong* sugar solution. Over the mouth tie firmly a piece of *bladder* membrane. Be certain that as the membrane is tied over the open end of the vial, the sugar solution fills it in order to keep out air bubbles.

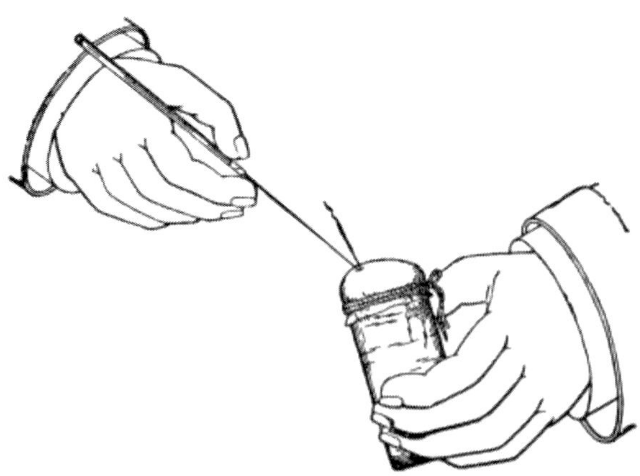

Figure 28. Puncturing a make-believe cell after it has been lying in water.

Figure 29. Same as Figure 28 after needle is removed.

Sink the vial in a vessel of fresh water and leave it there for twenty-four hours. Remove the vial and note that the membrane is arched outward. Thrust a sharp needle through the membrane when it is arched outward, and quickly pull it out. The liquid spurts out because of the inside pressure.

Diffusion Through an Animal Membrane

For this experiment we may use a thistle tube, across the larger end of which should be stretched and tied tightly a piece of a bladder membrane. A strong sugar solution (three parts sugar to one part water) is now placed in the tube so that the bulb is filled and the liquid extends part way in the neck of the tube. This is immersed in water within a wide-mouth bottle, the neck of the tube being supported in a perforated cork in such a way that the sugar solution in the tube is on a level with the water in the bottle or jar. In a short while the liquid begins to rise in the thistle tube, in the course of several hours having risen several centimeters. The diffusion current is thus stronger through the membrane in the direction of the sugar solution, so that this gains more water than it loses.

We have here two liquids separated by an animal membrane, water on the one hand which diffuses readily through the membrane, while on the other is a solution of sugar which diffuses through the animal membrane with difficulty. The water, therefore, not containing any solvent, according to a general law which has been found to obtain in such cases, diffuses more readily through the membrane into the sugar solution, which thus increases in volume, and also becomes more dilute. The bladder membrane is what is sometimes called a diffusion membrane, since the diffusion currents travel through it.

In this experiment then the bulk of the sugar solution is increased, and the liquid rises in the tube by this pressure above the level of the water in the jar outside of the thistle tube. The diffusion of liquids through a membrane is *osmosis*.

Importance of These Physical Processes in Plants

Now if we recur to our experiment with spirogyra we find that exactly the same processes take place. The protoplasmic membrane is the diffusion membrane, through which the diffusion takes place. The salt solution which is first used to bathe the threads of the plant is a stronger solution than that of the cell-sap within the cell. Water therefore is drawn out of the cell-sap, but the substances in solution in the cell-sap do not readily move out. As the bulk of the cell-sap diminishes the pressure from the outside pushes the protoplasmic membrane away from the wall. Now when we remove the salt solution and bathe the thread with water again, the cell-sap, being a solution of certain substances, diffuses with more difficulty than the water, and the diffusion current is inward, while the protoplasmic membrane moves out against the cell wall, and turgidity again results. Also in the experiments with salt and sugar solutions on the leaves of geranium, on the leaves and stems of the seedlings, on the tissues and cells of the beet and carrot, and on the root hairs of the seedlings, the same processes take place.

These experiments not only teach us that in the protoplasmic membrane, the cell wall, and the cell-sap of plants do we have structures which are capable of performing these physical processes, but they also show that these processes are of the utmost importance to the plant; not only in giving the plant the power to take up solutions of nutriment from the soil, but they serve also other purposes, as we shall see later.

Chapter 3

How Plants Obtain Water

In connection with the study of the means of absorption from the soil or water by plants, it will be found convenient to observe carefully the various forms of the plant. Without going into detail here, the suggestion is made that simple thread forms like spirogyra, œdogonium, and vaucheria; expanded masses of cells as are found in the thalloid liverworts, the duckweed, etc., be compared with those liverworts, and with the mosses, where leaf-like expansions of a central axis have been differentiated. We should then note how this differentiation, from the physiological standpoint, has been carried farther in the higher land plants.

Absorption by Algæ and Fungi

In the simpler forms of plant life, as in spirogyra and many of the algæ and fungi, the plant body is not differentiated into parts.[5] In many other cases the only differentiation is between the growing part and the fruiting part. In the algæ and fungi there is no differentiation into stem and leaf, though there is an approach to it in some of the higher forms. Where this simple plant body is flattened, as in the sea-wrack, or ulva, it is a frond. The Latin word for frond is thallus, and this name is applied to the plant body of all the lower plants, the algæ and fungi. The algæ and fungi together are sometimes called the thallophytes, or thallus plants. The word thallus is also sometimes applied to the flattened body of the liverworts. In the foliose liverworts and mosses there is an axis with leaf-like expansions. These are believed by some to represent true stems and leaves, by others to represent a flattened thallus in which the margins are deeply and regularly divided, or in which the expansion has only taken place at regular intervals.

[5] See Chapter 38 for organization of members of the plant body.

In nearly all of the algæ the plant body is submerged in water. In these cases absorption takes place through all portions of the surface in contact with the water, as in spirogyra, vaucheria, and all of the larger seaweeds. Comparatively few of the algæ grow on the surfaces of rocks or trees. In these cxamples it is likely that at times only portions of the plant body serve in the process of absorption of water from the substratum. A few of the algæ are parasitic, living in the tissues of higher plants, where they are surrounded by the water or liquids within the host. Absorption takes place in the same way in many of the fungi. The aquatic fungi are immersed in water. In other forms, like mucor, a portion of the mycelium is within the substratum, and being bathed by the water or watery solutions absorbs the same, while the fruiting portion and the aerial mycelium obtain their water and food solutions from the mycelium in the substratum. In higher fungi, like the mushrooms, the mycelium within the ground or decaying wood absorbs the water necessary for the fruiting portion; while in the case of the parasitic fungi the mycelium lies in the water or liquid within the host.

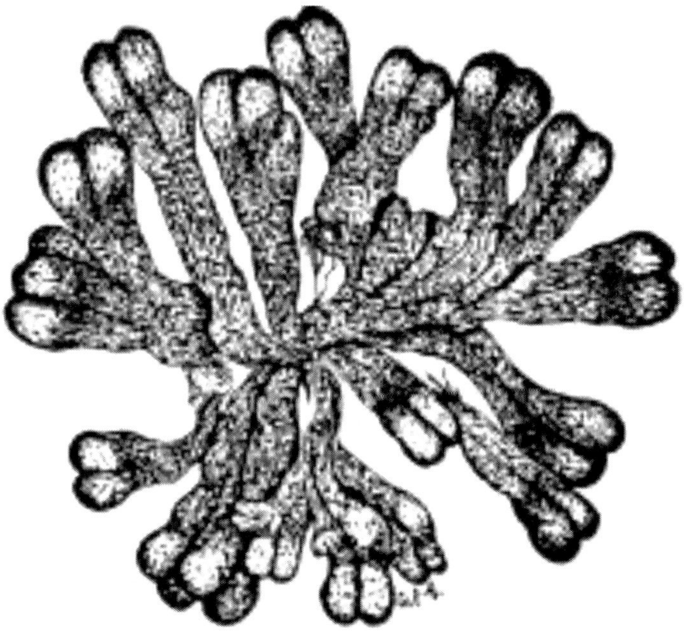

Figure 30. Thallus of Riccia lutescens.

Absorption by Liverworts

In many of the plants termed liverworts the vegetative part of the plant is a thin, flattened, more or less elongated green body known as a thallus.

Riccia

One of these, belonging to the genus riccia, is shown in Figure 30. Its shape is somewhat like that of a minute ribbon which is forked at intervals in a dichotomous manner, the characteristic kind of branching found in these thalloid liverworts. This riccia (known as R. lutescens) occurs on damp soil; long, slender, hair-like processes grow out from the under surface of the thallus which resemble root hairs and serve the same purpose in the processes of absorption. Another species of riccia (R. crystallina) is shown in Figure 252. This plant is quite circular in outline and occurs on muddy flats. Some species float on the water.

Marchantia

One of the larger and coarser liverworts is Figured at 31. This is a very common liverwort, growing in very damp and muddy places and also along the margins of streams, on the mud or upon the surfaces of rocks which are bathed with the water. This is known as *Marchantia polymorpha*. If we examine the under surface of the marchantia we see numerous hair-like processes which attach the plant to the soil. Under the microscope we see that some of these are similar to the root hairs of the seedlings which we have been studying, and they serve the purpose of absorption.

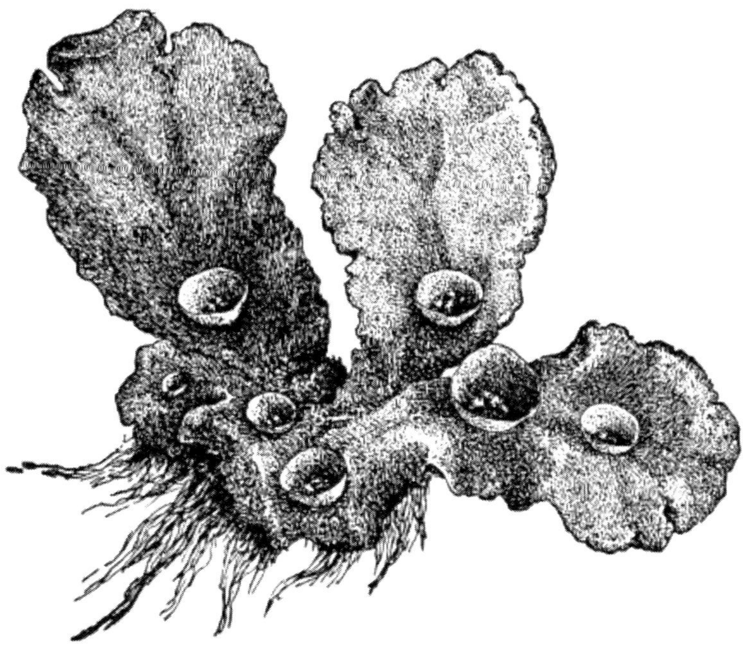

Figure 31. Marchantia plant with cupules and gemmæ; rhizoids below.

Since, however, there are no roots on the marchantia plant, these hair-like outgrowths are usually termed here *rhizoids*. In marchantia they are of two kinds, one kind the simple ones with smooth walls, and the other kind in which the inner surfaces of the walls are roughened by processes which extend inward in the form of irregular tooth-like points. Besides the hairs on the under side of the thallus we note especially near the growing end that there are two rows of leaf-like scales, those at the end of the thallus curving up over the growing end, thus serving to protect the delicate tissues at the growing point.

FRULLANIA

In Figure 32 is shown another liverwort, which differs greatly in form from the ones we have just been studying in that there is a well-defined axis with lateral leaf-like outgrowths. Such liverworts are called foliose liverworts. Besides these two quite prominent rows of leaves there is a third row of poorly developed leaves on the under surface. Also from the under surface of the axis we see here and there slender outgrowths, the rhizoids, through which much of the water is absorbed.

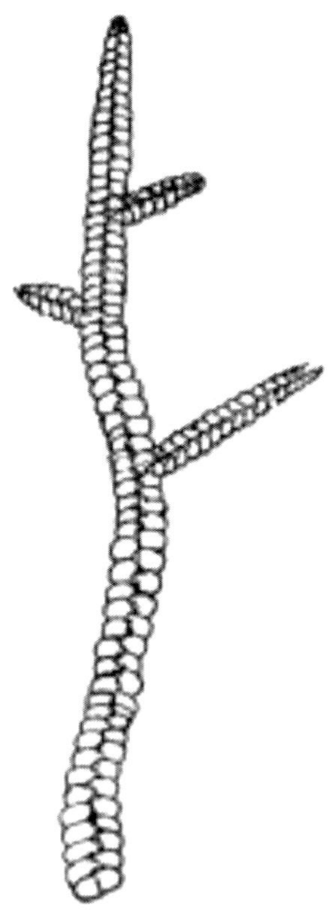

Figure 32. Portion of plant of Frullania, a foliose liverwort.

Figure 33. Portion of same more highly magnified, showingoverlapping leaves.

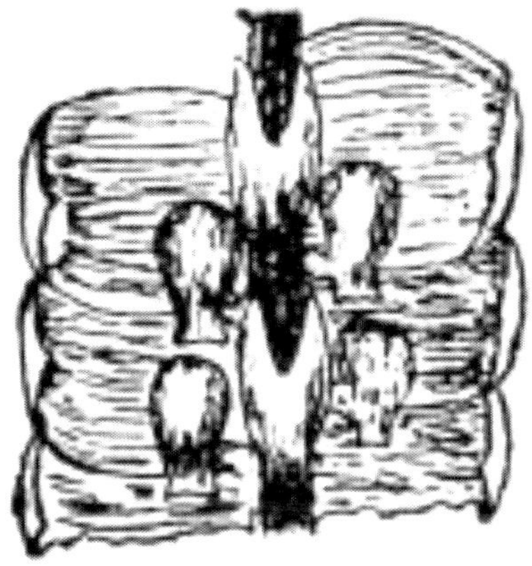

Figure 34. Under side, showing forked under row of leaves and lobes of lateral leaves.

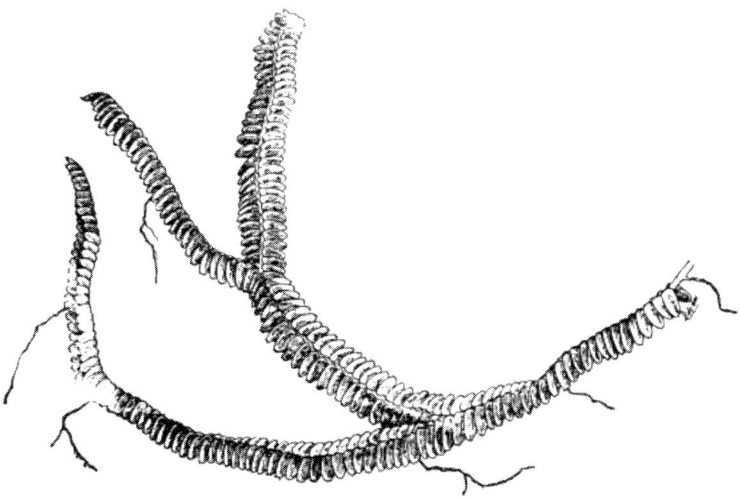

Figure 35. Foliose liverwort (bazzania) showing dichotomous branching and overlapping leaves.

Absorption by the Mosses

Among the mosses, which are usually common in moist and shaded situations, examples are abundant which are suitable for the study of the organs of absorption. If we take for example a plant of mnium (M. affine), which is illustrated in Figure 36, we note that it consists of a slender axis with thin flat, green, leaf-like expansions, Examining with the microscope the lower end of the axis, which is attached to the substratum, there are seen numerous brown-colored threads more or less branched.

Figure 36. Female plant (gametophyte) of a moss (mnium), showing rhizoids below, and the tuft of leaves above, which protect the archegonia.

Absorption by the Higher Aquatic Plants

Examples of the water plants which are entirely submerged in water are the water-crowfoots, some of the pondweeds, elodea or water-weeds, the tape-grass, vallisneria, etc. In these plants all parts of the body being submerged, they absorb water with which they are in contact. In other aquatic plants, like the water-lilies, some of the pondweeds, the duck-meats, etc., are only partially submerged in the water; the upper surface of the leaf or of the leaf-like expansion being exposed to the air, while the under surface lies in close contact with the water, and the stems and the petioles of the leaves are also immersed in water. In these plants absorption takes place through those parts in contact with the water.

Absorption by the Duck-Meats

These plants are very curious examples of the higher plants.

Lemna

One of these is illustrated in Figure 37. This is the common duckweed, *Lemna trisulca*. It is very peculiar in form and in its mode of growth. Each one of the lateral leaf-like expansions extends outwards by the elongation of the basal part, which becomes long and slender. Next, two new lateral expansions are formed on these by prolification from near the base, and thus the plant continues to extend. The plant occurs in ponds and ditches and is sometimes very common and abundant. It floats on the surface of the water. While the flattened part of the plant resembles a leaf, it is really the stem, no leaves being present. This expanded green body is usually termed a "frond." A single rootlet grows out from the under side and is destitute of root hairs. Absorption of water therefore takes place through this rootlet and through the under side of the "frond."

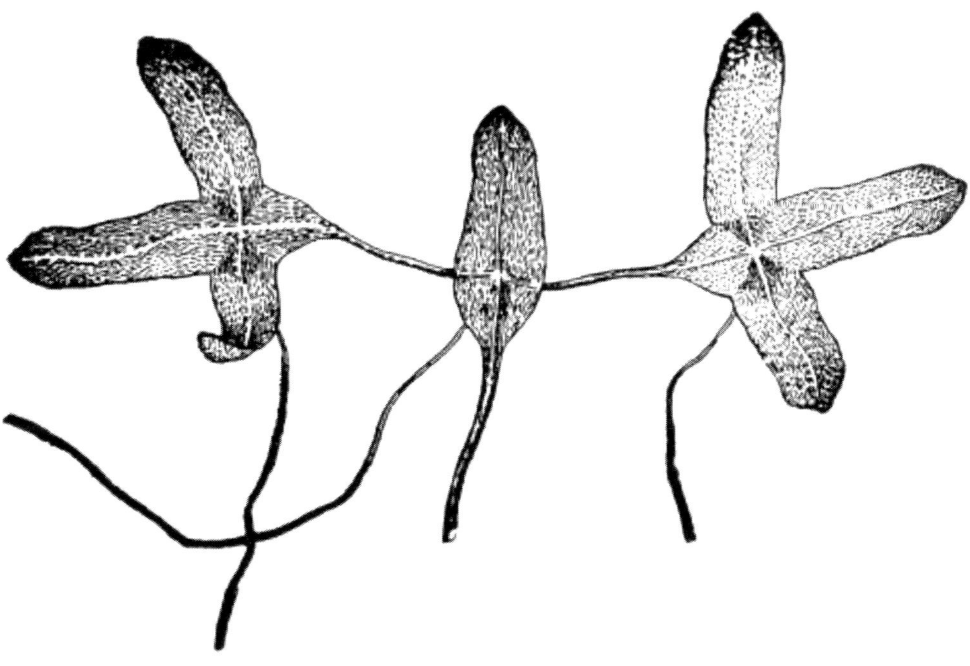

Figure 37. Fronds of the duckweed (Lemna trisculca).

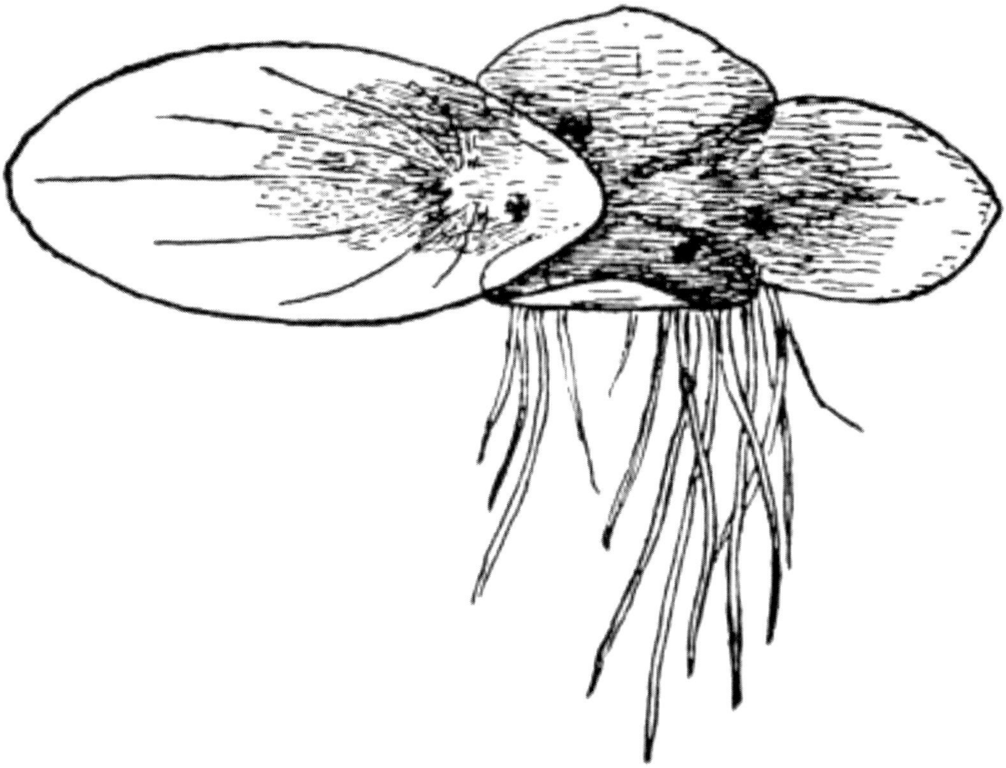

Figure 38. Spirodela polyrhiza.

Spirodela Polyrhiza

This is a very curious plant, closely related to the lemna and sometimes placed in the same genus. It occurs in similar situations, and is very readily grown in aquaria. It reminds one of a little insect as seen in Figure 38. There are several rootlets on the under side of the frond. Absorption of water takes place here in the same way as in lemna.

Absorption in Wolffia

Perhaps the most curious of these modified water plants is the little wolffia, which contains the smallest specimens of the flowering plants. Two species of this genus are shown in Figures 39-41. The plant body is reduced to nothing but a rounded or oval green body, which represents the stem. No leaves or roots are present. The plants multiply by "prolification," the new fronds growing out from a depression on the under side of one end. Absorption takes place through the under surface.

Absorption by Land Plants—Water Cultures

In connection with our inquiry as to how land plants obtain their water, it will be convenient to prepare some water cultures to illustrate this and which can also be used later in our study of nutrition (Chapter IX).

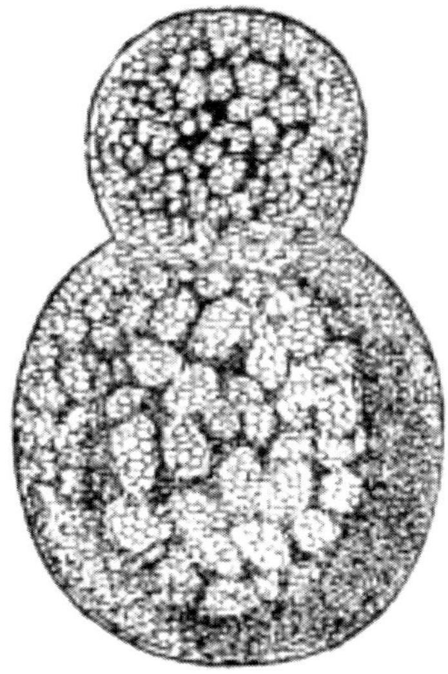

Figure 39. Young frond of wolffia growing out of older one.

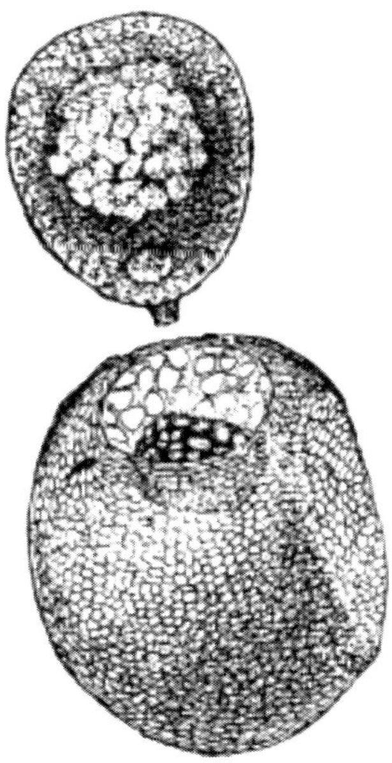

Figure 40. Young frond of wolffia separating from older one.

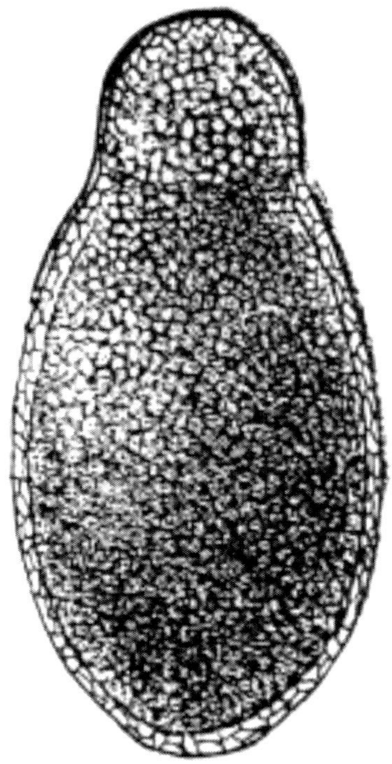

Figure 41. Another species of wolffia, the two fronds still connected.

Chemical analysis shows that certain mineral substances are common constituents of plants. By growing plants in different solutions of these various substances it has been possible to determine what ones are necessary constituents of plant food. While the proportion of the mineral elements which enter into the composition of plant food may vary considerably within certain limits, the concentration of the solutions should not exceed certain limits. A very useful solution is one recommended by Sachs, and is as follows:

Formula for water cultures:

Water	1000cc.
Potassium nitrate	0.5 gr.
Sodium chloride	0.5 "
Calcium sulphate	0.5 "
Magnesium sulphate	0.5 "
Calcium phosphate	0.5 "

The calcium phosphate is only partly soluble. The solution which is not in use should be kept in a dark cool place to prevent the growth of minute algæ.

Several different plants are useful for experiments in water cultures, as peas, corn, beans, buckwheat, etc. The seeds of these plants may be germinated, after soaking them for several hours in warm water, by placing them between the folds of wet paper on shallow trays, or in the folds of wet cloth. The seeds should not be kept immersed in water after they have imbibed enough to thoroughly soak and swell them. At the same time that the seeds are placed in damp paper or cloth for germination, one lot of the soaked seeds should be planted in good soil and kept under the same temperature conditions, for control. When the plants have germinated one series should be grown in distilled water, which possesses no plant food; another in the nutrient solution, and still another in the nutrient solution to which has been added a few drops of a solution of iron chloride or ferrous sulphate. There would then be four series of cultures which should be carried out with the same kind of seed in each series so that the comparisons can be made on the same species under the different conditions. The series should be numbered and recorded as follows:

- No. 1, soil.
- No. 2, distilled water.
- No. 3, nutrient solution.
- No. 4, nutrient solution with a few drops of iron solution added.

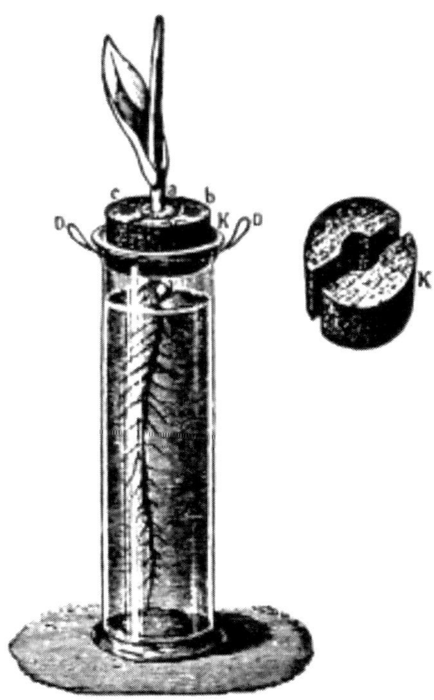

Figure 42. Culture cylinder to show position of corn seedling (Hansen).

Small jars or wide-mouth bottles, or crockery jars, can be used for the water cultures, and the cultures are set up as follows: A cork which will just fit in the mouth of the bottle, or which can be supported by pins, is perforated so that there is room to insert the seedling, with the root projecting below into the liquid. The seed can be fastened in position by inserting a pin through one side, if it is a large one, or in the case of small seeds a cloth of a coarse mesh can be tied over the mouth of the bottle instead of using the cork. After properly setting up the experiments the cultures should be arranged in a suitable place, and observed from time to time during several weeks. In order to obtain more satisfactory results several duplicate series should be set up to guard against the error which might arise from variation in individual plants and from accident. Where there are several students in a class, a single series set up by several will act as checks upon one another. If glass jars are used for the liquid cultures they should be wrapped with black paper or cloth to exclude the light from the liquid, otherwise numerous minute algæ are apt to grow and interfere with the experiment. Or the jars may be sunk in pots of earth to serve the same purpose. If crockery jars are used they will not need covering.

For some time all the plants grow equally well, until the nutriment stored in the seed is exhausted. The numbers 1, 3 and 4, in soil and nutrient solutions, should outstrip number 2, the plants in the distilled water. No. 4 in the nutrient solution with iron, having a perfect food, compares favorably with the plants in the soil.

Plants Take Liquid Food from the Soil

From these experiments then we judge that such plants take up the food they receive from the soil in the form of a liquid, the elements being in solution in water.

If we recur now to the experiments which were performed with the salt solution in producing plasmolysis in the cells of spirogyra, in the cells of the beet or corn, and in the root hairs of the corn and bean seedlings, and the way in which these cells become turgid again when the salt solution is removed and they arc again bathed with water, we shall have an explanation of the way in which plants take up nutrient solutions of food material through their roots.

How Food Solutions are Carried into the Plant

We can see how water and food solutions are carried into the plant, and we must next turn our attention to the way in which these solutions are carried farther into the plant. We should make a section across the root of a seedling in the region of the root hairs and examine it with the aid of a microscope. We here see that the root hairs are formed by the elongation of certain of the surface cells of the root. These cells elongate perpendicularly to the root, and become *3mm* to *6mm* long. They are flexuous or irregular in outline and cylindrical, as shown in Figure 43. The end of the hair next the root fits in between the adjacent superficial cells of the root and joins closely to the next deeper layer of cells. In studying the section of the young root we see that the root is made up of cells which lie closely side by side, each with its wall, its protoplasm and cell-sap, the protoplasmic membrane lying on the inside of each cell wall.

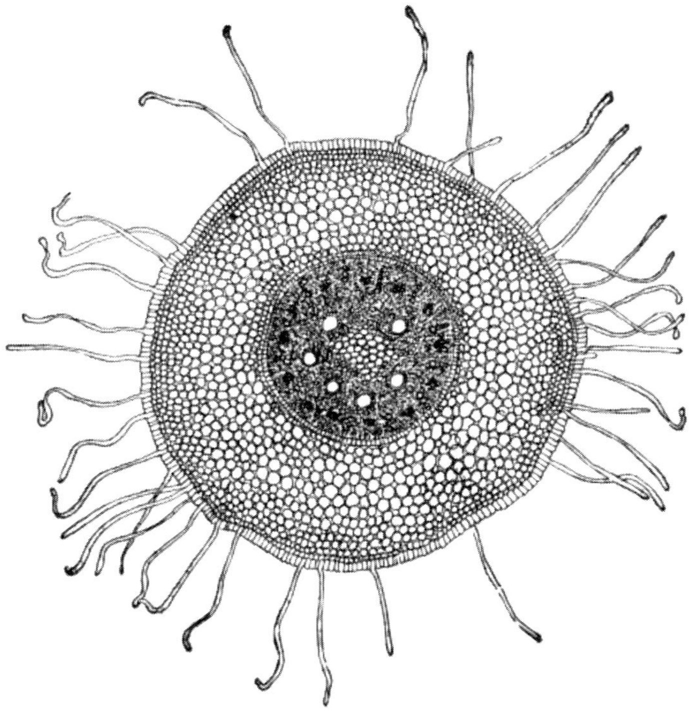

Figure 43. Section of corn root, showing rhizoids formed from elongated epidermal cells.

In the absorption of the watery solutions of plant food by the root hairs, the cell-sap, being a more concentrated solution, gains some of the former, since the liquid of less concentration flows through the protoplasmic membrane into the more concentrated cell-sap, increasing the bulk of the latter. This makes the root hairs turgid, and at the same time dilutes the cell-sap so that the concentration is not so great. The cells of the root lying inside and close to the base of the root hairs have a cell-sap which is now more concentrated than the diluted cell-sap of the hairs, and consequently gain some of the food solutions from the latter, which tends to lessen the content of the root hairs and also to increase the concentration of the cell-sap of the same. This makes it possible for the root hairs to draw on the soil for more of the food solutions, and thus, by a variation in the concentration of the substances in solution in the cell-sap of the different cells, the food solutions are carried along until they reach the *vascular bundles*, through which the solutions are carried to distant parts of the plant. Some believe that there is a rhythmic action of the elastic cell walls in these cells between the root hairs and the vascular bundles. This occurs in such a way that, after the cell becomes turgid, it contracts, thus reducing the size of the cell and forcing some of the food solutions into the adjacent cells, when by absorption of more food solutions, or water, the cell increases in turgidity again. This rhythmic action of the cells, if it does take place, would act as a pump to force the solutions along, and would form one of the causes of root pressure.

How the Root Hairs Get the Watery Solutions from the Soil

If we examine the root hairs of a number of seedlings which are growing in the soil under normal conditions, we shall see that a large quantity of soil readily clings to the roots. We should note also that unless the soil has been recently watered there is no free water in it; the soil is only moist. We are curious to know how plants can obtain water from soil which is not wet. If we attempt to wash off the soil from the roots, being careful not to break away the root hairs, we find, that small particles cling so tenaciously to the root hairs that they are not removed. Placing a few such root hairs under the microscope it appears as if here and there the root hairs were glued to the minute soil particles.

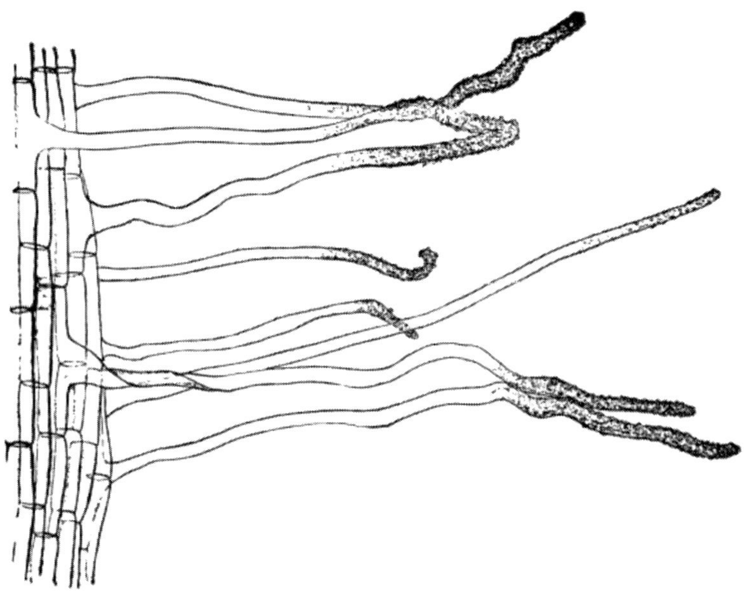

Figure 44. Root hairs of corn seedling with soil particles adhering closely.

If now we take some of the soil which is only moist, weigh it, and then permit it to become quite dry on exposure to dry air, and weigh again, we find that it loses weight in drying. Moisture has been given off. This moisture, it has been found, forms an exceedingly thin film on the surface of the minute soil particles. Where these soil particles lie closely together, as they usually do when massed together in the pot or elsewhere, this thin film of moisture is continuous from the surface of one particle to that of another. Thus the soil particles which are so closely attached to the root hairs connect the surface of the root hairs with this film of moisture. As the cell-sap of the root hairs draws on the moisture film with which they are in contact, the tension of this film is sufficient to draw moisture from distant particles. In this way the roots are supplied with water in soil which is only moist.

Plants Cannot Remove all the Moisture from the Soil

If we now take a potted plant, or a pot containing a number of seedlings, place it in a moderately dry room, and do not add water to the soil we find in a few days that the plant is wilting. The soil if examined will appear quite dry to the sense of touch. Let us weigh some of this soil, then dry it by artificial heat, and weigh again. It has lost in weight. This has been brought about by driving off the moisture which still remained in the soil after the plant began to wilt. This teaches that while plants can obtain water from soil which is only moist or which is even rather dry, they are not able to withdraw all the moisture from the soil.

Figure 45. Experiment to show root pressure (Detmer).

"Root Pressure" or Exudation Pressure

It is a very common thing to note, when certain shrubs or vines are pruned in the spring, the exudation of a watery fluid from the cut surfaces. In the case of the grape vine this has been known to continue for a number of days, and in some cases the amount of liquid, called "sap," which escapes is considerable. In many cases it is directly traceable to the activity of the roots, or root hairs, in the absorption of water from the soil. For this reason the term *root pressure* has been used to denote the force exerted in supplying the water from the soil. But there are some who object to the use of this term "root pressure." The principal objection is that the pressure which brings about the phenomenon known as "bleeding" by plants is not present in the roots alone. This pressure exists under certain conditions in all parts of the plant. The term exudation pressure has been proposed in lieu of root pressure. It should be remembered that the movement of water in the plant is started by the pressure which exists in the root. If the term "root pressure" is used, it should be borne clearly in mind that it does not express the phenomenon exactly in all cases.

Root Pressure may be Measured

It is possible to measure not only the amount of water which the roots will raise in a given time, but also to measure the force exerted by the roots during root pressure. It has been found that root pressure in the case of the nettle is sufficient to hold a column of water about 4.5 meters (15 ft.) high (Vines), while the root pressure of the vine (Hales, 1721) will hold a column of water about 10 meters (36.5 ft.) high, and the birch (Betula lutea) (Clark, 1873) has a root pressure sufficient to hold a column of water about 25 meters (84.7 ft.) high.

Experiment to Demonstrate Root Pressure

By a very simple method this lifting of water by root pressure is shown. During the summer season plants in the open may be used if it is preferred, but plants grown in pots are also very serviceable, and one may use a potted begonia or balsam, the latter being especially useful. The plants are usually convenient to obtain from the greenhouses, to illustrate this phenomenon. The stem is cut off rather close to the soil and a long glass tube is attached to the cut end of the stem, still connected with the roots, by the use of rubber tubing, as shown in Figure 45, and a very small quantity of water may be poured in to moisten the cut end of the stem. In a few minutes the water begins to rise in the glass tube. In some cases it rises quite rapidly, so that the column of water can readily be seen to extend higher and higher up in the tube when observed at quite short intervals. (To measure the force of root pressure is rather difficult for elementary work. To measure it see Ganong, Plant Physiology, pp. 67, 68, or some other book for advanced work.)

In either case where the experiment is continued for several days it is noticed that the column of water or of mercury rises and falls at different times during the same day, that is, the column stands at varying heights; or in other words the root pressure varies during the day. With some plants it has been found that the pressure is greatest at certain times of the day, or at certain seasons of the year. Such variation of root pressure exhibits what is termed a periodicity, and in the case of some plants there is a daily periodicity; while in others there is in addition an annual periodicity. With the grape vine the root pressure is greatest in the forenoon, and decreases from 12-6 p.m., while with the sunflower it is greatest before 10 a.m., when it begins to decrease. Temperature of the soil is one of the most important external conditions affecting the activity of root pressure.

Chapter 4

Transpiration, or the Loss of Water by Plants

We should now inquire if all the water which is taken up in excess of that which actually suffices for turgidity is used in the elaboration of new materials of construction. We notice when a leaf or shoot is cut away from a plant, unless it is kept in quite a moist condition, or in a damp, cool place, that it becomes flaccid, and droops. It wilts, as we say. The leaves and shoot lose their turgidity. This fact suggests that there has been a loss of water from the shoot or leaf. It can be readily seen that this loss is not in the form of drops of water which issue from the cut end of the shoot or petiole. What then becomes of the water in the cut leaf or shoot?

Loss of Water from Excised Leaves

Let us take a handful of fresh, green, rather succulent leaves, which are free from water on the surface, and place them under a glass bell jar, which is tightly closed below but which contains no water. Now place this in a brightly lighted window, or in sunlight. In the course of fifteen to thirty minutes we notice that a thin film of moisture is accumulating on the inner surface of the glass jar. After an hour or more the moisture has accumulated so that it appears in the form of small drops of condensed water. We should set up at the same time a bell jar in exactly the same way but which contains no leaves. In this jar there is no condensed moisture on the inner surface. We thus are justified in concluding that the moisture in the former jar comes from the leaves. Since there is no visible water on the surfaces of the leaves, or at the cut ends, before it may have condensed there, we infer that the water escapes from the leaves in the form of *water vapor*, and that this water vapor, when it comes in contact with the surface of the cold glass, condenses and forms the moisture film, and later the drops of water. The leaves of these cut shoots therefore lose water in the form of water vapor, and thus a loss of turgidity results.

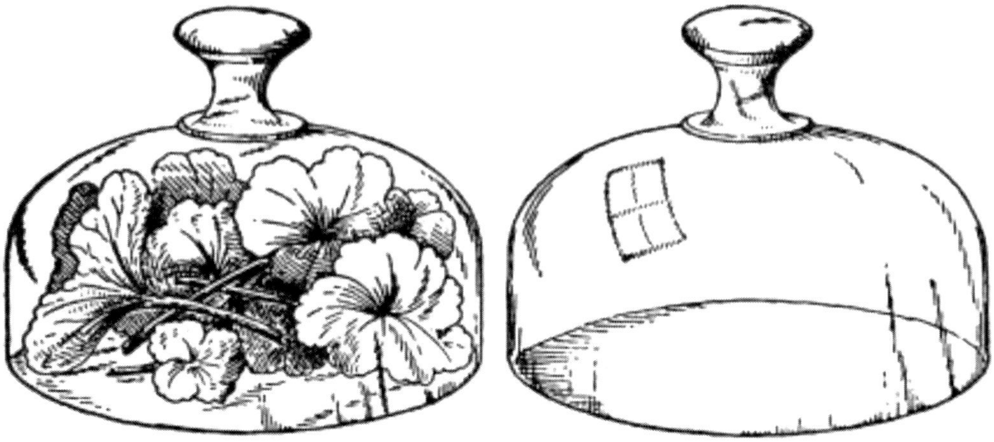

Figure 46. To show loss of water from leaves, the leaves just covered.

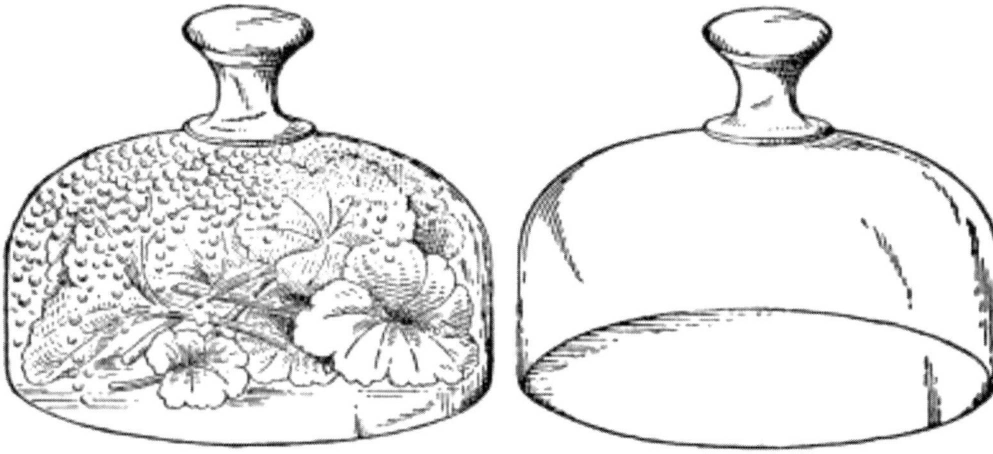

Figure 47. After a few hours drops of water have accumulated on the inside of the jar covering the leaves.

LOSS OF WATER FROM GROWING PLANTS

Suppose we now take a small and actively growing plant in a pot, and cover the pot and the soil with a sheet of rubber cloth or flexible oilcloth which fits tightly around the stem of the plant so that the moisture from the soil or from the surface of the pot cannot escape. Then place a bell jar over the plant, and set in a brightly lighted place, at a temperature suitable for growth. In the course of a few minutes on a dry day a moisture film forms on the inner surface of the glass, just as it did in the case of the glass jar containing the cut shoots and leaves. Later the moisture has condensed so that it is in the form of drops. If we have the same leaf surface here as we had with the cut shoots, we shall probably find that a larger amount of water accumulates on the surface of the jar from the plant that is still attached to its roots.

Water Escapes from the Surfaces of Living Leaves in the Form of Water Vapor

This living plant then has lost water, which also escapes in the form of water vapor. Since here there are no cut places on the shoots or leaves, we infer that the loss of water vapor takes place from the surfaces of the leaves and from the shoots. It is also to be noted that, while this plant is losing water from the surfaces of the leaves, it does not wilt or lose its turgidity. The roots by their activity and pressure supply water to take the place of that which is given off in the form of water vapor. This loss of water in the form of water vapor by plants is *transpiration*.

A Test for the Escape of Water Vapor from Plants

Make a solution of cobalt chloride in water. Saturate several pieces of filter paper with it. Allow them to dry. The water solution of cobalt chloride is red. The paper is also red when it is moist, but when it is thoroughly dry it is blue. It is very sensitive to moisture and the moisture of the air is often sufficient to redden it. Before using dry the paper in an oven or over a flame.

Take two bell jars, as shown in Figure 49. Under one place a potted plant, the pot and earth being covered by oiled paper. Or cover the plant with a fruit jar. To a stake in the pot pin a piece of the dried cobalt paper, and at the same time pin to a stake, in another jar covering no plant, another piece of cobalt paper. They should both be put under the jars at the same time. In a few moments the paper in the jar with the plant will begin to redden. In a short while, ten or fifteen minutes, probably, it will be entirely red, while the paper under the other jar will remain blue, or be only slightly reddened. The water vapor passing off from the living plant comes in contact with the sensitive cobalt chloride in the paper and reddens it before there is sufficient vapor present to condense as a film of moisture on the surface of the jar.

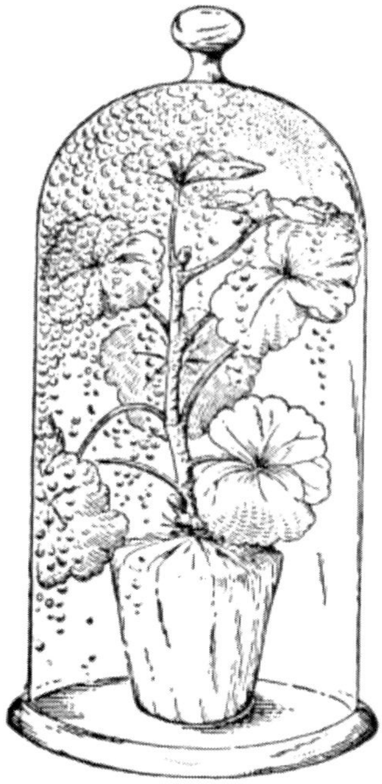

Figure 48. Water vapor is given off by the leaves when attached to the living plant. It condenses into drops of water on the cool surface of the glass covering the plant.

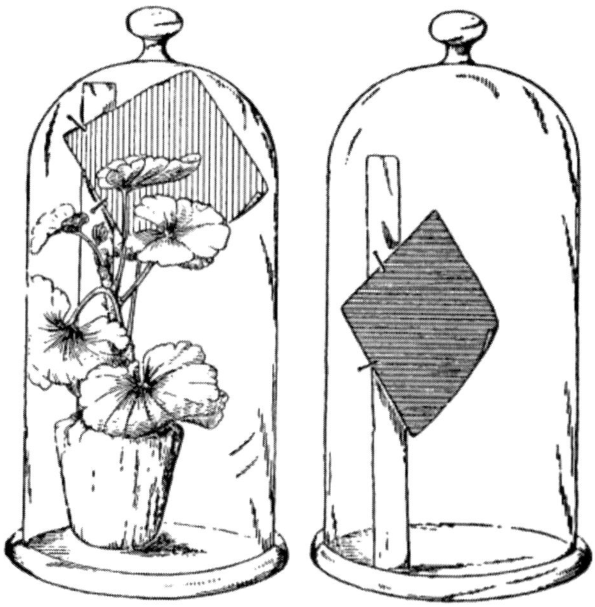

Figure 49. Figure 49.—A good way to show that the water passes off from the leaves in the form of water vapor.

Experiment to Compare Loss of Water in a Dry and a Humid Atmosphere

We should now compare the escape of water from the leaves of a plant covered by a bell jar, as in the last experiment, with that which takes place when the plant is exposed in a normal way in the air of the room or in the open. To do this we should select two plants of the same kind growing in pots, and of approximately the same leaf surface. The potted plants are placed one each on the arms of a scale. One of the plants is covered in this position with a bell jar. With weights placed on the pan of the other arm the two sides are balanced. In the course of an hour, if the air of the room is dry, moisture has probably accumulated on the inner surface of the glass jar which is used to cover one of the plants. This indicates that there has here been a loss of water. But there is no escape of water vapor into the surrounding air so that the weight on this arm is practically the same as at the beginning of the experiment. We see, however, that the other arm of the balance has risen. We infer that this is the result of the loss of water vapor from the plant on that arm. Now let us remove the bell jar from the other plant, and with a cloth wipe off all the moisture from the inner surface, and replace the jar over the plant. We note that the end of the scale which holds this plant is still lower than the other end.

The Loss of Water Is Greater in a Dry Than in a Humid Atmosphere

This teaches us that while water vapor escaped from the plant under the bell jar, the air in this receiver soon became saturated with the moisture, and thus the farther escape of moisture from the leaves was checked. It also teaches us another very important fact, viz., that plants lose water more rapidly through their leaves in a dry air than in a humid or moist atmosphere. We can now understand why it is that during the very hot and dry part of certain days plants often wilt, while at nightfall, when the atmosphere is more humid, they revive. They lose more water through their leaves during the dry part of the day, other things being equal, than at other times.

How Transpiration Takes Place

Since the water of transpiration passes off in the form of water vapor we are led to inquire if this process is simply *evaporation* of water through the surface of the leaves, or whether it is controlled to any appreciable extent by any condition of the living plant. An experiment which is instructive in this respect we shall find in a comparison between the transpiration of water from the leaves of a cut shoot, allowed to lie unprotected in a dry room, and a similar cut shoot the leaves of which have been killed.

Almost any plant will answer for the experiment. For this purpose I have used the following method. Small branches of the locust (*Robinia pseudacacia*), of sweet clover (*Melilotus alba*), and of a heliopsis were selected. One set of the shoots was immersed for a moment in hot water near the boiling point to kill them. The other set was immersed for the same length of time in cold water, so that the surfaces of the leaves might be well wetted, and thus the two sets of leaves at the beginning of the experiment would be similar, so far as the amount of water on their surfaces is concerned. All the shoots were then spread out on a table in a dry room, the leaves of the killed shoots being separated where they are inclined to cling together. In a short while all the water has evaporated from the surface of the living leaves, while the leaves of the dead shoots are still wet on the surface. In six hours the leaves of the dead shoots from which the surface water had now evaporated were beginning to dry up, while the leaves of the living plants were only becoming flaccid. In twenty-four hours the leaves of the dead shoots were crisp and brittle, while those of the living shoots were only wilted. In twenty-four hours more the leaves of the sweet clover and of the heliopsis were still soft and flexible, showing that they still contained more water than the killed shoots which had been crisp for more than a day.

It must be then that during what is termed transpiration the living plant is capable of holding back the water to some extent, which in a dead plant would escape more rapidly by evaporation. It is also known that a body of water with a surface equal to that of a given leaf surface of a plant loses more water by evaporation during the same length of time than the plant loses by transpiration.

Structure of a Leaf

We are now led to inquire why it is that a living leaf loses water less rapidly than dead ones, and why less water escapes from a given leaf surface than from an equal surface of water. To understand this it will be necessary to examine the minute structure of a leaf. For this purpose we may select the leaf of an ivy, though many other leaves will answer equally well. From a portion of the leaf we should make very thin cross-sections with a razor or other sharp instrument. These sections should be perpendicular to the surface of the leaf and should be then mounted in water for microscopic examination.[6]

[6] Demonstrations may be made with prepared sections of leaves.

EPIDERMIS OF THE LEAF

In this section we see that the green part of the leaf is bordered on what are its upper and lower surfaces by a row of cells which possess no green color. The walls of the cells of each row have nearly parallel sides, and the cross walls are perpendicular. These cells form a single layer over both surfaces of the leaf and are termed the *epidermis*. Their walls are quite stout and the outer walls are *cuticularized*.

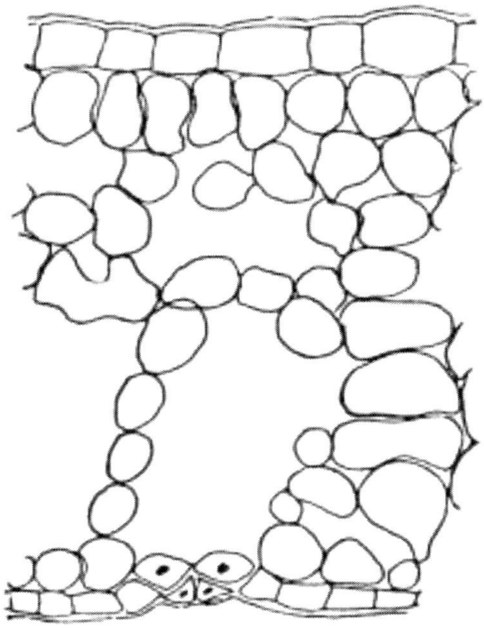

Figure 50. Section through ivy leaf showing communication between stomata and the large intercellular spaces of the leaf, stoma closed.

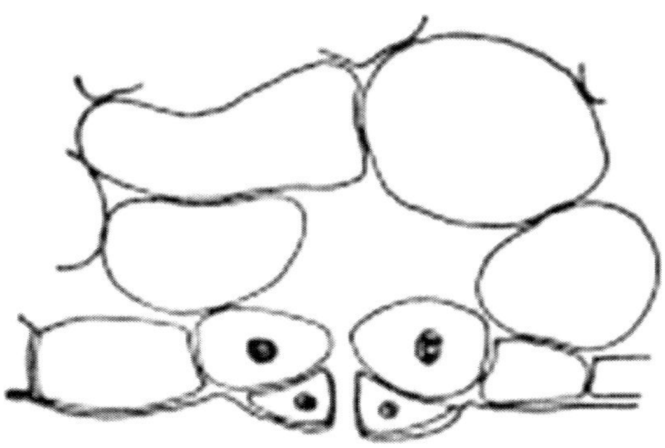

Figure 51. Stoma open.

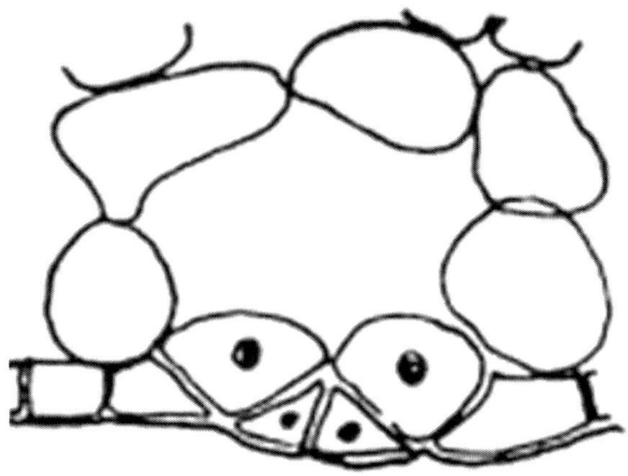

Figure 52. Stoma closed.

Figures 51, 52.—Section through stomata of ivy leaf.

Soft Tissue of the Leaf

The cells which contain the green chlorophyll bodies are arranged in two different ways. Those on the upper side of the leaf are usually long and prismatic in form and lie closely parallel to each other. Because of this arrangement of these cells they are termed the *palisade* cells, and form what is called the *palisade layer*. The other green cells, lying below, vary greatly in size in different plants and to some extent also in the same plant. Here we notice that they are elongated, or oval, or somewhat irregular in form. The most striking peculiarity, however, in their arrangement is that they are not usually packed closely together, but each cell touches the other adjacent cells only at certain points. This arrangement of these cells forms quite large spaces between them, the intercellular spaces. If we should examine such a section of a leaf before it is mounted in water we would see that the intercellular spaces are not filled with water or cell-sap, but are filled with air or some gas. Within the cells, on the other hand, we find the cell-sap and the protoplasm.

Stomata

If we examine carefully the row of epidermal cells on the under surface of the leaf, we find here and there a peculiar arrangement of cells shown at Figures 51, 52. This opening through the epidermal layer is a *stoma*. The cells which immediately surround the openings are the *guard cells*. The form of the guard cells can be better seen if we tear a leaf in such a way as to strip off a short piece of the lower epidermis, and mount this in water. The guard cells are nearly crescent-shaped, and the stoma is elliptical in outline. The epidermal cells are very irregular in outline in this view. We should also note that while the epidermal cells contain no chlorophyll, the guard cells do.

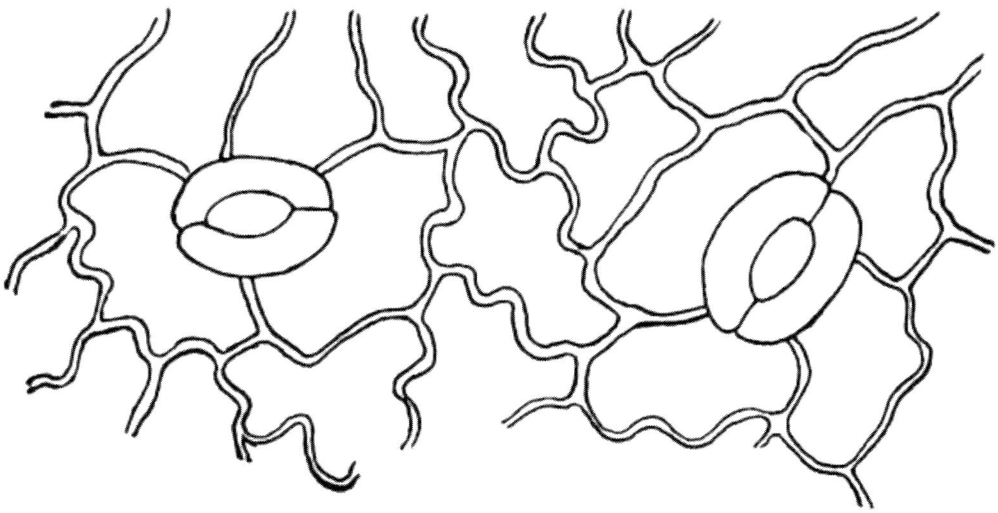

Figure 53. Portion of epidermis of ivy, showing irregular epidermal cells, stoma and guard cells.

82A

In the ivy leaf the guard cells are quite plain, but in most plants the form as seen in cross-section is irregular in outline, as shown in Figure 53*a*, which is from a section of a wintergreen leaf. This leaf is interesting because it shows the characteristic structure of leaves of many plants growing in soil where absorption of water by the roots is difficult owing to the cold water, acids, or salts in the water or soil, or in dry soil (see Chapters 47, 54, 55). The cuticle over the upper epidermis is quite thick. This lessens the loss of water by the leaf. The compact palisades of cells are in two to three cell layers, also reducing the loss of water.

The Living Protoplasm Retards the Evaporation of Water from the Leaf

If we now take into consideration a few facts which we have learned in a previous chapter, with reference to the physical properties of the living cell, we shall be able to give a partial explanation of the comparative slowness with which the water escapes from the leaves. The inner surfaces of the cell walls are lined with the membrane of protoplasm, and within this is the cell-sap. These cells have become turgid by the absorption of the water which has passed up to them from the roots. While the protoplasmic membrane of the cells does not readily permit the water to filter through, yet it is saturated with water, and the elastic cell wall with which it is in contact is also saturated. From the cell wall the water evaporates into the intercellular spaces. But the water is given up slowly through the protoplasmic membrane, so that the water vapor cannot be given off as rapidly from the cell walls as it could if the protoplasm were dead. The living protoplasmic membrane then which is only slowly permeable to the water of the cell-sap is here a very important factor in checking the too rapid loss of water from the leaves.

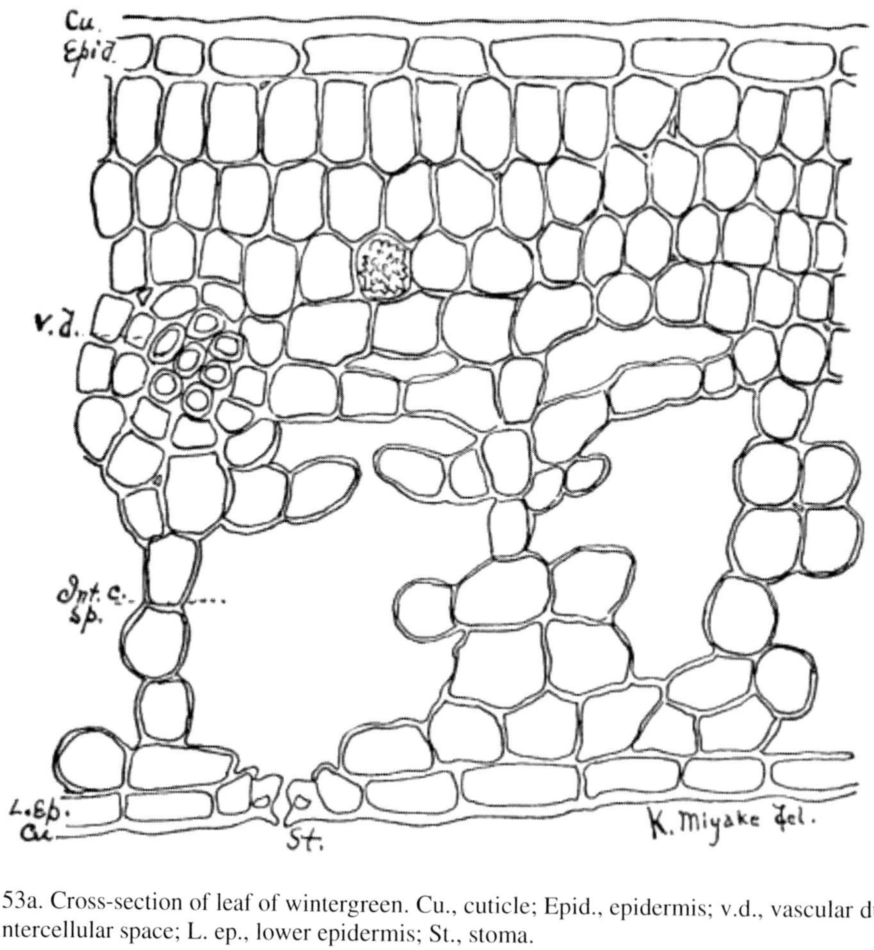

Figure 53a. Cross-section of leaf of wintergreen. Cu., cuticle; Epid., epidermis; v.d., vascular duct; Int. c. sp., intercellular space; L. ep., lower epidermis; St., stoma.

By an examination of our leaf section we see that the intercellular spaces are all connected, and that the stomata, where they occur, open also into intercellular spaces. There is here an opportunity for the water vapor in the intercellular spaces to escape when the stomata are open.

ACTION OF THE STOMATA

The guard cells serve an important function in regulating transpiration. During normal transpiration the guard cells are turgid and their peculiar form then causes them to arch away from each other, allowing the escape of water vapor. When the air becomes too dry transpiration is in excess of absorption by the roots. The guard cells lose some of their water, and collapse so that their inner faces meet in a straight line and close the stoma. Thus the rapid transpiration is checked. Some evaporation of water vapor, however, takes place through the epidermal cells, and if the air remains too dry, the leaves eventually become flaccid and droop. During the day the effect of sunlight is to increase certain sugars or salts in the guard cells so that they readily become turgid and open the stomates, but at night the cell-sap is less concentrated and the stomates are usually closed. Light therefore favors transpiration, while in darkness transpiration is checked.

COMPARE TRANSPIRATION FROM THE TWO SURFACES OF THE LEAF

This can be done by using the cobalt chloride paper. This paper can be kept from year to year and used repeatedly. It is thus a very simple matter to make these experiments. Provide two pieces of glass (discarded glass negatives, cleaned, are excellent), two pieces of cobalt chloride paper, and some geranium leaves entirely free from surface water. Dry the paper until it is blue. Place one piece of the paper on a glass plate; place the geranium leaf with the under side on the paper. On the upper side of the leaf now place the other cobalt paper, and next the second piece of glass. On the pile place a light weight to keep the parts well in contact. In fifteen or twenty minutes open and examine. The paper next the under side of the geranium leaf is red where it lies under the leaf. The paper on the upper side is only slightly reddened. The greater loss of water, then, is through the under side of the geranium leaf. This is true of a great many leaves, but it is not true of all.

NEGATIVE PRESSURE

This is not only indicated by the drooping of the leaves, but may be determined in another way. If the shoot of such a plant be cut underneath mercury, or underneath a strong solution of eosin, it will be found that some of the mercury or eosin, as the case may be, will be forcibly drawn up into the stem toward the roots. This is seen on quickly splitting the cut end of the stem. When plants in the open cannot be obtained in this condition, one may take a plant like a balsam plant from the greenhouse, or some other potted plant, knock it out of the pot, free the roots from the soil and allow to partly wilt. The stem may then be held under the eosin solution and cut.

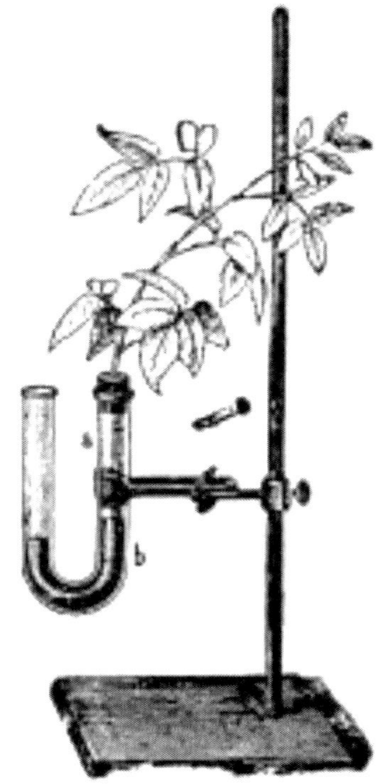

Figure 54. Experiment to show lifting power of transpiration.

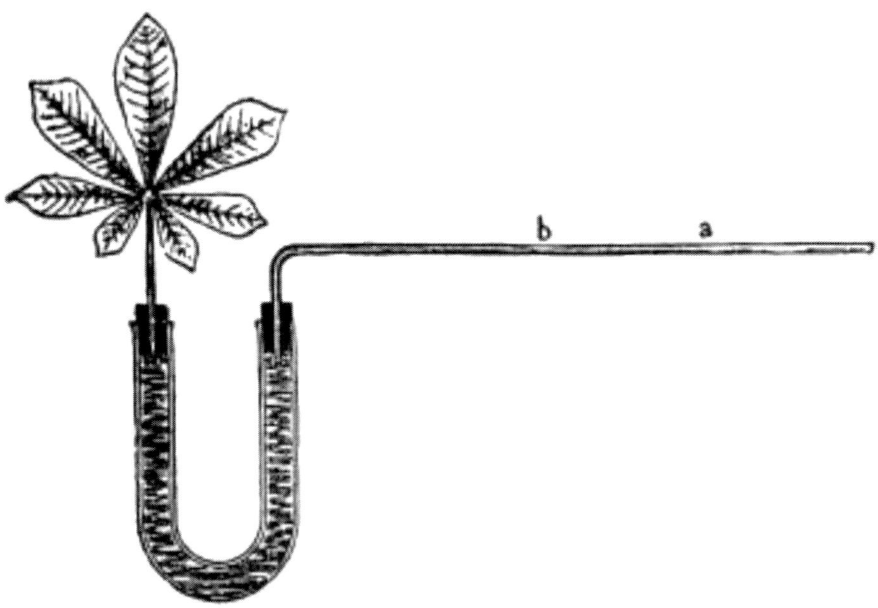

Figure 55. Estimation of the amount of transpiration. The tubes are filled with water, and as the water transpires from the leaf surface its movement in the tube from *a* to *b* can be measured. (After Mangin.)

Lifting Power Of Transpiration

Not only does transpiration go on quite independently of root pressure, as we have discovered from other experiments, but transpiration is capable of exerting a lifting power on the water in the plant. This may be demonstrated in the following way: Place the cut end of a leafy shoot in one end of a U tube and fit it water-tight. Partly fill this arm of the U tube with water, and add mercury to the other arm until it stands at a level in the two arms as in Figure 54. In a short time we note that the mercury is rising in the tube.

Root Pressure May Exceed Transpiration

If we cover small actively growing plants, such as the pea, corn, wheat, bean, etc., with a bell jar, and place them in the sunlight where the temperature is suitable for growth, in a few hours, if conditions are favorable, we shall see that there are drops of water standing out on the margins of the leaves. These drops of water have exuded through the ordinary stomata, or in other cases what are called water stomata, through the influence of root pressure. The plant being covered by the glass jar, the air soon becomes saturated with moisture and transpiration is checked. Root pressure still goes on, however, and the result is shown in the exuding drops. Root pressure is here in excess of transpiration. This phenomenon is often to be observed during the summer season in the case of low-growing plants. During the bright warm day transpiration equals, or may be in excess of, root pressure, and the leaves are consequently flaccid. As nightfall comes on the air becomes more moist, and the conditions of light are such also that transpiration is lessened. Root pressure, however, is still active because the soil is still warm. In these cases drops of water may be seen exuding from the margins of the leaves due to the excess of root pressure over transpiration. Were it not for this provision for the escape of the excess of water raised by root pressure, serious injury by lesions, as a result of the great pressure, might result. The plant is thus to some extent a self-regulatory piece of apparatus so far as root pressure and transpiration are concerned.

Injuries Caused by Excessive Root Pressure

Some varieties of tomatoes when grown in poorly lighted and poorly ventilated greenhouses suffer serious injury through lesions of the tissues. This is brought about by the cells at certain parts becoming charged so full with water through the activity of root pressure and lessened transpiration, assisted also probably by an accumulation of certain acids in the cell-sap which cannot be got rid of by transpiration. Under these conditions some of the cells here swell out, forming extensive cushions, and the cell walls become so weakened that they burst. It is possible to imitate the excess of root pressure in the case of some plants by connecting the stems with a system of water pressure, when very quickly the drops of water will begin to exude from the margins of the leaves.

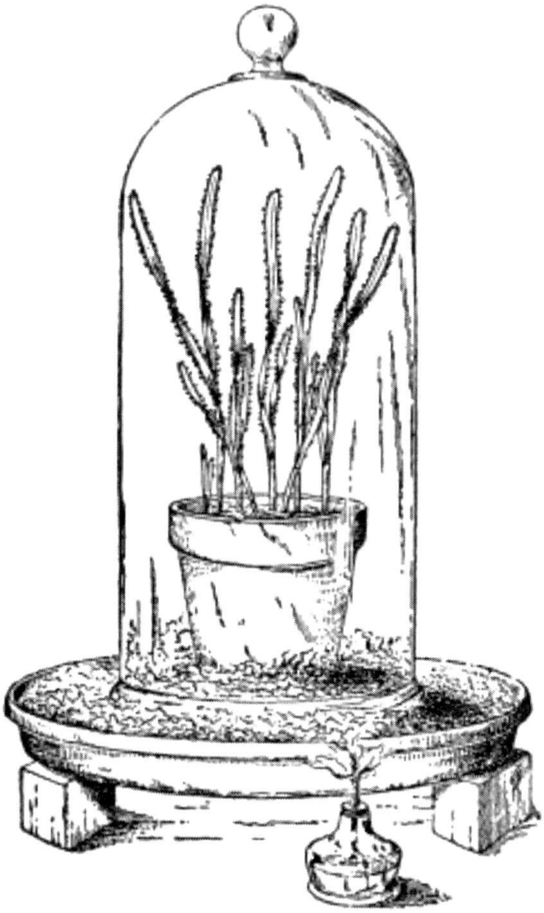

Figure 56. The roots are lifting more water into the plant than can be given off in the form of water vapor, so it is pressed out in drops. From "First Studies Plant Life."

It should be stated that in reality there is no difference between transpiration and evaporation, if we bear in mind that evaporation takes place more slowly from living plants than from dead ones, or from an equal surface of water.

The escape of water vapor is not the only function of the stomata. The exchange of gases takes place through them as we shall later see. A large number of experiments show that normally the stomata are open when the leaves are turgid. But when plants lose excessive quantities of water on dry and hot days, so that the leaves become flaccid, the guard cells automatically close the stomata to check the escape of water vapor. Some water escapes through the epidermis of many plants, though the cuticularized membrane of the epidermis largely prevents evaporation. In arid regions plants are usually provided with an epidermis of several layers of cells to more securely prevent evaporation there. In such cases the guard cells are often protected by being sunk deeply in the epidermal layer.

Demonstration of Stomates and Intercellular Air Spaces

A good demonstration of the presence of stomates in leaves, as well as the presence and intercommunication of the intercellular spaces, can be made by blowing into the cut end of the petiole of the leaf of a calla lily, the lamina being immersed in water. The air is forced out through the stomata and rises as bubbles to the surface of the water. At the close of the experiment some of the air bubbles will still be in contact with the leaf surface at the opening of the stomata. The pressure of the water gradually forces this back into the leaf. Other plants will answer for the experiment, but some are more suitable than others.

Number of Stomata

The larger number of stomata are on the under side of the leaf. (In leaves which float on the surface of the water all of the stomata are on the upper side of the leaf, as in the water-lily.) It has been estimated by investigation that in general there are 40-300 stomata to the square millimeter of surface. In some plants this number is exceeded, as in the olive, where there are 625. In an entire leaf of Brassica rapa there are about 11,000,000 stomata, and in an entire leaf of the sunflower there are about 13,000,000 stomata.

Amount of Water Transpired by Plants

The amount of water transpired by plants is very great. According to careful estimates a sunflower 6 feet high transpires on the average about one quart per day; an acre of cabbages 2,000,000 quarts in four months; an oak tree with 700,000 leaves transpires about 180 gallons of water per day. According to von Höhnel, a beech tree 110 years old transpired about 2250 gallons of water in one summer. A hectare of such trees (about 400 on 2½ acres) would at the same rate transpire about 900,000 gallons, or about 30,000 barrels in one summer.

Chapter 5

Path of Movement of Water in Plants

In our study of root pressure and transpiration we have seen that large quantities of water or solutions move upward through the stems of plants. We are now led to inquire through what part of the stems the liquid passes in this upward movement, or in other words, what is the path of the "sap" as it rises in the stem. This we can readily see by the following trial.

Place the cut ends of leafy shoots in a solution of some of the red dyes.—We may cut off leafy shoots of various plants and insert the cut ends in a vessel of water to which have been added a few crystals of the dye known as fuchsin to make a deep red color (other red dyes may be used, but this one is especially good). If the study is made during the summer, the "touch-me-not" (impatiens) will be found a very useful plant, or the garden balsam, which may also be had in the winter from conservatories. Almost any plant will do, however, but we should also select one like the corn plant (zea mays) if in the summer, or the petioles of a plant like caladium, which can be obtained from the conservatory. If seedlings of the castor-oil bean are at hand we may cut off some shoots which are 8-10 inches high, and place them in the solution also.

These solutions color the tracts in the stem and leaves through which they flow.—After a few hours in the case of the impatiens, or the more tender plants, we can see through the stem that certain tracts are colored red by the solution, and after 12 to 24 hours there may be seen a red coloration of the leaves of some of the plants used. After the shoots have been standing in the solution for a few hours, if we cut them at various places we will note that there are several points in the section where the tissues are colored red. In the impatiens perhaps from four to five, in the sunflower a larger number. In these plants the colored areas on a cross-section of the stem are situated in a concentric ring which separates more or less completely an outer ring of the stem from the central portion. If we now split portions of the stem lengthwise we see that these colored areas continue throughout the length of the stem, in some cases even up to the leaves and into them.

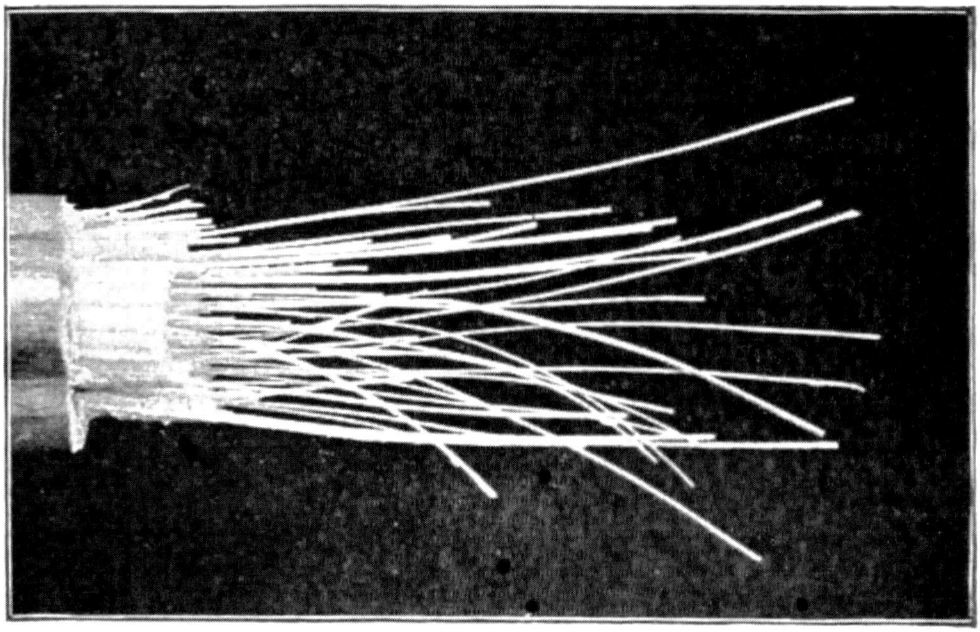

Figure 57. Broken corn stalk, showing fibrovascular bundles.

If we cut across the stem of a corn plant which has been in the solution, we see that instead of the colored areas being in a concentric ring they are irregularly scattered, and on splitting the stem we see here also that these colored areas extend for long distances through the stem. If we take a corn stem which is mature, or an old and dead one, cut around through the outer hard tissues, and then break the stem at this point, from the softer tissue long strings of tissue will pull out as shown in Figure 57. These strings of denser tissue correspond to the areas which are colored by the dye. They are in the form of minute bundles, and are called *vascular bundles*.

We thus see that instead of the liquids passing through the entire stem they are confined to definite courses. Now that we have discovered the path of the upward movement of water in the stem, we are curious to see what the structure of these definite portions of the stem is.

Structure of the Fibrovascular Bundles

We should now make quite thin cross-sections, either free hand and mount in water for microscopic examination, or they may be made with a microtome and mounted in Canada balsam, and in this condition will answer for future study.

To illustrate the structure of the bundle in one type we may take the stem of the castor-oil bean. On examining these cross-sections we see that there are groups of cells which are denser than the ground tissue. These groups correspond to the colored areas in the former experiments, and are the vascular bundles cut across. These groups are somewhat oval in outline, with the pointed end directed toward the center of the stem. If we look at the section as a whole we see that there is a narrow continuous ring[7] of small cells situated at the same distance from the center of the stem as the middle part of the bundles, and that it divides the bundles into two groups of cells.

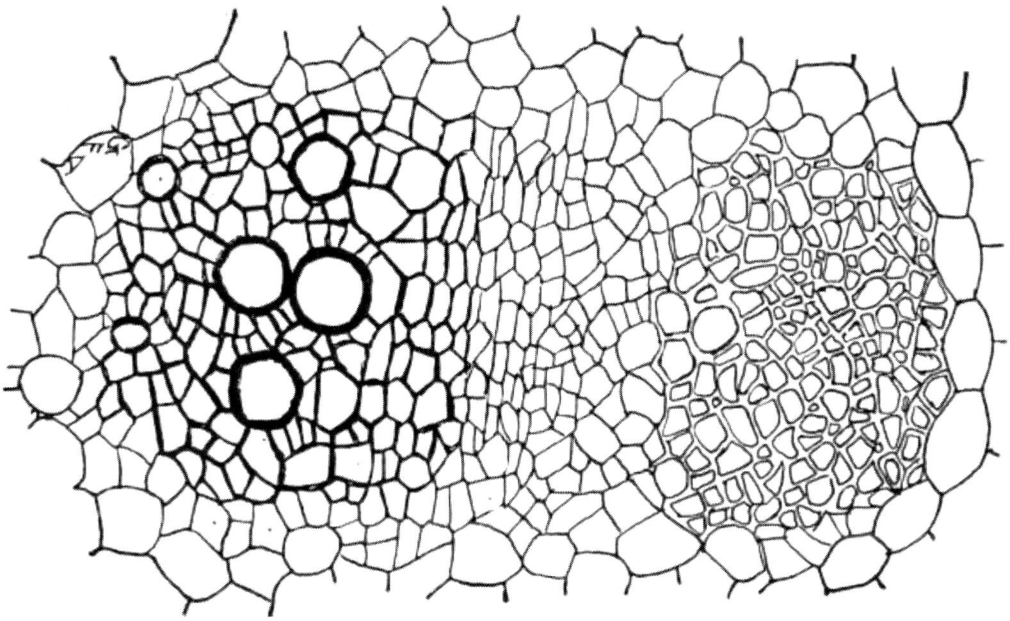

Figure 58. Xylem portion of bundle. Cambium portion of bundle. Bast portion of bundle. Section of vascular bundle of sunflower stem.

Woody Portion of the Bundle

In that portion of the bundle on the inside of the ring, i.e., toward the "pith," we note large, circular, or angular cavities. The walls of these cells are quite thick and woody. They are therefore called wood cells, and because they are continuous with cells above and below them in the stem in such a way that long tubes are formed, they are called woody vessels.

[7] This ring and the bundles separate the stem into two regions, an outer one composed of large cells with thin walls, known as the cortical cells, or collectively the *cortex*. The inner portion, corresponding to what is called the pith, is made up of the same kind of cells and is called the *medulla*, or *pith*. When the cells of the cortex, as well as of the pith, remain thin walled the tissue is called parenchyma. Parenchyma belongs to the group of tissues called fundamental.

Mixed in with these are smaller cells, some of which also have thick walls and are wood cells. Some of these cells may have thin walls. This is the case with all when they are young, and they are then classed with the fundamental tissue or soft tissue (parenchyma). This part of the bundle, since it contains woody vessels and fibres, is the *wood portion* of the bundle, or technically the *xylem*.

Bast Portion of the Bundle

If our section is through a part of the stem which is not too young, the tissues of the outer part of the bundle will show either one or several groups of cells which have white and shiny walls, that are thickened as much or more than those of the wood vessels. These cells are *bast* cells, and for this reason this part of the bundle is the *bast* portion, or the *phloem*. Intermingled with these, cells may often be found which have thin walls, unless the bundle is very old. Nearer the center of the bundle and still within the bast portion are cells with thin walls, angular and irregularly arranged. This is the softer portion of the bast, and some of these cells are what are called *sieve* tubes, which can be better seen and studied in a longitudinal section of the stem.

Cambium Region of the Bundle

Extending across the center of the bundle are several rows of small cells, the smallest of the bundle, and we can see that they are more regularly arranged, usually in quite regular rows, like bricks piled upon one another. These cells have thinner walls than any others of the bundle, and they usually take a deeper stain when treated with a solution of some of the dyes. This is because they are younger, and are therefore richer in protoplasmic contents. This zone of young cells across the bundle is the *cambium*. Its cells grow and divide, and thus increase the size of the bundle. By this increase in the number of the cells of the cambium layer, the outermost cells on either side are continually passing over into the phloem, on the one hand, and into the wood portion of the bundle, on the other hand.

Longitudinal Section of the Bundle

If we make thin longisections of the vascular bundle of the castor-oil seedling (or other dicotyledon) so that we have thin ones running through a bundle radially, as shown in Figure 59, we can see the structure of these parts of the bundle in side view.

We see here that the form of the cells is very different from what is presented in a cross-section of the same. The walls of the various ducts have peculiar markings on them. These markings are caused by the walls being thicker in some places than in others, and this thickening takes place so regularly in some instances as to form regular spiral thickenings. Others have the thickenings in the form of the rounds of a ladder, while still others have pitted walls or the thickenings are in the form of rings.

Figure 59. Longitudinal section of vascular bundle of sunflower stem; spiral, scalariform and pitted vessels at left; next are wood fibers with oblique cross walls; in middle are cambium cells with straight cross walls, next two sieve tubes, then phloem or bast cells.

Vessels or Ducts

One way in which the cells in side view differ greatly from an end view, in a cross-section in the bundle, is that they are much longer in the direction of the axis of the stem. The cells have become elongated greatly. If we search for the place where two of these large cells with spiral, or ladder-like, markings meet end to end, we see that the wall which formerly separated the cells has nearly or quite disappeared. In other words the two cells have now an open communication at the ends. This is so for long distances in the stem, so that long columns of these large cells form tubes or vessels through which the water rises in the stems of plants.

In the bast portion of the bundle we detect the cells of the bast fibers by their thick walls. They are very much elongated and the ends taper out to thin points so that they overlap. In this way they serve to strengthen the stem.

Sieve Tubes

Lying near the bast cells, usually toward the cambium, are elongated cells standing end to end, with delicate markings on their cross walls which appear like finely punctured plates or sieves. The protoplasm in such cells is usually quite distinct, and sometimes contracted away from the side walls, but attached to the cross walls, and this aids in the detection of the sieve tubes (Figure 59.) The granular appearance which these plates present is caused by minute perforations through the wall so that there is a communication between the cells. The tubes thus formed are therefore called sieve tubes and they extend for long distances through the tube so that there is communication throughout the entire length of the stem. (The function of the sieve tubes is supposed to be that for the downward transportation of substances elaborated in the leaves.)

If we section in like manner the stem of the sunflower we shall see similar bundles, but the number is greater than eight. In the garden balsam the number is from four to six in an ordinary stem 3-4*mm* diameter. Here we can see quite well the origin of the vascular bundle. Between the larger bundles we can see especially in free-hand sections of stems through which a colored solution has been lifted by transpiration, as in our former experiments, small groups of the minute cells in the cambial ring which are colored. These groups of cells which form strands running through the stem are *pro-cambium strands*. The cells divide and increase just like the cambium cells, and the older ones thrown off on either side change, those toward the center of the stem to wood vessels and fibers, and those on the outer side to bast cells and sieve tubes.

Fibrovascular Bundles in the Indian Corn

We should now make a thin transection of a portion of the center of the stem of Indian corn, in order to compare the structure of the bundle with that of the plants which we have just examined. In Figure 60 is represented a fibrovascular bundle of the stem of the Indian corn. The large cells are those of the spiral and reticulated and annular vessels. This is the woody portion of the bundle or xylem. Opposite this is the bast portion or phloem, marked by the lighter colored tissue at *i*. The larger of these cells are the sieve tubes, and intermingled with them are smaller cells with thin walls. Surrounding the entire bundle are small cells with thick walls. These are elongated and the tapering ends overlap. They are thus slender and long and form fibers. In such a bundle all of the cambium has passed over into permanent tissue and is said to be closed.

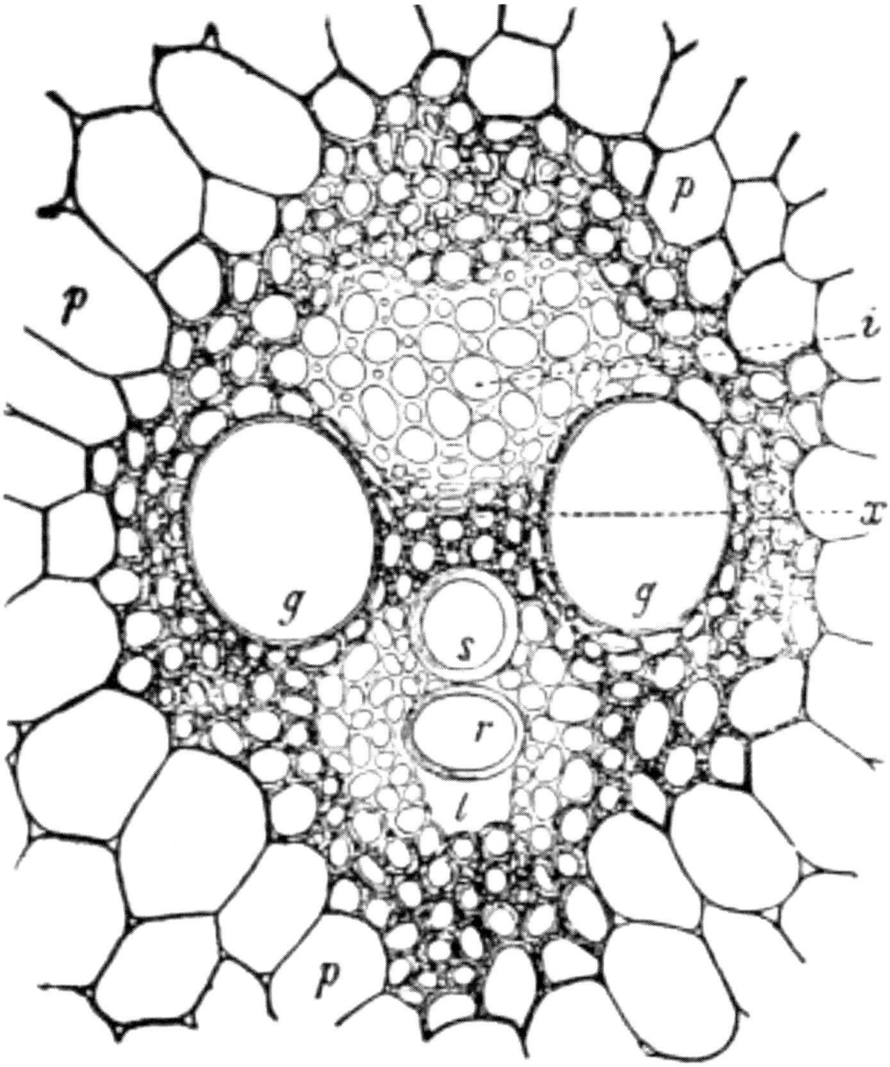

Figure 60. Transection of fibrovascular bundle of Indian corn. *a*, toward periphery of stem; *g*, large pitted vessels; *s*, spiral vessel; *r*, annular vessel; *l*, air cavity formed by breaking apart of the cells;*i*, soft bast, a form of sieve tissue; *p*, thin-walled parenchyma. (Sachs.)

RISE OF WATER IN THE VESSELS

During the movement of the water or nutrient solutions upward in the stem the vessels of the wood portion of the bundle in certain plants are nearly or quite filled, if root pressure is active and transpiration is not very rapid. If, however, on dry days transpiration is in excess of root pressure, as often happens, the vessels are not filled with the water, but are partly filled with certain gases because the air or other gases in the plant become rarefied as a result of the excessive loss of water. There are then successive rows of air or gas bubbles in the vessels separated by films of water which also line the walls of the vessels.

The condition of the vessel is much like that of a glass tube through which one might pass the "froth" which is formed on the surface of soapy water. This forms a chain of bubbles in the vessels. This chain has been called Jamin's chain because of the discoverer.

Why water or food solutions can be raised by the plant to the height attained by some trees has never been satisfactorily explained. There are several theories propounded which cannot be discussed here. It is probably a very complex process. Root pressure and transpiration both play a part, or at least can be shown, as we have seen, to be capable of lifting water to a considerable height. In addition to this, the walls of the vessels absorb water by diffusion, and in the other elements of the bundle capillarity comes also into play, as well as osmosis.

See Organization of Tissues, Chapter 38.

Flow of Sap in the Spring

The cause of the bleeding of trees and the flow of sap in the spring is little understood. One of the remarkable cases is the flow of sap in maple trees. It begins in early spring and ceases as the buds are opening, and seems to be initiated by alternation of high and low temperatures of day and night. It has been found that the pressures inside of the tree at this time are enormously increased during the day, when the temperature rises after a cold night. This has led to the belief that the pressure is caused by the expansion of the gases in the vascular ducts. The warming up of the twigs and branches of the tree would take place rapidly during the day, while the interior of the trunk would be only slightly affected. The pressures then would cause the sap to flow downward during the day, and at night the branches becoming cool, sap would flow back again from the roots and trunk.

Recent experiments by Jones *et al.* show that while some of the pressure is due to the expansion of gas in the tree by the rise of temperature, this cannot account for the enormous pressures which are often present, for example, when after a rise in the temperature of 2° C. there was an increase of 20 lbs. pressure.

Then again, after the cessation of the flow in late spring there are often as great differences between night and day temperatures. It therefore seems reasonable to conclude that the expansion of gases by a rise in temperature is not the direct cause.

Activities of the Cells

It has been suggested by some that the rise in temperature exercises an influence on the protoplasts, or living cells, so that they are stimulated to a special activity resulting in an exudation pressure from the individual cells, which is known to take place. With the fall of temperature at night this activity would cease and there might result a lessened pressure in the cells. Since the specific activities of cells are known to vary in different plants, and in the same plant at different seasons, some support is gained for this theory, though it is generally believed that the activities of the living cells in the stems are not necessary for the upward flow of water. It must be admitted, however, that at present we know very little about this interesting problem.

Chapter 6

Mechanical Uses of Water

Turgidity of Plant Parts

As we have seen by the experiments on the leaves, turgescence of the cells is one of the conditions which enables the leaves to stand out from the stem, and the lamina of the leaves to remain in an expanded position, so that they are better exposed to the light, and to the currents of air. Were it not for this turgidity the leaves would hang down close against the stem.

Figure 61. Restoration of turgidity (Sachs).

Restoration of Turgidity in Shoots

If we cut off a living stem of geranium, coleus, tomato, or "balsam," and allow the leaves to partly wilt so that the shoot loses its turgidity, it is possible for this shoot to regain turgidity. The end may be freshly cut again, placed in a vessel of water, covered with a bell jar and kept in a room where the temperature is suitable for the growth of the plant. The shoot will usually become turgid again from the water which is absorbed through the cut end of the stem and is carried into the leaves where the individual cells become turgid, and the leaves are again expanded. Such shoots, and the excised leaves also, may often be made turgid again by simply immersing them in water, as one of the experiments with the salt solution would teach.

Turgidity may be restored more certainly and quickly in a partially wilted shoot in another way. The cut end of the shoot may be inserted in a U tube as shown in Figure 61, the end of the tube around the stem of the plant being made air-tight. The arm of the tube in which the stem is inserted is filled with water and the water is allowed to partly fill the other arm. Into this other arm is then poured mercury. The greater weight of the mercury causes such pressure upon the water that it is pushed into the stem, where it passes up through the vessels in the stems and leaves, and is brought more quickly and surely to the cells which contain the protoplasm and cell-sap, so that turgidity is more quickly and certainly attained.

Tissue Tensions

Besides the turgescence of the cells of the leaves and shoots there are certain tissue tensions without which certain tender and succulent shoots, etc., would be limp, and would droop. There are a number of plants usually accessible, some at one season and some at others, which may be used to illustrate tissue tension.

Longitudinal Tissue Tension

For this in early summer one may use the young and succulent shoots of the elder (sambucus); or the petioles of rhubarb during the summer and early autumn; or the petioles of richardia. Petioles of caladium are excellent for this purpose, and these may be had at almost any season of the year from the greenhouses, and are thus especially advantageous for work during late autumn or winter. The tension is so strong that a portion of such a petiole 10-15*cm* long is ample to demonstrate it. As we grasp the lower end of the petiole of a caladium, or rhubarb leaf, we observe how rigid it is, and how well it supports the heavy expanded lamina of the leaf.

The ends of a portion of such a petiole or other object which may be used are cut off squarely. With a knife a strip from 2-3*mm* in thickness is removed from one side the full length of the object. This strip we now find is shorter than the larger part from which it was removed. The outer tissue then exerts a tension upon the petiole which tends to shorten it. Let us remove another strip lying next this one, and another, and so on until the outer tissues remain only upon one side. The object will now bend toward that side. Now remove this strip and compare the length of the strips removed with the central portion. We find that they are much shorter now. In other words there is also a tension in the tissue of the central portion of the petiole, the direction of which is opposite to that of the superficial tissue. The parts of the petiole now are not rigid, and they easily bend. These two longitudinal tissue tensions acting in opposition to each other therefore give rigidity to the succulent shoot. It is only when the individual cells of such shoots or petioles are turgid that these tissue tensions in succulent shoots manifest themselves or are prominent.

Figure 62. Strip from dandelion stem made to imitate a plant tendril.

To demonstrate the efficiency of this tension in giving support, let us take a long petiole of caladium or of rhubarb. Hold it by one end in a horizontal position. It is firm and rigid, and does not droop, or but little. Remove all of the outer portion of the tissues, as described above, leaving only the central portion. Now attempt to hold it in a horizontal position by one end. It is flabby and droops downward because the longitudinal tension is removed.

Longitudinal Tension in Dandelion Stems

Take long and fresh dandelion stems. Split them. Note that they coil. The longitudinal tension is very great. Place some of these strips in fresh water. They coil up into close curls because by the absorption of water by the cells the turgescence of the individual cells is increased, and this increases the tension in the stem. Now place them in salt water (a 5 per cent solution). Why do they uncoil?

To Imitate the Coiling of a Tendril

Cut out a narrow strip from a long dandelion stem. Fasten to a piece of soft wood, with the ends close together, as shown in Figure 62. Now place it in fresh water and watch it coil. Part of it coils one way and part another way, just as a tendril does after the free end has caught hold of some place for support.

Transverse Tissue Tension

To illustrate this one may take a willow shoot 3-5 cm diameter and saw off sections about 2 cm long. Cut through the bark on one side and peel it off in a single strip. Now attempt to replace it. The bark will not quite cover the wood again, since the ends will not meet. It must then have been held in transverse tension by the woody part of the shoot.

Chapter 7

STARCH AND SUGAR FORMATION

THE GASES CONCERNED

Gas Given Off by Green Plants in the Sunlight

Let us take some green alga, like spirogyra, which is in a fresh condition, and place one lot in a beaker or tall glass vessel of water and set this in the direct sunlight or in a well lighted place. At the same time cover a similar vessel of spirogyra with black cloth so that it will be in the dark, or at least in very weak light.

Figure 63. Oxygen gas given off by spirogyra.

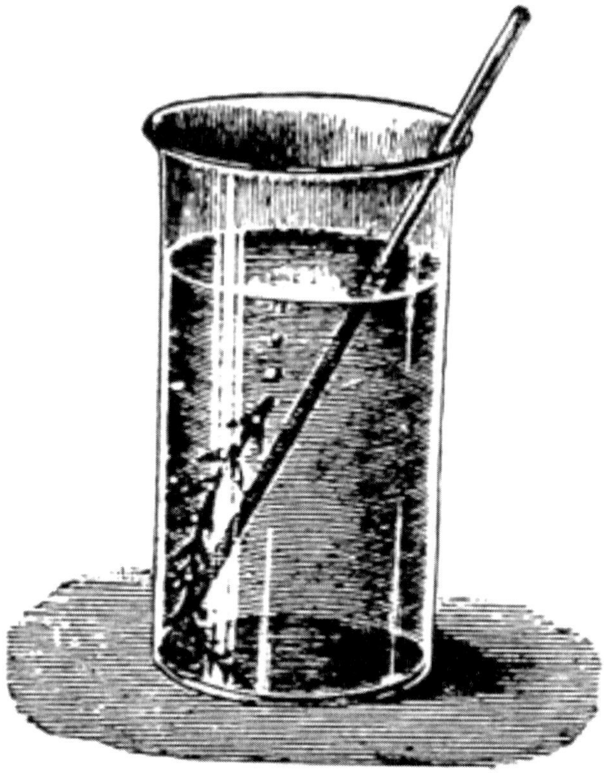

Figure 64. Bubbles of oxygen gas given off from elodea in presence of sunlight. (Oels.)

In a short time we note that in the first vessel small bubbles of gas are accumulating on the surface of the threads of the spirogyra, and now and then some free themselves and rise to the surface of the water. Where there is quite a tangle of the threads the gas is apt to become caught and held back in larger bubbles, which on agitation of the vessel are freed.

If we now examine the second vessel we see that there are no bubbles, or only a very few of them. We are led to believe then that sunlight has had something to do with the setting free of this gas from the plant.

We may now take another alga-like vaucheria and perform the experiment in the same way, or to save time the two may be set up at once. In fact if we take any of the green algæ and treat them as described above gas will be given off in a similar manner.

We may now take one of the higher green plants, an aquatic plant like elodea, callitriche, etc. Place the plant in the water with the cut end of the stem uppermost, but still immersed, the plant being weighted down by a glass rod or other suitable object. If we place the vessel of water containing these leafy stems in the bright sunlight, in a short time bubbles of gas will pass off quite rapidly from the cut end of the stem. If in the same vessel we place another stem, from which the leaves have been cut, the number of bubbles of gas given off will be very few. This indicates that a large part of the gas is furnished by the leaves.

Another vessel fitted up in the same way should be placed in the dark or shaded by covering with a box or black cloth. It will be seen here, as in the case of spirogyra, that very few or no bubbles of gas will be set free. Sunlight here also is necessary for the rapid escape of the gas.

We may easily compare the rapidity with which light of varying intensity effects the setting free of this gas. After cutting the end of the stem let us plunge the cut surface several times in melted paraffine, or spread over the cut surface a coat of varnish. Then prick with a needle a small hole through the paraffine or varnish. Immerse the plant in water and place in sunlight as before. The gas now comes from the puncture through the coating of the cut end, and the number of bubbles given off during a given period can be ascertained by counting. If we duplicate this experiment by placing one plant in weak light or diffused sunlight, and another in the shade, we can easily compare the rapidity of the escape of the gas under the different conditions, which represent varying intensities of light. We see then that not only is sunlight necessary for the setting free of this gas, but that in diffused light or in the shade the activity of the plant in this respect is less than in direct sunlight.

What This Gas Is

If we take quite a quantity of the plants of elodea and place them under an inverted funnel which is immersed in water, the gas will be given off in quite large quantities and will rise into the narrow exit of the funnel. The funnel should be one with a short tube, or the vessel one which is quite deep so that a small test tube which is filled with water may in this condition be inverted over the opening of the funnel tube. With this arrangement of the experiment the gas will rise in the inverted test tube, slowly displace a portion of the water, and become collected in a sufficient quantity to afford us a test. When a considerable quantity has accumulated in the test tube, we may close the end of the tube in the water with the thumb, lift it from the water and invert. The gas will rise against the thumb. A dry soft-pine splinter should be then lighted, and after it has burned a short time, extinguish the flame by blowing upon it, when the still burning end of the splinter should be brought to the mouth of the tube as the thumb is quickly moved to one side. The glowing of the splinter shows that the gas is *oxygen*.

It is better to allow the apparatus to stand several days in the sunlight in order to catch a full tube of the gas. Or on a sunny day carbon dioxide gas can be led into the water in the jar from a generator, such an one as is used for the evolution of CO_2. The CO_2 can be produced by the action of hydrochloric acid on bits of marble. The CO_2 should not be run below the funnel. The test tube should be fastened so that the light oxygen gas will not raise it off the funnel. With the tube full of gas the test for oxygen can be made by lifting the tube with one hand and quickly thrusting the glowing end of the splinter in with the other hand. If properly handled, the splinter will flame again. If it is necessary to keep the apparatus standing for more than one day it is well to add fresh water in the place of most of the water in the jar. Do not use leaves of land plants in this experiment, since the bubbles which rise when these leaves are placed in water are not evidence that this process is taking place.

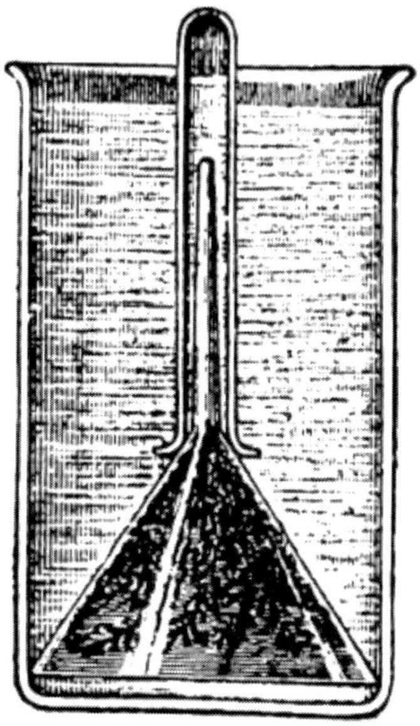

Figure 65. Apparatus for collecting quantity of oxygen from elodea. (Detmer.)

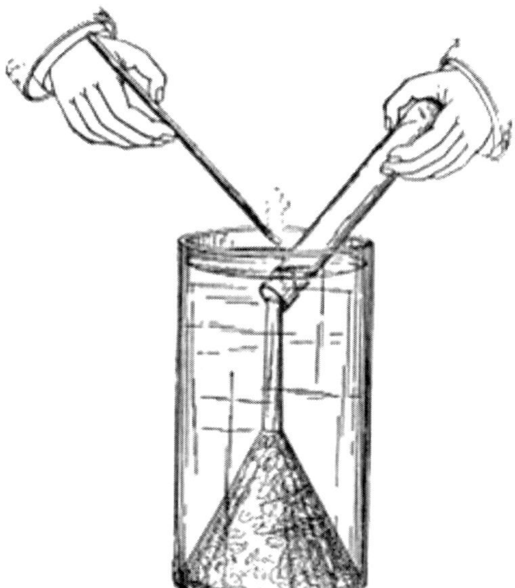

Figure 66. Ready to see what the gas is.

Figure 67. The splinter lights again in the presence of oxygen gas.

Oxygen Given Off By Green Land Plants Also

If we should extend our experiments to land plants we should find that oxygen is given off by them under these conditions of light. Land plants, however, will not do this when they are immersed in water, but it is necessary to set up rather complicated apparatus and to make analyses of the gases at the beginning and at the close of the experiments. This has been done, however, in a sufficiently large number of cases so that we know that all green plants in the sunlight, if temperature and other conditions are favorable, give off oxygen.

Absorption of Carbon Dioxide

We have next to inquire where the oxygen comes from which is given off by green plants when exposed to the sunlight, and also to learn something more of the conditions necessary for the process. We know that water which has been for some time exposed to the air and soil, and has been agitated, like running water of streams, or the water of springs, has mixed with it a considerable quantity of oxygen and carbon dioxide.

If we boil spring water or hydrant water which comes from a stream containing oxygen and carbon dioxide, for about 20 minutes, these gases are driven off. We should set this aside where it will not be agitated, until it has cooled sufficiently to receive plants without injury. Let us now place some spirogyra or vaucheria, and elodea, or other green water plant, in this boiled water and set the vessel in the bright sunlight under the same conditions which were employed in the experiments for the evolution of oxygen. No oxygen is given off.

Can it be that this is because the oxygen was driven from the water in boiling? We shall see. Let us take the vessel containing the water, or some other boiled water, and agitate it so that the air will be thoroughly mixed with it. In this way oxygen is again mixed with the water. Now place the plant again in the water, set in the sunlight, and in several minutes observe the result. No oxygen or but little is given off. There must be then some other requisite for the evolution of the oxygen.

The Gases Are Interchanged in the Plants

We will now introduce carbon dioxide again in the water. This can be done by leading CO_2 from a gas generator into the water. Broken bits of marble are placed in the generator, acted upon by hydrochloric acid, and the gas is led over by glass tubing. Now if we place the plant in the water and set the vessel in the sunlight, in a few minutes the oxygen is given off rapidly.

A Chemical Change of the Gas Takes Place within the Plant Cell

This leads us to believe then that CO_2 is in some way necessary for the plant in this process. Since oxygen is given off while carbon dioxide, a different gas, is necessary, it would seem that a chemical change takes place in the gases within the plant. Since the process takes place in such simple plants as spirogyra as well as in the more bulky and higher plants, it appears that the changes go on within the cell, in fact within the protoplasm.

Gases As Well As Water Can Diffuse Through the Protoplasmic Membrane

Carbon dioxide then is absorbed by the plant while oxygen is given off. We see therefore that gases as well as water can diffuse through the protoplasmic membrane of plants under certain conditions.

Where Starch Is Formed

We have found by these simple experiments that some chemical change takes place within the protoplasm of the green cells of plants during the absorption of carbon dioxide and the giving off of oxygen. We should examine some of the green parts of those plants used in the experiments, or if they are not at hand we should set up others in order to make this examination.

Starch Formed as a Result of this Process

We may take spirogyra which has been standing in water in the bright sunlight for several hours. A few of the threads should be placed in alcohol for a short time to kill the protoplasm. From the alcohol we transfer the threads to a solution of iodine in potassium iodide. We find that at certain points in the chlorophyll band a bluish tinge, or color, is imparted to the ring or sphere which surrounds the pyrenoid. In our first study of the spirogyra cell we noted this sphere as being composed of numerous small grains of starch which surround the pyrenoid.

Iodine Used as a Test for Starch

This color reaction which we have obtained in treating the threads with iodine is the well-known reaction, or test, for starch. We have demonstrated then that starch is present in spirogyra threads which have stood in the sunlight with free access to carbon dioxide.

If we examine in the same way some threads which have stood in the dark for a few days we obtain no reaction for starch, or at best only a slight reaction. This gives us some evidence that a chemical change does take place during this process (absorption of CO_2 and giving off of oxygen), and that starch is a product of that chemical change.

Schimper's Method of Testing for the Presence of Starch

Another convenient and quick method of testing for the presence of starch is what is known as Schimper's method. A strong solution of chloral hydrate is made by taking 8 grams of chloral hydrate for every 5*cc* of water. To this solution is added a little of an alcoholic tincture of iodine. The threads of spirogyra may be placed directly in this solution, and in a few moments mounted in water on the glass slip and examined with the microscope. The reaction is strong and easily seen.

We should also examine the leaves of elodea, or one of the higher green plants which has been for some time in the sunlight. We may use here Schimper's method by placing the leaves directly in the solution of chloral hydrate and iodine. The leaves are made transparent by the chloral hydrate so that the starch reaction from the iodine is easily detected.

The following is a convenient and safe method of extracting chlorophyll from leaves. Fill a large pan, preferably a dishpan, half full of hot water. This may be kept hot by a small flame. On the water float an evaporating dish partly filled with alcohol. The leaves should be first immersed in the hot water for several minutes, then placed in the alcohol, which will quickly remove the chlorophyll. Now immerse the leaves in the iodine solution.

Figure 68. Leaf of coleus showing green and white areas, before treatment with iodine.

Figure 69. Similar leaf treated with iodine, the starch reaction only showing where the leaf was green.

Green Parts of Plants Form Starch When Exposed to Light

Thus we find that in the case of all the green plants we have examined, starch is present in the green cells of those which have been standing for some time in the sunlight where the process of the absorption of CO_2 and the giving off of oxygen can go on, and that in the case of plants grown in the dark, or in leaves of plants which have stood for some time in the dark, starch is absent. We reason from this that starch is the product of the chemical change which takes place in the green cells under these conditions. The CO_2 which is absorbed by the plant mixes with the water (H_2O) in the cell and immediately forms carbonic acid. The chlorophyll in the leaf absorbs radiant energy from the sun which splits up the carbonic acid, and its elements then are put together into a more complex compound, starch. This process of putting together the elements of an organic compound is a *synthesis*, or a *synthetic assimilation*, since it is done by the living plant. It is therefore a synthetic assimilation of carbon dioxide. Since the sunlight supplies the energy it is also called *photosynthesis*, or *photosynthetic assimilation*. We can also say carbon dioxide assimilation, or CO_2 assimilation (see paragraph on assimilation at close of Chapter 10).

Starch is Formed Only in the Green Parts of Variegated Leaves

If we test for starch in variegated leaves like the leaf of a coleus plant, we shall have an interesting demonstration of the fact that the green parts of plants only form starch. We may take a leaf which is partly green and partly white, from a plant which has been standing for some time in bright light. Figure 68 is from a photograph of such a leaf. We should first boil it in alcohol to remove the green color. Now immerse it in the potassium iodide of iodine solution for a short time. The parts which were formerly green are now dark blue or nearly black, showing the presence of starch in those portions of the leaf, while the white part of the leaf is still uncolored. This is well shown in Figure 69, which is from a photograph of another coleus leaf treated with the iodine solution.

3. Chlorophyll and the Formation of Starch

In our experiments thus far in treating of the absorption of carbon dioxide and the evolution of oxygen, with the accompanying formation of starch, we have used green plants.

Fungi Cannot Form Starch

If we should extend our experiments to the fungi, which lack the green color so characteristic of the majority of plants, we should find that photosynthesis does not take place even though the plants are exposed to direct sunlight. These plants cannot then form starch, but obtain carbohydrates for food from other sources.

Photosynthesis Cannot Take Place in Etiolated Plants

Moreover photosynthesis is usually confined to the green plants, and if by any means one of the ordinary green plants loses its green color this process cannot take place in that plant, even when brought into the sunlight, until the green color has appeared under the influence of light.

This may be very easily demonstrated by growing seedlings of the bean, squash, corn, pea, etc. (pine seedlings are green even when grown in the dark), in a dark room, or in a dark receiver of some kind which will shut out the rays of light. The room or receiver must be quite dark. As the seedlings are "coming up," and as long as they remain in the dark chamber, they will present some other color than green; usually they are somewhat yellowed. Such plants are said to be *etiolated*. If they are brought into the sunlight now for a few hours and then tested for the presence of starch the result will be negative. But if the plant is left in the light, in a few days the leaves begin to take on a green color, and then we find that carbon dioxide assimilation begins.

Chlorophyll and Chloroplasts

The green substance in plants is then one of the important factors in this complicated process of forming starch. This green substance is *chlorophyll*, and it usually occurs in definite bodies, the chlorophyll bodies, or *chloroplasts*.

The material for new growth of plants grown in the dark is derived from the seed. Plants grown in the dark consist largely of water and protoplasm, the walls being very thin.

Form of the Chlorophyll Bodies

Chlorophyll bodies vary in form in some different plants, especially in some of the lower plants. This we have already seen in the case of spirogyra, where the chlorophyll body is in the form of a very irregular band, which courses around the inner side of the cell wall in a spiral manner. In zygnema, which is related to spirogyra, the chlorophyll bodies are star-shaped. In the desmids the form varies greatly. In œdogonium, another of the thread-like algæ, illustrated in Figure 144, the chlorophyll bodies are more or less flattened oval disks. In vaucheria, too, a branched thread-like alga shown in Figure 138, the chlorophyll bodies are oval in outline. These two plants, œdogonium and vaucheria, should be examined here if possible, in order to become familiar with their form, since they will be studied later under morphology (see chapters on œdogonium and vaucheria, for the occurrence and form of these plants). The form of the chlorophyll body found in œdogonium and vaucheria is that which is common to many of the green algæ, and also occurs in the mosses, liverworts, ferns, and the higher plants. It is a more or less rounded, oval, flattened body.

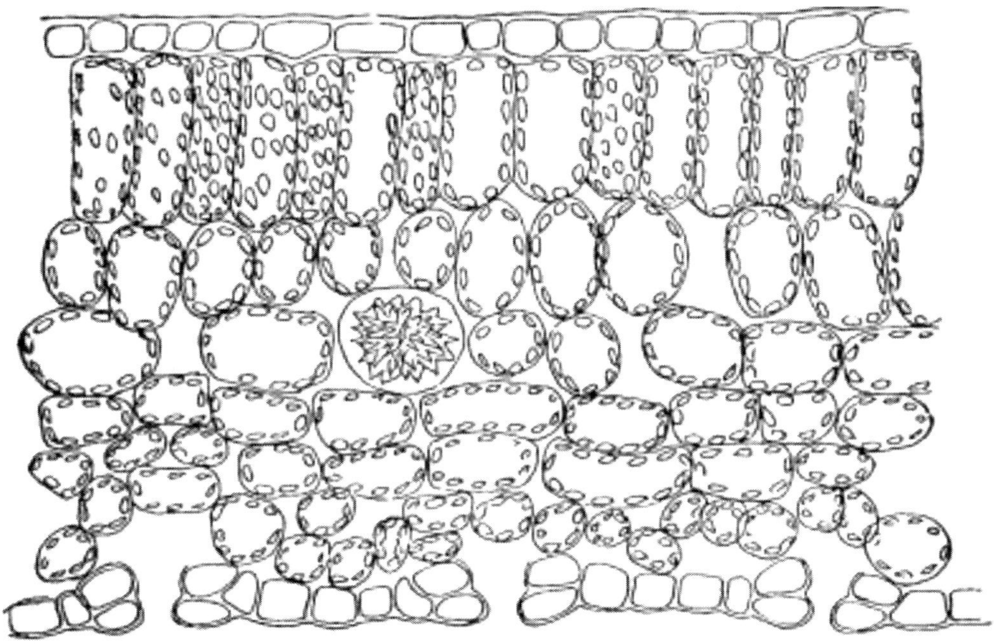

Figure 69*a*. Section of ivy leaf, palisade cells above, loose parenchyma, with large intercellular spaces in center. Epidermal cells on either edge, with no chlorophyll bodies.

Chlorophyll is a Pigment Which Resides in the Chloroplast

That the chlorophyll is a coloring substance which resides in the chloroplastid, and does not form the body itself, can be demonstrated by dissolving out the chlorophyll when the framework of the chloroplastid is apparent. The green parts of plants which have been placed for some time in alcohol lose their green color. The alcohol at the same time becomes tinged with green. In sectioning such plant tissue we find that the chlorophyll bodies, or chloroplastids as they are more properly called, are still intact, though the green color is absent. From this we know that chlorophyll is a substance distinct from that of the chloroplastid.

Chlorophyll Absorbs Energy from Sunlight for Photosynthesis

It has been found by analysis with the spectroscope that chlorophyll absorbs certain of the rays of the sunlight. The energy which is thus obtained from the sun, called *kinetic* energy, acts on the molecules of CH_2O_3, separating them into molecules of C, H, and O. (When the CO_2 from the air enters the plant cell it immediately unites with some of the water, forming carbonic acid = CH_2O_3.) After a series of complicated chemical changes starch is formed by the union of carbon, oxygen, and hydrogen. In this process of the reduction of the CH_2O_3 and the formation of starch there is a surplus of oxygen, which accounts for the giving off of oxygen during the process.

Rays of Light Concerned in Photosynthesis

If a solution of chlorophyll be made, and light be passed through it, and this light be examined with the spectroscope, there appear what are called absorption bands. These are dark bands which lie across certain portions of the spectrum. These bands lie in the red, orange, yellow, green, blue, and violet, but the bands are stronger in the red, which shows that chlorophyll absorbs more of the red rays of light than of the other rays. These are the rays of low refrangibility. The kinetic energy derived by the absorption of these rays of light is transformed into potential energy. That is, the molecule of CH_2O_3 is broken up, and then by a different combination of certain elements starch is formed.[8]

Starch Grains Formed in the Chloroplasts

During photosynthesis the starch formed is deposited generally in small grains within the green chloroplast in the leaf. We can see this easily by examining the leaves of some moss-like funaria which has been in the light, or in the chloroplasts of the prothallia of ferns, etc. Starch grains may also be formed in the chloroplasts from starch which was

[8] In the formation of starch during photosynthesis the separated molecules from the carbon dioxide and water unite in such a way that carbon, hydrogen, and oxygen are united into a molecule of starch. This result is usually represented by the following equation: $CO_2 + H_2O = CH_2O + O_2$. Then by polymerization $6(CH_2O) = C_6H_{12}O_6$ = grape sugar. Then $C_6H_{12}O_6\text{-}H_2O = C_6H_{10}O_5$ = starch. It is believed, however, that the process is much more complicated than this, that several different compounds are formed before starch finally appears, and that the formula for starch is much higher numerically than is represented by $C_6H_{10}O_5$.

formed in some other part of the plant, but which has passed in solution. Thus the functions of the chloroplast are twofold, that of photosynthesis and the formation of starch grains.

In the translocation of starch when it becomes stored up in various parts of the plant, it passes from the state of solution into starch grains in connection with plastids similar to the chloroplasts, but which are not green. The green ones are sometimes called *chloroplasts*, while the colorless ones are termed *leucoplasts*, and those possessing other colors, as red and yellow, in floral leaves, the root of the carrot, etc., are called *chromoplasts*.

Photosynthesis in Other than Green Plants

While carbohydrates are usually only formed by green plants, there are some exceptions. Apparent exceptions are found in the blue-green algæ, like oscillatoria, nostoc, or in the brown and red sea weeds like fucus, rhabdonia, etc. These plants, however, possess chlorophyll, but it is disguised by another pigment or color. There are plants, however, which do not have chlorophyll and yet form carbohydrates with evolution of oxygen in the presence of light, as for example a purple bacterium, in which the purple coloring substance absorbs light, though the rays absorbed most energetically are not the red.

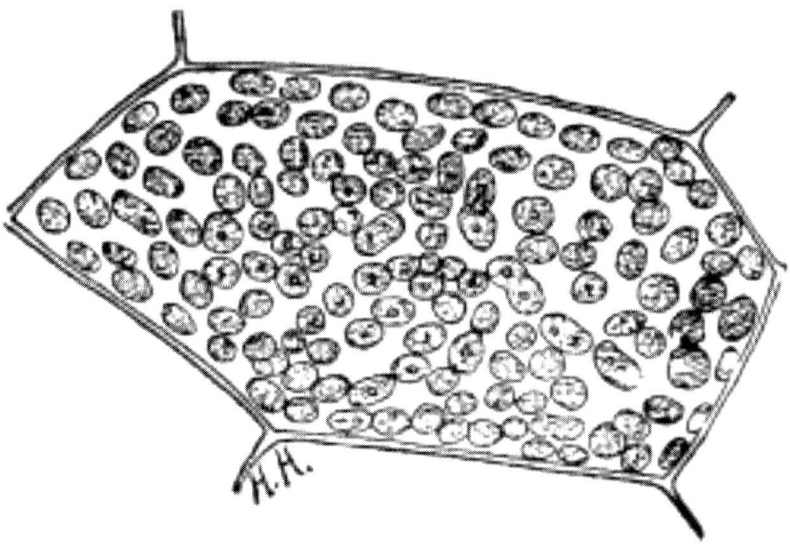

Figure 70. Cell exposed to weak diffused light showing chlorophyll bodies along the horizontal walls.

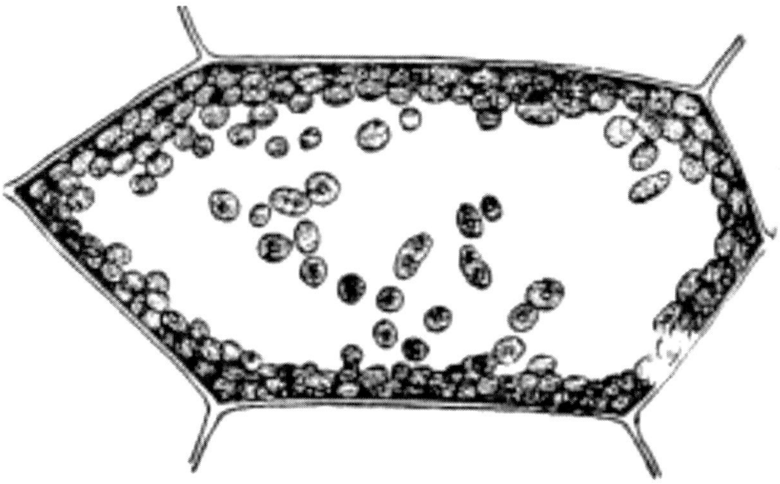

Figure 71. Same cell exposed to strong light, showing chlorophyll bodies have moved to perpendicular walls.

Figures 70, 71.—Cell of prothallium of fern.

Influence of Light on the Movement of Chlorophyll Bodies—In Fern Prothallia

If we place fern prothallia in weak light for a few hours, and then examine them under the microscope, we find that the most of the chlorophyll bodies in the cells are arranged along the inner surface of the horizontal wall. If now the same prothallia are placed in a brightly lighted place for a short time most of the chlorophyll bodies move so that they are arranged along the surfaces of the perpendicular walls, and instead of having the flattened surfaces exposed to the light as in the former case, the edges of the chlorophyll bodies are now turned toward the light. (See Figures 70, 71.) The same phenomenon has been observed in many plants. Light then has an influence on chlorophyll bodies, to some extent determining their position. In weak light they are arranged so that the flattened surfaces are exposed to the incidence of the rays of light, so that the chlorophyll will absorb as great an amount as possible of kinetic energy; but intense light is stronger than necessary, and the chlorophyll bodies move so that their edges are exposed to the incidence of the rays. This movement of the chlorophyll bodies is different from that which takes place in some water plants like elodea. The chlorophyll bodies in elodea are free in the protoplasm. The protoplasm in the cells of elodea streams around the inside of the cell wall much as it does in nitella and the chlorophyll bodies are carried along in the currents, while in nitella they are stationary.

Chapter 8

Starch and Sugar Concluded. Analysis of Plant Substance

Translocation of Starch

Translocation of Starch

It has been found that leaves of many plants grown in the sunlight contain starch when examined after being in the sunlight for several hours. But when the plants are left in the dark for a day or two the leaves contain no starch, or a much smaller amount. This suggests that starch after it has been formed may be transferred from the leaves, or from those areas of the leaves where it has been formed.

Figure 72. Leaf of tropæolum with portion covered with corks to prevent the formation of starch. (After Detmer.)

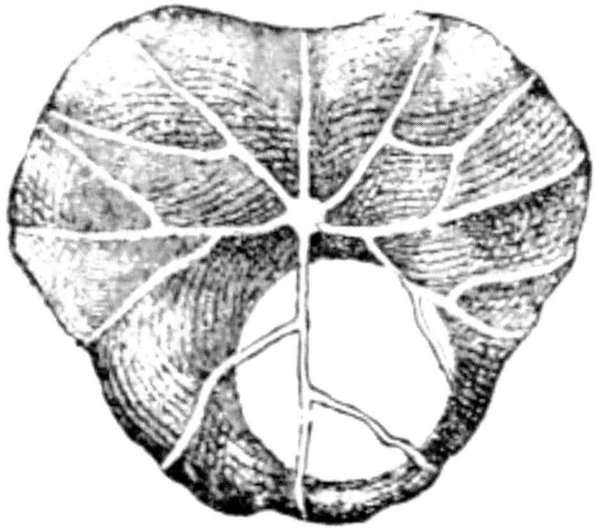

Figure 73. Leaf of tropæolum treated with iodine after removal of cork, to show that starch is removed from the leaf during the night.

To test this let us perform an experiment which is often made. We may take a plant such as a garden tropæolum or a clover plant, or other land plant in which it is easy to test for the presence of starch. Pin a piece of circular cork, which is smaller than the area of the leaf, on either side of the leaf, as in Figure 72, but allow free circulation of air between the cork and the under side of the leaf. Place the plant where it will be in the sunlight. On the afternoon of the following day, if the sun has been shining, test the entire leaf for starch. The part covered by the cork will not give the reaction for starch, as shown by the absence of the bluish color, while the other parts of the leaf will show it. The starch which was in that part of the leaf the day before was dissolved and removed during the night, and then during the following day, the parts being covered from the light, no starch was formed in them.

Starch in Other Parts of Plants than the Leaves

We may use the iodine test to search for starch in other parts of plants than the leaves. If we cut a potato tuber, scrape some of the cut surface into a pulp, and apply the iodine test, we obtain a beautiful and distinct reaction showing the presence of starch. Now we have learned that starch is only formed in the parts containing chlorophyll. We have also learned that the starch which has been formed in the leaves disappears from the leaf or is transferred from the leaf. We judge therefore that the starch which we have found in the tuber of the potato was formed first in the green leaves of the plant, as a result of photosynthesis. From the leaves it is transferred in solution to the underground stems, and stored in the tubers. The starch is stored here by the plant to provide food for the growth of new plants from the tubers, which are thus much more vigorous than the plants would be if grown from the seed.

Form of Starch Grains

Where starch is stored as a reserve material it occurs in grains which usually have certain characters peculiar to the species of plant in which they are found. They vary in size in many different plants, and to some extent in form also. If we scrape some of the cut surface of the potato tuber into a pulp and mount a small quantity in water, or make a thin section for microscopic examination, we find large starch grains of a beautiful structure. The grains are oval in form and more or less irregular in outline. But the striking peculiarity is the presence of what seem to be alternating dark and light lines in the starch grain. We note that the lines form irregular rings, which are smaller and smaller until we come to the small central spot termed the "hilum" of the starch grain. It is supposed that these apparent lines in the starch grain are caused by the starch substance being deposited in alternating dense and dilute layers, the dilute layers containing more water than the dense ones; others think that the successive layers from the hilum outward are regularly of diminishing density, and that this gives the appearance of alternating lines. The starch formed by plants is one of the organic substances which are manufactured by plants, and it (or glucose) is the basis for the formation of other organic substances in the plant. Without such organic substances green plants cannot make any appreciable increase of plant substance, though a considerable increase in size of the plant may take place.

Note.—The organic compounds resulting from photosynthesis, since they are formed by the union of carbon, hydrogen, and oxygen in such a way that the hydrogen and oxygen are usually present in the same proportion as in water, are called *carbohydrates*. The most common carbohydrates are sugars (cane sugar, $C_{12}H_{22}O_{11}$ for example, in beet roots, sugar cane, sugar maple, etc.), starch, and cellulose.

Vaucheria

The result of carbon dioxide assimilation in the threads of Vaucheria is not clearly understood. Starch is absent or difficult to find in all except a few species, while oil globules are present in most species. These oil globules are spherical, colorless, globose and highly refringent. Often small ones are seen lying against chlorophyll bodies. Oil is a *hydrocarbon* (containing C, H, and O, but the H and O are in different proportions from what they are in H_2O) and until recently it was supposed that this oil in Vaucheria was the direct result of photosynthesis. But the oil does not disappear when the plant is kept for a long time in the dark, which seems to show that it is not the direct product of carbon dioxide assimilation, and indicates that it comes either from a temporary starch body or from glucose. Schimper found glucose in several species of Vaucheria, and Waltz says that some starch is present in Vaucheria sericea, while in V. tuberosa starch is abundant and replaces the oil. To test for oil bodies in Vaucheria treat the threads with weak osmic acid, or allow them to stand for twenty-four hours in Fleming's solution (which contains osmic acid). Mount some threads and examine with microscope. The oil globules are stained black.

Sugar, and Digestion of Starch[9]

It is probable that some form of sugar is always produced as the result of photosynthesis. The sugar thus formed may be stored as such or changed to starch. In general it may be said that sugar is most common in the green parts of monocotyledonous plants, while starch is most frequent in dicotyledons. Plant sugars are of three general kinds: cane sugar abundant in the sugar cane, sugar beet, sugar maple, etc.; glucose and fruit sugar, found in the fruits of a majority of plants, and abundant in some, as in apples, pears, grapes, etc.; and maltose, a variety produced in germinating seeds, as in malted barley.

Test for Sugar

A very pretty experiment maybe made by taking two test tubes, placing in one a solution of commercial grape sugar (glucose), in the other one of granulated cane sugar, and adding to each a few drops of Fehling's solution.[10] After these tubes have stood in a warm place for half an hour, it will be found that a bright orange brown or cinnabar-colored precipitate of copper and cuprous oxide has formed in the tube containing grape sugar, while the other solution is unchanged. Grape sugar or glucose, therefore, reduces Fehling's solution, while cane sugar as such has no effect upon it.

Cane sugar may be changed or converted to glucose by being boiled for a short time with a dilute acid, or by adding Fehling's solution to the sugar solution and boiling. In the

---

[9] Paragraphs 156-160 were prepared by Dr. E. J. Durand.

[10] Make up three stock solutions as follows:

(1)

Copper sulphate	9 grams
Water	250 cc.

(2)

Caustic potash	30 grams
Water	250 cc.

(3)

Rochelle salts	49 grams
Water	250 cc.

For Fehling's solution take one volume of each of (1), (2), and (3), and to the mixture add two volumes of water.

latter case the change is brought about by the alkali and the precipitate of copper and cuprous oxide forms.

Tests for Sugar in Plant Tissue

(*a*) Scrape out a little of the tissue from the inside of a ripe apple or pear, place it with a little water in a test tube, and add a few drops of Fehling's solution. After standing half an hour the characteristic precipitate of copper and cuprous oxide appears, showing that grape sugar is present in quantity.

Make thin sections of the apple and mount in a drop of Fehling's solution on a slide. After half an hour examine with the microscope. The granules of cuprous oxide are present in the cells of the tissue in great abundance.

(*b*) Cut up several leaves of a young vigorous corn seedling, cover with water in a test tube and boil for a minute. After the decoction has cooled add the Fehling's solution and allow to stand. The precipitate will appear. For comparison take similar corn leaves, remove the chlorophyll with alcohol and test with iodine. No starch reaction appears. The carbohydrate in corn leaves is therefore glucose and not starch. If now the corn seed be examined the cells will be found to be full of starch grains which give the beautiful blue reaction with iodine. This experiment shows that grape sugar is formed in the leaves of the corn plant, but is changed to starch when stored in the seed.

(*c*) Take two leaves of bean seedling or coleus, test one for sugar and the other for starch. Both are present.

(*d*) Procure some maple sap in the spring, or in the winter months make a decoction of the broken tips of young branches of the sugar maple by boiling them in water in a test tube. To the sap or cool decoction add Fehling's solution. No precipitate appears after standing. Now heat the same solution to the boiling point, and the precipitate forms, showing the presence of cane sugar in the maple sap which was converted to glucose and fruit sugar by boiling in the presence of an alkali.

(*e*) Scrape out some of the tissue from a sugar beet root, cover with water in a test tube and add Fehling's solution. No change takes place after standing. Boil the same solution and the precipitate forms, showing the presence of cane sugar, inverted to grape sugar and fruit sugar by the hot alkali.

How Starch Is Changed To Sugar

We have seen that in many plants the carbohydrate formed as the result of carbon dioxide assimilation is stored as starch. This substance being insoluble in water must be changed to sugar, which is soluble before it can be used as food or transported to other parts of the plant. This is accomplished through the action of certain enzymes, principally diastase. This substance has the power of acting upon starch under proper conditions of temperature and moisture, causing it to take up the elements of water, and so to become sugar.

This process takes place commonly in the leaves where starch is formed, but especially in seeds, tubers (during the sprouting, etc.), and other parts which the plant uses as storehouses for starch food. It is probable that the same conditions of temperature and moisture which favor germination or active growth are also favorable to the production of diastase.

Experiments to Show the Action of Diastase

(*a*) Place a bit of starch half as large as a pea in a test tube, and cover with a weak solution[11] (about ⅕ per cent) of commercial taka diastase. After it has stood in a warm place for five or ten minutes test with Fehling's solution. The precipitate of cuprous oxide appears showing that some of the starch has been changed to sugar. By using measured quantities, and by testing with iodine at frequent intervals, it can be determined just how long it takes a given quantity of diastase to change a known quantity of starch. In this connection one should first test a portion of the same starch with Fehling's solution to show that no sugar is present.

(*b*) Repeat the above experiment using a little tissue from a potato, and some from a corn seed.

(*c*) Take 25 germinating barley seeds in which the radicle is just appearing. Grind up thoroughly in a mortar with about three parts of water. After this has stood for ten or fifteen minutes, filter. Fill a test tube one-third full of water, add a piece of starch half the size of a pea or less, and boil the mixture to make starch-paste. Add the barley extract. Put in a warm place and test from time to time with iodine. The first samples so treated will be blue, later ones violet, brown, and finally colorless, showing that the starch has all disappeared. This is due to the action of the diastase which was present in the germinating seeds, and which was dissolved out and added to the starch mixture. The office of this diastase is to change the starch in the seeds to sugar. Germinating wheat is sweet, and it is a matter of common observation that bread made from sprouted wheat is sweet.

(*d*) Put a little starch-paste in a test tube and cover it with saliva from the mouth. After ten or fifteen minutes test with Fehling's solution. A strong reaction appears showing how quickly and effectively saliva acts in converting starch to sugar. Successive tests with iodine will show the gradual disappearance of the starch.

---

[11] This solution of taka diastase should be made up cold. If it is heated to 60° C. or over it is destroyed.

These experiments have shown us that diastase from three different sources can act upon starch converting it into sugar. The active principle in the saliva is an *animal* diastase (*ptyalin*), which is necessary as one step in the digestion of starch food in animals. The *taka* diastase is derived from a fungus (Eurotium oryzæ) which feeds on the starch in rice grains converting it into sugar which the fungus absorbs for food. The *malt* diastase and *leaf* diastase are formed by the seed plants. That in seeds converts the starch to sugar which is absorbed by the embryo for food. That in the leaf converts the starch into sugar so that it can be transported to other parts of the plant to be used in building new tissue, or to be stored again in the form of starch (example, the potato, in seeds, etc.). The starch is formed in the leaf during the daylight. The light renders the leaf diastase inactive. But at night the leaf diastase becomes active and converts the starch made during the day. Starch is not soluble in water, while the sugar is, and the sugar in solution is thus easily transported throughout the plant. In those green plants which do not form starch in their leaves (sugar beet, corn, and many monocotyledons), grape sugar and fruit sugar are formed in the green parts as the result of photosynthesis. In some, like the corn, the grape sugar formed in the leaves is transported to other parts of the plant, and some of it is stored up in the seed as starch. In others like the sugar beet the glucose and fruit sugar formed in the leaves flow to other parts of the plant, and much of it is stored up as cane sugar in the beet root. The process of photosynthesis probably proceeds in the same way in all cases up to the formation of the grape sugar and fruit sugar in the leaves. In the beet, corn, etc., the process stops here, while in the bean, clover, and most dicotyledons the process is carried one step farther in the leaf and starch is formed.

3. Rough Analysis of Plant Substance

Some Simple Experiments to Indicate the Nature of Plant Substance

After these building-up processes of the plant, it is instructive to perform some simple experiments which indicate roughly the nature of the plant substance, and serve to show how it can be separated into other substances, some of them being reduced to the form in which they existed when the plant took them as food. For exact experiments and results it would be necessary to make chemical analyses.

The Water in the Plant

Take fresh leaves or leafy shoots or other fresh plant parts. Weigh. Permit them to remain in a dry room until they are what we call "dry." Now weigh. The plants have lost weight, and from what we have learned in studies of transpiration this loss in weight we know to result from the loss of water from the plant.

The Dry Plant Material Contains Water

Take air-dry leaves, shavings, or other dry parts of plants. Place them in a test tube. With a holder rest the tube in a nearly horizontal position, with the bottom of the tube in the flame of a Bunsen burner. Very soon, before the plant parts begin to "burn," note that moisture is accumulating on the inner surface of the test tube. This is water driven off which could not escape by drying in air, without the addition of artificial heat, and is called "hygroscopic water."

Water Formed On Burning the Dry Plant Material

Light a soft-pine or basswood splinter. Hold a thistle tube in one hand with the bulb downward and above the flame of the splinter. Carbon will be deposited over the inner surface of the bulb. After a time hold the tube toward the window and look through it above the carbon. Drops of water have accumulated on the inside of the tube. This water is formed by the rearrangement of some of the hydrogen and oxygen, which is set free by the burning of the plant material, where they were combined with carbon, as in the cellulose, and with other elements.

Formation of Charcoal by Burning

Take dried leaves, and shavings from some soft wood. Place in a porcelain crucible, and cover about 3 cm. deep with dry fine earth. Place the crucible in the flame of a Bunsen burner and let it remain for about fifteen minutes. Remove and empty the contents. If the flame was hot the plant material will be reduced to a good quality of charcoal. The charcoal consists largely of carbon.

The Ash of the Plant

Place in the porcelain crucible dried leaves and shavings as before. Do not cover with earth. Place the crucible in the flame of the Bunsen burner, and for a moment place on the porcelain cover; then remove the cover, and note the moisture on the under surface from the escaping water.

Permit the plant material to burn; it may even flame for a time. In the course of fifteen minutes it is reduced to a whitish powder, much smaller in bulk than the charcoal in the former experiment. This is the ash of the plant.

What Has Become of the Carbon?

In this experiment the air was not excluded from the plant material, so that oxygen combined with carbon as the water was freed, and formed carbon dioxide, passing off into the air in this form. This it will be remembered is the form in which the plant took the carbon-food in through the leaves. Here the carbon dioxide met the water coming from the soil, and the two united to form, ultimately, starch, cellulose, and other compounds of carbon; while with the addition of nitrogen, sulphur, etc., coming also from the soil, still other plant substances were formed.

The carbohydrates are classed among the non-nitrogenous substances. Other non-nitrogenous plant substances are the organic acids like oxalic acid ($H_2C_2O_4$), malic acid ($H_2C_4H_4O_5$), etc.; the fats and fixed oils, which occur in the seeds and fruits of many plants. Of the nitrogenous substances the proteids have a very complex chemical formula and contain carbon, hydrogen, oxygen, nitrogen, sulphur, etc. (example, *aleuron*, or proteid grains, found in seeds). The proteids are the source of nitrogenous food for the seedling during germination. Of the amides, *asparagin* ($C_4H_8N_2O_3$) is an example of a nitrogenous substance; and of the alkaloids, nicotin ($C_{10}H_{14}N_2$) from tobacco.

All living plants contain a large per cent of water. According to Vines "ripe seeds dried in the air contain 12 to 15 per cent of water, herbaceous plants 60 to 80 per cent, and many water plants and fungi as much as 95 per cent of their weight." When heated to 100 °C. the water is driven off. The dry matter remaining is made up partly of organic compounds, examples of which are given above, and inorganic compounds. By burning this dry residue the organic substances are mostly changed into volatile products, principally carbonic acid, water, and nitrogen. The inorganic substances as a result of combustion remain as a white or gray powder, the *ash*.

The amount of the ash increases with the age of the plant, though the percentage of ash may vary at different times in the different members of the plant. The following table taken from Vines will give an idea of the amount and composition of the ash in the dry solid of a few plants:

Content of 1000 Parts of Dry Solid Matter

	Clover, in blossom	Wheat, grain	Wheat, straw	Potato tubers	Apples	Peas (the seed)
Ash	68.3	19.7	53.7	37.7	14.4	27.3
Potash.	21.96	6.14	7.33	22.76	5.14	11.41
Soda.	1.39	0.44	0.74	0.99	3.76	0.26
Lime.	24.06	0.66	3.09	0.97	0.59	1.36
Magnesium.	7.44	2.36	1.33	1.77	1.26	2.17
Ferric Oxide.	0.72	0.26	0.33	0.45	0.2	0.16
Phosphoric Acid.	6.74	9.26	2.58	6.53	1.96	9.95
Sulphuric Acid.	2.06	0.07	1.32	2.45	0.88	0.95
Silica.	1.62	0.42	36.25	0.8	0.62	0.24
Chlorine.	2.66	0.04	0.9	1.17	....	0.42

Chapter 9

HOW PLANTS OBTAIN THEIR FOOD - I

SOURCES OF PLANT FOOD

The Necessary Constituents of Plant Food

As indicated in Chapter 3, investigation has taught us the principal constituents of plant food. Some suggestion as to the food substances is derived by a chemical analysis of various plants. In Chapter 8 it was noted that there are two principal kinds of compounds in plant substances, the organic compounds and the inorganic compounds or mineral substances. The principal elements in the organic compounds are *hydrogen*, *carbon*, *oxygen* and *nitrogen*. The elements in the inorganic compounds which have been found indispensable to plant growth are *calcium*,[12] *potassium*, *magnesium*, *phosphorus*, *sulphur* and *iron*. (See paragraphs 54-58, and complete observations on water cultures.) Other elements are found in the ash of plants; and while they are not absolutely necessary for growth, some[13] of them are beneficial in one way or another.

The carbohydrates are derived, as we have learned, from the CO_2 of the air, and water in the plant tissue drawn from the soil; though in the case of aquatic plants entirely submerged, all the constituents are absorbed from the surrounding water.

Food Substances in the Soil

Land plants derive their mineral food from the soil, the soil received the mineral substances from dissolving and disintegrating rocks. Nitrogenous food is chiefly derived

[12] Calcium is not essential for the growth of the fungi.

[13] For example, silicon is used by some plants in strengthening supporting tissues. Buckwheat thrives better when supplied with a chloride.

from the same source, but under a variety of conditions which will be discussed in later paragraphs, but the nitrogen comes primarily from the air. Some of the mineral substances, those which are soluble as well as some of the nitrogenous substances, are found in solution in the soil. These are absorbed by the plant, as needed, along with water, through the root hairs.

Absorption of Soluble Substances

Since these substances are dissolved in the water of the soil, it is not necessary for us to dwell on the process of absorption. This in general is dwelt upon in Chapter 3. It should be noted, however, that food substances in solution, during absorption, diffuse through the protoplasmic membrane independently of each other and also independently of the rate of movement of the water from the soil into the root hairs and cells of the root.

When the cells have absorbed a certain amount of a given substance, no more is absorbed until the concentration of the cell-sap in that particular substance is reduced. This, however, does not interfere with the absorption of water, or of other substances in solution by the same cells. Plants have therefore a certain selective power in the absorption of food substances.

Action of Root Hairs on Insoluble Substances. Acidity of Root Hairs

If we take a seedling which has been grown in a germinator, or in the folds of cloths or paper, so that the roots are free from the soil, and touch the moist root hairs to blue litmus paper, the paper becomes red in color where the root hairs have come in contact. This is the reaction for the presence of an acid salt, and indicates that the root hairs excrete certain acid substances. This acid property of the root hairs serves a very important function in the preparation of certain of the elements of plant food in the soil. Certain of the chemical compounds of potash, phosphoric acid, etc., become deposited on the soil particles, and are not soluble in water. The acid of the root hairs dissolves some of these compounds where the particles of soil are in close contact with them, and the solutions can then be taken up by the roots. Carbonic acid and other acids are also formed in the soil, and aid in bringing these substances into solution.

This corrosive action of the roots can be shown by the well-known experiment of growing a plant on a marble plate which is covered by soil. In lieu of the marble plate, the peas may be planted in clam or oyster shells, which are then buried in the soil of the pot, so that the roots of the seedlings will come in contact with the smooth surface of the shell. After a few weeks, if the soil be washed from the marble where the roots have been in close contact, there will be an outline of this part of the root system. Several different acid substances are excreted from the roots of plants which have been found to redden blue litmus paper by contact. Experiments by Czapek show, however, that the carbonic acid excreted by the roots has the power of directly bringing about these corrosion phenomena. The acid salts are the substances which are most actively concerned in reddening the blue litmus paper. They do not directly aid in the corrosion phenomena. In the soil, however, where these compounds of potash, phosphoric acid, etc., are which are not soluble in water, the acid salt (primary acid potassium phosphate) which is most actively concerned in reddening the blue litmus paper may act indirectly on these mineral substances, making them available for plant food. This salt soon unites with certain chlorides in the soil, making among other things small quantities of hydrochloric acid.

Note.—It is a general rule that plants cannot take solid food into their bodies, but obtain all food in either a liquid or gaseous state. The only exception to this is in the case of the plasmodia of certain *Myxomycetes* (Slime Moulds), and also perhaps some of the Flagellates and other very low forms, which engulf solid particles of food. It is uncertain, however, whether these organisms belong to the plant or animal kingdom, and they probably occupy a more or less intermediate position.

Action of Nitrite and Nitrate Bacteria

Many of the higher green plants prefer their nitrogenous food in the form of nitrates. (Example, nitrate of soda, potassium nitrate, saltpetre.) Nitrates are constantly being formed in soil by the action of certain bacteria. The nitrite bacteria (Nitromonas) convert ammonia in the soil to *nitrous acid* (a *nitrite*), while at this point the nitrate bacteria (Nitrobacter) convert the nitrites into nitrates. The fact that this nitrification is going on constantly in soil is of the utmost importance, for while commercial nitrates are often applied to the soil, the nitrates are easily washed from the soil by heavy rains. These nitrite and nitrate bacteria require oxygen for their activity, and they arc able to obtain their carbohydrates by decomposing organic matter in the soil, or directly by assimilating the CO_2 in the soil, deriving the energy for the assimilation of the carbon dioxide from the chemical process of nitrification. This kind of carbon dioxide assimilation is called *chemosynthetic* assimilation.

Parasites and Saprophytes

Parasites Among the Fungi

A parasite is an organism which derives all or a part of its food directly from another living organism (its host) and at the latter's expense. The larger number of plant parasites are found among the fungi (rusts, smuts, mildews, etc.). (See Nutrition of the Fungi, paragraph 185.) Some of these are not capable of development unless upon their host, and are called *obligate* parasites. Others can grow not only as parasites but at other times can also grow on dead organic matter, and are called *facultative* parasites, i.e. they can choose either a parasitic life or a saprophytic one.

Parasites among the Seed Plants—Cuscuta

There are, however, parasites among the seed plants; for example, the dodder (Cuscuta), parasitic on clover, and a great variety of other plants. There is food enough in the seed for the young plant to take root and develop a slender stem until it takes hold of its host. It then twines around the stem of its host sending wedge-shaped haustoria into the stem to obtain food. The part then in connection with the ground dies.

Figure 74. Dodder.

The haustoria of the dodder form a complete junction with the vascular bundles of its host so that through the vessels water and salts are obtained, while through the junction of sieve tubes the elaborated organic food is obtained. The union of the dodder with its host is like that between a graft and the graft stock. The beech drops (Epiphegus) is another example of a parasitic seed plant. It is parasitic on the roots of the beech.

The mistletoe (*Viscum album*), which grows on the branches of trees, sends its roots into the branches, and only the vessels of the vascular system are fused according to some. If this is true then it probably obtains only water and salts from its host. But the mistletoe has green leaves and is thus able to assimilate carbon dioxide and manufacture its own organic substances. It is claimed by some, however, that the host derives some food from the parasite during the winter when the host has shed its leaves, and if this is true it would seem that organic food could also be derived during the summer from the host by the mistletoe.

Saprophytes

A saprophyte is a plant which is enabled to obtain its food, especially its organic food, directly from dead animals or plants or from dead organic substances. Many fungi are saprophytes, as the moulds, mushrooms, etc. (See Nutrition of the Fungi.)

Humus Saprophytes

The action of fungi as described in the preceding chapter, as well as of certain bacteria, gradually converts the dead plants or plant parts into the finely powdered brown substance known as *humus*. In general the green plants cannot absorb organic food from humus directly. But plants which are devoid of chlorophyll can live saprophytically on this humus. They are known as *humus saprophytes*. Many of the mushrooms and other fungi, as well as some seed plants which lack chlorophyll or possess only a small quantity, are able to absorb all their organic food from humus. It is uncertain whether any seed plants can obtain all of their organic food directly from humus, though it is believed that many can so obtain a portion of it. But a number of seed plants, like the Indian-pipe (Monotropa) and certain orchids, obtain organic food from humus. These plants lack chlorophyll and cannot therefore manufacture their own carbohydrate food. Not being parasitic on plants which can, as in the case of the dodder and beech drops mentioned above, they undoubtedly derive their organic food from the humus. But fungus mycelium growing in the humus is attached to their roots, and in some orchids enters the roots and forms a nutritive connection. The fungus mycelium can absorb organic food from the humus and in some cases at least can transfer it over to the roots of the higher plant (see Mycorhiza).

Autotrophic, Heterotrophic, and Mixotrophic Plants

An *autotrophic* plant is one which is self-nourishing, i.e. it is provided with an abundant chlorophyll apparatus for carbon dioxide assimilation and with absorbing organs for obtaining water and salts. Heterotrophic plants are not provided with a chlorophyll apparatus sufficient to assimilate all the carbon dioxide necessary, so they nourish themselves by other means. *Mixotrophic* plants are those which are intermediate between the other two, i.e. they have some chlorophyll but not enough to provide all the organic food necessary, so they obtain a portion of it by other means. Evidently there are all gradations of mixotrophic plants between the two other kinds (example, the mistletoe).

Symbiosis

Symbiosis means a living with or living together, and is said of those organisms which live so closely in connection with each other as to be influenced for better or worse, especially from a nutrition standpoint. *Conjunctive* symbiosis has reference to those cases where there is a direct interchange of food material between the two organisms (lichens, mycorhiza, etc.). *Disjunctive* symbiosis has reference to an inter-life relation without any fixed union between them (example, the relations between flowers and insects, ants and plants, and even in a broad sense the relation between saprophytic plants in reducing organic matter to a condition in which it may be used for food by the green plants, and these in turn provide organic matter for the saprophytes to feed upon, etc.). *Antagonistic* symbiosis is shown in the relation of parasite to its host, *reciprocal* symbiosis, or *mutualistic* symbiosis is shown in those cases where both symbionts derive food as a result of the union (lichens, mycorhiza, etc.)

How Fungi Obtain their Food

Nutrition of Moulds

In our study of mucor, as we have seen, the growing or vegetative part of the plant, the mycelium, lies within the substratum, which contains the food materials in solution, and the slender threads are thus bathed on all sides by them. The mycelium absorbs the watery solutions throughout the entire system of ramifications. When the upright fruiting threads are developed they derive the materials for their growth directly from the mycelium with which they are in connection. The moulds which grow on decaying fruit or on other organic matter derive their nutrient materials in the same way. The portion of the mould which we usually see on the surface of these substances is in general the fruiting part. The larger part of the mycelium lies hidden within the substratum.

Figure 75. Carnation rust on leaf and flower stem. From photograph.

Nutrition of Parasitic Fungi

Certain of the fungi grow on or within the higher plants and derive their food materials from them and at their expense. Such a fungus is called a *parasite*, and there are a large number of these plants which are known as *parasitic fungi*. The plant at whose expense they grow is called the "*host*."

One of these parasitic fungi, which it is quite easy to obtain in greenhouses or conservatories during the autumn and winter, is the carnation rust (*Uromyces caryophyllinus*), since it breaks out in rusty dark brown patches on the leaves and stems of the carnation (see Figure 75). If we make thin cross-sections through one of these spots on a leaf, and place them for a few minutes in a solution of chloral hydrate, portions of the tissues of the leaf will be dissolved. After a few minutes we wash the sections in water on a glass slip, and stain them with a solution of eosin. If the sections were carefully made, and thin, the threads of the mycelium will be seen coursing between the cells of the leaf as slender threads. Here and there will be seen short branches of these threads which penetrate the cell wall of the host and project into the interior of the cell in the form of an irregular knob. Such a branch is a *haustorium*.

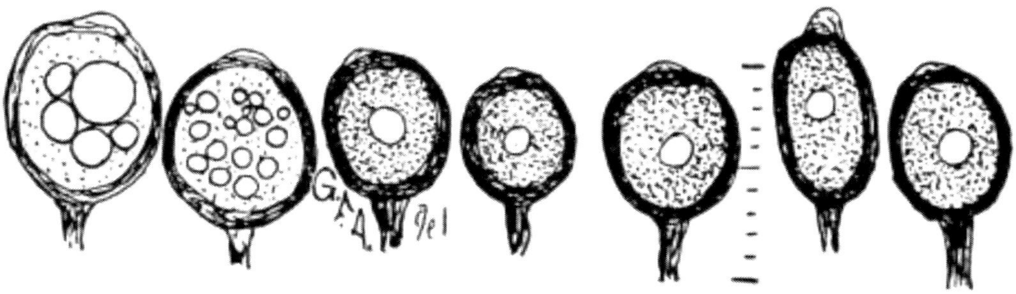

Figure 76. Several teleutospores, showing the variations in form.

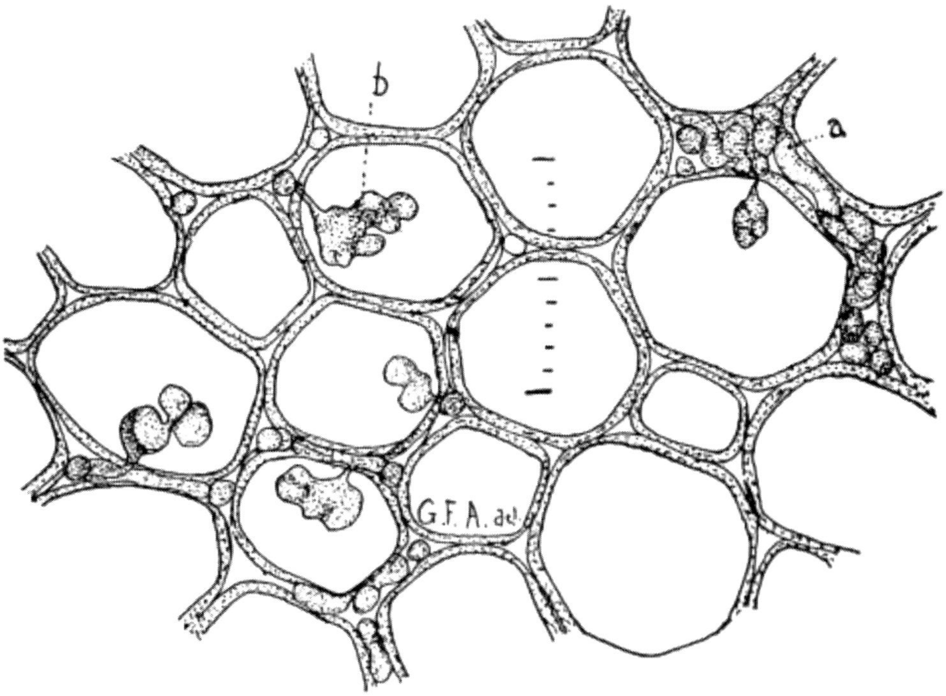

Figure 77. Cells from the stem of a rusted carnation, showing the intercellular mycelium and haustoria. Object magnified 30 times more than the scale.

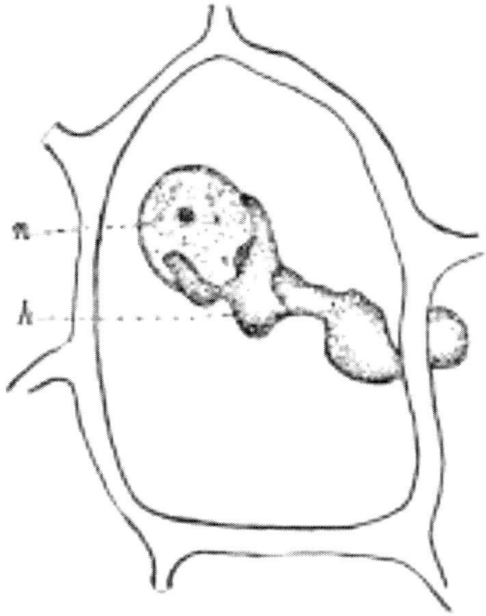

Figure 78. Cell from carnation leaf, showing haustorium of rust mycelium grasping the nucleus of the host. *h*, haustorium; *n*, nucleus of host.

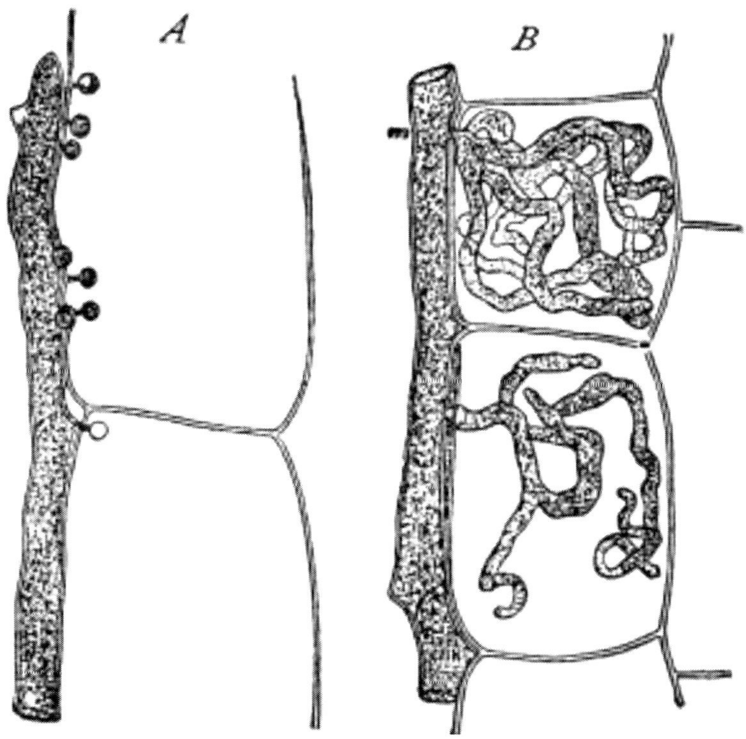

Figure 79. Intercellular mycelium with haustoria entering the cells. *A*, of Cystopus candidus (white rust); *B*, of Peronospora calotheca. (De Bary.)

By means of this haustorium, which is here only a short branch of the mycelium, nutritive substances are taken by the fungus from the protoplasm or cell-sap of the carnation. From here it passes to the threads of the mycelium. These in turn supply food material for the development of the dark brown gonidia, which we see form the dark-looking powder on the spots. Many other fungi form haustoria, which take up nutrient matters in the way described for the carnation rust. In the case of other parasitic fungi the threads of the mycelium themselves penetrate the cells of the host, while in still others the mycelium courses only between the cells of the host (fungus of peach leaf curl for example) and derives food materials from the protoplasm or cell-sap of the host by the process of osmosis.

Nutrition of the Larger Fungi

If we select some one of the larger fungi, the majority of which belong to the mushroom family and its relatives, which is growing on a decaying log or in the soil, we shall see on tearing open the log, or on removing the bark or part of the soil, as the case may be, that the stem of the plant, if it have one, is connected with whitish strands. During the spring, summer, or autumn months, examples of the mushrooms connected with these strands may usually be found readily in the fields or woods, but during the winter and colder parts of the year often they may be seen in forcing houses, especially those cellars devoted to the propagation of the mushroom of commerce.

These strands are made up of numerous threads of the mycelium which are closely twisted and interwoven into a cord or strand, which is called a mycelium strand, or *rhizomorph*. These are well shown in Figure 236, which is from a photograph of the mycelium strands, or "spawn" as the grower of mushrooms calls it, of Agaricus campestris. The little knobs or enlargements on the strands are the young fruit bodies, or "buttons."

While these threads or strands of the mycelium in the decaying wood or in the decaying organic matter of the soil are not true roots, they function as roots, or root hairs, in the absorption of food materials. In old cellars and on damp soil in moist places we sometimes see fine examples of this vegetative part of the fungi, the mycelium. But most magnificent examples are to be seen in abandoned mines where timber has been taken down into the tunnels far below the surface of the ground to support the rock roof above the mining operations. I have visited some of the coal mines at Wilkesbarre, Pa., and here on the wood props and doors, several hundred feet below the surface, and in blackest darkness, in an atmosphere almost completely saturated at all times, the mycelium of some of the wood destroying fungi grows in a profusion and magnificence which is almost beyond belief. Figure 80 is from a flash-light photograph of a beautiful example 400 feet below the surface of the ground. This was growing over the surface of a wood prop or post, and the picture is much reduced. On the doors in the mine one can see the strands of the mycelium which radiate in fan-like Figures at certain places near the margin of growth, and farther back the delicate tassels of mycelium which hang down in fantastic Figures, all in spotless white and rivalling the most beautiful fabric in the exquisiteness of its construction.

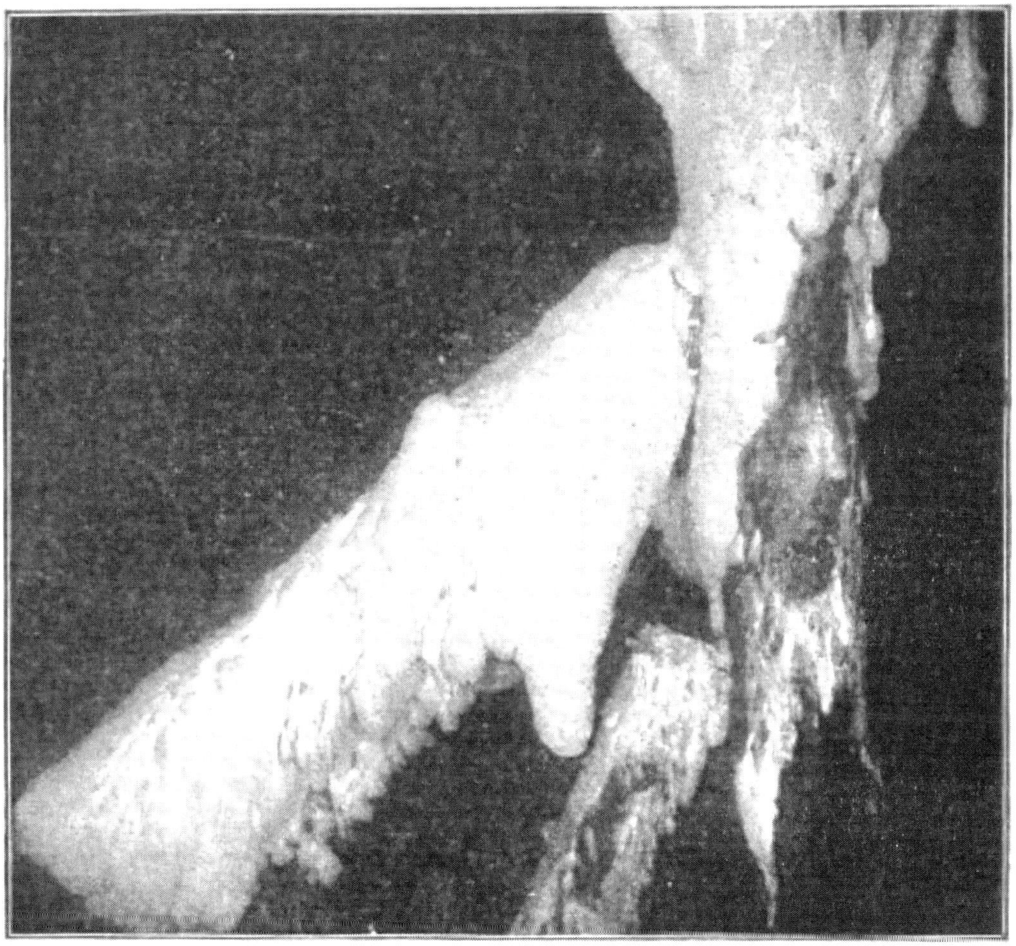

Figure 80. Sterile mycelium on wood props in coal mine, 400 feet below surface. (Photographed by the author.)

How Fungi Derive Carbohydrate Food

The fungi being devoid of chlorophyll cannot assimilate the CO_2 from the air. They are therefore dependent on the green plants for their carbohydrate food. Among the saprophytes, the leaf and wood destroying fungi excrete certain substances (known as *enzymes*) which dissolve the carbohydrates and certain other organic compounds in the woody or leafy substratum in which they grow. They thus produce a sort of extracellular digestion of carbohydrates, converting them into a soluble form which can be absorbed by the mycelium. The parasitic fungi also obtain their carbohydrates and other organic food from the host. The mycelium of certain parasitic, and of wood destroying fungi, excretes enzymes (*cytase*) which dissolve minute perforations in the cell walls of the host and thus aid the hypha during its boring action in penetrating cell walls.

Note.—Certain wood destroying fungi growing in oaks absorb tannin directly, i.e. in an unchanged form. One of the pine destroying fungi (*Trametes pini*) absorbs the xylogen from the wood cells, leaving the pure cellulose in which the xylogen was filtrated; while *Polyporus mollis* absorbs the cellulose, leaving behind only the wood element.

Mycorhiza

While such plants as the Indian-pipe (Monotropa), some of the orchids, etc., are *humus saprophytes* and some of them are possibly able to absorb organic food from the humus, many of them have fungus mycelium in close connection with their roots, and these fungus threads aid in the absorption of organic food. The roots of plants which have fungus mycelium intimately associated in connection with the process of nutrition, are termed *mycorhiza*. There is a mutual interchange of food between the fungus and the host, a *reciprocal symbiosis*.

Mycorhiza are of two kinds as regards the relation of the fungus to the root; *ectotrophic* (or *epiphytic*), where the mycelium is chiefly on the outside of the root, and *endotrophic* (or *endophytic*) where the mycelium is chiefly within the tissue of the root.

Ectotrophic Mycorrhiza

Ectotrophic mycorhiza occur on the roots of the oak, beech, hornbean, etc., in forests where there is a great deal of humus from decaying leaves and other vegetation. The young growing roots of these trees become closely covered with a thick felt of the mycelium, so that no root hairs can develop. The terminal roots also branch profusely and are considerably thickened. The fungus serves here as the absorbent organ for the tree. It also acts on the humus, converting some of it into available plant food and transferring it over to the tree.

Endotrophic Mycorrhiza

These are found on many of the humus saprophytes, which are devoid of chlorophyll, as well as on those possessing little or even on some plants possessing an abundance, of chlorophyll. Examples are found in many orchids (see the coral root orchid, for example), some of the ferns (Botrychium), the pines, leguminous plants, etc. In endotrophic mycorhiza the mycelium is more abundant within the tissues of the root, though some of the threads extend to the outside. In the case of the mycorhiza on the humus saprophytes which have no chlorophyll, or but little, it is thought by some that the fungus mycelium in the humus assists in converting organic substances and carbohydrates into a form available for food by the higher plant and then conducts it into the root, thus aiding also in the process of absorption, since there are few or no root hairs on the short and fleshy mycorhiza. The roots, however, of some of these humus saprophytes have the power of absorbing a portion of their organic compounds from the humus. It is thought by some, though not definitely demonstrated, that in the case of the oaks, beeches, hornbeans, and other chlorophyll-bearing symbionts, the fungus threads do not absorb any carbohydrates for the higher symbiont, but that they actually derive their carbohydrates from it.[14] But it is reasonably certain that the fungus threads do assimilate from the humus certain unoxidized, or feebly oxidized, nitrogenous substances (ammonia, for example), and transfer them over to the host, for the higher plants with difficulty absorb these substances, while they readily absorb nitrates which are not abundant in humus. This is especially important in the forest. It is likely therefore.

NITROGEN GATHERERS

How Clovers, Peas, and Other Legumes Gather Nitrogen

It has long been known that clover plants, peas, beans, and many other leguminous plants are often able to thrive in soil where the cereals do but poorly. Soil poor in nitrogenous plant food becomes richer in this substance where clovers, peas, etc., are grown, and they are often planted for the purpose of enriching the soil. Leguminous plants, especially in poor soil, are almost certain to have enlargements, in the form of nodules, or "root-tubercles." A root of the common vetch with some of these root-tubercles is shown in Figure 81.

---

[14] Evidence points to the belief that certain cells of the host form substances which attract, chemitropically, the fungus threads, and that in these cells the fungus threads are more abundant than in others. Furthermore in the vicinity of the nucleus of the host seems to be the place where these activities are more marked.

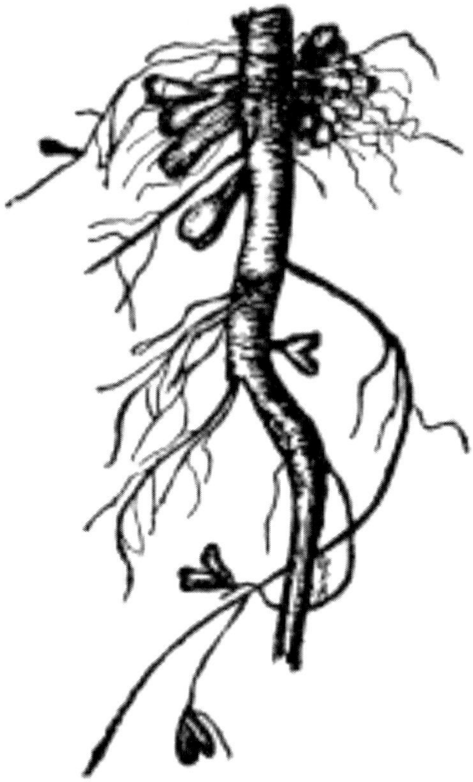

Figure 81. Root of the common vetch, showing root tubercles.

A Fungal or Bacterial Organism in These Root-Tubercles

If we cut one of these root tubercles open, and mount a small portion of the interior in water for examination with the microscope, we shall find small rod-shaped bodies, some of which resemble bacteria, while others are more or less forked into forms like the letter Y, as shown in Figure 82. These bodies are rich in nitrogenous substances, or proteids. They are portions of a minute organism, of a fungus or bacterial nature, which attacks the roots of leguminous plants and causes these nodular outgrowths. The organism (Phytomyxa leguminosarum) exists in the soil and is widely distributed where legumes grow.

How the Organism Gets into the Roots of the Legumes

This minute organism in the soil makes its way through the wall of a root hair near the end. It then grows down the interior of the root hair in the form of a thread. When it reaches the cell walls it makes a minute perforation, through which it grows to enter the adjacent cell, when it enlarges again. In this way it passes from the root hair to the cells of the root and down to near the center of the root. As soon as it begins to enter the cells of the root it stimulates the cells of that portion to greater activity. So the root here develops a large lateral nodule, or "root-tubercle." As this "root-tubercle" increases in size, the fungus threads branch in all directions, entering many cells. The threads are very irregular in form, and from certain enlargements it appears that the rod-like bodies are formed, or the thread later breaks into myriads of these small "bacteroids."

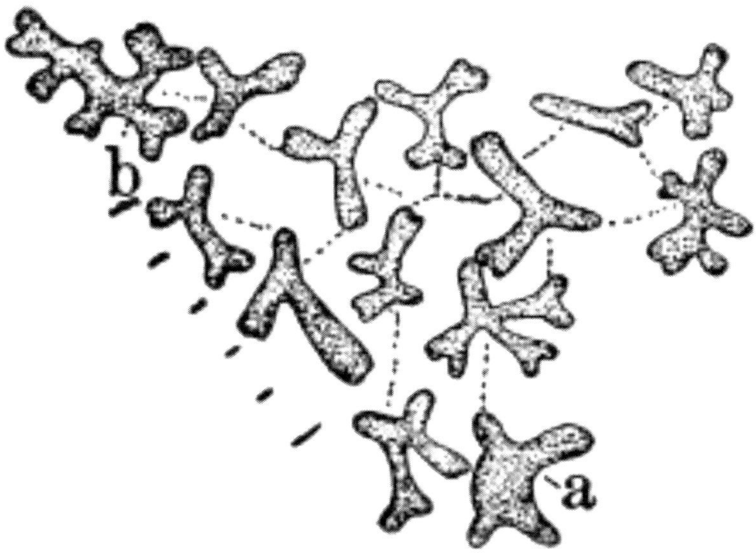

Figure 82. Root-tubercle organism from vetch, old condition.

Figure 83. Root-tubercle organism from Medicago denticulata.

The Root Organism Assimilates Free Nitrogen for Its Host

This organism assimilates the free nitrogen from the air in the soil, to make the proteid substance which is found stored in the bacteroids in large quantities. Some of the bacteroids, rich in proteids, are dissolved, and the proteid substance is made use of by the clover or pea, as the case may be. This is why such plants can thrive in soil with a poor nitrogen content. Later in the season some of the root-tubercles die and decay. In this way some of the proteid substance is set free in the soil. The soil thus becomes richer in nitrogenous plant food.

The forms of the bacteroids vary. In some of the clovers they are oval, in vetch they are rod-like or forked, and other forms occur in some of the other genera.

Note.—So far as we know the legume tubercle organism does not assimilate free nitrogen of the air unless it is within the root of the legume. But there are microörganisms in the soil which are capable of assimilating free nitrogen independently. Example, a bacterium, *Clostridium pasteurianum*. Certain bacteria and algæ live in *contact symbiosis* in the soil, the bacteria fixing free nitrogen, while in return for the combined nitrogen, the algæ furnish the bacteria with carbohydrates. It seems that these bacteria cannot fix the free nitrogen of the air unless they are supplied with carbohydrates, and it is known that *Clostridium pasteurianum* cannot assimilate free nitrogen unless sugar is present.

Lichens

Nutrition of Lichens

Lichens are very curious plants which grow on rocks, on the trunks and branches of trees, and on the soil. They form leaf-like expansions more or less green in color, or brownish, or gray, or they occur in the form of threads, or small tree-like formations. Sometimes the plant fits so closely to the rock on which it grows that it seems merely to paint the rock a slightly different color, and in the case of many which occur on trees there appears to be to the eye only a very slight discoloration of the bark of the trunk, with here and there the darker colored points where fruit bodies are formed. The most curious thing about them is, however, that while they form plant bodies of various form, these bodies are of a "dual nature" as regards the organisms composing them. The plant bodies, in other words, are formed of two different organisms which, woven together, exist apparently as one. A fungus on the one hand grows around and encloses in the meshes of its mycelium the cells or threads of an alga, as the case may be.

If we take one of the leaf-like forms known as peltigera, which grows on damp soil or on the surfaces of badly decayed logs, we see that the plant body is flattened, thin, crumpled, and irregularly lobed. The color is dull greenish on the upper side, while the under side is white or light gray, and mottled with brown, especially the older portions. Here and there on the under surface are quite long slender blackish strands. These are composed entirely of fungus threads and serve as organs of attachment or holdfasts, and for the purpose of supplying the plant body with mineral substances which are in solution in the water of the soil. If we make a thin section of the leaf-like portion of a lichen as shown in Figure 85, we shall see that it is composed of a mesh of colorless threads which in certain definite portions contain entangled green cells. The colorless threads are those of the fungus, while the green cells are those of the alga. These green cells of the alga perform the function of chlorophyll bodies for the dual organism, while the threads of the fungus provide the mineral constituents of plant food. The alga, while it is not killed in the embrace of the fungus, does not reach the perfect state of development which it attains when not in connection with the fungus. On the other hand the fungus profits more than the alga by this association. It forms fruit bodies, and perfects spores in the special fruit bodies, which are so very distinct in the case of so many of the species of the lichens. These plants have lived for so long a time in this close association that the fungi are rarely found separate from the algæ in nature, but in a number of cases they have been induced to grow in artificial cultures separate from the alga. This fact, and also the fact that the algæ are often found to occur separate from the fungus in nature, is regarded by many as an indication that the plant body of the lichens is composed of two distinct organisms, and that the fungus is parasitic on the alga.

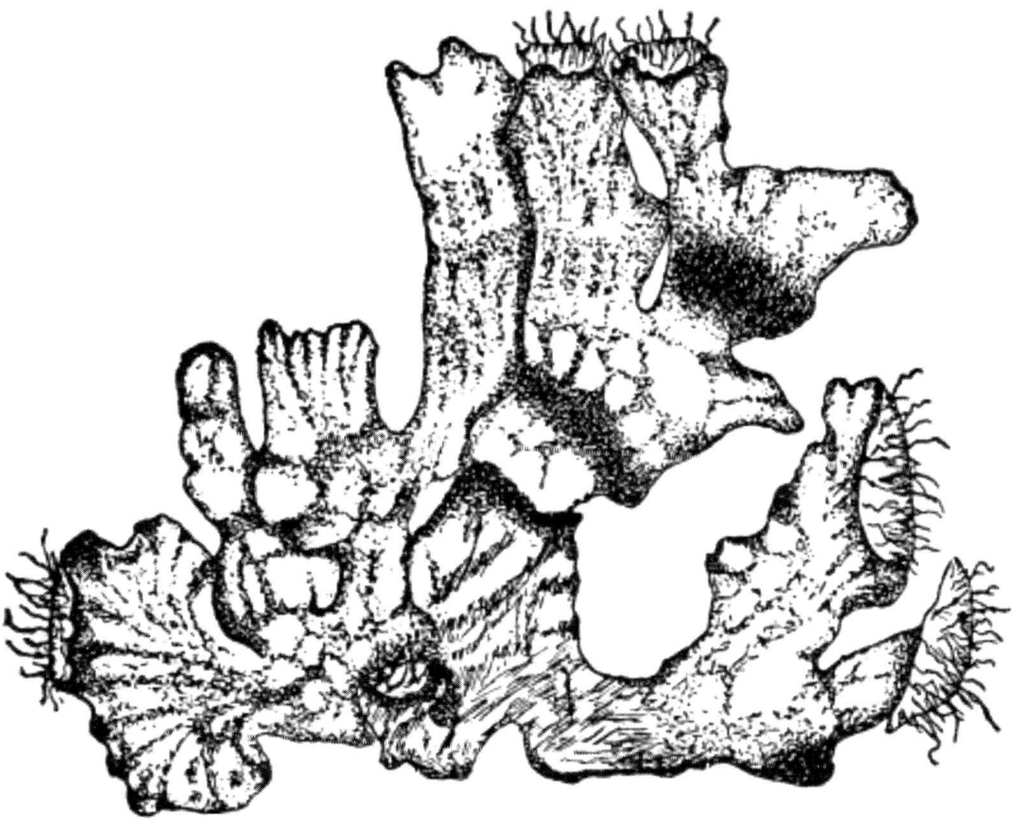

Figure 84. Frond of lichen (peltigera), showing rhizoids.

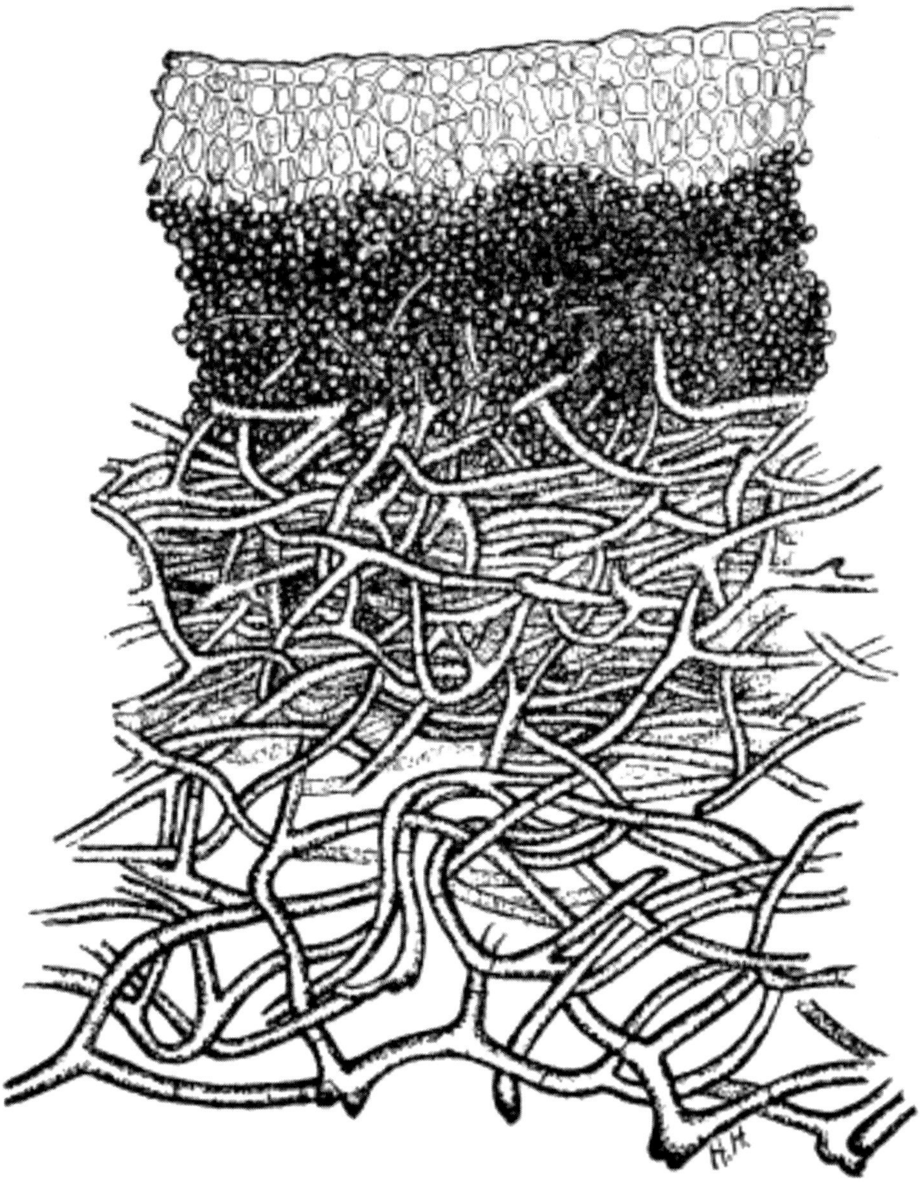

Figure 85. Lichen (peltigera), section of thallus; dark zone of rounded bodies made up largely of the algal cells. Fungus cells above, and threads beneath and among the algal cells.

Others regard the lichens as autonomous plants, that is, the two organisms have by this long-continued community of existence become unified into an individualized organism, which possesses a habit and mode of life distinct from that of either of the organisms forming the component parts. This community of existence between two different organisms is called by some *mutualism*, or *symbiosis*. While the alga enclosed within the meshes of the fungus is not so free to develop, and probably does not attain the full development which it would alone under favorable conditions, still it is very likely that it is often preserved from destruction during very dry periods, within the tough thallus, on the surface of bare rocks.

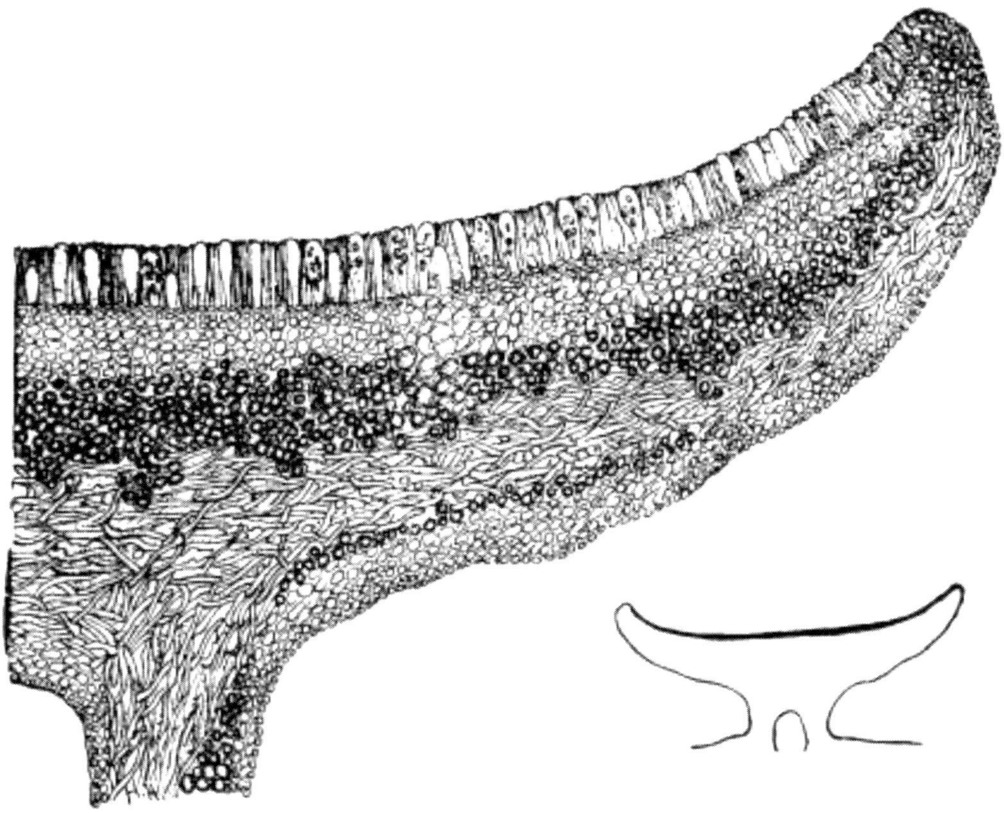

Figure 86. Section of fruit body or apothecium of lichen (parmelia), showing asci and spores of the fungus.

Chapter 10

HOW PLANTS OBTAIN THEIR FOOD, II

SEEDLINGS

It is evident from some of the studies which we have made in connection with germination of seeds and nutrition of the plant that there is a period in the life of the seed plants in which they are able to grow if supplied with moisture, but may entirely lack any supply of food substance from the outside, though we understand that growth finally comes to a standstill unless they are supplied with food from the outside. In connection with the study of the nutrition of the plant, therefore, it will be well to study some of the representative seeds and seedlings to learn more accurately the method of germination and nutrition in seedlings during the germinating period.

To Prepare Seeds for Germination

Soak a handful of seeds (or more if the class is large) in water for 12 to 24 hours. Take shallow crockery plates, or ordinary plates, or a germinator with a fluted bottom. Place in the bottom some sheets of paper, and if sphagnum moss is at hand scatter some over the paper. If the moss is not at hand, throw the upper layer of paper into numerous folds. Thoroughly wet the paper and moss, but do not have an excess of water. Scatter the seeds among the moss or the folds of the paper. Cover with some more wet paper and keep in a room where the temperature is about 20°C. to 25°C. The germinator should be looked after to see that the paper does not become dry. It may be necessary to cover it with another vessel to prevent the too rapid evaporation of the water. The germinator should be started about a week before the seedlings are wanted for study. Some of the soaked seeds should be planted in soil in pots and kept at the same temperature, for comparison with those grown in the germinator.

Structure of the Grain of Corn

Take grains of corn that have been soaked in water for 24 hours and note the form and difference in the two sides (in all of these studies the form and structure of the seed, as well as the stages in germination, should be illustrated by the student). Make a longisection of a grain of corn through the middle line, if necessary making several in order to obtain one which shows the structures well near the smaller end of the grain. Note the following structures: 1st, the hard outer "wall" (formed of the consolidated wall of the ovary with the integuments of the ovules—see Chapters 35 and 36); 2d, the greater mass of starch and other plant food (the endosperm) in the centre; 3d, a somewhat crescent-shaped body (the *scutellum*) lying next the endosperm and near the smaller end of the grain; 4th, the remaining portion of the young embryo lying between the scutellum and the seed coat in the depression. When good sections are made one can make out the radicle at the smaller end of the seed, and a few successive leaves (the plumule) which lie at the opposite end of the embryo shown by sharply curved parallel lines. Observe the attachment of the scutellum to the caulicle at the point of junction of the plumule and the radicle. The scutellum is a part of the embryo and represents a cotyledon. The endosperm is also called *albumen*, and such a seed is *albuminous*.

Figure 87. Section of corn seed; at upper right of each is the plantlet, next the cotyledon, at left the endosperm.

Dissect out an embryo from another seed, and compare with that seen in the section.

In the germination of the grain of corn the endosperm supplies the food for the growth of the embryo until the roots are well established in the soil and the leaves have become expanded and green, in which stage the plant has become able to obtain its food from the soil and air and live independently. The starch in the endosperm cannot of course be used for food by the embryo in the form of starch. It is first converted into a soluble form and then absorbed through the surface of the scutellum or cotyledon and carried to all parts of the embryo. An enzyme developed by the embryo acts upon the starch, converting it into a form of sugar which is in solution and can thus be absorbed. This enzyme is one of the so-called diastatic "ferments" which are formed during the germination of all seeds which contain food stored in the form of starch. In some seedlings, this diastase formed is developed in much greater abundance than in others, for example, in barley. Examine grains of corn still attached to seedlings several weeks old and note that a large part of their content has been used up. The action of diastase on starch is described in Chapter 8.

Structure of the Pumpkin Seed

The pumpkin seed has a tough papery outer covering for the protection of the embryo plant within. This covering is made up of the seed coats. When the seed is opened by slitting off these coats there is seen within the "meat" of the pumpkin seed. This is nothing more than the embryo plant. The larger part of this embryo consists of two flattened bodies which are more prominent than any other part of the plantlet at this time. These two flattened bodies are the two first leaves, usually called *cotyledons*. If we spread these cotyledons apart we see that they are connected at one end. Lying between them at this point of attachment is a small bud. This is the *plumule*. The plumule consists of the very young leaves at the end of the stem which will grow as the seed germinates. At the other end where the cotyledons are joined is a small projection, the young root, often termed the *radicle*.

How the Embryo Gets Out of a Pumpkin Seed

To see how the embryo gets out of the pumpkin seed we should examine seeds germinated in the folds of damp paper or on damp sphagnum, as well as some which have been germinated in earth. Seeds should be selected which represent several different stages of germination.

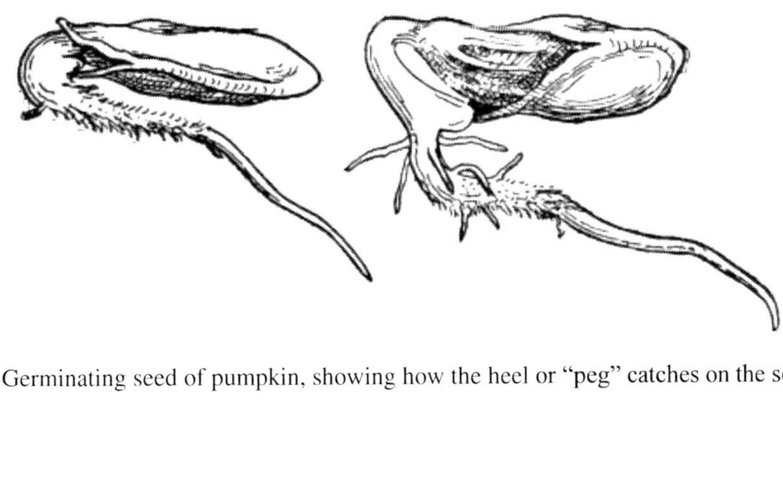

Figure 88. Germinating seed of pumpkin, showing how the heel or "peg" catches on the seed coat to cast it off.

Figure 89. Escape of the pumpkin seedling from the seed coats.

The Peg Helps To Pull the Seed Coats Apart

The root pushes its way out from between the stout seed coats at the smaller end, and then turns downward unless prevented from so doing by a hard surface. After the root is 2-4 *cm* long, and the two halves of the seed coats have begun to be pried apart, if we look in this rift at the junction of the root and stem, we shall see that one end of the seed coat is caught against a heel, or "peg," which has grown out from the stem for this purpose. Now if we examine one which is a little more advanced, we shall see this heel more distinctly, and also that the stem is arching out away from the seed coats. As the stem arches up its back in this way it pries with the cotyledons against the upper seed coat, but the lower seed coat is caught against this heel, and the two are pulled gradually apart. In this way the embryo plant pulls itself out from between the seed coats. In the case of seeds which are planted deeply in the soil we do not see this contrivance unless we dig down into the earth. The stem of the seedling arches through the soil, pulling the cotyledons up at one end. Then it straightens up, the green cotyledons part, and open out their inner faces to the sunlight, as shown in Figure 90. If we dig into the soil we shall see that this same heel is formed on the stem, and that the seed coats are cast off into the soil.

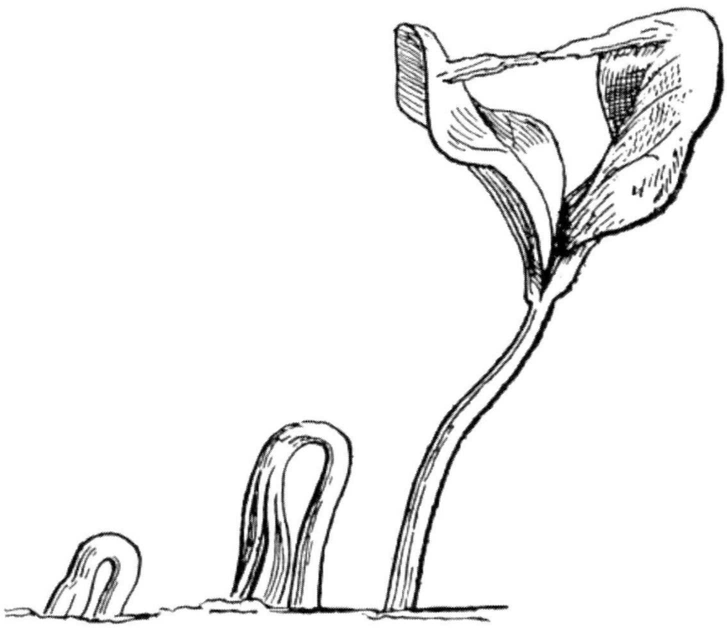

Figure 90. Pumpkin seedling rising from the ground.

Parts of the Pumpkin Seedling

During the germination of the seed all parts of the embryo have enlarged. This increase in size of a plant is one of the peculiarities of growth. The cotyledons have elongated and expanded somewhat, though not to such a great extent as the root and the stem. The cotyledons also have become green on exposure to the light. Very soon after the main root has emerged from the seed coats, other lateral roots begin to form, so that the root soon becomes very much branched. The main root with its branches makes up the root system of the seedling. Between the expanded cotyledons is seen the plumule. This has enlarged somewhat, but not nearly so much as the root, or the part of the stem which extends below the cotyledons. This part of the stem, i.e., that part below the cotyledons and extending to the beginning of the root, is called in all seedlings the *hypocotyl*, which means "below the cotyledon."

The Common Garden Bean

The common garden bean, or the lima bean, may be used for study. The garden bean is not so flattened or broadened as the lima bean. It is rounded compressed, elongate slightly curved, slightly concave on one side and convex on the other, and the ends are rounded. At the middle of the concave side note the distinct scar (the hilum) formed where the bean seed separates from its attachment to the wall of the pod. Upon one side of this scar is a slight prominence which is continued for a short distance toward the end of the bean in the form of a slight ridge. This is the *raphe*, and represents that part of the stalk of the ovule which is joined to the side of the ovule when the latter is curved around against it (see Chapter 36), and at the outer end of the raphe is the *chalaza*, the point where the stalk is joined to the end of the ovule, best understood in a straight ovule. Upon the opposite side of the scar and close to it can be seen a minute depression, the *micropyle*. Underneath the seed coat and lying between this point and the end of the seed is the *embryo*, which gives greater prominence to the bean at this point, but it is especially more prominent after the bean has been soaked in water. Soak the beans in water and as they are swelling note how the seed coats swell faster than the inner portion of the seed, which causes them to wrinkle in a curious way, but finally the inner portion swells and fills the seed coat out smooth again. Sketch a bean showing all the external features both in side view and in front. Split one lengthwise and sketch the half to which the embryo clings, noting the young root, stem, and the small leaves which were lying between the cotyledons. There is no endosperm here now, since it was all used up in the growth of the embryo, and a large part of its substance was stored up in the cotyledons.

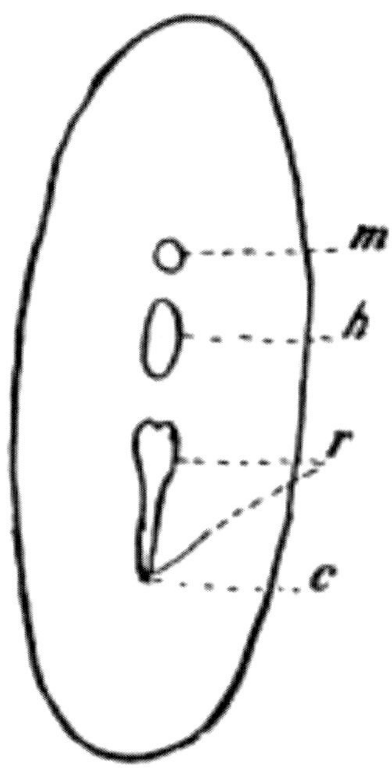

Figure 91. Garden bean. *m*, micropyle; *h*, hilum or scar; *r*, raphe; *c*, point where chalaza lies.

As the seed germinates the young plant gets its first food from that stored in the cotyledons. The hypocotyl elongates, becomes strongly arched, and at last straightens up, lifting the cotyledons from the soil. As the cotyledons become exposed to the light they assume a green color. Some of the stored food in them goes to nourish the embryo during germination, and they therefore become smaller, shrivel somewhat, and at last fall off.

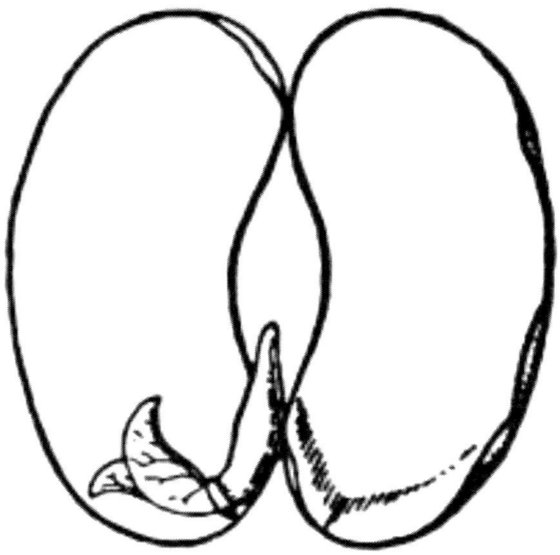

Figure 92. Bean seed split open to show plantlet.

The Castor-Oil Bean

This is not a true bean, since it belongs to a very different family of plants (Euphorbiaceæ). In the germination of this seed a very interesting comparison can be made with that of the garden bean. As the "bean" swells the very hard outer coat generally breaks open at the free end and slips off at the stem end. The next coat within, which is also hard and shining black, splits open at the opposite end, that is at the stem end. It usually splits open in the form of three ribs. Next within the inner coat is a very thin, whitish film (the remains of the nucellus, and corresponding to the perisperm) which shrivels up and loosens from the white mass, the endosperm, within. In the castor-oil bean, then, the endosperm is not all absorbed by the embryo during the formation of the seed. As the plant becomes older we should note that the fleshy endosperm becomes thinner and thinner, and at last there is nothing but a thin, whitish film covering the green faces of the cotyledons. The endosperm has been gradually absorbed by the germinating plant through its cotyledons and used for food.

Figure 93. How the garden bean comes out of the ground. First the looped hypocotyl, then the cotyledons pulled out, next casting off the seed coat, last the plant erect, bearing thick cotyledons, the expanding leaves, and the plumule between them.

ARISÆMA TRIPHYLLUM[15]

Germination of Seeds of Jack-In-The-Pulpit

The ovaries of jack-in-the-pulpit form large, bright red berries with a soft pulp enclosing one to several large seeds. The seeds are oval in form. Their germination is interesting, and illustrates one type of germination of seeds common among monocotyledonous plants. If the seeds are covered with sand, and kept in a moist place, they will germinate readily.

---

[15] In lieu of Arisæma make a practical study of the pea. See paragraph 216*a*.

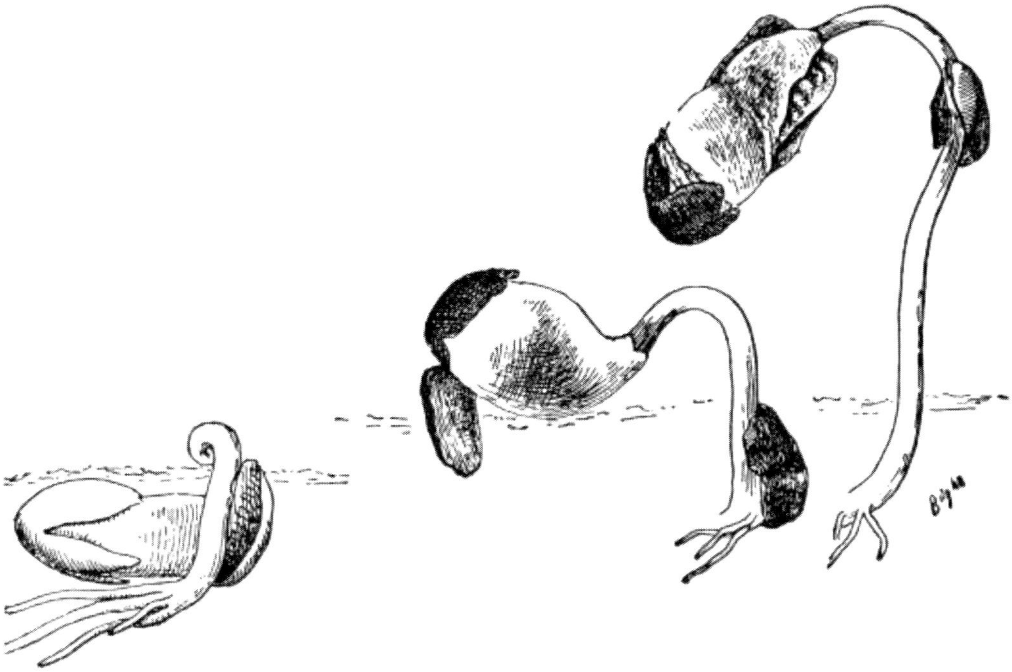

Figure 94. Germination of castor-oil bean.

How the Embryo Backs Out of the Seed

The embryo lies within the mass of the endosperm; the root end, near the smaller end of the seed. The club-shaped cotyledon lies near the middle of the seed, surrounded firmly on all sides by the endosperm. The stalk, or petiole, of the cotyledon, like the lower part of the petiole of the leaves, is a hollow cylinder, and contains the younger leaves, and the growing end of the stem or bud. When germination begins, the stalk, or petiole, of the cotyledon elongates. This pushes the root end of the embryo out at the small end of the seed. The free end of the embryo now enlarges somewhat, as seen in the Figures, and becomes the bulb, or corm, of the young plant. At first no roots are visible, but in a short time one, two, or more roots appear on the enlarged end.

Section of an Embryo

If we make a longisection of the embryo and seed at this time we can see how the club-shaped cotyledon is closely surrounded by the endosperm. Through the cotyledon, then, the nourishment from the endosperm is readily passed over to the growing embryo. In the hollow part of the petiole near the bulb can be seen the first leaf.

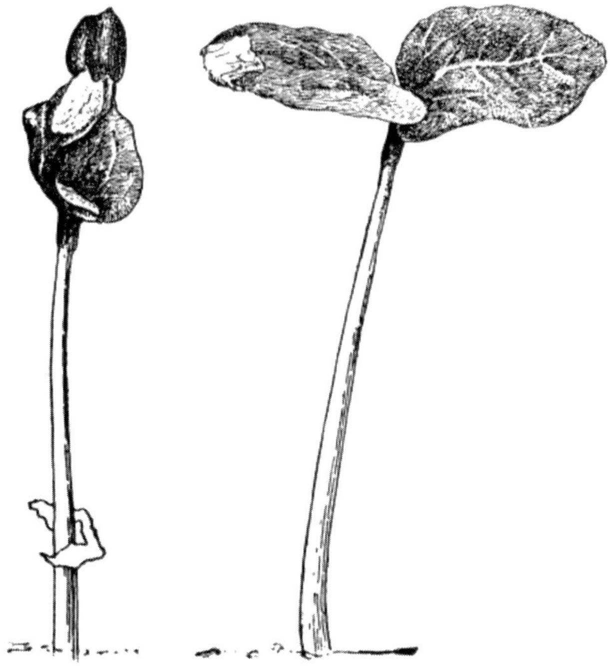

Figure 95. Seedlings of castor-oil bean casting the seed coats, and showing papery remnant of the endosperm.

Figure 96. Seedlings of jack-in-the-pulpit; embryo backing out of the seed.

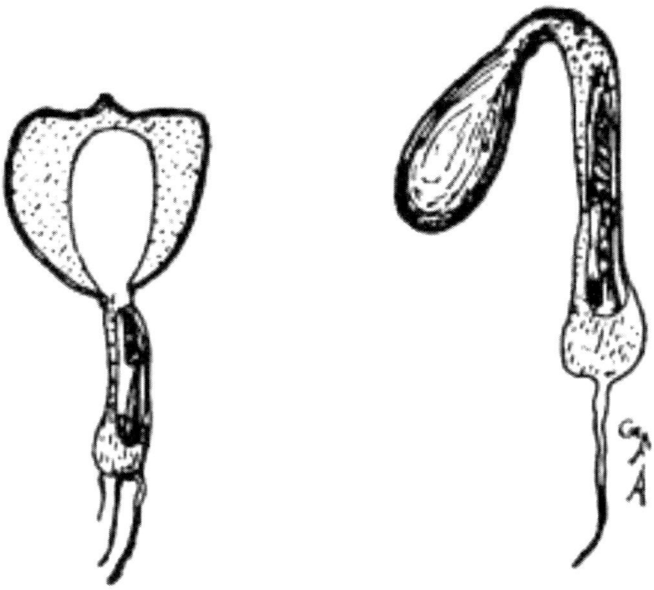

Figure 97. Section of germinating embryos of jack-in-the-pulpit, showing young leaves inside the petiole of the cotyledon. At the left cotyledon shown surrounded by the endosperm in the seed; at right endosperm removed to show the club-shaped cotyledon.

How the First Leaf Appears

As the embryo backs out of the seed, it turns downward into the soil, unless the seed is so lying that it pushes straight downward.

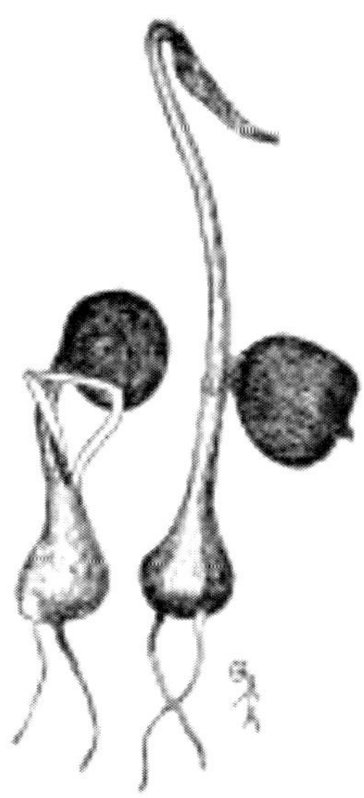

Figure 98. Seedlings of jack-in-the-pulpit, first leaf arching out of the petiole of the cotyledon.

On the upper side of the arch thus formed, in the petiole of the cotyledon, a slit appears, and through this opening the first leaf arches its way out. The loop of the petiole comes out first, and the leaf later, as shown in Figure 98. The petiole now gradually straightens up, and as it elongates the leaf expands.

Figure 99. Embryos of jack-in-the-pulpit still attached to the endosperm in seed coats, and showing the simple first leaf.

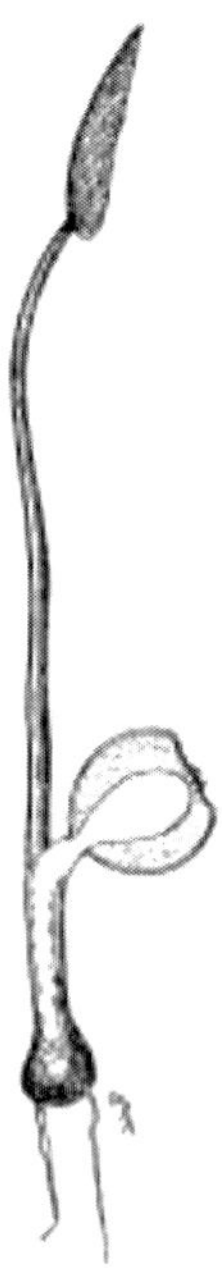

Figure 100. Seedling of jack-in-the-pulpit; section of the endosperm and cotyledon.

The First Leaf of the Jack-In-The-Pulpit Is a Simple One

The first leaf of the embryo jack-in-the-pulpit is very different in form from the leaves which we are accustomed to see on mature plants. If we did not know that it came from the seed of this plant we would not recognize it. It is simple, that is it consists of one lamina or blade, and not of three leaflets as in the compound leaf of the mature plant. The simple leaf is ovate and with a broad heart-shaped base. The jack-in-the-pulpit, then, as trillium, and some other monocotyledonous plants which have compound leaves on the mature plants, have simple leaves during embryonic development. The ancestral monocotyledons are supposed to have had simple leaves. Thus there is in the embryonic development of the jack-in-the-pulpit, and others with compound leaves, a sort of recapitulation of the evolutionary history of the leaf in these forms.

Germination of the Pea

Compare with the bean. Note especially that the cotyledons are not lifted above the soil as in the beans. Compare germination of acorns.

Digestion

To Test For Stored Food Substance in the Seedlings Studied

The pumpkin, squash, and castor-oil bean are examples of what are called oily seeds, since considerable oil is stored up in the protoplasm in the cotyledons. To test for this, remove a small portion of the substance from the cotyledon of the squash and crush it on a glass slip in a drop or two of osmic acid.[16] Put on a cover glass and examine with a microscope. The black amorphous matter shows the presence of oil in the protoplasm. The small bodies which are stained yellow are *aleurone* grains, a form of protein or albuminous substance. Both the oil and the protein substance are used by the seedling during germination. The oil is converted into an available food form by the action of an enzyme called *lipase*, which splits up the fatty oil into glucose and other substances. Lipase has been found in the endosperm of the castor-oil, cocoanut, and in the cotyledons of the pumpkin, as well as in other seeds containing oil as a stored product. The aleurone is made available by an enzyme of the nature of trypsin. Test the endosperm of the castor-oil bean in the same way. Make another test of both the squash and castor-oil seeds with iodine to show that starch is not present.

---

[16] Dissolve a half gram of osmic acid in 50 *cc*. of water and keep tightly corked when not using.

Test the cotyledon of the bean with iodine for the presence of starch. If the endosperm of corn seed has not been tested do so now with iodine. The endosperm consists largely of starch. The starch is converted to glucose by a diastatic "ferment" formed by the seedling as it germinates. Make a thin cross-section of a grain of wheat, including the seed coat and a portion of the interior, treat with iodine and mount for microscopic examination. Note the abundance of starch in the internal portion of endosperm. Note a layer of cells on the outside of the starch portions filled with small bodies which stain yellow. These are aleurone grains. The cellulose in the cell walls of the endosperm is dissolved by another enzyme called *cytase*, and some plants store up cellulose for food. For example, in the endosperm of the *date* the cell walls are very much thickened and pitted. The cell walls consist of reserve cellulose and the seedling makes use of it for food during growth.

Albuminous and Exalbuminous Seeds

In seeds where the food is stored outside of the embryo they are called *albuminous*; examples, corn, wheat and other cereals, Indian turnip, etc. In those seeds where the food is stored up in the embryo they are called *exalbuminous*; examples, bean, pea, pumpkin, squash, etc.

Digestion has a well-defined meaning in animal physiology and relates to the conversion of solid food, usually within the stomach, into a soluble form by the action of certain gastric juices, so that the liquid food may be absorbed into the circulatory system. The term is not often applied in plant physiology, since the method of obtaining food is in general fundamentally different in plants and animals. It is usually applied to the process of the conversion of starch into some form of sugar in solution, as glucose, etc. This we have found takes place in the leaf, especially at night, through the action of a diastatic ferment developed more abundantly in darkness. As a result, the starch formed during the day in the leaves is digested at night and converted into sugar, in which form it is transferred to the growing parts to be employed in the making of new tissues, or it is stored for future use; in other cases it unites with certain inorganic substances, absorbed by the roots and raised to the leaf, to form proteids and other organic substances. In tubers, seeds, parts of stems or leaves where starch is stored, it must first be "digested" by the action of some enzyme before it can be used as food by the sprouting tubers or germinating seeds.

For example, starch is converted to a glucose by the action of a diastase. Cellulose is converted to a glucose by cytase. Albuminoids are converted into available food by a tryptic ferment. Fatty oils are converted into glucose and other products by lipase.

Inulin, a carbohydrate closely related to starch, is stored up for food in solution in many composite plants, as in the artichoke, the root tuber of dahlia, etc. When used for food by the growing plant it is converted into glucose by an enzyme, inulase. Make a section of a portion of a dahlia tuber or artichoke and treat with alcohol. The inulin is precipitated into sphæro crystals. (See also paragraphs 156-161 and 216*b*.)

Then there are certain fungi which feed on starch or other organic substances whether in the host or not, which excrete certain enzymes to dissolve the starch, etc., to bring it into a soluble form before they can absorb it as food. Such a process is a sort of *extracellular digestion*, i.e., the organism excretes the enzyme and digests the solid outside, since it cannot take the food within its cells in the solid form. To a certain degree the higher plants perform also extracellular digestion in the action of root hair excretion on insoluble substances, and in the case of the humus saprophytes. But for them soluble food is largely prepared by the action of acids, etc., in the soil or water, or by the work of fungi and bacteria as described in Chapter 9.

Assimilation

In plant physiology the term assimilation has been chiefly used for the process of carbon dioxide assimilation (= photosynthesis). Some objections have been raised against the use of assimilation here as one of the life processes of the plant, since its inception stages are due to the combined action of light, an external factor, and chlorophyll in the plant along with the living chloroplastid. So long, however, as it is not known that this process can take place without the aid of the living plant, it does not seem proper to deny that it is altogether not a process of assimilation. It is not necessary to restrict the term assimilation to the formation of new living matter in the plant cell; it can be applied also to the synthetic processes in the formation of carbohydrates, proteids, etc., and called synthetic assimilation. The sun supplies the energy, which is absorbed by the chlorophyll, for splitting up the carbonic acid, and the living chloroplast then assimilates by a synthetic process the carbon, hydrogen, and oxygen. This process then can be called *photosynthetic assimilation*. The nitrite and nitrate bacteria derive energy in the process of nitrification, which enables them to assimilate CO_2 from the air, and this is called *chemosynthetic assimilation*. The inorganic material in the form of mineral salts, nitrates, etc., absorbed by the root, and carried up to the leaves, here meets with the carbohydrates manufactured in the leaf. Under the influence of the protoplasm synthesis takes place, and proteids and other organic compounds are built up by the union of the salts, nitrates, etc., with the carbohydrates. This is also a process of synthetic assimilation. These are afterward stored as food, or assimilated by the protoplasm in the making of new living matter, or perhaps without the first process of synthetic assimilation some of the inorganic salts, nitrates, and carbohydrates meeting in the protoplasm are assimilated into new living matter directly.

Chapter 11

RESPIRATION

One of the life processes in plants which is extremely interesting, and which is exactly the same as one of the life processes of animals, is easily demonstrated in several ways.

SIMPLE EXPERIMENT TO DEMONSTRATE THE EVOLUTION OF CO_2 DURING GERMINATION

Where there are a number of students and a number of large cylinders are not at hand, take bottles of a pint capacity and place in the bottom some peas soaked for 12 to 24 hours. Cover with a glass plate which has been smeared with vaseline to make a tight joint with the mouth of the bottle. Set aside in a warm place for 24 hours. Then slide the glass plate a little to one side and quickly pour in a little baryta water so that it will run down on the inside of the bottle. Cover the bottle again. Note the precipitate of barium carbonate which demonstrates the presence of CO_2 in the bottle. Lower a lighted taper. It is extinguished because of the great quantity of CO_2. If flower buds are accessible, place a small handful in each of several jars and treat the same as in the case of the peas. Young growing mushrooms are excellent also for this experiment, and serve to show that respiration takes place in the fungi.

If we now take some of the baryta water and blow our "breath" upon it the same film will be formed. The carbon dioxide which we exhale is absorbed by the baryta water, and forms barium carbonate, just as in the case of the peas. In the case of animals the process by which oxygen is taken into the body and carbon dioxide is given off is *respiration*. The process in plants which we are now studying is the same, and also is respiration. The oxygen in the vessel was partly used up in the process, and carbon dioxide was given off. (It will be seen that this process is exactly the opposite of that which takes place in carbon dioxide assimilation.)

Figure 101. Test for presence of carbon dioxide in vessel with germinating peas. (Sachs.)

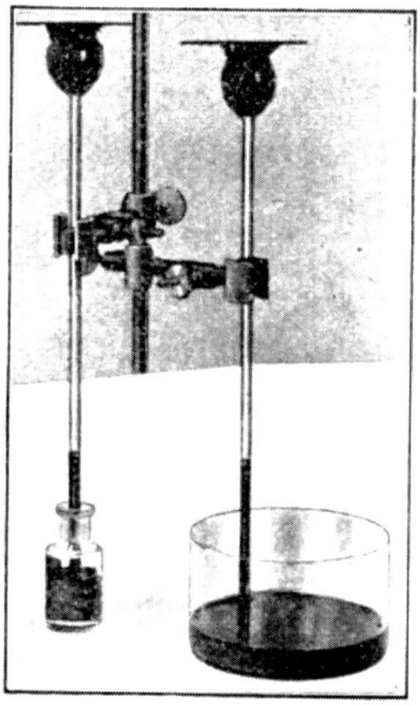

Figure 102. Apparatus to show respiration of germinating wheat.

To Show that Oxygen from the Air is Used Up While Plants Respire

Soak some wheat for 24 hours in water. Remove it from the water and place it in the folds of damp cloth or paper in a moist vessel. Let it remain until it begins to germinate. Fill the bulb of a thistle tube with the germinating wheat. By the aid of a stand and clamp, support the tube upright, as shown in Figure 102. Let the small end of the tube rest in a strong solution of caustic potash (one stick caustic potash in two-thirds tumbler of water) to which red ink has been added to give a deep red color. Place a small glass plate over the rim of the bulb and seal it air-tight with an abundance of vaseline. Two tubes can be set up in one vessel, or a second one can be set up in strong baryta water colored in the same way.

The Result

It will be seen that the solution of caustic potash rises slowly in the tube; the baryta water will also, if that is used. The solution is colored so that it can be plainly seen at some distance from the table as it rises in the tube. In the experiment from which the Figure was made for the accompanying illustration, the solution had risen in 6 hours to the height shown in Figure 102. In 24 hours it had risen to the height shown in Figure 103.

Why the Solution of Caustic Potash Rises in the Tube

Since no air can get into the thistle tube from above or below, it must be that some part of the air which is inside of the tube is used up while the wheat is germinating.

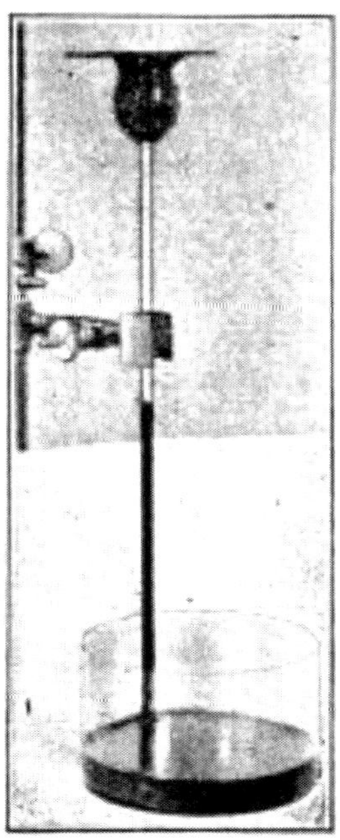

Figure 103. Apparatus to show respiration of germinating wheat.

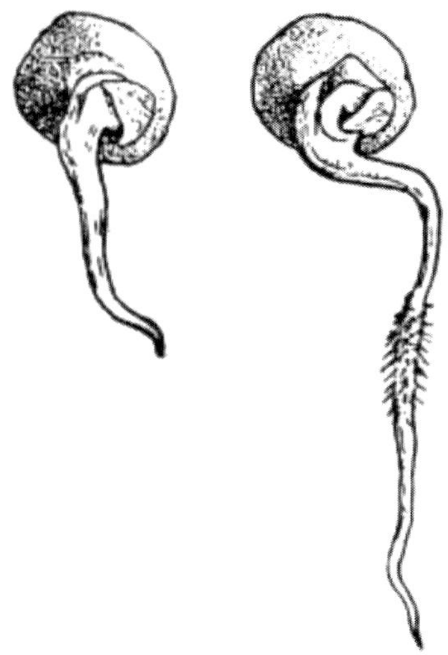

Figure 104. Pea seedlings; the one at the left had no oxygen and little growth took place, the one at the right in oxygen and growth was evident.

From our study of germinating peas, we know that a suffocating gas, carbon dioxide, is given off while respiration takes place. The caustic potash solution, or the baryta water, whichever is used, absorbs the carbon dioxide. The carbon dioxide is heavier than air, and so it settles down in the tube where it can be absorbed.

Where Does the Carbon Dioxide Come From?

We know it comes from the growing seedlings. The symbol for carbon dioxide is CO_2. The carbon comes from the plant, because there is not enough in the air. Nitrogen could not join with the carbon to make CO_2. Some oxygen from the air or from the protoplasm of the growing seedlings (more probably the latter) joins with some of the carbon of the plant. These break away from their association with the living substance and unite, making CO_2. The oxygen absorbed by the plant from the air unites with the living substance, or perhaps first with food substances, and from these the plant is replenished with carbon and oxygen. After the demonstration has been made, remove the glass plate which seals the thistle tube above, and pour in a small quantity of baryta water. The white precipitate formed affords another illustration that carbon dioxide is released.

Respiration is Necessary for Growth

After performing experiment in paragraph 221, if the vessel has not been open too long so that oxygen has entered, we may use the vessel for another experiment, or set up a new one to be used in the course of 12 to 24 hours, after some oxygen has been consumed. Place some folded damp filter paper on the germinating peas in the jar. Upon this place one-half dozen peas which have just been germinated, and in which the roots are about 20-25 *mm* long. The vessel should be covered tightly again and set aside in a warm room. A second jar with water in the bottom instead of the germinating peas should be set up as a check. Damp folded filter paper should be supported above the water, and on this should be placed one-half dozen peas with roots of the same length as those in the jar containing carbon dioxide.

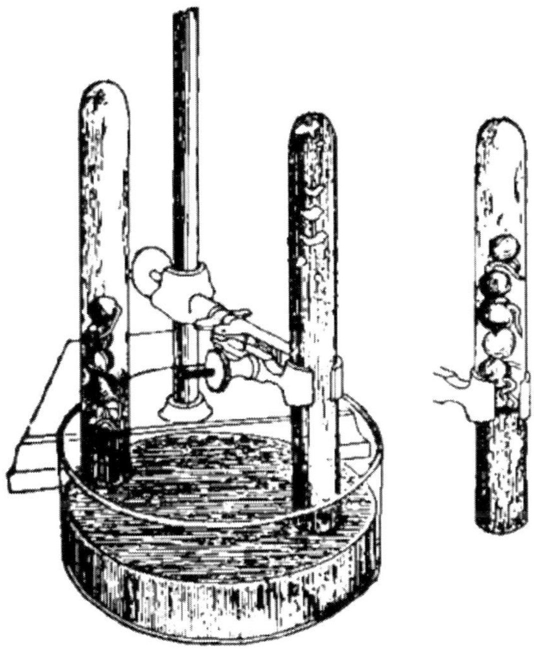

Figure 105. Experiment to show that growth takes place more rapidly in presence of oxygen than in absence of oxygen. The two tubes in the vessel represent the condition at the beginning of the experiment. At the close of the experiment the roots in the tube at the left were longer than those in the tube filled at the start with mercury. The tube outside of the vessel represents the condition of things where the peas grew in absence of oxygen; the carbon dioxide given off has displaced a portion of the mercury. This also shows *anaerobic* respiration.

In 24 hours examine and note how much growth has taken place. It will be seen that the roots have elongated but very little or none in the first jar, while in the second one we see that the roots have elongated considerably, if the experiment has been carried on carefully. Therefore in an atmosphere devoid of oxygen very little growth will take place, which shows that normal respiration with access of oxygen (aerobic respiration) is necessary for growth.

Another Way of Performing the Experiment

If we wish we may use the following experiment instead of the simple one indicated above. Soak a handful of peas in water for 12-24 hours, and germinate so that twelve with the radicles 20-25 *mm* long may be selected. Fill a test tube with mercury and carefully invert it in a vessel of mercury so that there will be no air in the upper end. Now nearly fill another tube and invert in the same way. In the latter there will be some air. Remove the outer coats from the peas so that no air will be introduced in the tube filled with the mercury, and insert them one at a time under the edge of the tube beneath the mercury, six in each tube, having first measured the length of the radicles. Place in a warm room. In 24 hours measure the roots. Those in the air will have grown considerably, while those in the other tube will have grown but little or none.

Anaerobic Respiration

The last experiment is also an excellent one to show *anaerobic* respiration. In the tube filled with mercury so that when inverted there will be no air, it will be seen after 24 hours that a gas has accumulated in the tube which has crowded out some of the mercury. With a wash bottle which has an exit tube properly curved, some water may be introduced in the tube.

Figure 106. Test for liberation of carbon dioxide from leafy plant during respiration. Baryta water in smaller vessel. (Sachs.)

Then insert underneath a small stick of caustic potash. This will form a solution of potash, and the gas will be partly or completely absorbed. This shows that the gas was carbon dioxide. This evolution of carbon dioxide by living plants when there is no access of oxygen is anaerobic respiration (sometimes called intramolecular respiration). It occurs markedly in oily seeds and especially in the yeast plant.

Energy Set Free During Respiration

From what we have learned of the exchange of gases during respiration we infer that the plant loses carbon during this process. If the process of respiration is of any benefit to the plant, there must be some gain in some direction to compensate the plant for the loss of carbon which takes place.

It can be shown by an experiment that during respiration there is a slight elevation of the temperature in the plant tissues. The plant then gains some heat during respiration. Energy is also manifested by growth.

Respiration in a Leafy Plant

We may take a potted plant which has a well-developed leaf surface and place it under a tightly fitting bell jar. Under the bell jar there also should be placed a small vessel containing baryta water. A similar apparatus should be set up, but with no plant, to serve as a check. The experiment must be set up in a room which is not frequented by persons, or the carbon dioxide in the room from respiration will vitiate the experiment. The bell jar containing the plant should be covered with a black cloth to prevent carbon assimilation. In the course of 10 or 12 hours, if everything has worked properly, the baryta water under the jar with the plant will show the film of barium carbonate, while the other one will show none. Respiration, therefore, takes place in a leafy plant as well as in germinating seeds.

Respiration in Fungi

If several large actively growing mushrooms are accessible, place them in a tall glass jar as described for determining respiration in germinating peas. In the course of 12 hours test with the lighted taper and the baryta water. Respiration takes place in fungi as well as in green plants.

Respiration in Plants in General

Respiration is general in all plants, though not universal. There are some exceptions in the lower plants, notably in certain of the bacteria, which can only grow and thrive in the absence of oxygen.

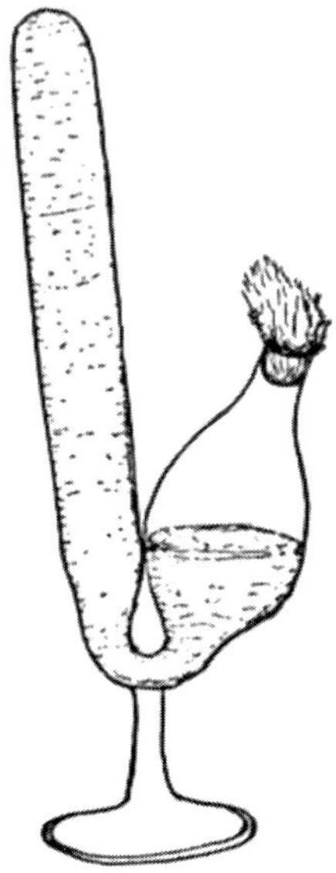

Figure 107. Fermentation tube with culture of yeast.

Figure 108. Fermentation tube filled with CO_2 from action of yeast in a sugar solution.

Respiration a Breaking-Down Process

We have seen that in respiration the plant absorbs oxygen and gives off carbon dioxide. We should endeavor to note some of the effects of respiration on the plant. Let us take, say, two dozen dry peas, weigh them, soak for 12-24 hours in water, and, in the folds of a cloth kept moist by covering with wet paper or sphagnum, germinate them. When well germinated and before the green color appears dry well in the sun, or with artificial heat, being careful not to burn or scorch them. The aim should be to get them about as dry as the seeds were before germination. Now weigh. The germinated seeds weigh less than the dry peas. There has then been a loss of plant substance during respiration.

Fermentation of Yeast

Take two fermentation tubes. Fill the closed tubular parts of each with a weak solution of grape sugar, or with potato decoction, leaving the open bulb nearly empty. Into the liquid of one of the tubes place a piece of compressed yeast as large as a pea. If the tubes are kept in a warm place for 24 hours bubbles of gas may be noticed rising in the one in which the yeast was placed, while in the second tube no such bubbles appear, especially if the filled tubes are first sterilized. The tubes may be kept until the first is entirely filled with the gas. Now dissolve in the liquid a small piece of caustic potash.

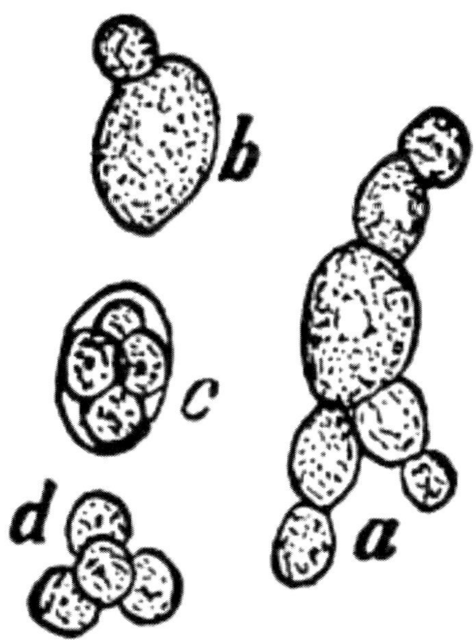

Figure 108*a*. Yeast. Saccharomyces ceriviseæ. *a*, small colony; *b*, single cell budding; *c*, single cell forming an ascus with four spores; *d*, spores free from the ascus. (After Rees.)

Soon the gas will begin to be absorbed, and the liquid will rise until it again fills the tube. The gas was carbon dioxide, which was chiefly produced during the anaerobic respiration of the rapidly growing yeast cells. In bread making this gas is produced in considerable quantities, and rising through the dough fills it with numerous cavities containing gas, so that the bread "rises." When it is baked the heat causes the gas in the cavities to expand greatly. This causes the bread to "rise" more, and baked in this condition it is "light." There are two special processes accompanying the fermentation by yeast: 1st, the evolution of carbon dioxide as shown above; and, 2d, the formation of alcohol. The best illustration of this second process is the brewing of beer, where a form of the same organism which is employed in "bread rising" is used to "brew beer."

The Yeast Plant

Before the caustic potash is placed in the tube some of the fermented liquid should be taken for study of the yeast plant, unless separate cultures are made for this purpose. Place a drop of the fermented liquid on a glass slip, place on this a cover glass, and examine with the microscope. Note the minute oval cells with granular protoplasm. These are the yeast plant. Note in some a small "bud" at one side of the end. These buds increase in size and separate from the parent plant. The yeast plant is one-celled, and multiplies by "budding" or "sprouting." It is a fungus, and some species of yeast like the present one do not form any mycelium. Under certain conditions, which are not very favorable for growth (example, when the yeast is grown in a weak nutrient substance on a thin layer of a plaster Paris slab), several spores are formed in many of the yeast cells. After a period of rest these spores will sprout and produce the yeast plant again. Because of this peculiar spore formation some place the yeast among the sac fungi. (See classification of the fungi.)

Organized Ferments and Unorganized Ferments

An organism like the yeast plant which produces a fermentation of a liquid with evolution of gas and alcohol is sometimes called a *ferment*, or *ferment organism*, or an *organized* ferment. On the other hand the diastatic ferments or enzymes like diastase, taka diastase, animal diastase (ptyalin in the saliva), cytase, etc., are *unorganized* ferments. In the case of these it is better to say *enzyme* and leave the word ferment for the ferment organisms.

Importance of Green Plants in Maintaining Purity of Air

By respiration, especially of animals, the air tends to become "foul" by the increase of CO_2. Green plants, i.e., plants with chlorophyll, purify the air during photosynthesis by absorbing CO_2 and giving off oxygen. Animals absorb in respiration large quantities of oxygen and exhale large quantities of CO_2. Plants absorb a comparatively small amount of oxygen in respiration and give off a comparatively small amount of CO_2. But they absorb during photosynthesis large quantities of CO_2 and give off large quantities of oxygen. In this way a balance is maintained between the two processes, so that the percentage of CO_2 in the air remains approximately the same, viz., about four-tenths of one per cent, while there are approximately 21 parts oxygen and 79 parts nitrogen.

Comparison of respiration and photosynthesis

Starch formation or Photosynthesis	Carbon dioxide is taken in by the plant and oxygen is liberated.
	Starch is formed as a result of the metabolism, or chemical change.
	The process takes place only in green plants, and in the green parts of plants, that is, in the presence of the chlorophyll. (Exception in purple bacterium.)
	The process only takes place under the influence of sunlight.
	It is a building-up process, because new plant substance is formed.
Respiration	Oxygen is taken in by the plant and carbon dioxide is liberated.
	Carbon dioxide is formed as a result of the metabolism, or chemical change.
	The process takes place in all plants whether they possess chlorophyll or not. (Exceptions in anaerobic bacteria).
	The process takes place in the dark as well as in the sunlight.
	It is a breaking-down process, because disintegration of plant substance occurs.

Chapter 12

GROWTH

By growth is usually meant an increase in the bulk of the plant accompanied generally by an increase in plant substance. Among the lower plants growth is easily studied in some of the fungi.

GROWTH IN MUCOR

Some of the gonidia (often called spores) may be sown in nutrient gelatine or agar, or even in prune juice. If the culture has been placed in a warm room, in the course of 24 hours, or even less, the preparation will be ready for study.

Form of the Gonidia

It will be instructive if we first examine some of the gonidia which have not been sown in the culture medium. We should note their rounded or globose form, as well as their markings if they belong to one of the species with spiny walls. Particularly should we note the size, and if possible measure them with the micrometer, though this would not be absolutely necessary for a comparison, if the comparison can be made immediately. Now examine some of the gonidia which were sown in the nutrient medium. If they have not already germinated we note at once that they are much larger than those which have not been immersed in a moist medium.

The Gonidia Absorb Water and Increase in Size Before Germinating

From our study of the absorption of water or watery solutions of nutriment by living cells, we can easily understand the cause of this enlargement of the gonidium of the mucor when surrounded by the moist nutrient medium. The cell-sap in the spore takes up more water than it loses by diffusion, thus drawing water forcibly through the protoplasmic membrane. Since it does not filter out readily, the increase in quantity of the water in the cell produces a pressure from within which stretches the membrane, and the elastic cell wall yields. Thus the gonidium becomes larger.

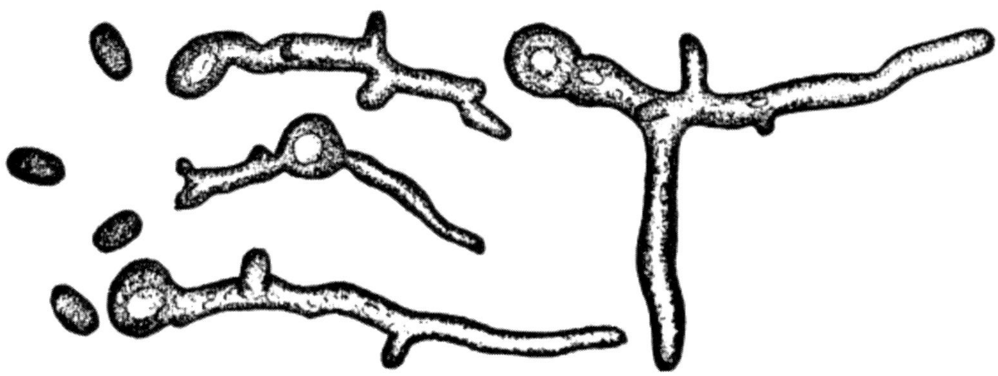

Figure 109. Spores of mucor, and different stages of germination.

How the Gonidia Germinate

We should find at this time many of the gonidia extended on one side into a tube-like process the length of which varies according to time and temperature. The short process thus begun continues to elongate. This elongation of the plant is *growth*, or, more properly speaking, one of the phenomena of growth.

The Germ Tube Branches and Forms the Mycelium

In the course of a day or so branches from the tube will appear. This branched form of the threads of the fungus is, as we remember, the mycelium. We can still see the point where growth started from the gonidium. Perhaps by this time several tubes have grown from a single one. The threads of the mycelium near the gonidium, that is, the older portions of them, have increased in diameter as they have elongated, though this increase in diameter is by no means so great as the increase in length. After increasing to a certain extent in diameter, growth in this direction ceases, while apical growth is practically unlimited, being limited only by the supply of nutriment.

Growth in Length Takes Place Only at the End of the Thread

If there were any branches on the mycelium when the culture was first examined, we can now see that they remain practically the same distance from the gonidium as when they were first formed. That is, the older portions of the mycelium do not elongate. Growth in length of the mycelium is confined to the ends of the threads.

Protoplasm Increases by Assimilation of Nutrient Substances

As the plant increases in bulk we note that there is an increase in the protoplasm, for the protoplasm is very easily detected in these cultures of mucor. This increase in the quantity of the protoplasm has come about by the assimilation of the nutrient substance, which the plant has absorbed. The increase in the protoplasm, or the formation of additional plant substance, is another phenomenon of growth quite different from that of elongation, or increase in bulk.

GROWTH OF ROOTS

For the study of the growth of roots we may take any one of many different plants. The seedlings of such plants as peas, beans, corn, squash, pumpkin, etc., serve excellently for this purpose.

Roots of the Pumpkin

The seeds, a handful or so, are soaked in water for about 12 hours, and then placed between layers of paper or between the folds of cloth, which must be kept quite moist but not very wet, and should be kept in a warm place. A shallow crockery plate, with the seeds lying on wet filter paper, and covered with additional filter paper, or with a bell jar, answers the purpose well.

The primary or first root (radicle) of the embryo pushes its way out between the seed coats at the small end. When the seeds are well germinated, select several which have the root 4-5 *cm* long. With a crow-quill pen we may now mark the terminal portion of the root off into very short sections as in Figure 110. The first mark should be not more than 1 *mm* from the tip, and the others not more than 1mm apart. Now place the seedlings down on damp filter paper, and cover with a bell jar so that they will remain moist, and if the season is cold place them in a warm room. At intervals of 8 or 10 hours, if convenient, observe them and note the farther growth of the root.

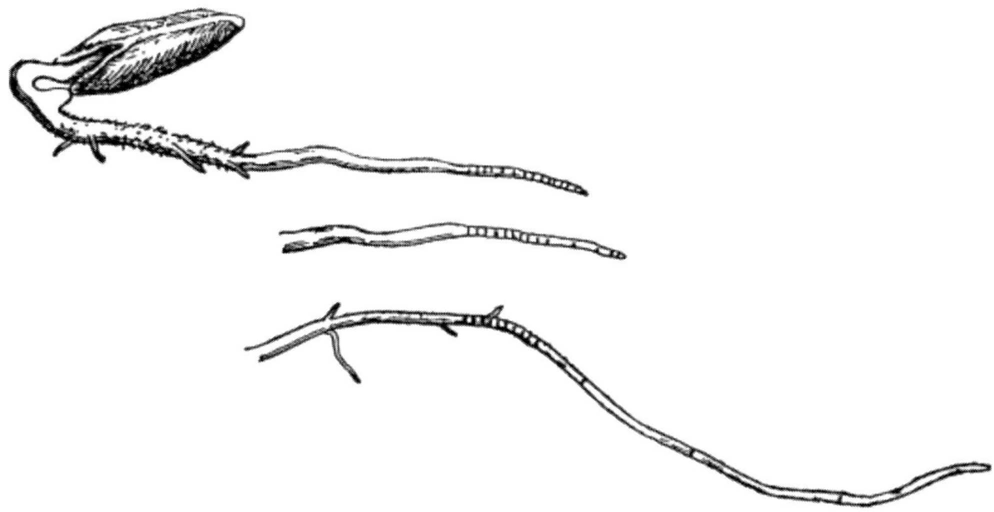

Figure 110. Root of germinating pumpkin, showing region of elongation just back of the tip.

The Region of Elongation

While the root has elongated, the region of elongation *is not at the tip of the root. It lies a little distance back from the tip*, beginning at about 2mm from the tip and extending over an area represented by from 4-5 of the millimeter marks. The root shown in Figure 110 was marked at 10 a.m. on July 5. At 6 p.m. of the same day, 8 hours later, growth had taken place as shown in the middle Figure. At 9 a.m. on the following day, 15 hours later, the growth is represented in the lower one. Similar experiments upon a number of seedlings give the same result: the region of elongation in the growth of the root is situated a little distance back from the tip. Farther back very little or no elongation takes place, but growth in diameter continues for some time, as we should discover if we examined the roots of growing pumpkins, or other plants, at different periods.

Movement of Region of Greatest Elongation

In the region of elongation the areas marked off do not all elongate equally at the same time. The middle spaces elongate most rapidly and the spaces marked off by the 6, 7, and

8 *mm* marks elongate slowly, those farthest from the tip more slowly than the others, since elongation has nearly ceased here. The spaces marked off between the 2-4 *mm* marks also elongate slowly, but soon begin to elongate more rapidly, since that region is becoming the region of greatest elongation. Thus the region of greatest elongation moves forward as the root grows, and remains approximately at the same distance behind the tip.

Formative Region

If we make a longitudinal section of the tip of a growing root of the pumpkin or other seedling, and examine it with the microscope, we see that there is a great difference in the character of the cells of the tip and those in the region of elongation of the root. First there is in the section a V-shaped cap of loose cells which are constantly being sloughed off. Just back of this tip the cells are quite regularly isodiametric, that is, of equal diameter in all directions. They are also very rich in protoplasm, and have thin walls. This is the region of the root where new cells are formed by division. It is the *formative region*. The cells on the outside of this area are the older, and pass over into the older parts of the root and root cap. If we examine successively the cells back from this *formative* region we find that they become more and more elongated in the direction of the axis of the root. The elongation of the cells in this older portion of the root explains then why it is that this region of the root elongates more rapidly than the tip.

Growth of the Stem

We may use a bean seedling growing in the soil. At the junction of the leaves with the stem there are enlargements. These are the *nodes*, and the spaces on the stem between successive nodes are the *internodes*. We should mark off several of these internodes, especially the younger ones, into sections about 5 *mm* long. Now observe these at several times for two or three days, or more. The region of elongation is greater than in the case of the roots, and extends back farther from the end of the stem. In some young garden bean plants the region of elongation extended over an area of 40 *mm* in one internode. See also Chapters 38, 39.

Force Exerted by Growth

One of the marvelous things connected with the growth of plants is the force which is exerted by various members of the plant under certain conditions. Observations on seedlings as they are pushing their way through the soil to the air often show us that considerable force is required to lift the hard soil and turn it to one side. A very striking illustration may be had in the case of mushrooms which sometimes make their way through the hard and packed soil of walks or roads. That succulent and tender plants should be capable of lifting such comparatively heavy weights seems incredible until we have witnessed it. Very striking illustrations of the force of roots are seen in the case of trees which grow in rocky situations, where rocks of considerable weight are lifted, or small rifts in large rocks are widened by the lateral pressure exerted by the growth of a root, which entered when it was small and wedged its way in.

Zone of Maximum Growth

Great variation exists in the rapidity of growth even when not influenced by outside conditions. In our study of the elongation of the root we found that the cells just back of the formative region elongated slowly at first. The rapidity of the elongation of these cells increases until it reaches the maximum. Then the rapidity of elongation lessens as the cells come to lie farther from the tip. The period of maximum elongation here is the *zone of maximum growth* of these cells.

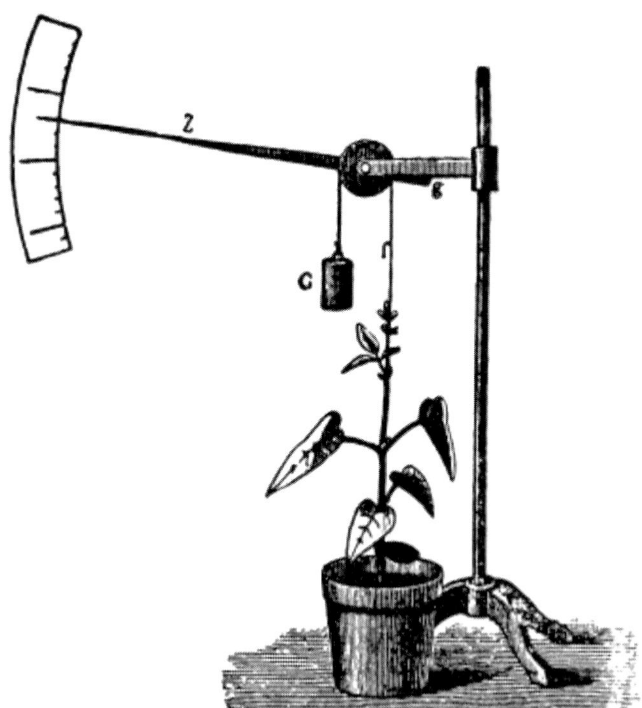

Figure 111. Lever auxanometer (Oels) for measuring elongation of the stem during growth.

Just as the cells exhibit a zone of maximum growth, so the members of the plant exhibit a similar zone of maximum growth. In the case of leaves, when they are young the rapidity of growth is comparatively slow, then it increases, and finally diminishes in rapidity again. So it is with the stem. When the plant is young the growth is not so rapid; as it approaches middle age the rapidity of growth increases; then it declines in rapidity at the close of the season.

Energy of Growth

Closely related to the zone of maximum growth is what is termed the energy of growth. This is manifested in the comparative size of the members of a given plant. To take the sunflower for example, the lower and first leaves are comparatively small. As the plant grows larger the leaves are larger, and this increase in size of the leaves increases up to a maximum period, when the size decreases until we reach the small leaves at the top of the stem. The zone of maximum growth of the leaves corresponds with the maximum size of the leaves on the stem. The rapidity and energy of growth of the stem is also correlated with that of the leaves, and the zone of maximum growth is coincident with that of the leaves. It would be instructive to note it in the case of other plants and also in the case of fruits.

Nutation

During the growth of the stem all of the cells of a given section of the stem do not elongate simultaneously. For example the cells at a given moment on the south side are elongating more rapidly than the cells on the other side. This will cause the stem to bend slightly to the north. In a few moments later the cells on the west side are elongating more rapidly, and the stem is turned to the east; and so on, groups of cells in succession around the stem elongate more rapidly than the others. This causes the stem to describe a circle or ellipse about a central point. Since the region of greatest elongation of the cells of the stem is gradually moving toward the apex of the growing stem, this line of elongation of the cells which is traveling around the stem does so in a spiral manner. In the same way, while the end of the stem is moving upward by the elongation of the cells, and at the same time is slowly moved around, the line which the end of the stem describes must be a spiral one. This movement of the stem, which is common to all stems, leaves, and roots, is nutation.

The importance of nutation to twining stems in their search for a place of support, as well as for the tendrils on leaves or stems, will be seen. In the case of the root it is of the utmost importance, as the root makes its way through the soil, since the particles of soil are more easily thrust aside. The same is also true in the case of many stems before they emerge from the soil.

Chapter 13

IRRITABILITY

We should now examine the movements of plant parts in response to the influence of certain stimuli. By this time we have probably observed that the direction which the root and stem take upon germination of the seed is not due to the position in which the seed happens to lie. Under normal conditions we have seen that the root grows downward and the stem upward.

INFLUENCE OF THE EARTH ON THE DIRECTION OF GROWTH

When the stem and root have been growing in these directions for a short time let us place the seedling in a horizontal position, so that the end of the root extends over an object of support in such a way that it will be free to go in any direction. It should be pinned to a cork and placed in a moist chamber. In the course of twelve to twenty-four hours the root which was formerly horizontal has turned the tip downward again. If we should mark off millimeter spaces beginning at the tip of the root, we should find that the motor zone, or region of curvature, lies in the same region as that of the elongation of the root.

Knight found that the stimulus which influences the root to turn downward is the force of gravity. The reaction of the root in response to this stimulus is geotropism, a turning influenced by the earth. This term is applied to the growth movements of plants influenced by the earth with regard to direction. While the motor zone lies back of the root-tip, the latter receives the stimulus and is the perceptive zone. If the root-tip is cut off, the root is no longer geotropic, and will not turn downward when placed in a horizontal position. Growth toward the earth is *progeotropism*. The lateral growth of secondary roots is *diageotropism*.

Figure 112. Germinating pea placed in a horizontal position.

Figure 113. In 24 hours gravity has caused the root to turn downward.

Figures 112, 113.—Progeotropism of the pea root.

The stem, on the other hand, which was placed in a horizontal position has become again erect. This turning of the stem in the upward direction takes place in the dark as well as in the light, as we can see if we start the experiment at nightfall, or place the plant in the dark. This upward growth of the stem is also influenced by the earth, and therefore is a case of geotropism. The special designation in the case of upright stems is *negative geotropism*, or *apogeotropism*, or the stems are said to be *apogeotropic*. If we place a rapidly growing potted plant in a horizontal position by laying the pot on its side, the ends of the shoots will soon turn upward again when placed in a horizontal position. Young bean plants growing in a pot began within two hours to turn the ends of the shoots upward.

Figure 114. Pumpkin seedling showing apogeotropism. Seedling at the left placed horizontally, in 24 hours the stem has become erect.

Horizontal leaves and shoots can be shown to be subject to the same influence, and are therefore diageotropic.

INFLUENCE OF LIGHT

Figure 115. Radish seedlings grown in the dark, long, slender, not green.

Figure 116. Radish seedlings grown in the light, shorter, stouter, and green in color. Growth retarded by light.

Not only is light a very important factor for plants during photosynthesis, it exerts great influence on plant growth and movement.

Growth in the Absence of Light

Plants grown in the dark are subject to a number of changes. The stems are often longer, more slender and weaker since they contain a larger amount of water in proportion to building material which the plant obtains from carbohydrates manufactured in the light. On many plants the leaves are very small when grown in the dark.

Influence of Light on Direction of Growth

While we are growing seedlings, the pots or boxes of some of them should be placed so that the plants will have a one-sided illumination. This can be done by placing them near an open window, in a room with a one-sided illumination, or they may be placed in a box closed on all sides but one which is facing the window or light. In 12-24 hours, or even in a much shorter time in some cases, the stems of the seedlings will be directed toward the source of light. This influence exerted by the rays of light is *heliotropism*, a turning influenced by the sun or sunlight.

Figure 117. Seedling of castor-oil bean, before and after a one-sided illumination.

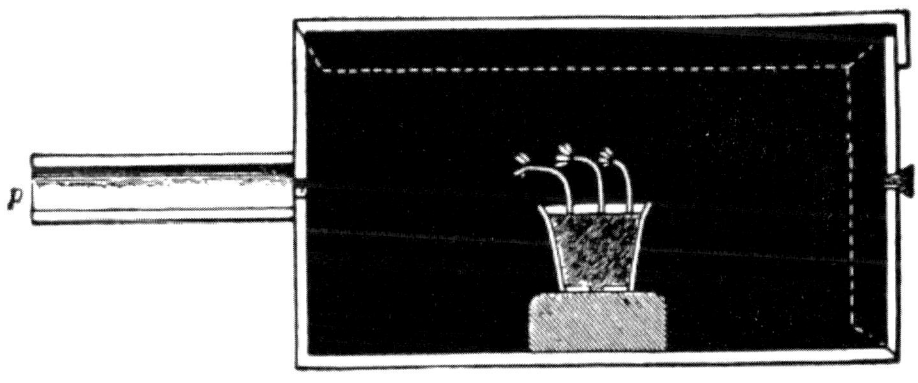

Figure 118. Dark chamber with opening at one side to show heliotropism. (After Schleichert.)

DIAHELIOTROPISM

Horizontal leaves and shoots are *diaheliotropic* as well as *diageotropic*. The general direction which leaves assume under this influence is that of placing them with the upper surface perpendicular to the rays of light which fall upon them. Leaves, then, exposed to the brightly lighted sky are, in general, horizontal. This position is taken in direct response to the stimulus of light. The leaves of plants with a one-sided illumination, as can be seen by trial, are turned with their upper surfaces toward the source of light, or perpendicular to the incidence of the light rays. In this way light overcomes for the time being the direction which growth gives to the leaves. The so-called "sleep" of plants is of course not sleep, though the leaves "nod," or hang downward, in many cases. There are many plants in which we can note this drooping of the leaves at nightfall, and in order to prove that it is not determined by the time of day we can resort to a well-known experiment to induce this condition during the day. The plant which has been used to illustrate this is the sunflower. Some of these plants, which were grown in a box, when they were about 35 *cm* high were covered for nearly two days, so that the light was excluded. At midday on the second day the box was removed, and the leaves on the covered plants are well represented by Figure 119, which was made from one of them. The leaves of the other plants in the box which were not covered were horizontal, as shown by Figure 120. Now on leaving these plants, which had exhibited induced "sleep" movements, exposed to the light they gradually assumed the horizontal position again.

Figure 119. Sunflower plant. Epinastic condition of leaves induced during the day in darkness.

Figure 120. Sunflower plant removed from darkness, leaves extending under influence of light (diaheliotropism.)

EPINASTY AND HYPONASTY

During the early stages of growth of many leaves, as in the sunflower plant, the direction of growth is different from what it is at a later period. The under surface of the young leaves grows more rapidly in a longitudinal direction than the upper side, so that the leaves are held upward close against the bud at the end of the stem. This is termed *hyponasty*, or the leaves are said to be *hyponastic*. Later the growth is more rapid on the upper side and the leaves turn downward or away from the bud. This is termed *epinasty*, or the leaves are said to be *epinastic*. This is shown by the night position of the leaves, or in the induced "sleep" of the sunflower plant in the experiment detailed above. The day position of the leaves on the other hand, which is more or less horizontal, is induced because of their irritability under the influence of light, the inherent downward or epinastic growth is overcome for the time. Then at nightfall or in darkness, the stimulus of light being removed, the leaves assume the position induced by the direction of growth.

Figure 121. Squash seedling. Position of cotyledons in light.

Figure 122. Squash seedling. Position of cotyledons in the dark.

In the case of the cotyledons of some plants it would seem that the growth was hyponastic even after they have opened. The day position of the cotyledons of the pumpkin is more or less horizontal, as shown in Figure 121. At night, or if we darken the plant by covering with a tight box, the leaves assume the position shown in Figure 122.

While the horizontal position is the general one which is assumed by plants under the influence of light, their position is dependent to a certain extent on the intensity of the light as well as on the incidence of the light rays. Some plants are so strongly heliotropic that they change their positions all during the day.

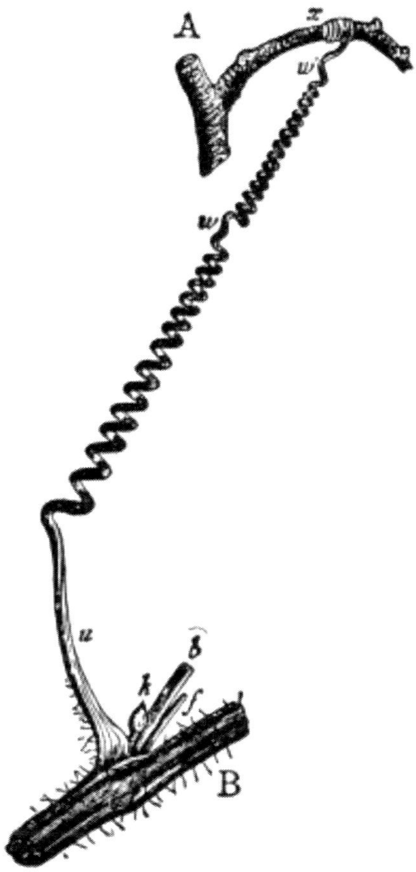

Figure 123. Coiling tendril of bryony.

LEAVES WITH A FIXED DIURNAL POSITION

Leaves of some plants when they are developed have a fixed diurnal position and are not subject to variation. Such leaves tend to arrange themselves in a vertical or paraheliotropic position, in which the surfaces are not exposed to the incidence of light of the greatest intensity, but to the incidence of the rays of diffused light. Interesting cases of the fixed position of leaves are found in the so-called compass plants (like Silphium laciniatum, Lactuca scariola, etc.). In these the horizontal leaves arrange themselves with the surfaces vertical, and also pointing north and south, so that the surfaces face east and west.

Importance of These Movements

Not only are the leaves placed in a position favorable for the absorption of the rays of light which are concerned in making carbon available for food, but they derive other forms of energy from the light, as heat, which is absorbed during the day. Then with the nocturnal position, the leaves being drooped down toward the stem, or with the margin toward the sky, or with the cotyledons as in the pumpkin, castor-oil bean, etc., clasped upward together, the loss of heat by radiation is less than it would be if the upper surfaces of the leaves were exposed to the sky.

Influence of Light on the Structure of the Leaf

In our study of the structure of a leaf we found that in the ivy leaf the palisade cells were on the upper surface. This is the case with a great many leaves, and is the normal arrangement of "dorsiventral" leaves which are diaheliotropic. Leaves which are paraheliotropic tend to have palisade cells on both surfaces. The palisade layer of cells as we have seen is made up of cells lying very close together, and they thus prevent rapid evaporation. They also check to some extent the entrance of the rays of light, at least more so than the loose spongy parenchyma cells do. Leaves developed in the shade have looser palisade and parenchyma cells. In the case of some plants, if we turn over a very young leaf, so that the under side will be uppermost, this side will develop the palisade layer. This shows that light has a great influence on the structure of the leaf.

Movement Influenced by Contact

In the case of tendrils, twining leaves, or stems, the irritability to contact is shown in a movement of the tendril, etc., toward the object in touch. This causes the tendril or stem to coil around the object for support. The stimulus is also extended down the part of the tendril below the point of contact (see Figure 123), and that part coils up like a wire coil spring, thus drawing the leaf or branch from which the tendril grows closer to the object of support. This coil between the object of support and the plant is also very important in easing up the plant when subject to violent gusts of wind which might tear the plant from its support were it not for the yielding and springing motion of this coil.

Figure 124. Sensitive plant leaf in normal position.

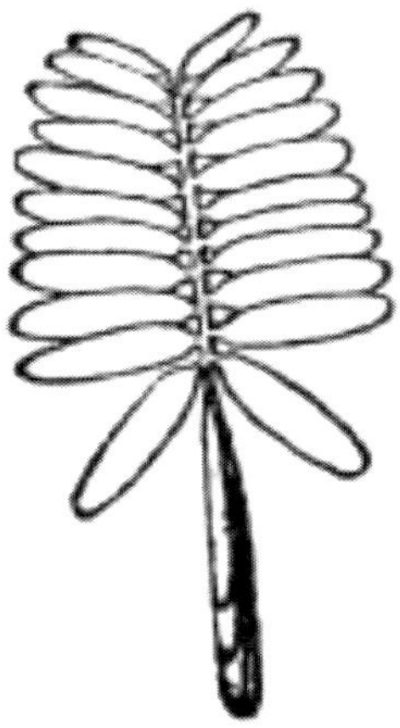

Figure 125. Pinnæ folding up after stimulus.

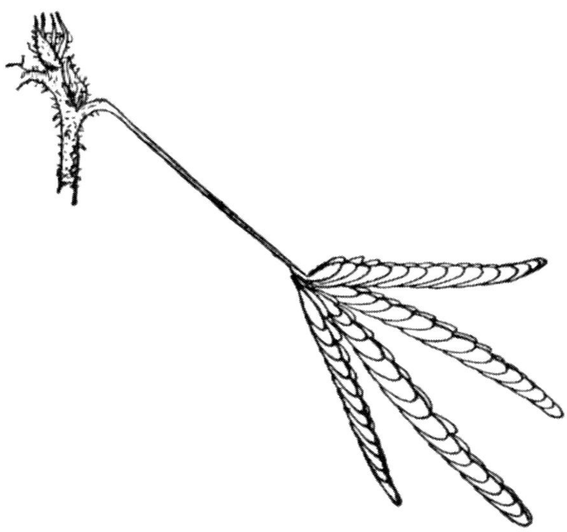

Figure 126. Later all the pinnæ folded and leaf drooped.

Sensitive Plants

These plants are remarkable for the rapid response to stimuli. Mimosa pudica is an excellent plant to study for this purpose.

Movement in Response to Stimuli

If we pinch with the forceps one of the terminal leaflets, or tap it with a pencil, the two end leaflets fold above the "vein" of the pinna. This is immediately followed by the movement of the next pair, and so on as shown in Figure 125, until all the leaflets on this pinna are closed, then the stimulus travels down the other pinnæ in a similar manner, and soon the pinnæ approximate each other and the leaf then drops downward as shown in Figure 126. The normal position of the leaf is shown in Figure 124. If we jar the plant by striking it or by jarring the pot in which it is grown all the leaves quickly collapse into the position shown in Figure 126. If we examine the leaf now we see minute cushions at the base of each leaflet, at the junction of the pinnæ with the petiole, and a larger one at the junction of the petiole with the stem. We shall also note that the movement resides in these cushions.

Transmission of the Stimulus

The transmission of the stimulus in this mimosa from one part of the plant has been found to be along thc cells of the bast.

Cause of the Movement

The movement is caused by a sudden loss of turgidity on the part of the cells in one portion of the pulvinus, as the cushion is called. In the case of the large pulvinus at the base of the petiole this loss of turgidity is in the cells of the lower surface. There is a sudden change in the condition of the protoplasm of the cells here so that they lose a large part of their water. This can be seen if with a sharp knife we cut off the petiole just above the pulvinus before movement takes place. A drop of liquid exudes from the cells of the lower side.

Paraheliotropism of the Leaves of the Sensitive Plant

If the mimosa plant is placed in very intense light the leaflets will turn their edges toward the incidence of the rays of light. This is also true of other plants in intense light, and is *paraheliotropism*. Transpiration is thus lessened, and chlorophyll is protected from too intense light.

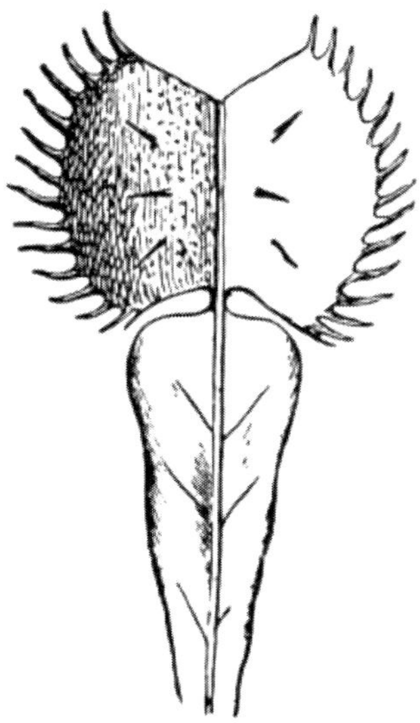

Figure 126*a*. Leaf of Venus fly-trap (Dionæa muscipula), showing winged petiole and toothed lobes.

Figure 127. Leaf of Drosera rotundifolia, some of the glandular hairs folding inward as a result of a stimulus.

We thus see that variations in the intensity of light have an important influence in modifying movements. Variations in temperature also exert a considerable influence, rapid elevation of temperature causing certain flowers to open, and falling temperature causing them to close.

Sensitiveness of Insectivorous Plants

The Venus fly-trap (Dionæa muscipula) and the sundew (drosera) are interesting examples of sensitive plants, since the leaves close in response to the stimulus from insects.

Hydrotropism

Roots are sensitive to moisture. They will turn toward moisture. This is of the greatest importance for the well-being of the plant, since the roots will seek those places in the soil where suitable moisture is present. On the other hand, if the soil is too wet there is a tendency for the roots to grow away from the soil which is saturated with water. In such cases roots are often seen growing upon the surface of the soil so that they may obtain oxygen, which is important for the root in the processes of absorption and growth. Plants then may be injured by an excess of water as well as by a lack of water in the soil.

Temperature

In the experiments on germination thus far made it has probably been noted that the temperature has much to do with the length of time taken for seeds to germinate. It also influences the rate of growth. The effect of different temperatures on the germination of seed can be very well noted by attempting to germinate some in rooms at various temperatures. It will be found, other conditions being equal, that in a moderately warm room, or even in one quite warm, 25-30 degrees centigrade, germination and growth goes on more rapidly than in a cool room, and here more rapidly than in one which is decidedly cold. In the case of most plants in temperate climates, growth may go on at a temperature but little above freezing, but few will thrive at this temperature.

If we place dry peas or beans in a temperature of about 70 °C. for 15 minutes they will not be killed, but if they have been thoroughly soaked in water and then placed at this temperature they will be killed, or even at a somewhat lower temperature. The same seeds in the dry condition will withstand a temperature of 10 °C. below, but if they are first soaked in water this low temperature will kill them.

In order to see the effect of freezing we may thoroughly freeze a section of a beet root, and after thawing it out place it in water. The water is colored by the cell-sap which escapes from the cells, just as we have seen it does as a result of a high temperature, while a section of an unfrozen beet placed in water will not color it if it was previously washed.

If the slice of the beet is placed at about -6 °C. in a shallow glass vessel, and covered, ice will be formed over the surface. If we examine it with the microscope ice crystals will be seen formed on the outside, and these will not be colored. The water for the formation of the crystals came from the cell-sap, but the concentrated solutions in the sap were not withdrawn by the freezing over the surface.

If too much water is not withdrawn from the cells of many plants in freezing, and they are thawed out slowly, the water which was withdrawn from the cells will be absorbed again and the plant will not be killed. But if the plant is thawed out quickly the water will not be absorbed, but will remain on the surface and evaporate. Some will also remain in the intercellular spaces, and the plant will die. Some plants, however, no matter how slowly they are thawed out, are killed after freezing, as the leaves of the pumpkin, dahlia, or the tubers of the potato.

It has been found that as a general rule when plants, or plant parts, contain little moisture they will withstand quite high degrees of temperature, as well as quite low degrees, but when the parts are filled with sap or water they are much more easily killed. For this reason dry seeds and the winter buds of trees, and other plants, because they contain but little water, are better able to resist the cold of winters. But when growth begins in the spring, and the tissues of these same parts become turgid and filled with water, they are quite easily killed by frosts. It should be borne in mind, however, that there is great individual variation in plants in this respect, some being more susceptible to cold than others. There is also great variation in plants as to their resistance to the cold of winters, and of arctic climates, the plants of the latter regions being able to resist very low temperatures. We have examples also in the arctic plants, and those which grow in arctic climates on high mountains, of plants which are able to carry on all the life functions at temperatures but little above freezing.

For further discussion as to relation of plants to temperature, see Chapters 46, 48, 49, and 53.

PART II. MORPHOLOGY AND LIFE HISTORY OF REPRESENTATIVE PLANTS

Chapter 14

SPIROGYRA

In our study of protoplasm and some of the processes of plant life we became acquainted with the general appearance of the plant spirogyra. It is now a familiar object to us. And in taking up the study of representative plants of the different groups, we shall find that in knowing some of these lower plants the difficulties of understanding methods of reproduction and relationship are not so great as they would be if we were entirely ignorant of any members of the lower groups.

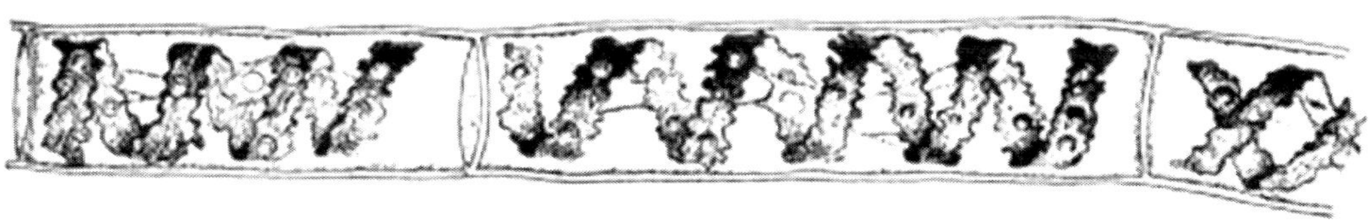

Figure 128. Thread of spirogyra, showing long cells, chlorophyll band, nucleus, strands of protoplasm, and the granular wall layer of protoplasm.

FORM OF SPIROGYRA

We have found that the plant spirogyra consists of simple threads, with cylindrical cells attached end to end. We have also noted that each cell of the thread is exactly alike, with the exception of certain "holdfasts" on some of the species. If we should examine threads in different stages of growth we should find that each cell is capable of growth and division, just as it is capable of performing all the functions of nutrition and assimilation. The cells of spirogyra then multiply by division. Not simply the cells at the ends of the threads but any and all of the cells divide as they grow, and in this way the threads increase in length.

Multiplication of the Threads

In studying living material of this plant we have probably noted that the threads often become broken by two of the adjacent cells of a thread becoming separated. This may be and is accomplished in many cases without any injury to the cells. In this manner the threads or plants of spirogyra, if we choose to call a thread a plant, multiply, or increase. In this breaking of a thread the cell wall which separates any two cells splits. If we should examine several species of spirogyra we would probably find threads which present two types as regards the character of the walls at the ends of the cells. In Figure 128 we see that the ends are plain, that is, the cross walls are all straight. But in some other species the inner wall of the cells presents a peculiar appearance. This inner wall at the end of the cell is at first straight across. But it soon becomes folded back into the interior of its cell, just as the end of an empty glove finger may be pushed in. Then the infolded end is pushed partly out again, so that a peculiar Figure is the result.

How Some of the Threads Break

In the separation of the cells of a thread this peculiarity is often of advantage to the plant. The cell-sap within the protoplasmic membrane absorbs water and the pressure pushes on the ends of the infolded cell walls. The inner wall being so much longer than the outer wall, a pull is exerted on the latter at the junction of the cells. Being weaker at this point the outer wall is ruptured. The turgidity of the two cells causes these infolded inner walls to push out suddenly as the outer wall is ruptured, and the thread is snapped apart as quickly as a pipe-stem may be broken.

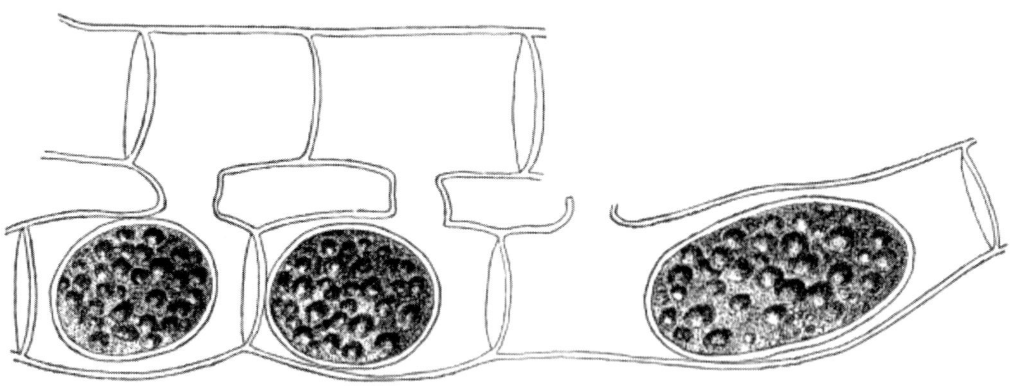

Figure 129. Zygospores of spirogyra.

Conjugation of Spirogyra

Under certain conditions, when vegetative growth and multiplication cease, a process of reproduction takes place which is of a kind termed sexual reproduction. If we select mats of spirogyra which have lost their deep green color, we are likely to find different stages of this sexual process, which in the case of spirogyra and related plants is called *conjugation*. A few threads of such a mat we should examine with the microscope. If the material is in the right condition we see in certain of the cells an oval or elliptical body. If we note carefully the cells in which these oval bodies are situated, there will be seen a tube at one side which connects with an empty cell of a thread which lies near as shown in Figure 129. If we search through the material we may see other threads connected in this ladder fashion, in which the contents of the cells are in various stages of collapse from what we have seen in the growing cell. In some the protoplasm and chlorophyll band have moved but little from the wall; in others it forms a mass near the center of the cell, and again in others we will see that the contents of the cell of one of the threads has moved partly through the tube into the cell of the thread with which it is connected.

This suggests to us that the oval bodies found in the cells of one thread of the ladder, while the cells of the other thread were empty, are formed by the union of the contents of the two cells. In fact that is what does take place. This kind of union of the contents of two similar or nearly similar cells is *conjugation*. The oval bodies which are the result of this conjugation are *zygotes*, or *zygospores*. When we are examining living material of spirogyra in this stage it is possible to watch this process of conjugation. Figure 130 represents the different stages of conjugation of spirogyra.

How the Threads Conjugate, or Join

The cells of two threads lying parallel put out short processes. The tubes from two opposite cells meet and join. The walls separating the contents of the two tubes dissolve so that there is an open communication between the two cells. The content of each one of these cells which take part in the conjugation is a *gamete*. The one which passes through the tube to the receiving cell is the *supplying gamete*, while that of the receiving cell is the *receiving gamete*.

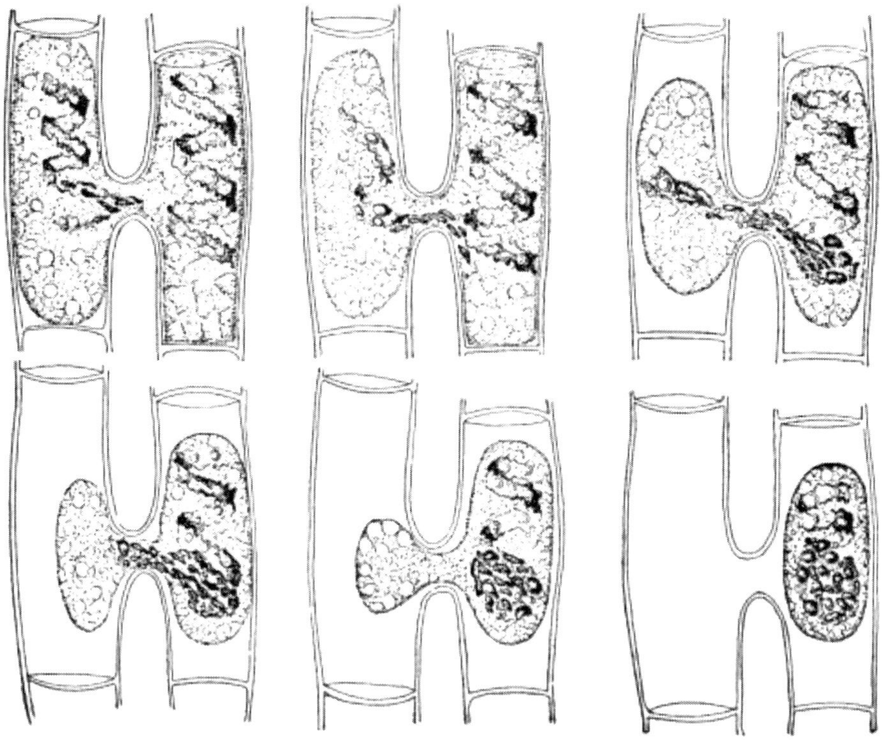

Figure 130. Conjugation in spirogyra; from left to right beginning in the upper row is shown the gradual passage of the protoplasm from the supplying gamete to the receiving gamete.

How the Protoplasm Moves from One Cell to Another

Before any movement of the protoplasm of the supplying cell takes place we can see that there is great activity in its protoplasm. Rounded vacuoles appear which increase in size, are filled with a watery fluid, and swell up like a vesicle, and then suddenly contract and disappear. As the vacuole disappears it causes a sudden movement or contraction of the protoplasm around it to take its place. Simultaneously with the disappearance of the vacuole the membrane of the protoplasm is separated from a part of the wall. This is probably brought about by a sudden loss of some of the water in the cell-sap. These activities go on, and the protoplasmic membrane continues to slip away from the wall. Every now and then there is a movement by which the protoplasm is moved a short distance. It is moved toward the tube and finally a portion of it with one end of the chlorophyll band begins to move into the tube. About this time the vacuoles can be seen in an active condition in the receptive cell. At short intervals movement continues until the content of the supplying cell has passed over into that of the receptive cell. The protoplasm of this one is now slipping away from the cell wall, until finally the two masses round up into the one zygospore.

The Zygospore

This zygospore now acquires a thick wall which eventually becomes brown in color. The chlorophyll color fades out, and a large part of the protoplasm passes into an oily substance which makes it more resistant to conditions which would be fatal to the vegetative threads. The zygospores are capable therefore of enduring extremes of cold and dryness which would destroy the threads. They pass through a "resting" period, in which the water in the pond may be frozen, or dried, and with the oncoming of favorable conditions for growth in the spring or in the autumn they germinate and produce the green thread again.

Life Cycle

The growth of the spirogyra thread, the conjugation of the gametes and formation of the zygospore, and the growth of the thread from the zygospore again, makes what is called a complete *life cycle*.

Fertilization

While conjugation results in the fusion of the two masses of protoplasm, fertilization is accomplished when the nuclei of the two cells come together in the zygospore and fuse into a single nucleus. The different stages in the fusion of the two nuclei of a recently formed zygospore are shown in Figure 131.

In the conjugation of the two cells, the chlorophyll band of the supplying cell is said to degenerate, so that in the new plant the number of chlorophyll bands in a cell is not increased by the union of the two cells.

Simplicity of the Process

In spirogyra any cell of the thread may form a gamete (excepting the holdfasts of some species). Since all of the cells of a thread are practically alike, there is no structural difference between a vegetative cell and a cell about to conjugate. The difference is a physiological one. All the cells are capable of conjugation if the physiological conditions are present. All the cells therefore are potential gametes. (Strictly speaking the wall of the cell is the *garnetangium*, while the content forms the gamete.)

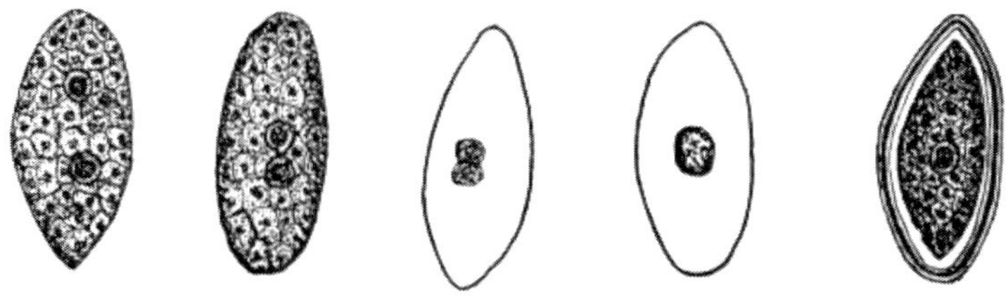

Figure 131. Fertilization in spirogyra; shows different stages of fusion of the two nuclei, with mature zygospore at right. (After Overton.)

While there is sometimes a slight difference in size between the conjugating cells, and the supplying cell may be the smaller, this is not general. We say, therefore, that there is no differentiation among the gametes, so that usually before the protoplasm begins to move one cannot say which is to be the supplying and which the receiving gamete.

Position of the Plant Spirogyra

From our study then we see that there is practically no differentiation among the vegetative cells, except where holdfasts grow out from some of the cells for support. They are all alike in form, in capacity for growth, division, or multiplication of the threads.

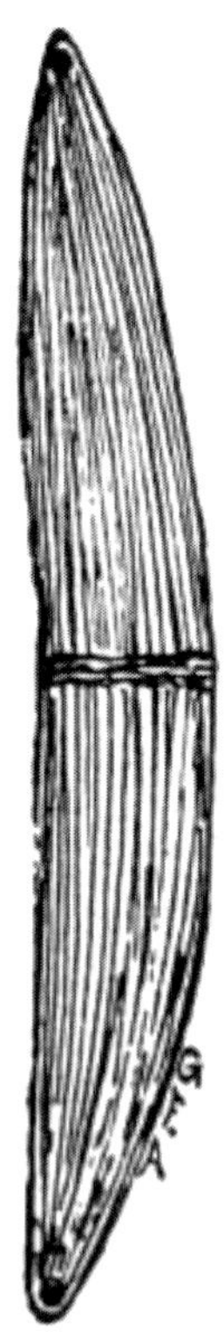

Figure 132. Closterium.

Each cell is practically an independent plant. There is no differentiation between vegetative cell and conjugating cell. All the cells are potential gametes. Finally there is no structural differentiation between the gametes. This indicates then a simple condition of things, a low grade of organization.The alga spirogyra is one of the representatives of the lower algæ belonging to the group called *Conjugatæ*. Zygnema with star-shaped chloroplasts, mougeotia with straight or sometimes twisted chlorophyll bands, belong to the same group. In the latter genus only a portion of the protoplasm of each cell unites to form the zygospore, which is located in the tube between the cells.

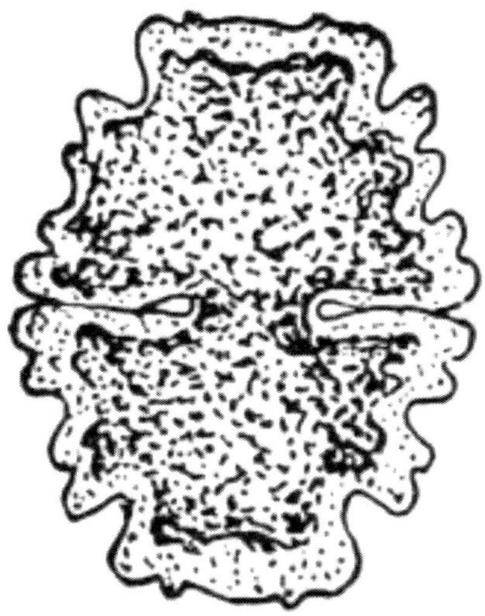

Figure 133. Micrasterias.

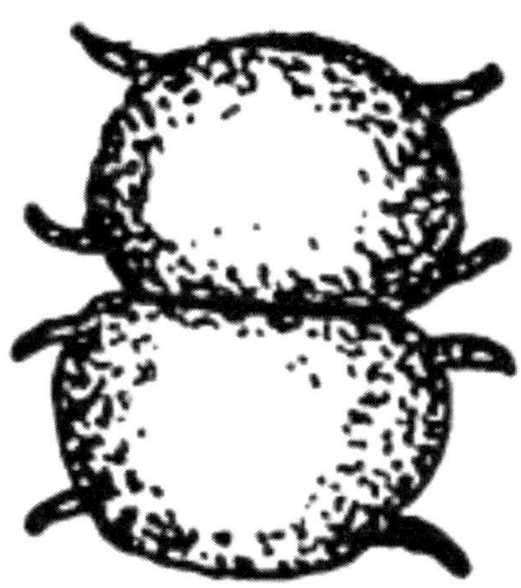

Figure 134. Xanthidium.

Figure 135. Staurastrum.

Figure 136. Euastrum.

Figure 137. Cosmarium.

The desmids also belong to the same group. The desmids usually live as separate cells. Many of them are beautiful in form. They grow entangled among other algæ, or on the surface of aquatic plants, or on wet soil. Several genera are illustrated in Figures 132-137.

Chapter 15

VAUCHERIA

The plant vaucheria we remember from our study in an earlier chapter. It usually occurs in dense mats floating on the water or lying on damp soil. The texture and feeling of these mats remind one of "felt," and the species are sometimes called the "green felts." The branched threads are continuous, that is there are no cross walls in the vegetative threads. This plant multiplies itself in several ways which would be too tedious to detail here. But when fresh bright green mats can be obtained they should be placed in a large vessel of water and set in a cool place. Only a small amount of the alga should be placed in a vessel, since decay will set in more rapidly with a large quantity. For several days one should look for small green bodies which may be floating at the side of the vessel next the lighted window.

ZOOGONIDIA OF VAUCHERIA

If these minute floating green bodies are found, a small drop of water containing them should be mounted for examination. If they are rounded, with slender hair-like appendages over the surface, which vibrate and cause motion, they very likely are one of the kinds of reproductive bodies of vaucheria. The hair-like appendages are *cilia*, and they occur in pairs, several of them distributed over the surface. These rounded bodies are *gonidia*, and because they are motile they are called *zoogonidia*.

By examining some of the threads in the vessel where they occurred we may have perhaps an opportunity to see how they are produced. Short branches are formed on the threads, and the contents are separated from those of the main thread by a septum. The protoplasm and other contents of this branch separate from the wall, round up into a mass, and escape through an opening which is formed in the end. Here they swim around in the water for a time, then come to rest, and germinate by growing out into a tube which forms another vaucheria plant. It will be observed that this kind of reproduction is not the result of the union of two different parts of the plant. It thus differs from that which is termed sexual reproduction. A small part of the plant simply becomes separated from it as a special body, and then grows into a new plant, a sort of multiplication. This kind of reproduction has been termed *asexual reproduction*.

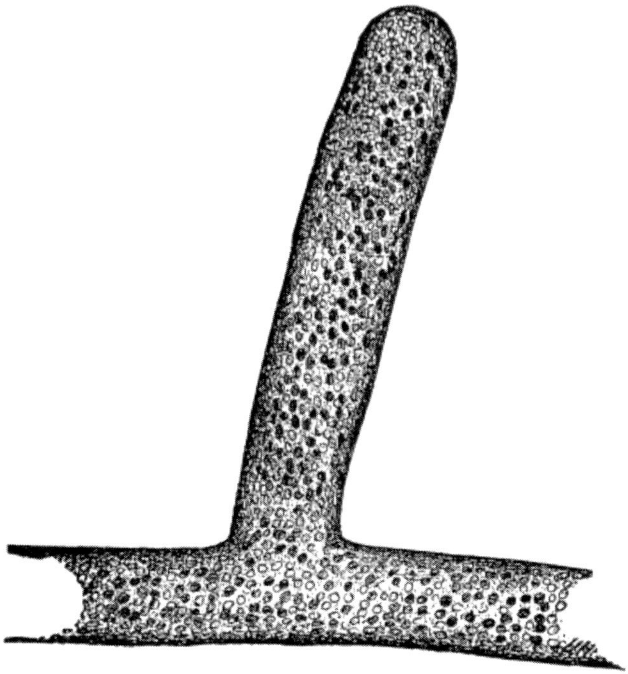

Figure 138. Portion of branched thread of vaucheria.

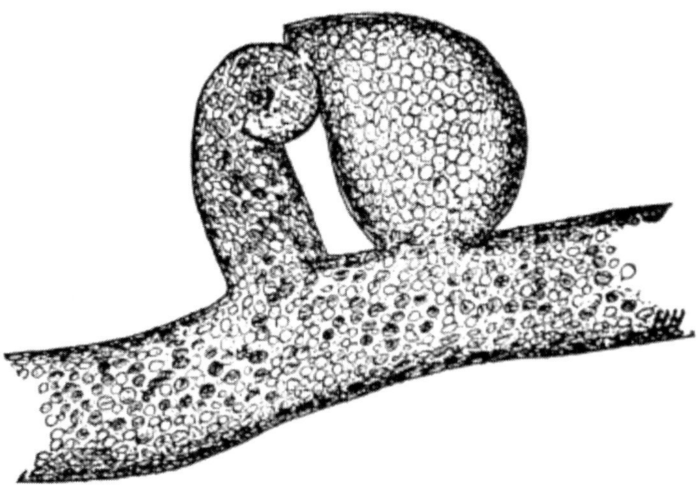

Figure 139. Young antheridium and oogonium of Vaucheria sessilis, before separation from contents of thread by a septum.

Sexual Reproduction in Vaucheria

The organs which are concerned in sexual reproduction in vaucheria are very readily obtained for study if one collects the material at the right season. They are found quite readily during the spring and autumn, and may be preserved in formalin for study at any season, if the material cannot be collected fresh at the time it is desired for study. Fine material for study often occurs on the soil of pots in greenhouses during the winter. While the zoogonidia are more apt to be found in material which is quite green and freshly growing, the sexual organs are usually more abundant when the threads appear somewhat yellowish, or yellow green.

Vaucheria Sessilis; The Sessile Vaucheria

In this plant the sexual organs are sessile, that is they are not borne on a stalk as in some other species. The sexual organs usually occur several in a group. Figure 139 represents a portion of a fruiting plant.

Sexual Organs of Vaucheria. Antheridium

The antheridia are short, slender, curved branches from a main thread. A septum is formed which separates an end portion from the stalk. This end cell is the *antheridium*. Frequently it is collapsed or empty as shown in Figure 140. The protoplasm in the antheridium forms numerous small oval bodies each with two slender lashes, the cilia. When these are formed the antheridium opens at the end and they escape. It is after the escape of these spermatozoids that the antheridium is collapsed. Each spermatozoid is a male gamete.

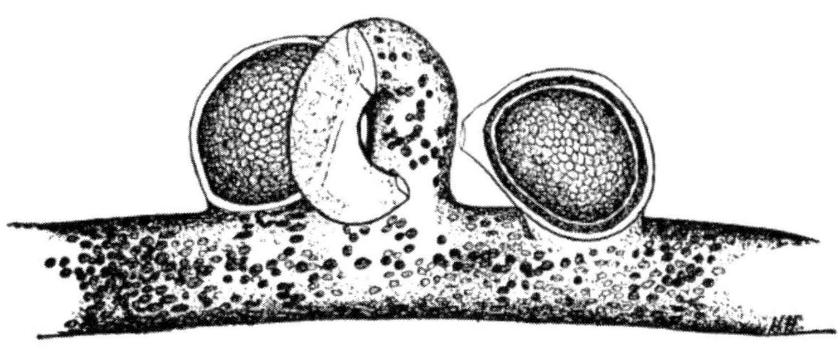

Figure 140. Vaucheria sessilis, one antheridium between two oogonia.

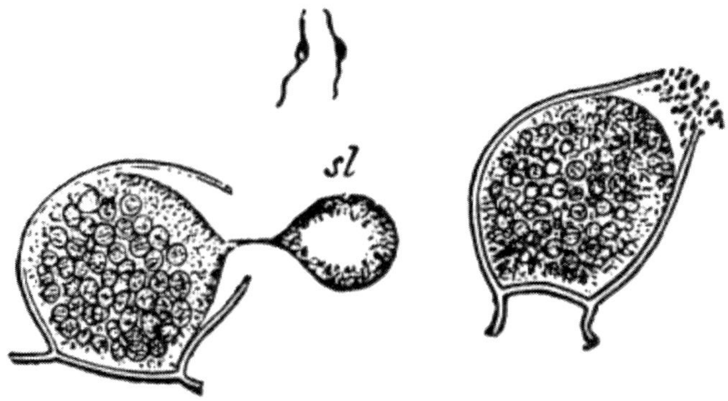

Figure 141. Vaucheria sessilis; oogonium opening and emitting a bit of protoplasm; spermatozoids; spermatozoids entering oogonium. (After Pringsheim and Goebel.)

Oogonium

The oogonia are short branches also, but they become large and somewhat oval. The septum which separates the protoplasm from that of the main thread is as we see near the junction of the branch with the main thread. The oogonium, as shown in the Figure, is usually turned somewhat to one side. When mature the pointed end opens and a bit of the protoplasm escapes. The remaining protoplasm forms the large rounded egg-cell which fills the wall of the oogonium. In some of the oogonia which we examine this egg is surrounded by a thick brown wall, with starchy and oily contents. This is the fertilized egg (sometimes called here the oospore). It is freed from the oogonium by the disintegration of the latter, sinks into the mud, and remains here until the following autumn or spring, when it grows directly into a new plant.

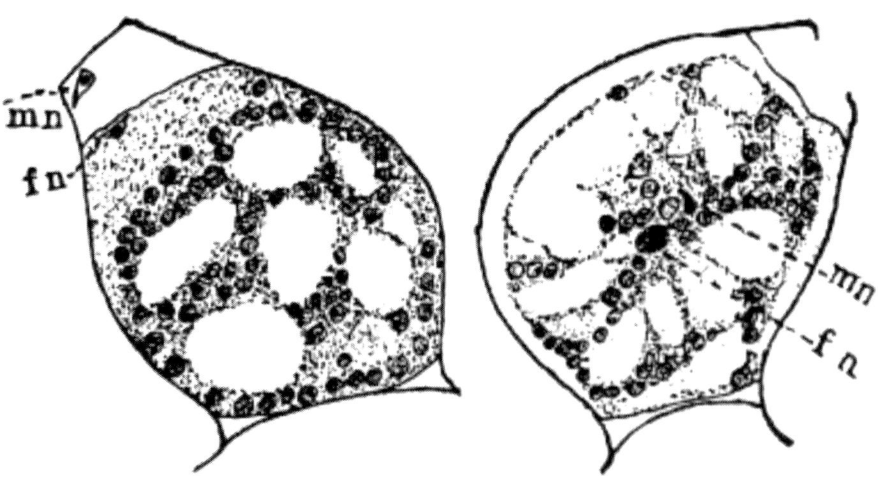

Figure 142. Fertilization in vaucheria, *mn*, male nucleus; *fn*, female nucleus. Male nucleus entering the egg and approaching the female nucleus. (After Oltmans.)

FERTILIZATION

Fertilization is accomplished by the spermatozoids swimming in at the open end of the oogonium, when one of them makes its way down into the egg and fuses with the nucleus of the egg.

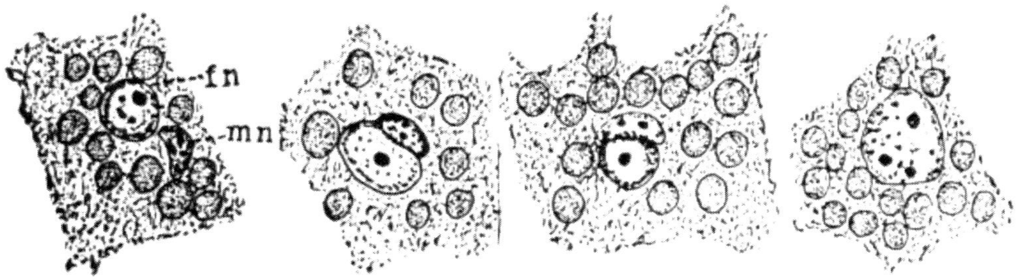

Figure 143. Fertilization of vaucheria. *fn*, female nucleus; *mn*, male nucleus. The different Figures show various stages in the fusion of the nuclei.

THE TWIN VAUCHERIA (V. GEMINATA)

Another species of vaucheria is the twin vaucheria. This is also a common one, and may be used for study instead of the sessile vaucheria if the latter cannot be obtained. The sexual organs are borne at the end of a club-shaped branch. There are usually two oogonia, and one antheridium between them which terminates the branch. In a closely related species, instead of the two oogonia there is a whorl of them with the antheridium in the center.

VAUCHERIA COMPARED WITH SPIROGYRA

In vaucheria we have a plant which is very interesting to compare with spirogyra in several respects. Growth takes place, not in all parts of the thread, but is localized at the ends of the thread and its branches. This represents a distinct advance on such a plant as spirogyra. Again, only specialized parts of the plant in vaucheria form the sexual organs. These are short branches. Farther there is a great difference in the size of the two organs, and especially in the size of the gametes, the supplying gametes (spermatozoids) being very minute, while the receptive gamete is large and contains all the nutriment for the fertilized egg. In spirogyra, on the other hand, there is usually no difference in size of the gametes, as we have seen, and each contributes equally in the matter of nutriment for the fertilized egg. Vaucheria, therefore, represents a distinct advance, not only in the vegetative condition of the plant, but in the specialization of the sexual organs. Vaucheria, with other related algæ, belongs to a group known as the *Siphoneæ*, so called because the plants are tube-like or *siphon*-like.

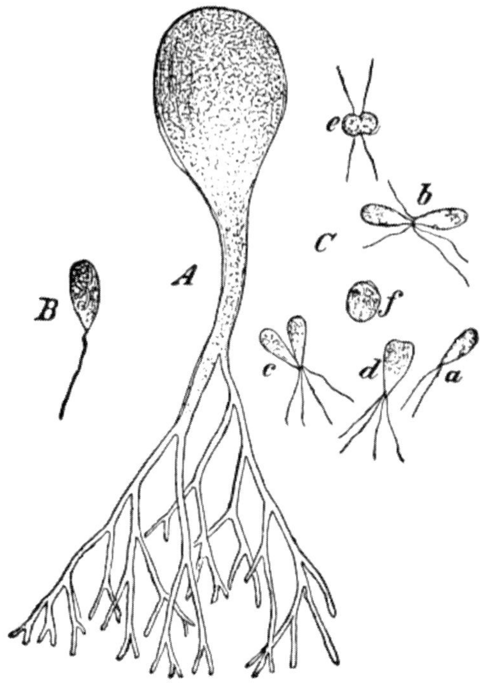

Figure 143*a*. Botrydium granulatum. *A*, the whole plant; *B*, swarm spore; *C*, planogametes; *a*, a single gamete; *b-e*, two gametes in process of fusion; *f*, zygote.

Botrydium Granulatum

An example of one of the simpler members of the Siphoneæ is Botrydium granulatum. It is found sometimes in abundance on wet ground which is colored green or red by its presence, according to the stage of development. The plant body is long pear-shaped, the smaller end attached to the ground by slender branched rhizoids (Figure 143). The protoplasm contains many nuclei and lines the inside of the wall. When multiplication takes place large numbers of small zoospores with one cilium each are formed in the protoplasm, and escape at free end. Reproduction takes place by two-ciliated gametes, which fuse in pairs to form zygospores. In dry seasons the protoplasm in the pear-shaped plant passes down into the rhizoids and forms small rounded *planospores*. All the stages of development are too complicated to describe here.

Chapter 16

ŒDOGONIUM

Œdogonium is also an alga. The plant is sometimes associated with spirogyra, and occurs in similar situations. Our attention was called to it in the study of chlorophyll bodies. These we recollect are, in this plant, small oval disks, and thus differ from those in spirogyra.

FORM OF ŒDOGONIUM

Like spirogyra, œdogonium forms simple threads which are made up of cylindrical cells placed end to end. But the plant is very different from any member of the group to which spirogyra belongs. In the first place each cell is not the equivalent of an individual plant as in spirogyra. Growth is localized or confined to certain cells of the thread which divide at one end in such a way as to leave a peculiar overlapping of the cell walls in the form of a series of shallow caps or vessels (Figure 144), and this is one of the characteristics of this genus. Other differences we find in the manner of reproduction.

FRUITING STAGE OF ŒDOGONIUM

Material in the fruiting stage is quite easily obtainable, and may be preserved for study in formalin if there is any doubt about obtaining it at the time we need it for study. This condition of the plant is easily detected because of the swollen condition of some of the cells, or by the presence of brown bodies with a thick wall in some of the cells.

Sexual Organs of Œdogonium. Oogonium and Egg

The enlarged cell is the oogonium, the wall of the cell being the wall of the oogonium. (See Figure 145.) The protoplasm inside, before fertilization, is the egg-cell. In those cases where the brown body with a thick wall is present fertilization has taken place, and this body is the *fertilized egg*, or *oospore*. It contains large quantities of an oily substance, and, like the fertilized egg of spirogyra and vaucheria, is able to withstand greater changes in temperature than the vegetative stage, and can endure drying and freezing for some time without injury.

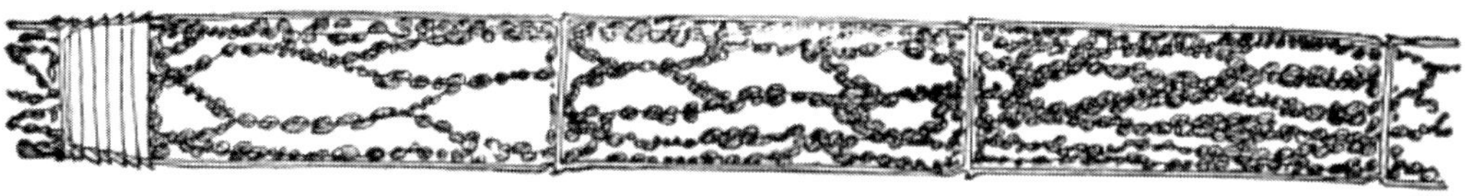

Figure 144. Portion of thread of œdogonium, showing chlorophyll grains, and peculiar cap cell walls.

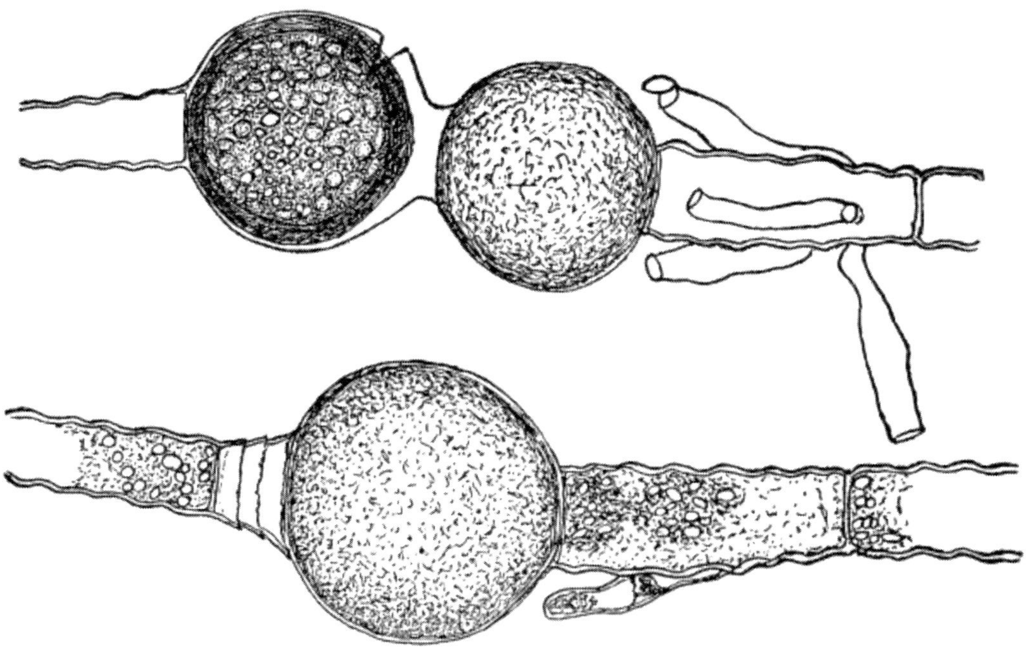

Figure 145. Œdogonium undulatum, with oogonia and dwarf males; the upper oogonium at the right has a mature oospore.

In the oogonium wall there can frequently be seen a rift near the middle of one side, or near the upper end. This is the opening through which the spermatozoid entered to fecundate the egg.

Dwarf Male Plants

In some species there will also be seen peculiar club-shaped dwarf plants attached to the side of the oogonium, or near it, and in many cases the end of this dwarf plant has an open lid on the end.

Antheridium

The end cell of the dwarf male in such species is the *antheridium*. In other species the spermatozoids are developed in different cells (antheridia) of the same thread which bears the oogonium, or on a different thread.

Zoospore Stage of Œdogonium

The egg after a period of rest starts into active life again. In doing so it does not develop the thread-like plant directly as in the case of vaucheria and spirogyra. It first divides into four zoospores which are exactly like the zoogonidia in form. (See Figure 152.)

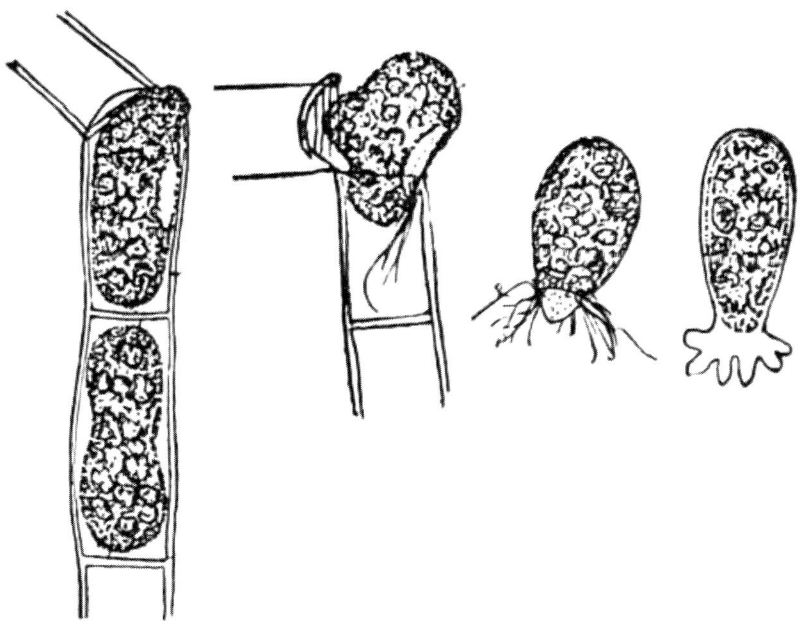

Figure 146. Zoogonidia of œdogonium escaping. At the right one is germinating and forming the holdfasts, by means of which these algæ attach themselves to objects for support. (After Pringsheim.)

These germinate and develop the thread form again. This is a quite remarkable peculiarity of œdogonium when compared with either vaucheria or spirogyra. It is the introduction of an intermediate stage between the fertilized egg and that form of the plant which bears the sexual organs, and should be kept well in mind.

Asexual Reproduction

Material for the study of this stage of œdogonium is not readily obtainable just when we wish it for study. But fresh plants brought in and placed in a quantity of fresh water may yield suitable material, and it should be examined at intervals for several days. This kind of reproduction takes place by the formation of *zoogonidia*. The entire contents of a cell round off into an oval body, the wall of the cell breaks, and the zoogonidium escapes. It has a clear space at the small end, and around this clear space is a row or crown of cilia as shown in Figure 146. By the vibration of these cilia the zoogonidium swims around for a time, then settles down on some object of support, and several slender holdfasts grow out in the form of short rhizoids which attach the young plant.

Sexual Reproduction. Antheridia

The antheridia are short cells which are formed by one of the ordinary cells dividing into a number of disk-shaped ones as shown in Figure 147.

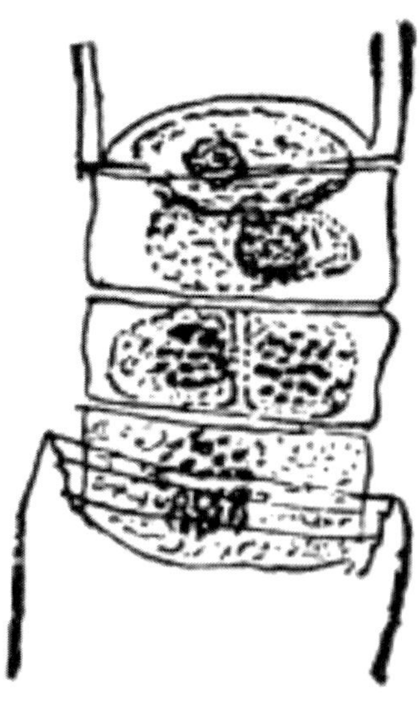

Figure 147. Portion of thread of œdogonium showing antheridia.

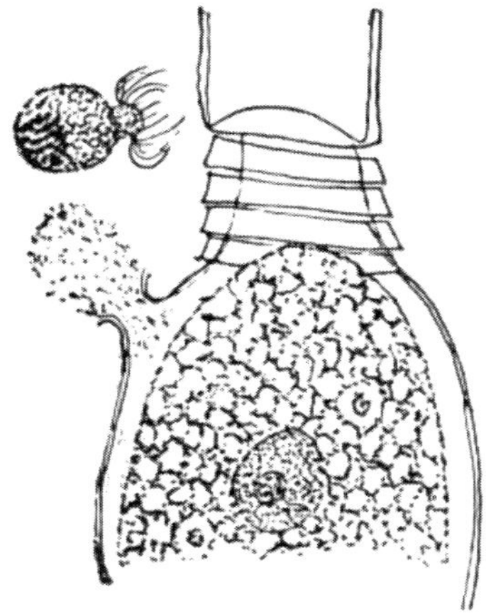

Figure 148. Portion of thread of œdogonium showing upper half of egg open, and a spermatozoid ready to enter. (After Klebahn).

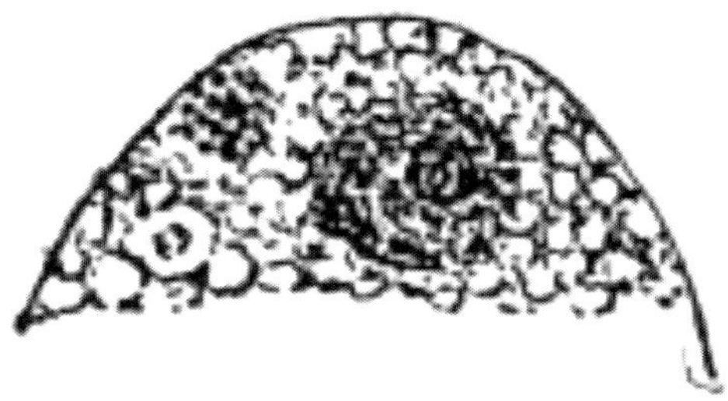

Figure 149. Male nucleus just entering egg at left side.

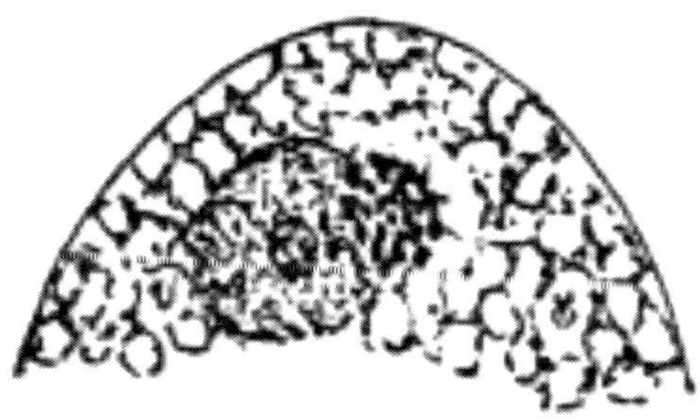

Figure 150. Male nucleus fusing with female nucleus.

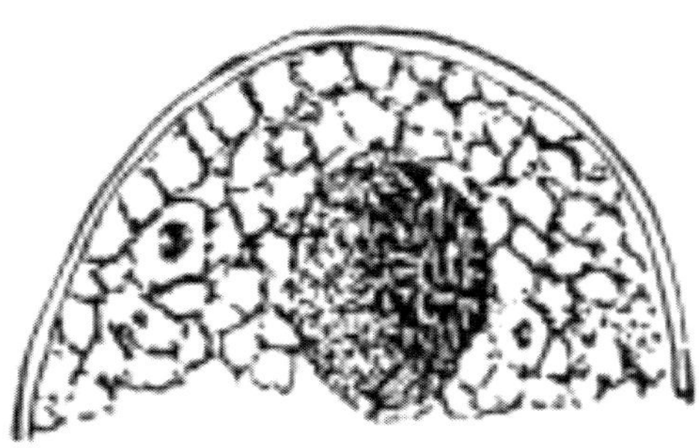

Figure 151. The two nuclei fused, and fertilization complete.

The protoplasm in each antheridium forms two spermatozoids (sometimes only one) which are of the same form as the zoogonidia but smaller, and yellowish instead of green. In some species a motile body intermediate in size and color between the spermatozoids and zoogonidia is first formed, which after swimming around comes to rest on the oogonium, or near it, and develops what is called a "dwarf male plant" from which the real spermatozoid is produced.

Figures 149-151.—Fertilization in œdogonium. (After Klebahn).

Oogonia

The oogonia are formed directly from one of the vegetative cells. In most species this cell first enlarges in diameter, so that it is easily detected. The protoplasm inside is the egg-cell. The oogonium wall opens, a bit of the protoplasm is emitted, and the spermatozoid then enters and fertilizes it (Figure 148). Now a hard brown wall is formed around it, and, just as in spirogyra and vaucheria, it passes through a resting period. At the time of germination it does not produce the thread-like plant again directly, but first forms four zoospores exactly like the zoogonidia (Figure 152). These zoospores then germinate and form the plant.

Œdogonium Compared with Spirogyra

Now if we compare œdogonium with spirogyra, as we did in the case of vaucheria, we find here also that there is an advance upon the simple condition which exists in spirogyra. Growth and division of the thread is limited to certain portions. The sexual organs are differentiated. They usually differ in form and size from the vegetative cells, though the oogonium is simply a changed vegetative cell. The sexual organs are differentiated among themselves, the antheridium is small, and the oogonium large. The gametes are also differentiated in size, and the male gamete is motile, and carries in its body the nucleus which fuses with the nucleus of the egg-cell.

But a more striking advance is the fact that the fertilized egg does not produce the vegetative thread of œdogonium directly, but first forms four zoospores, each of which is then capable of developing into the thread. On the other hand we found that in spirogyra the zygospore develops directly into the thread form of the plant.

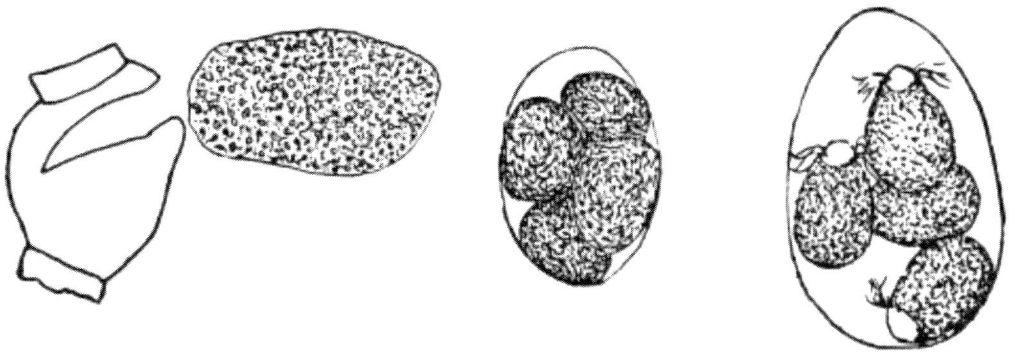

Figure 152. Fertilized egg of œdogonium after a period of rest escaping from the wall of the oogonium, and dividing into the four zoospores. (After Juranyi.)

Figure 153. Tuft of chætophora, natural size.

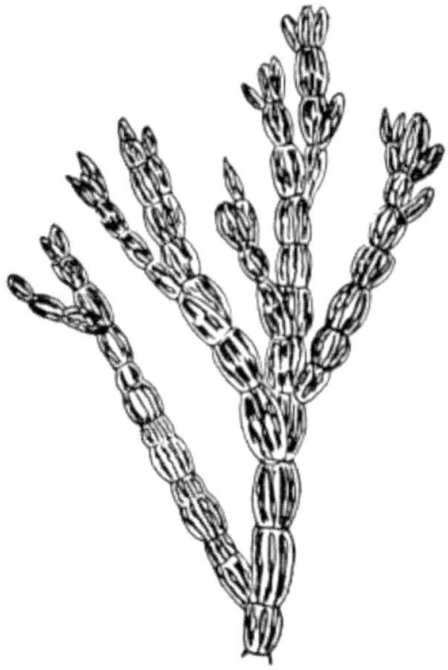

Figure 154. Portion of chætophora showing branching.

Position of Œdogonium

Œdogonium is one of the true thread-like algæ, green in color, and the threads are divided into distinct cells. It, along with many relatives, was once placed in the old genus conferva. These are all now placed in the group *Confervoideæ*, that is, the *conferva-like algæ*.

Relatives of Œdogonium

Many other genera are related to œdogonium. Some consist of simple threads, and others of branched threads. An example of the branched forms is found in chætophora, represented in Figures 153, 154. This plant grows in quiet pools or in slow-running water. It is attached to sticks, rocks, or to larger aquatic plants. Many threads spring from the same point of attachment and radiate in all directions. This, together with the branching of the threads, makes a small, compact, greenish, rounded mass, which is held firmly together by a gelatinous substance. The masses in this species are about the size of a small pea, or smaller. Growth takes place in chætophora at the ends of the threads and branches. That is, growth is apical. This, together with the branched threads and the tendency to form cell masses, is a great advance of the vegetative condition of the plant upon that which we find in the simple threads of œdogonium.

Chapter 17

COLEOCHÆTE

Among the green algæ coleochæte is one of the most interesting. Several species are known in this country. One of these at least should be examined if it is possible to obtain it. It occurs in the water of fresh lakes and ponds, attached to aquatic plants.

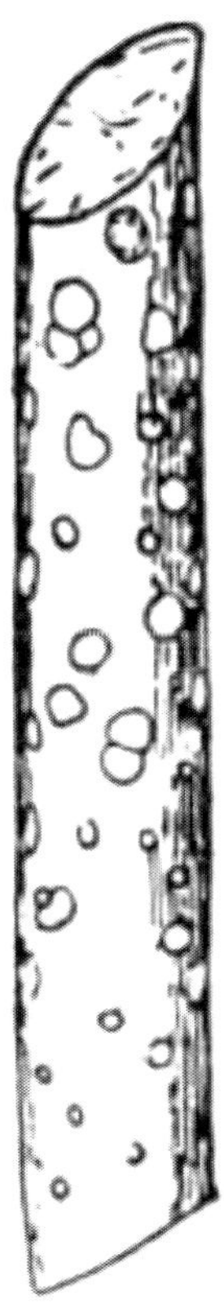

Figure 155. Stem of aquatic plant showing coleochæte, natural size.

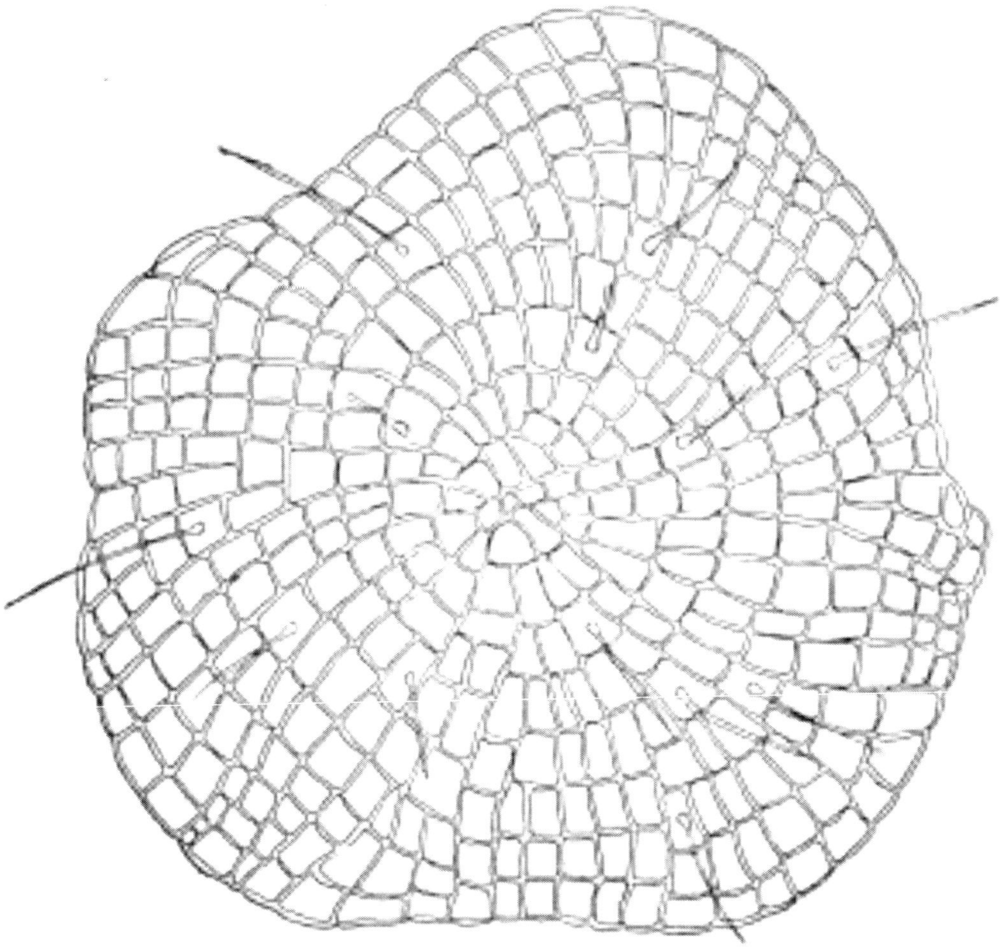

Figure 156. Thallus of Coleochæte scutata.

The Shield-Shaped Coleochæte

This plant (C. scutata) is in the form of a flattened, circular, green plate, as shown in Figure 156. It is attached near the center on one side to rushes and other plants, and has been found quite abundantly for several years in the waters of Cayuga Lake at its southern extremity. As will be seen it consists of a single layer of green cells which radiate from the center in branched rows to the outside, the cells lying so close together as to form a continuous plate. The plant started its growth from a single cell at the central point, and grew at the margin in all directions. Sometimes they are quite irregular in outline, when they lie quite closely side by side and interfere with one another by pressure.

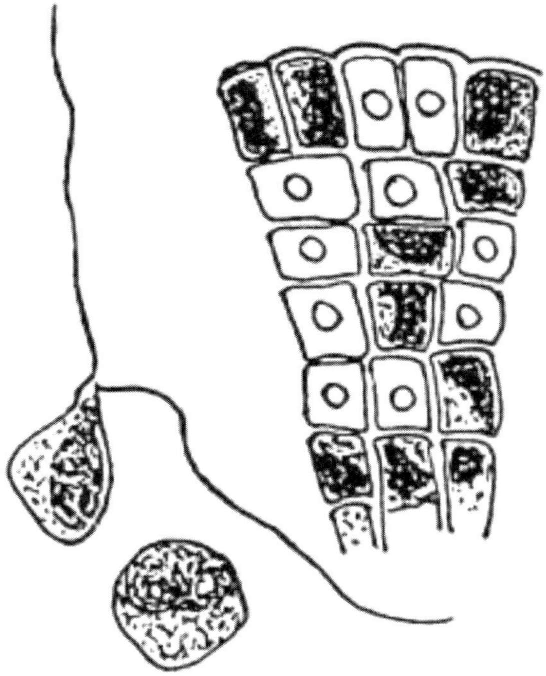

Figure 157. Portion of thallus of Coleochæte scutata, showing empty cells from which zoogonidia have escaped, one from each cell; zoogonidia at the left. (After Pringsheim.)

Figure 158. Portion of thallus of Coleochæte scutata, showing four antheridia formed from one thallus cell; a single spermatozoid at the right. (After Pringsheim.)

If the surface is examined carefully there will be found long hairs, the base of which is enclosed in a narrow sheath. It is from this character that the genus takes its name of coleochæte (sheathed hair).

Fruiting Stage of Coleochæte

It is possible at some seasons of the year to find rounded masses of cells situated near the margin of this green disk. These have developed from a fertilized egg which remained attached to the plant, and probably by this time the parent plant has lost its color.

Zoospore Stage

This mass of tissue does not develop directly into the circular green disk, but each of the cells forms a zoospore. Here then, as in œdogonium, we have another stage of the plant interpolated between the fertilized egg and that stage of the plant which bears the gametes. But in coleochæte we have a distinct advance in this stage upon what is present in œdogonium, for in coleochæte the fertilized egg develops first into a several-celled mass of tissue before the zoospores are formed, while in œdogonium only four zoospores are formed directly from the egg.

Asexual Reproduction

In asexual reproduction any of the green cells on the plant may form zoogonida. The contents of a cell round off and form a single zoogonidium which has two cilia at the smaller end of the oval body, Figure 157. After swimming around for a time they come to rest, germinate, and produce another plant.

Sexual Reproduction—Oogonium

The oogonium is formed by the enlargement of a cell at the end of one of the threads, and then the end of the cell elongates into a slender tube which opens at the end to form a channel through which the spermatozoid may pass down to the egg. The egg is formed of the contents of the cell (Figure 159). Several oogonia are formed on one plant, and in such a plant as C. scutata they are formed in a ring near the margin of the disk.

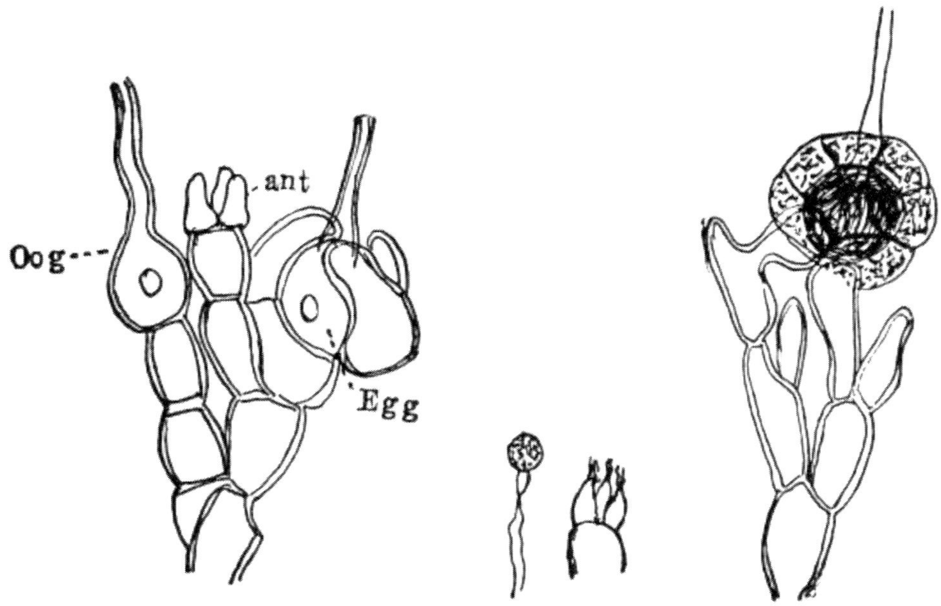

Figure 159. Coleochæte soluta; at left branch bearing oogonium (*oog*); antheridia (*ant*); egg in oogonium and surrounded by enveloping threads; at center three antheridia open, and one spermatozoid; at right sporocarp, mature egg inside sporocarp wall.

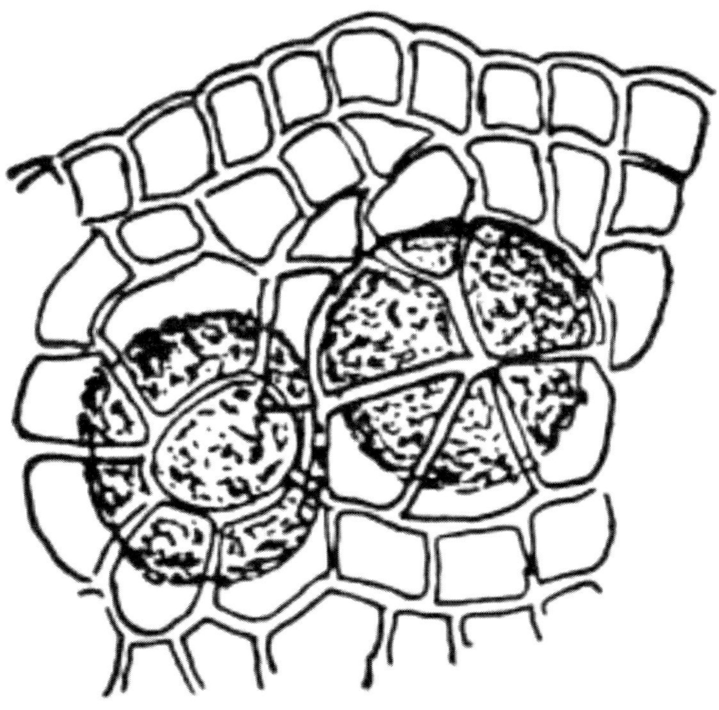

Figure 160. Two sporocarps still surrounded by thallus. Thallus finally decays and sets sporocarp free.

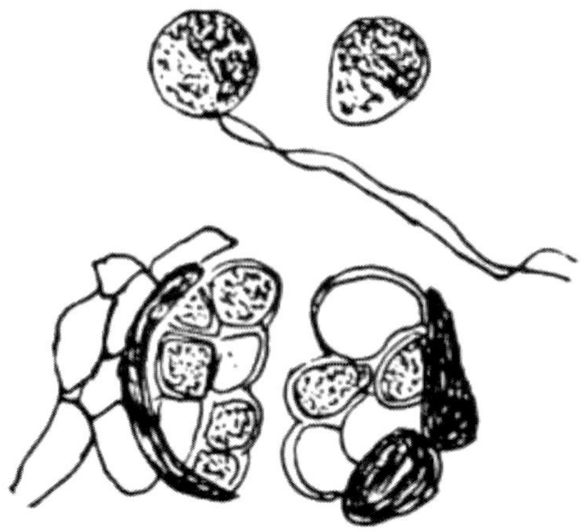

Figure 161. Sporocarp ruptured by growth of egg to form cell mass.Cells of this sporophyte forming zoospores.

Figures 160, 161. C. scutata.

Antheridia

In C. scutata certain of the cells of the plant divide into four smaller cells, and each one of these becomes an antheridium. In C. soluta the antheridia grow out from the end of terminal cells in the form of short flasks, sometimes four in number or less (Figure 159). A single spermatozoid is formed from the contents. It is oval and possesses two long cilia. After swimming around it passes down the tube of the oogonium and fertilizes the egg.

Sporocarp

After the egg is fertilized the cells of the threads near the egg grow up around it and form a firm covering one cell in thickness. This envelope becomes brown and hard, and serves to protect the egg. This is the "fruit" of the coleochæte, and is sometimes called a sporocarp (spore-fruit). The development of the cell mass and the zoospores from the egg has been described above. Some of the species of coleochæte consist of branched threads, while others form circular cushions several layers in thickness. These forms together with the form of our plant C. scutata make an interesting series of transitional forms from filamentous structures to an expanded plant body formed of a mass of cells.

Comparative Table for Spirogyra, Vaucheria, Œdogonium, Coleochæte

<table>
<tr><th rowspan="3"></th><th colspan="7">GAMETOPHYTE. (Bears the sexual organs and gonidia.)</th><th rowspan="2">SPOROPHYTE
Bears spores</th><th rowspan="3">How Veg. Phase of Gametophyte is Developed.</th></tr>
<tr><th rowspan="2">Vegetative Phase</th><th rowspan="2">Growth.</th><th rowspan="2">Mulitiplication.</th><th colspan="4">Sexual Reproduction.</th></tr>
<tr><th colspan="2">Sexual Organs.</th><th colspan="2">Gametes.</th><th>Fruit.</th></tr>
<tr><td rowspan="2">Spirogyra.</td><td rowspan="2">Simple threads of cylindrical cells.</td><td rowspan="2">All cells divide and grow.</td><td rowspan="2">By breaking up of threads.</td><td colspan="2">Undifferentiated.</td><td colspan="2">Undifferentiated.</td><td rowspan="2">Zygospore Rests.</td><td rowspan="2">Develops vegetative phase directly.</td></tr>
<tr><td colspan="2">Any cell of thread. Conjugate by tube.</td><td colspan="2">Entire contents of cojugating cells.</td></tr>
<tr><td rowspan="2">Vaucheria.</td><td rowspan="2">Branched threads, continuous.</td><td rowspan="2">Limited to ends of threads and branches.</td><td rowspan="2">By multiciliate zoogonidia, and other cells, from terminal portions.</td><td colspan="2">Differentiated.</td><td colspan="2">Differentiated.</td><td rowspan="2">Egg (or oospore). Rests.</td><td rowspan="2">Develops vegetative phase directly.</td></tr>
<tr><td>Antheridia slender cells on special branches.</td><td>Oogonium, large rounded cell on special branch, opens and emits bit of protoplasm.</td><td>Small two-ciliated spermatozoids.</td><td>Large egg cell.</td></tr>
<tr><td rowspan="2">Œdogonium.</td><td rowspan="2">Simple threads of cylindrical cells.</td><td rowspan="2">Limited to certain portions of thread.</td><td rowspan="2">By oval zoogonidia, with crown of cilia. Any cell may form a single zoogonidium.</td><td colspan="2">Differentiated.</td><td colspan="2">Differentiated.</td><td rowspan="2">Egg (or oospore). Rests.</td><td rowspan="2">Divides into four cells; each forms zoospore which develops veg.phase again.</td></tr>
<tr><td>Antheridia disk-shaped, several from one vegetative cell. Sometimes on dwarf males.</td><td>Oogonium, changed vegetative cell, opens and emits bit of protoplasm.</td><td>Oval spermatozoids with crown of cilia. Two from each antheridium.</td><td>Large egg cell.</td></tr>
<tr><td rowspan="2">Coleochæte.</td><td rowspan="2">Branched threads, or compact circular plates.</td><td rowspan="2">Terminal or marginal.</td><td rowspan="2">By zoogonidia with two cilia. Any cell may form a single zoogonidium.</td><td colspan="2">Differentiated.</td><td colspan="2">Differentiated.</td><td rowspan="2">Egg (surrounded by wall from gametophyte). Rests. Divides and grows to form a mass of cells.</td><td rowspan="2">Each forms a zoospore. Zoospore develops veg. phase again.</td></tr>
<tr><td>Antheridia, four or several from single veg. cell.</td><td>Oogonium, enlarged veg. cell, with long tube through opening of which spermatozoid</td><td>Oval, biciliate spermatozoid, one from each ntheridium.</td><td>Large egg cell.</td></tr>
</table>

Comparative Table (Continued)

	GAMETOPHYTE. (Bears the sexual organs and gonidia.)							SPOROPHYTE Bears spores	How Veg. Phase of Gametophyte is Developed.
	Vegetative Phase	Growth.	Mulitiplication.	Sexual Reproduction.					
				Sexual Organs.		Gametes.		Fruit.	
					enters. After fertilization wall of enveloping threads surrounds oogonium.				

Chapter 18

Classification and Additional Studies of the Algæ

In order to show the general relationship of the algæ studied, the principal classes are here enumerated as well as some of the families. In some of the groups not represented by the examples studied above, a few species are described which may serve as the basis of additional studies if desired. The principal classes[17] of algæ are as follows:

Class Chlorophyceæ

These are the green algæ, so called because the chlorophyll green is usually not masked by other pigments, though in some forms it is. There are three subclasses.

Subclass Protococcoideæ

In the Protococcoideæ are found the simplest green plants. Many of them consist of single cells which live an independent life. Others form "colonies," loose aggregations of individuals not yet having attained the permanency of even a simple plant body, for the cells often separate readily and are able to form new colonies. The colonies are often held together by a gelatinous membrane, or matrix. Some are motile, while others are non-motile. A few of the families are here enumerated.

[17] In Engler & Prantl's Pflanzenfamilien, Wille uses the term class for these principal subdivisions of the algæ. Systematists are not yet agreed upon a uniform use of the terms.

Family Volvocaceæ

These are all motile, during the vegetative stage. The individuals are single or form more or less globose colonies.

The "Red Snow" Plant (Sphærella nivalis)

This is often found in arctic and alpine regions forming a red covering over more or less large areas of snow or ice. For this reason it is called the "red snow plant."

Sphærella lacustris, a closely related species, is very widely distributed in temperate regions along streams or on the borders of lakes and ponds. Here in dry weather it is often found closely adhering to the dry rock surface, and giving it a reddish color as if the rock were painted. This is especially the case in the shallow basins formed over the uneven surface of the rock near the water's edge. These places during heavy rains or in high water are provided with sufficient water to fill the basins. During such times the red snow plant grows and multiplies, loses its red color and becomes green, and, being motile, is free swimming. It is a single-celled plant, oval in form, surrounded by a gelatinous sheath and with two cilia or flagella at the smaller end, by the vibration of which it moves (Figure 162). The single cell multiplies by dividing into two cells. When the water dries out of the basin, the motile plant comes to rest, and many of the cells assume the red color. To obtain the plant for study, scrape some of the red covering from these rock basins and place it in fresh spring water, and in a day or so the swarmers are likely to be found. Under certain conditions small microzoids are formed.

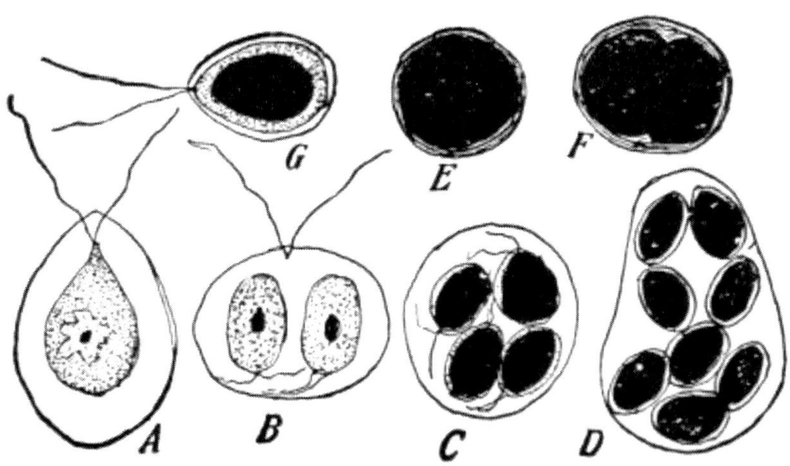

Figure 162. Sphærella lacustris (Girod.) Wittrock. *A*, mature free swimming individual with central red spot. *B*, division of mother individual to form two. *C*, division of a red one to form four. *D*, division into eight. *E*, a typical resting cell, red. *F*, same beginning to divide. *G*, one of four daughter zoospores after swimming around for a time losing its red color and becoming green. (After Hazen.)

Chlamydomonas is a very interesting genus of motile one-celled green algæ, because the species are closely related to the Flagellates among the lower animals. The plant is oval, with a single chloroplast and surrounded by a gelatinous envelope through which the two cilia or flagella extend. One-celled organisms of this kind are sometimes called *monads*, i.e., a one-celled being. This one has a gelatinous cloak and is, therefore, a *cloaked monad* (*Chlamydomonas*). The species often are found as a very thin green film on fresh water. C. pulvisculus is shown in Figure 163. When it multiplies the single cell divides into two, as shown in *B*. Sometimes a non-motile palmella stage is formed, as shown in *C* and *D*. Reproduction takes place by gametes which are of unequal size, the smaller one representing the sperm and the larger one the egg, as in *E* and *F*. These conjugate as in *G* and *H*, the protoplasm of the smaller one passing over into the larger one, and a zygospore is thus formed.

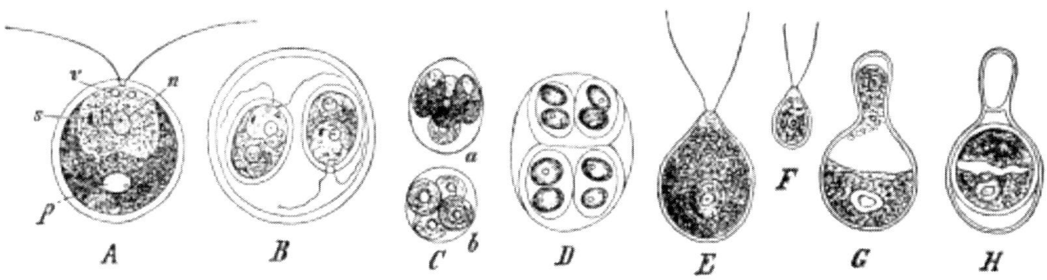

Figure 163. *Chlamydomonas pulvisculus* (Müll.) Ehrb. *A*, an old motile individual; *n*, nucleus; *p*, pyrenoid; *s*, red eye spot; *v*, contractile vacuole; *B*, motile individual has drawn in its cilia and divided into two; *C*, mother plant has drawn in its cilia and divided into four non-motile cells; *D*, pamella stage; *E*, female gamete—egg; *F*, male gamete—sperm; *G*, early stage of conjugation; *H*, zygospore with conjugating tube and empty male cell attached. (After Wille.)

Of those which form colonies, Pandorina morum is widely distributed and not rare. It consists of a sphere formed of sixteen individuals enclosed in a thin gelatinous membrane. Each cell possesses two cilia (or flagella), which extend from the broader end out through the enveloping membrane. By the movement of these flagella the colony goes rolling around in the water. When the plant multiplies each individual cell divides into sixteen small cells, which then grow and form new colonies. Reproduction takes place when the individual cells of the young colonies separate, and usually a small individual unites with a larger one and a zygospore is formed (see Figure 164). Eudorina elegans is somewhat similar, but when the gametes are formed certain mother cells divide into sixteen small motile males or sperms, and certain other mother cells divide into sixteen large motile females or eggs. These separate from the colonies, and the sperms pair with the eggs and fuse to form zygospores. This plant as well as Chlamydomonas pulvisculus foreshadows the early differentiation of sex in plants.

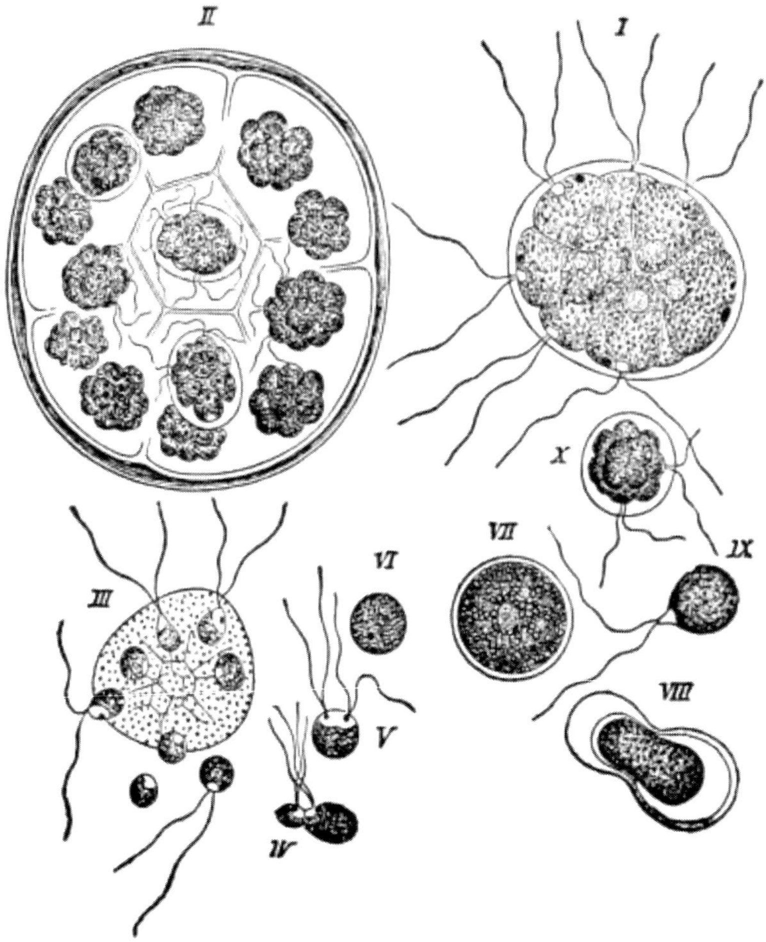

Figure 164. Pandorina morum (Müll.) Bory.

I, motile colony;
II, colony divided into 16 daughter colonies;
III, sexual colony, gametes escaping;
IV, V, conjugating gametes;
VI, VII, young and old zygospore;
VIII, zygospore forming a large swarm spore, which is free in IX;
X, same large swarm spore divided to form young colony.
(After Pringsheim.)

Figure 165. Pleurococcus (protococcus) vulgaris.

Family Tetrasporaceæ

This family is well represented by Tetraspora lubrica forming slimy green net-like sheets attached to objects in slow-running water. It is really a single-celled plant. The rounded cells divide by cross walls into four cells, and these again, and so on, large numbers being held in loose sheets by the slime in which they are imbedded.

Family Pleurococcaceæ

The members of this family are all non-motile in the vegetative stage. They consist of single individuals, or of colonies. Pleurococcus vulgaris (Protococcus vulgaris) is a single-celled alga, usually obtained with little difficulty. It is often found on the shaded, and cool, or moist side of trees, rocks, walls, etc., in damp places. This plant is not motile. It multiplies by fission (Figure 165) into two, then four, etc. These cells remain united for a time, then separate. Sometimes the cells are found growing out into filaments, and it is thought by some that P. vulgaris may be only a simple stage of a higher alga. Eremosphæra viridis is another single-celled alga found in fresh water among filamentous forms. The cells are large and globose.

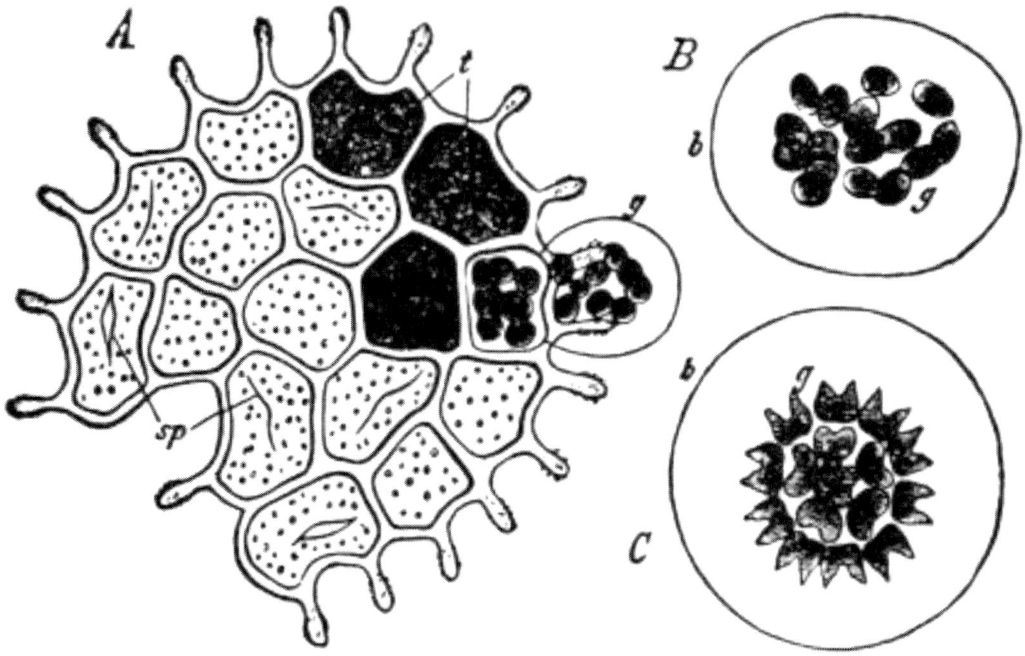

Figure 166. Pediastrum boryanum. *A*, mature colony, most of the young colonies have escaped from their mother cells; at *g*, a young colony is escaping; *sp*, empty mother cells; *B*, young colony; *C*, same colony with spores arranged in order. (After Braun.)

Family Hydrodictyaceæ

These plants form colonies of cells. Hydrodictyon reticulatum, the water net, is made up of large numbers of cylindrical cells so joined at their ends as to form a large open mesh or net. Pediastrum forms circular flat colonies, as shown in Figure 166. Both of these plants are rather common in fresh-water pools, the latter one intermingled with filamentous algæ, while the former forms large sheets or nets. Multiplication in Hydrodictyon takes place by the protoplasm in one of the cells dividing into thousands of minute cells, which gradually arrange themselves in the form of a net, escape together from the mother cell, and grow into a large net. In Pediastrum multiplication takes place in a similar way, but the protoplasm in each cell usually divides into sixteen small cells, and escaping together from the mother cell arrange themselves and grow to full size (Figure 166).

The Conjugateæ include several families of green algæ, which probably should be included among the Chlorophyceæ. They have probably had their origin from some of the more simple members of the Protococcoideæ. They are represented by Spirogyra, Zygnema, and the desmids, studied in Chapter 14.

Subclass Confervoideæ

These are mostly filamentous algæ, the filaments being composed of cells firmly united, and, with the exception of the simplest forms, there is a definite growing point. A few of the families are as follows:

Family Ulvaceæ

These contain the sea wracks, or sea lettuce, like Ulva, forming expanded green, ribbon-like growths in the sea.

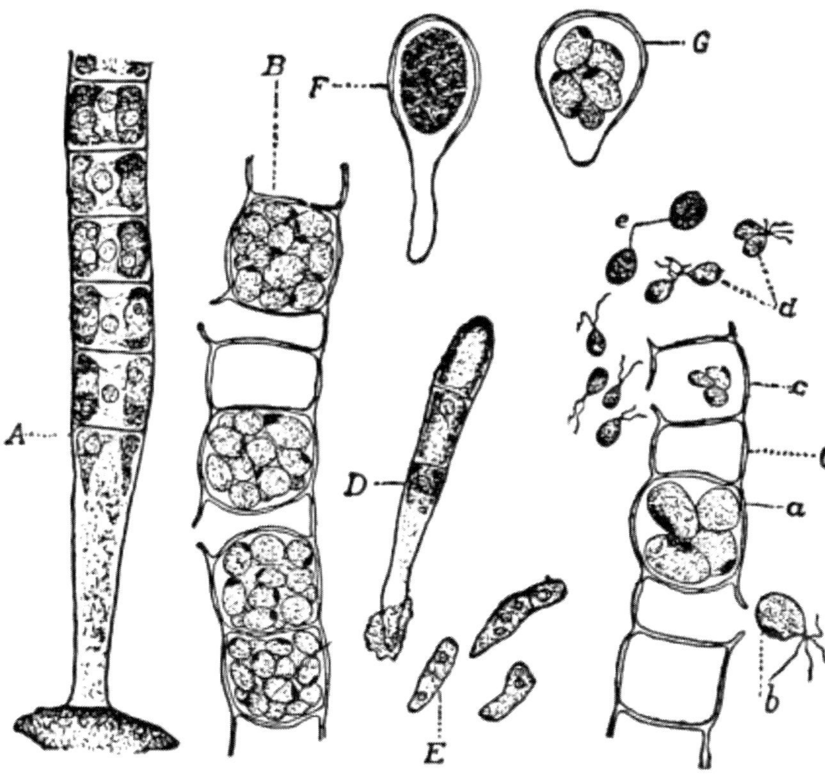

Figure 167. Ulothrix zonata. *A*, base of thread. *B*, cells with zoospores, *C*, one cell with zoospores escaping another cell with small biciliate gametes escaping and some fusing to form zygospores, *E*, zoospores germinating and forming threads: *F*, *G*, zygospore growing and forming zoospores. (After Caldwell and Dodel-Port.)

Family Ulotrichaceæ

Represented by Ulothrix zonata, not uncommon in slow-running water or in ponds of fresh water attached to rocks or wood. It consists of simple threads of short cells. Multiplication takes place by zoospores. Reproduction takes place by motile sexual cells (gametes) which fuse to form a zygospore (Figure 167).

Family Chætophoraceæ

Represented by Chætophora (in Chapter 15) and Drapernaudia in fresh water.

Family Œdogoniaceæ

Rrepresented by Œdogonium (Chapter 16).

Family Coleochætaceæ

Represented by Coleochæte (Chapter 17).

Subclass Siphoneæ

There are several families.

Family Botrydiaceæ

This is represented by Botrydium granulatum (Chapter 15, p. 146).

Family Vaucheriaceæ

Represented by Vaucheria (Chapter 15), with quite a large number of species, is widely distributed.

CLASS SCHIZOPHYCEÆ (= CYANOPHYCEÆ)

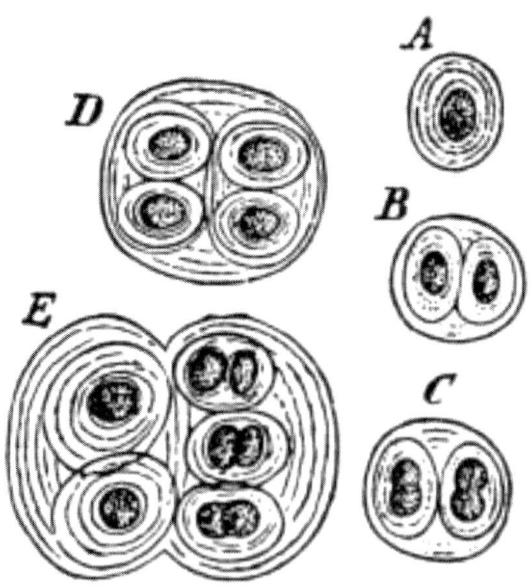

Figure 168. Glœocapsa.

The Blue-Green Algæ, or Cyanophyceæ form slimy looking thin mats on damp wood or the ground, or floating mats or scum on the water. The color is usually bluish green, but in some species it is purple, red or brown. All have chlorophyll, but it is not in distinct chloroplasts and is more or less completely guised by the presence of other pigments. Two orders and eight families are recognized. The following include some of our common forms:

Order Coccogonales (Coccogoneæ)

Single-celled plants, occurring singly or in colonies, in some forms forming short threads. One of the two families is mentioned.

Family Chroococcaceæ

The plants multiply only through cell division. Chroococcus, forms rounded, blue-green cells enclosed in a thick gelatinous coat, in fresh water and in damp places; certain species form "lichen-gonidia" in some genera of lichens. Glœocapsa is similar to Chroococcus, but the colonies are surrounded by an additional common gelatinous envelope (Figure 168); on damp rocks, etc.

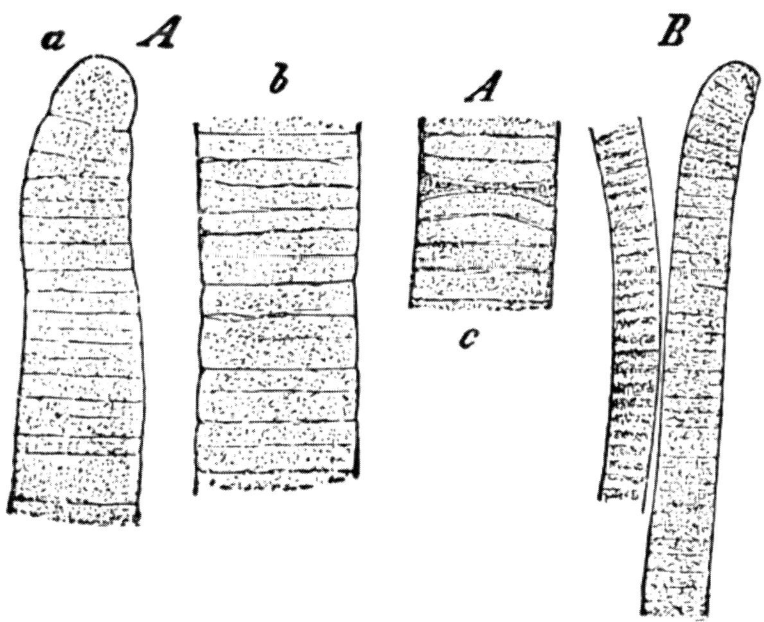

Figure 169. *A*, Oscillatoria princeps, *a*, terminal cell; *b*, *c*, portions from the middle of a filament. In *c*, a dead cell is shown between the living cells; *B*, Oscillatoria froelichii, *b*, with granules along the partition walls.

Order Hormogonales (Hormogoneæ)

Plants filamentous, simple celled or with false or true branching, usually several celled (Spirulina is single celled). Multiplication takes place through *hormogones*, short sections of the threads becoming free; also through resting cells. Two of the six families are mentioned.

Family Oscillatoriaceæ

This family is represented by the genus Oscillatoria, and by several other genera common and widely distributed. Oscillatoria contains many species. They are found on the damp ground or wood, or floating in mats in the water. They often form on the soil at the bottom of the pool, and as gas becomes entangled in the mat of threads, it is lifted from the bottom and floated to the surface of the water. The plant is thread-like, and divided up into many short cells. The threads often show an oscillating movement, whence the name *Oscillatoria*.

Family Nostocaceæ

This family is represented by Nostoc, which forms rounded, slimy, blue-green masses on wet rocks. The individual plants in the slimy ball resemble strings of beads, each cell being rounded, and several of these arranged in chains as shown in Figure 170.

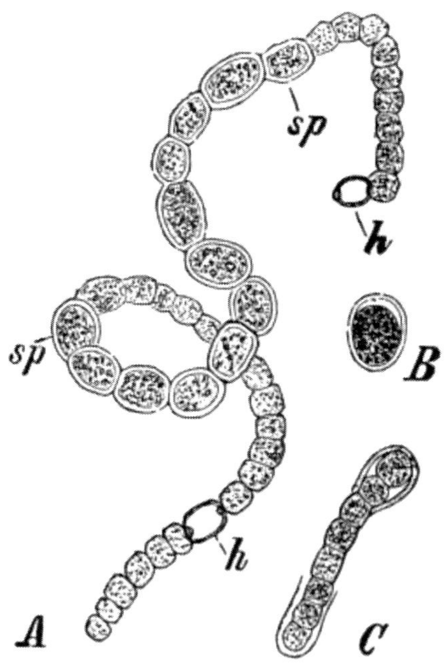

Figure 170. Nostoc linckii. *A*, filament with two heterocysts (*h*), and a large number of spores (*sp*); *B*, isolated spore beginning to germinate; *C*, young filament developed from spore. (After Bornet.)

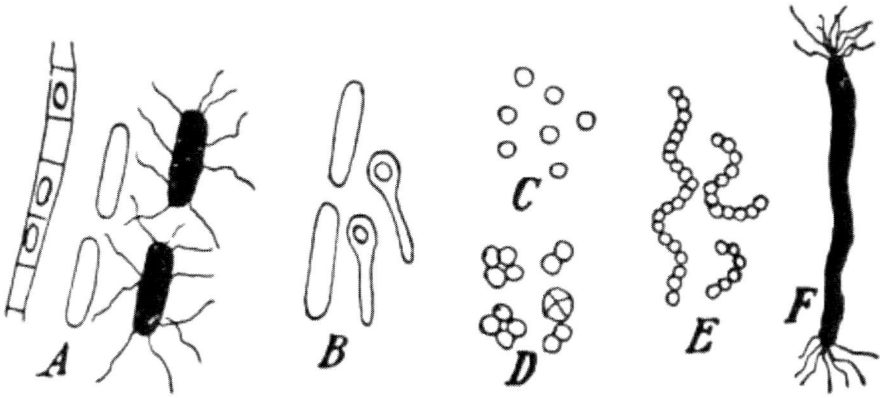

Figure 171. Bacteria. *A*, Bacillus subtilis. Spores in threads, unstained rods, and stained rods showing cilia; *B*, Bacillus tetani, the tetanus or lockjaw bacillus, found in garden soil and on old rusty nails. Spores in club-shaped ends. *C*, Micrococcus; *D*, Sarcina; *E*, Streptococcus; *F*, Spirillum. (After Migula.)

Here and there are often found larger cells (heterocysts) in the chain. Nostoc punctiforme lives in the intercellular spaces of the roots of cycads (often found in greenhouses), and in the stems of Gunnera. N. sphæricum lives in the spaces between the cells in many species of liverworts (in the genera Anthoceros, Blasia, Pellia, Aneura, Riccia, etc.), and in the perforated cells of Sphagnum acutifolium. Anabæna is another common and widely distributed genus. The species occur in fresh or salt water, singly or in slimy masses. Anabæna azollæ lives endophytically in the leaves of the water fern, Azolla.

Class Schizomycetes

Bacteriales

The bacteria are sometimes classified with the Cyanophyceæ, under the name Schizophyta, and represent the subdivision Schizomycetes, or fission fungi, because many of them multiply by a division of the cells just as the blue-green algæ do. For example, Bacillus forms rods which increase in length and divide into two rods, or it may grow into a long thread of many short rods. Micrococcus consists of single rounded cells. Streptococcus forms chains of rounded cells, Sarcina forms irregular cubes of rounded cells, while others like Spirillum are spiral in form. Bacillus subtilis may be obtained by making an infusion from hay and allowing it to stand for several days. Bacillus tetani occurs in the soil, on old rusty nails, etc. It is called the tetanus bacillus because it causes a permanent spasm of certain muscles, as in "lockjaw." This bacillus grows and produces this result on the muscles when it occurs in deep and closed wounds such as are caused by stepping on an old nail or other object which pierces the flesh deeply. In such a deep wound oxygen is deficient, and in this condition the bacillus is virulent. Opening the wounds to admit oxygen and washing them out with a solution of bichloride of mercury prevents the tetanus. Many bacteria are of great importance in bringing about the decay of dead animal and plant matter, returning it to a condition for plant food. (See also nitrate and nitrite bacteria, Chapter IX.) While most bacteria are harmless there are many which cause very serious diseases of man and animals, as typhoid fever, diphtheria, tuberculosis, etc., while some others produce disease in plants. Others aid in certain fermentations of liquids and are employed for making certain kinds of wines or other beverages. Some work in symbiosis with yeasts, as in the kephir yeast, used in fermenting certain crude beverages by natives of some countries.

Myxobacteriales (Myxobacteriaceæ Thaxter[18])

These plants consist of colonies of bacteria-like organisms, motile rods, which multiply by cross-division and secrete a gelatinous substance or matrix which surrounds the colonies. They form plasmodium-like masses which superficially resemble the slime moulds. In the fruiting stage some species become elevated from the substratum into cylindrical, clavate, or branched forms, which bear cysts of various shapes containing the rods in a resting stage, or the rods are converted into spore-like masses. Ex., Chondromyces crocatus on decaying plant parts, Myxobacter aureus on wet wood and bark, Myxococcus rubescens on dung, decaying lichens, paper, etc.

Class Flagellata

The flagellates are organisms of very low organization resembling animals as much as they do plants. They are single celled and possess two cilia or flagella, by the vibration of which they move. Some are without a cell wall, while others have a well-defined membrane, but it rarely consists of cellulose. Some have chromatophores and are able to manufacture carbohydrates like ordinary green plants. These are green in Euglena, and brown in Hydrurus. Some possess a mouth-like opening and are able to ingest solid particles of food (more like animals), while others have no such opening and absorb food substances dissolved in water (more like plants). The Euglena viridis is not uncommon in stagnant water, often forming a greenish film on the water.

[18] See Bot. Gaz., 17, 389, 1892.

Class Peridineæ

These are peculiar one-celled organisms provided with two flagella and show some relationship to the Flagellates. They usually are provided with a cellulose membrane, which in some forms consists of curiously sculptured plates. In the higher forms this cellulose membrane consists of two valves fitting together in such a way as to resemble some of the diatoms. Like the Flagellates, some have green chromatophores, which in some are obscured by a yellow or brown pigment (resembling the diatoms), while still others have no chlorophyll. The Peridineæ are abundant in the sea, while some are found in fresh water.

Class Diatomaphyceæ (Bacillariales, Diatomaceæ)

The diatoms are minute and peculiar organisms believed to be algæ. They live in fresh, brackish, and salt water. They are often found covering the surface of rocks, sticks, or the soil in thin sheets. They occur singly and free, or several individuals may be joined into long threads, or other species may be attached to objects by slender gelatinous stalks. Each protoplast is enclosed in a silicified skeleton in the form of a box with two halves, often shaped like an old-fashioned pill box, one half fitting over the other like the lid of a box. It is evident that in this condition the plant cannot increase much in size.

They multiply by fission. This takes place longitudinally, i.e., in the direction of the two halves or *valves* of the box. Each new plant then has a valve only on one side. A new valve is now formed over the naked half, and fits inside the old valve. At each division the individuals thus become smaller and smaller until they reach a certain point, when the valves are cast off and the cell forms an *auxospore*, i.e., it grows alone, or after conjugation with another, to the full size again, and eventually provides itself with new valves. The valves are often marked, with numerous and fine lines, often making beautiful Figures, and some are used for test objects for microscopes.

The free forms are capable of movement. The movement takes place in the longitudinal direction of the valves. They glide for some time in one direction, and then stop and move back again. It is not a difficult thing to mount them in fresh water and observe this movement.

The diatoms have small chlorophyll plates, but the green color is disguised by a brownish pigment called diatomin. The relationships of the diatoms are uncertain, but some, because of the color, think they are related to the Phæophyceæ.

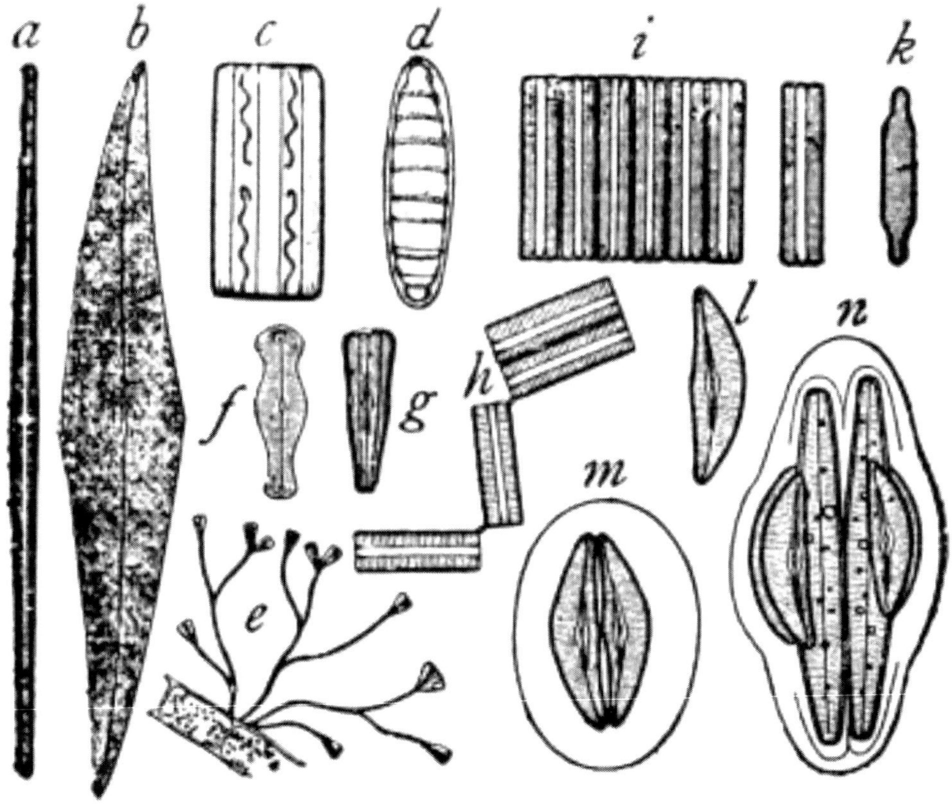

Figure 171*a*. A group of Diatoms: *c* and *d*, top and side views of the same form; *e*, colony of stalked forms attached to an alga; *f* and *g*, top and side views of the form shown at *e*; *h*, a colony; *i*, a colony, the top and side view shown at *k* and *n*, forming auxospores. (After Kerner.)

Class Phæophyceæ

The Brown Algæ. (Phæophyceæ)

The members of this class possess chlorophyll, but it is obscured by a brown pigment. The plants are accessible at the seashore, and for inland laboratories may be preserved in formalin (2½ per cent). (See also Chapter LVI.)

Figure 172. *A*, Ectocarpus siliculosus; *B*, branch with a young and a ripe plurilocular sporangium; *E*, gametes fusing to form zygospore, (*B*, after Thuret; *E*, after Berthold.)

Ectocarpus

The genus Ectocarpus represents well some of the simpler forms of the brown algæ (Figure 172). They are slender, filamentous branched algæ growing in tufts, either epiphytic on other marine algæ (often on Fucaceæ), or on stones. The slender threads are only divided crosswise, and thus consist of long series of short cells. The sporangia are usually plurilocular (sometimes unilocular), and usually occur in the place of lateral branches. The zoospores escape from the apex of the sporangium and are biciliate, and they fuse to form zygospores.

Sphacelaria

The species of this genus represent an advance in the development of the thallus. While they are filamentous and branched, division takes place longitudinally as well as crosswise (Figure 173).

Leathesia difformis represents an interesting type because the plant body is small, globose, later irregular and hollow, and consists of short radiately arranged branches, the surface ones in the form of short, crowded, but free, trichome-like green branches. This trichothallic body recalls the similar form of Chætophora pisiformis (Chapter 16) among the Chlorophyceæ.

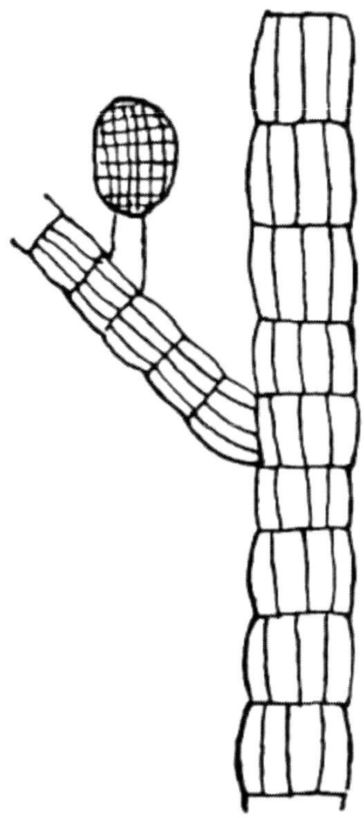

Figure 173. Sphacelaria, portion of plant showing longitudinal division of cells, and brood bud (plurilocular sporangium).

Figure 174. Laminaria digitata, forma cloustoni, North Sea. (Reduced.)

The Giant Kelps

Among the brown algæ are found the largest specimens, some of the laminarias or giant kelps, rivaling in size the largest land plants, and some of them have highly developed tissues. *Postelsia palmæformis* has a long, stout stem, from the free end of which extend numerous large and long blades, while the stem is attached to the rocks by numerous "root" like outgrowths, the holdfasts. It occurs along the northern Pacific coast, and appears to flourish where it receives the shock of the surf beating on the shore. Several species of Laminaria occur on our north Atlantic coast. In L. digitata, the stem expands at the end into a broad blade, which becomes split into several smaller blades (Figure 174). *Macrocystis pyrifera* inhabits the ocean in the southern hemisphere, and sometimes is found along the north American coast. It is said to reach a length of 200-300 meters.

Fucus, or Rockweed

This plant is a more or less branched and flattened thallus or "frond." One of them, illustrated in Figure 119, measures 15-30 *cm* (6-12 inches) in length. It is attached to rocks and stones which are more or less exposed at low tide. From the base of the plant are developed several short and more or less branched expansions called "holdfasts," which, as their name implies, are organs of attachment. Some species (F. vesiculosus) have vesicular swellings in the thallus.

The fruiting portions are somewhat thickened as shown in the Figure. Within these portions are numerous oval cavities opening by a circular pore, which gives a punctate appearance to these fruiting cushions. Tufts of hairs frequently project through them.

Figure 175. Portion of plant of Fucus showing conceptacles in enlarged ends; and below the vesicles (Fucus vesiculosus).

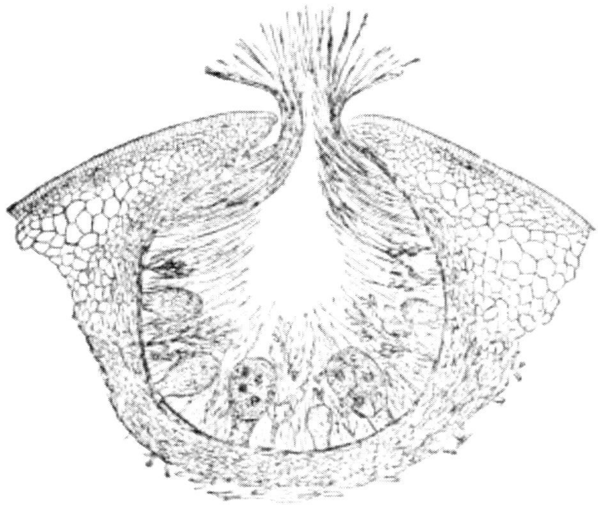

Figure 176. Section of conceptacle of Fucus, showing oogonia, and tufts of antheridia.

Structure of the Conceptacles

On making sections of the fruiting portions one finds the walls of the cavities covered with outgrowths. Some of these are short branches which bear a large rounded terminal sac, the oogonium, at maturity containing eight egg-cells. More slender and much-branched threads bear narrowly oval antheridia. In these are developed several two-ciliated spermatozoids.

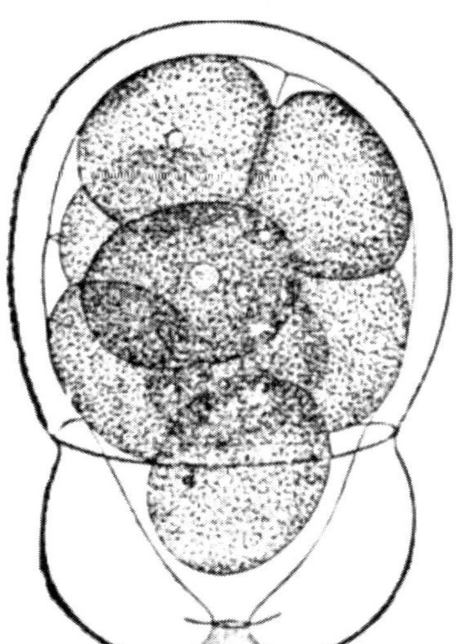

Figure 177. Oogonium of Fucus with ripe eggs.

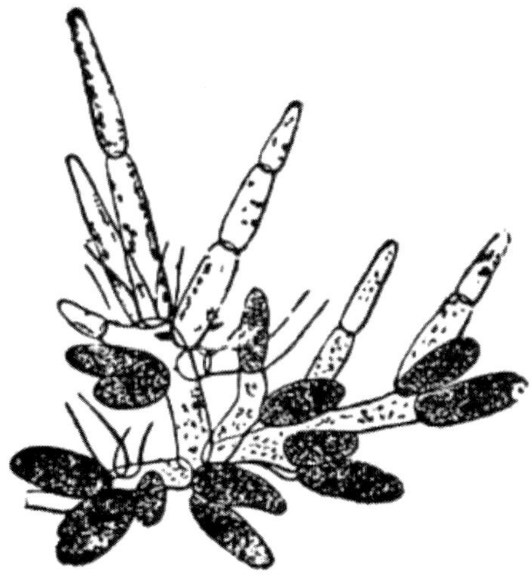

Figure 178. Antheridia of Fucus, on branched threads.

Fertilization

At maturity the spermatozoids and egg-cells float outside of the oval cavities, where fertilization takes place. The spermatozoid sinks into the protoplasm of the egg-cell, makes its way to the nucleus of the egg, and fuses with it as shown in Figure 181. The fertilized egg then grows into a new plant. Nearly all the brown algæ are marine.

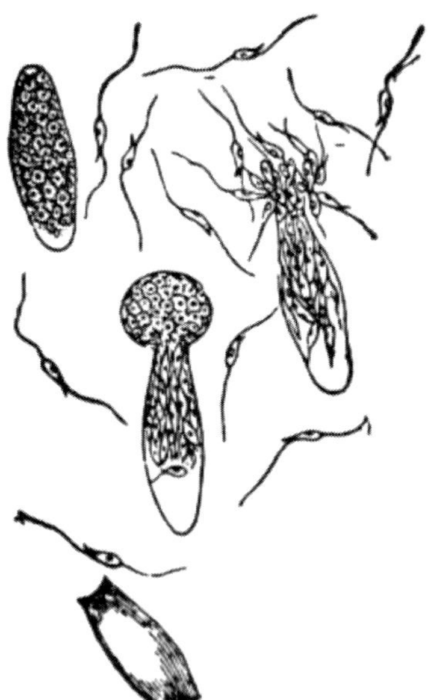

Figure 179. Antheridia of Fucus with escaping spermatozoids.

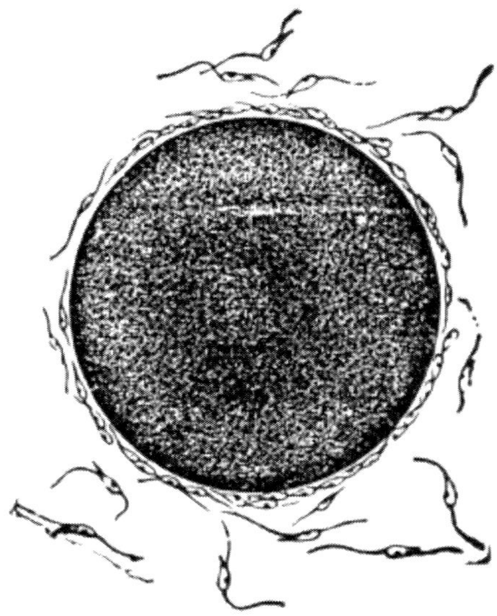

Figure 180. Eggs of Fucus surrounded by spermatozoids.

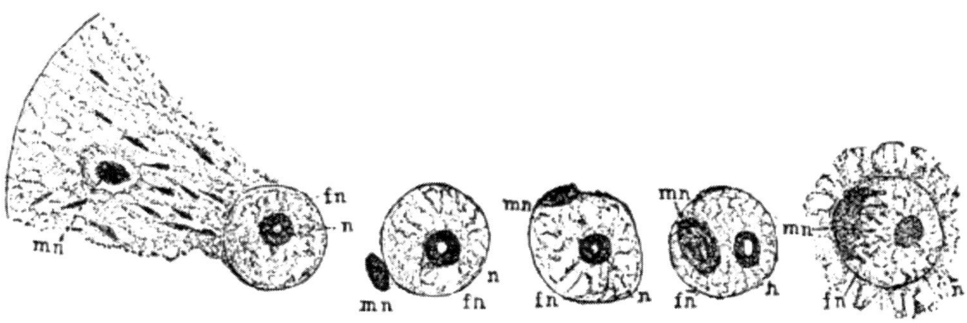

Figure 181. Fertilization in Fucus; *fn*, female nucleus; *mn*, male nucleus; *n*, nucleolus. In the left Figure the male nucleus is shown moving down through the cytoplasm of the egg; in the remaining Figures the cytoplasm of the egg is omitted. (After Strasburger.)

The Gulf weed (Sargassum bacciferum) in the warmer Atlantic ocean unites in great masses which float on the water, whence comes the name "Sargassum Sea." The Sargassum grows on the coast where it is attached to the rocks, but the beating of the waves breaks many specimens loose and these float out into the more quiet waters, where they continue to grow and multiply vegetatively.

Uses

Laminaria japonica and L. angustata are used as food by the Chinese and Japanese. Some species of the Laminariaceæ are used as food for cattle and are also used for fertilizers, while L. digitata is sometimes employed in surgery.

Classification

Kjellman divides the Phæophyceæ into two orders.

Order Phæosporales (Phæosporeæ)

Including 18 families. One of the most conspicuous families is the Laminariaceæ, including among others the Giant Kelps mentioned above (Laminaria, Postelsia, Macrocystis, etc.).

Order Cyclosporales (Cyclosporeæ)

This includes one family, the *Fucaceæ* with Ectocarpus, Sphacelaria, Læathesia, Fucus, Sargassum, etc.

CLASS RHODOPHYCEÆ

The Red Algæ (Rhodophyceæ)

The larger number of the so-called red algæ occur in salt water, though a few genera occur in fresh water. The plants possess chlorophyll, but it is usually obscured by a reddish or purple pigment.

Nemalion

This is one of the lower marine forms, though its thallus is not one of the simplest in structure. The plant body consists of a slender cylindrical branched shoot, sometimes very profusely branched. The central strand is rather firm, while the cortex is composed of rather loose filaments.

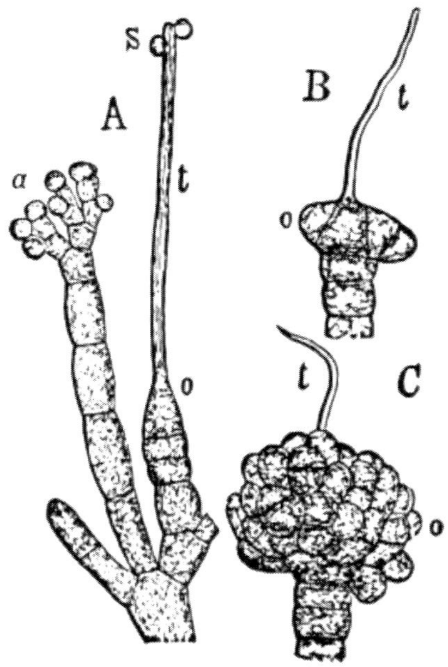

Figure 182. A red alga (Nemalion). *A*, sexual branches, showing antheridia (*a*); carpogonium or procarp (*o*) with its trichogyne (*t*), to which are attached two spermatia (*s*); *B*, beginning of a cystocarp (*o*), the trichogyne (*t*) still showing; *C*, an almost mature cystocarp (*o*), with the disorganizing trichogyne (*t*). (After Vines.)

Batrachospermum

This genus occurs in fresh water, and the species are found in slow-running water of shallow streams or ditches. There is a central slender strand which is more or less branched, and on these branches are whorls of densely crowded slender branches occurring at regular intervals. The plants are usually very slippery. Gonidia are formed on the ends of some of these branches in globose sporangia, called monosporangia, since but a single spore or gonidium is developed in each. Other branches often terminate in long slender hyaline setæ.

Lemanea

This genus also occurs in fresh water. The species develop only during the cold winter months in rapids of streams or where the water from falls strikes the rocks and is thoroughly aerated. They form tufts of greenish threads, cylindrical or whiplike, which in the summer are usually much broken down. The threads are hollow and have a firm cortex. These are the sexual shoots, and they arise as branches from a sterile filamentous-branched, Chantransia-like form.

Fertilization in the Lower Red Algæ

The sexual organs in the red algæ consist of antheridia and carpogonia. The antheridia are usually borne in crowded clusters, or surfaces, and bear terminally the small non-motile sperm cells. The carpogonium is a branch of one or several cells, the terminal cell (procarp) extending into a long slender process, the trichogyne. The sperm cell comes in contact with the trichogyne, and in the case of Nemalion and some others the nucleus has been found to pass down the inside and fuse with the nucleus of the procarp.

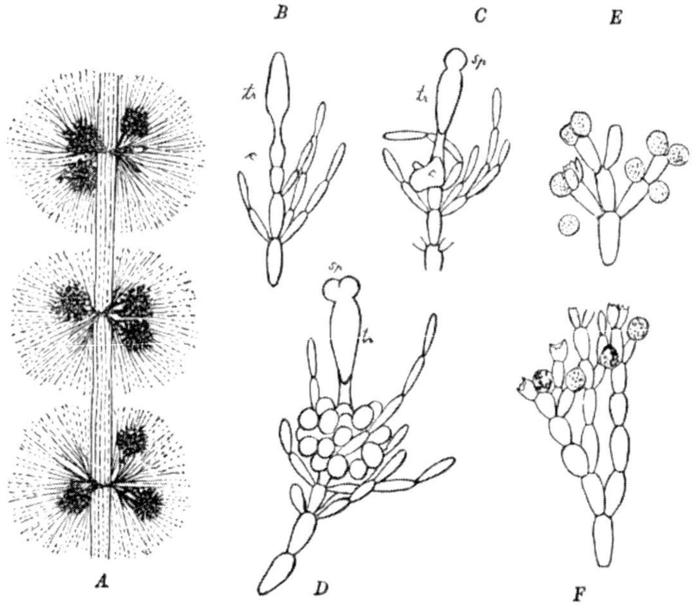

Figure 183. *A*, part of a shoot showing whorls of branches with clusters of carpospores. *B*, carpogonic branch or procarp. *c*, procarp cell; *tr*, trichogyne. *C*, same with sperm (*sp*) uniting with trichogyne. *D*, same with carpospores developing from procarp cell. *E*, male branch with one-celled antheridia. *F*, same with some of antheridia empty. (After Schmitz.)

From this point in the lower red algæ like Nemalion, Batrachospermum and Lemanea the formation of the spores is very simple. The procarp is stimulated to growth, and buds in different directions, producing branched chains of spores (carpospores). The carpospores form a rather compact cluster called the sporocarp, which means spore-fruit or spore-fruit body. In Batrachospermum it is seen as a compact tuft in the loose branching, in Nemalion it lies in the surface of the cortex, while in Lemanea the sporocarps lie at different positions in the hollow tube of the sexual shoot.

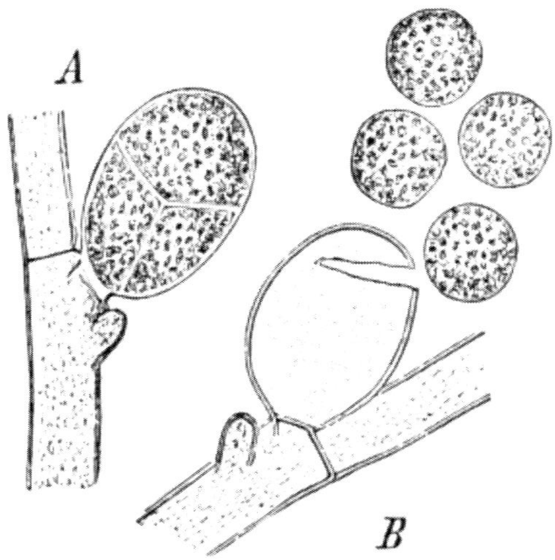

Figure 184. A red alga (Callithamnion), showing sporangium *A*, and the tetraspores discharged *B*. (After Thuret.)

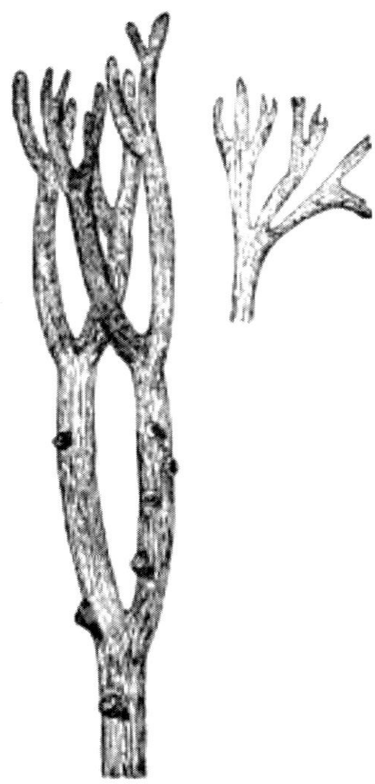

Figure 185. Gracilaria, portion of frond, showing position of cystocarps.

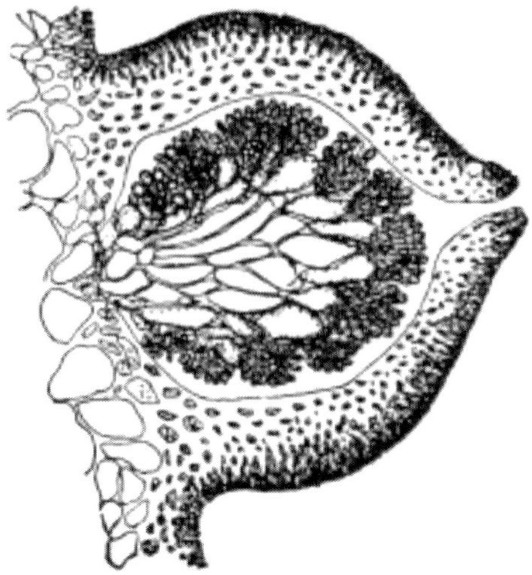

Figure 186. Gracilaria, section of cystocarp showing spores.

Gonidia in the Red Algæ

The common type of gonidium in the red algæ is found in the *tetraspores*. A single mother cell divides into four cells arranged usually in the form of tetrads within the *tetrasporangium*. In Callithamnion the tetrasporangium is exposed. In Polysiphonia, Rhabdonia, Gracilaria, etc., it is imbedded in the cortex. In Batrachospermum there are monosporangia, each monosporangium containing a single gonidium, while in Lemanea, and according to some also in Nemalion, gonidia are wanting.

Gracilaria

Gracilaria is one of the marine forms, and one species is illustrated in Figure 185. It measures 15-20*cm* or more long, and is profusely branched in a palmate manner. The parts of the thallus are more or less flattened. The fruit is a cystocarp, which is characteristic of the Rhodophyceæ (Florideæ). In Gracilaria these fruit bodies occur scattered over the thallus. They are somewhat flask-shaped, are partly sunk in the thallus, and the conical end projects strongly above the surface. The carpospores are grouped in radiating threads within the oval cavity of the cystocarp. These cystocarps are developed as a result of fertilization. Other plants bear gonidia in groups of four, the so-called *tetraspores*.

Rhabdonia

This plant is about the same size as the gracilaria, though it possesses more filiform branches. The cystocarps form prominent elevations, while the carpospores lie in separated groups around the periphery of a sterile tissue within the cavity. (See Figures 187, 188.) Gonidia in the form of tetraspores are also developed in Rhabdonia.

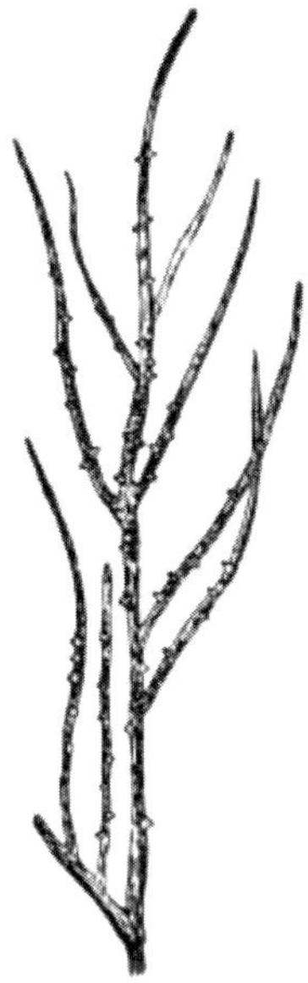

Figure 187. Rhabdonia, branched portion of frond showing cystocarps.

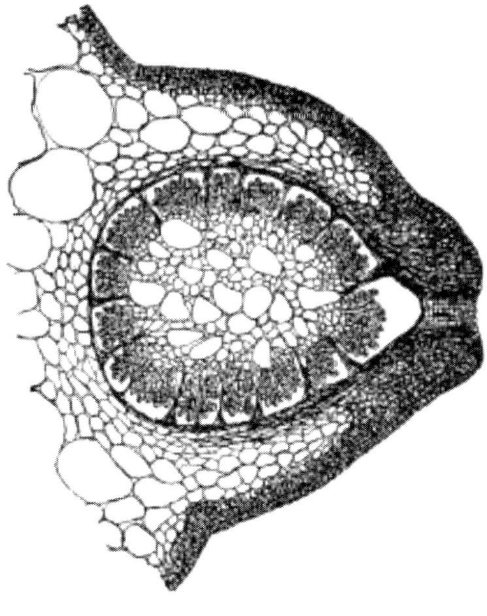

Figure 188. Section of cystocarp of rhabdonia, showing spores.

Fertilization of the Higher Red Algæ

The process of fertilization in most of the red algæ is very complicated, chiefly because the fertilized egg-cell (procarp) does not develop the spores directly, as in Nemalion, Lemanea, etc., but fuses directly, or by a short cell or long filament with one or more auxiliary cells before the sporocarp is finally formed. Examples are Rhabdonia, Polysiphonia, Callithamnion, Dudresnaya, etc. (Figure 189). The auxiliary cell then develops the sporocarp. See Figure 189 for conjugation of a filament from the fertilized procarp with an auxiliary cell.

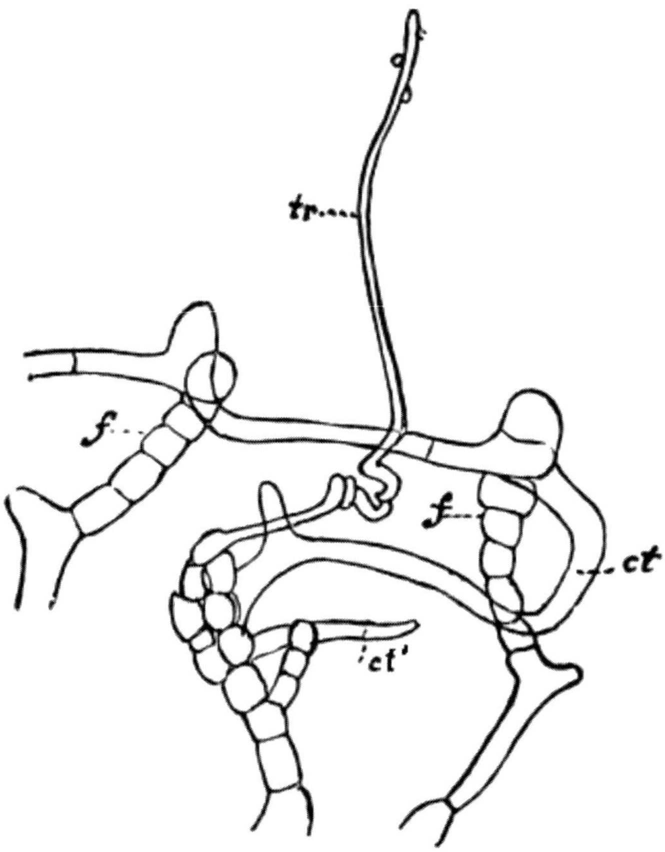

Figure 189. Dudresnaya purpurifera. *tr*, trichogyne, with spermatozoids attached; *ct*, connecting-tube which grows out from below the base of the trichogyne, and comes in contact with the fertile branches *f, f*; *ct'*, young connecting-tube. (After Thuret and Bornet.)

Uses of the Red Algæ

Many species produce a great amount of gelatinous substance in their tissues, and several of these are used for food, for the manufacture of gelatines and agar-agar. Some of these are Gracilaria lichenoides and wrightii, the former species occurring along the coast of India and China. The plant is easily converted into gelatinous substance (agar-agar). Chondrus crispus, widely distributed in the northern Atlantic is known as "Irish" moss and is used for food and for certain medicinal purposes. Gigartina mamillosa in the Atlantic and Arctic oceans is similarly employed. The following orders are recognized in the red algæ:

Order Bangiales

Example, Bangia atropurpurea (= Conferva atropurpurea) in springs and brooks in North America and Europe. Porphyra contains a number of species forming broad, thin, leaf-like purple sheets in the sea.

Order Nemalionales

Including Lemanea, Batrachospermum, Nemalion, described above, and many others.

Order Gigartinales

In this order occurs the common Iceland moss (Chondrus crispus) in the sea, and Rhabdonia and Gigartina mentioned above.

Order Rhodomeniales

In this order occurs Gracilaria and Polysiphonia mentioned above, also the beautiful marine forms like Ceramium.

Order Cryptonemiales

Examples are Dudresnaya, Melobesia, Corallina, etc., the last two genera include many species with a wide distribution.

Class Charophyceæ, Order Charales

The Charales are by some thought to represent a distinct class of algæ standing near the mosses, perhaps, because of the biciliate character of the spermatozoids. There is one family, the Characeæ. The plants occur in fresh and brackish water. Aside from the peculiarity of the reproductive organs they are remarkable for the large size of the cells of the internodes and of the "leaves," and the protoplasm exhibits to a remarkable degree the phenomenon of "cyclosis" (see paragraphs 17-20). Three of the genera are found in North America (Chara, Nitella (Figure 8) and Tolypella).

The complicated structure of the sexual organs shows a higher state of organization than any of the other living algæ known. While the internodes in Nitella are composed of a single, stout cell, some times a foot or more in length, the nodes in all are composed of a group of smaller cells. From the lateral cells of this group lateral axes (sometimes called leaves) arise in whorls.

In Nitella the internodes are naked, but in most species of Chara they are *corticated*, i.e., they are covered by a layer of numerous elongated cells which grow downward from the nodes at the base of the whorl of lateral shoots.

The sexual organs are situated at the nodes of the whorled lateral shoots, and consist of antheridia and carpogonia. Most of the plants are monœcious, and both antheridia and carpogonia are often attached to the same node, the antheridium projecting downward while the carpogonium is more or less ascending. The sexual organs are visible to the unaided eye. The antheridium is a globose red body of an exceedingly complicated structure. The sperms are borne in several very long coiled slender threads which are divided transversely into numerous cells. The carpogonium is oval or elliptical in outline, the wall of which is composed of several closely coiled spiral threads enclosing the large egg.

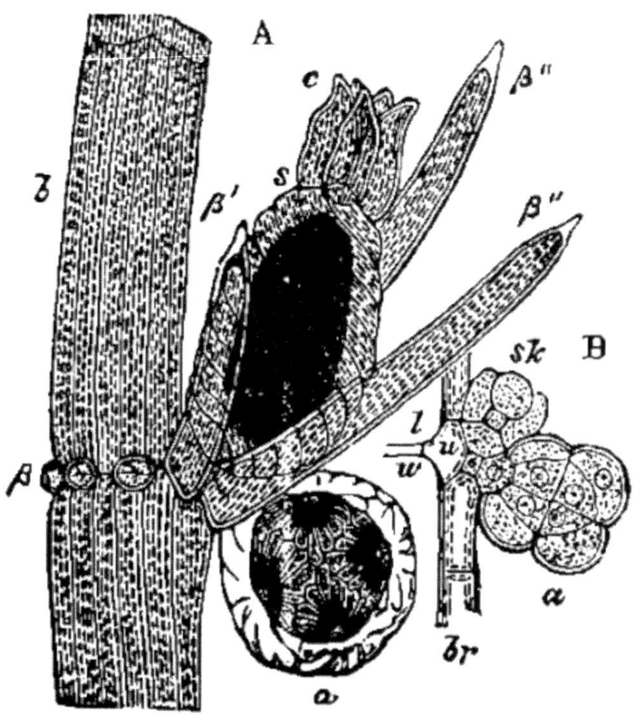

Figure 172*a*. Reproductive organs of *Chara fragilis*. *A*, a central portion of a leaf, *b*, with an antheridium, *a*, and a carpogonium, *s*, surrounded by the spirally twisted enveloping cells; *c*, crown of five cells at apex; β, sterile lateral leaflets; β′, large lateral leaflet near the fruit; β″, bracteoles springing from the basal node of the reproductive organs. *B*, a young antheridium, *a*, and a young carpogonium, *sk*; *w*, nodal cell of leaf; *u*, intermediate cell between *w* and the basal-node cell of the antheridium; *l*, cavity of the internode of the leaf; *br*, cortical cells of the leaf. *A* × about 33; *B* × 240. (After Sachs.)

Chapter 19

FUNGI: MUCOR AND SAPROLEGNIA

MUCOR

In the chapter on growth, and in our study of protoplasm, we have become familiar with the vegetative condition of mucor. We now wish to learn how the plant multiplies and reproduces itself. For this study we may take one of the mucors. Any one of several species will answer. This plant may be grown by placing partially decayed fruits, lemons, or oranges, from which the greater part of the juice has been removed, in a moist chamber; or often it occurs on animal excrement when placed under similar conditions. In growing the mucor in this way we are likely to obtain Mucor mucedo, or another plant sometimes known as Mucor stolonifer, or Rhizopus nigricans, which is illustrated in Figure 191. This latter one is sometimes very injurious to stored fruits or vegetables, especially sweet potatoes or rutabagas. Figure 190 is from a photograph of this fungus on a banana.

Asexual Reproduction

On the decaying surface of the vegetable matter where the mucor is growing there will be seen numerous small rounded bodies borne on very slender stalks. These heads contain the gonidia, and if we sow some of them in nutrient gelatine or agar in a Petrie dish the material can be taken out very readily for examination under the microscope. Or we may place glass slips close to the growing fungus in the moist chamber, so that the fungus will develop on them, though cultures in a nutrient medium are much better. Or we may take the material directly from the substance on which it is growing. After mounting a small quantity of the mycelium bearing these heads, if we have been careful to take it where the heads appear quite young, it may be possible to study the early stages of their development.

We shall probably note at once that the stalks or upright threads which support the heads are stouter than the threads of the mycelium.

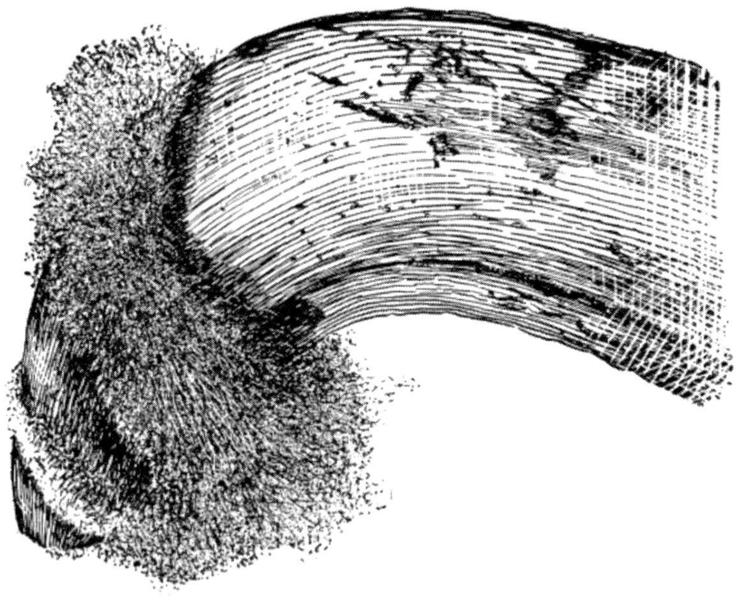

Figure 190. Portion of banana with a mould (Rhizopus nigricans) growing on one end.

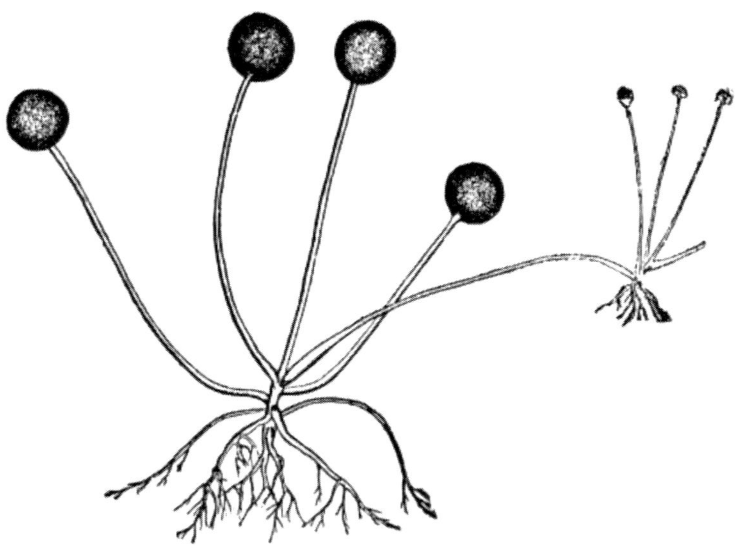

Figure 191. Group of sporangia of a mucor (Rhizopus nigricans) showing rhizoids and the stolon extending from an older group.

These upright threads soon have formed near the end a cross wall which separates the protoplasm in the end from the remainder. This end cell now enlarges into a vesicle of considerable size, the head as it appears, but to which is applied the name of *sporangium* (sometimes called gonidangium), because it encloses the *gonidia*.

At the same time that this end cell is enlarging the cross wall is arching up into the interior. This forms the *columella*. All the protoplasm in the sporangium now divides into gonidia. These are small-rounded or oval bodies. The wall of the sporangium becomes dissolved, except a small collar around the stalk which remains attached below the columella (Figure 192). By this means the gonidia are freed. These gonidia germinate and produce the mycelium again.

Sexual Stage

This stage is not so frequently found, but may sometimes be obtained by growing the fungus on bread.

Conjugation takes place in this way. Two threads of the mycelium which lie near each other put out each a short branch which is clavate in form. The ends of these branches meet, and in each a septum is formed which cuts off a portion of the protoplasm in the end from that of the rest of the mycelium. The meeting walls of the branches now dissolve and the protoplasm of each gamete fuses into one mass. A thick wall is now formed around this mass, and the outer layer becomes rough and brown. This is the *zygote* or *zygospore*. The mycelium dies and it becomes free often with the suspensors, as the stalks of these sexual branches are called, still attached. This zygospore passes through a period of rest, when with the entrance of favorable conditions of growth it germinates, and usually produces directly a sporangium with gonidia. This completes the normal life cycle of the plant.

Gemmæ

Gemmæ, as they arc sometimes called, are often formed on the mycelium. A short cell with a stout wall is formed on the side of a thread of the mycelium. In other cases large portions of the threads of the mycelium may separate into chains of cells. Both these kinds of cells are capable of growing and forming the mycelium again. They are sometimes called *chlamydospores*.

The Mucorineæ according to their manner of zygospore formation are of two kinds: 1st, the *homothallic* (monœcious), in which all of the colonies of thalli developed from different spores are the same, and both gametes may be developed from the mycelium from a single spore, as in Sporodinia grandis, a mould common on old mushrooms; 2d, the *heterothallic* (diœcious), in which certain plants are of a male nature and small in comparison with those of perhaps a female nature which are larger or more vigorous. When grown separately each of these two kinds of thalli, or colonies of mycelium, produce their own kind but only sporangia. If the two kinds are brought together, however, branches from one conjugate with branches from the other and zygospores are produced, as in

Rhizopus nigricans, the common bread or fruit mould. This is one reason why we rarely find this fungus forming zygospores. (See Blakeslee, Sexual Reproduction in the Mucorineæ, Proc. Am. Acad. Arts and Sci., 40, 205-319, pl. 1-4, 1904.)

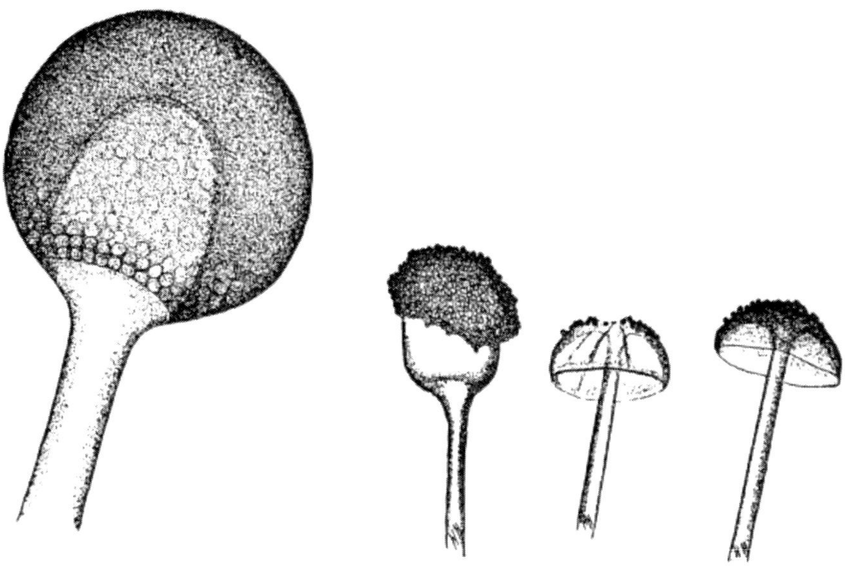

Figure 194. A mucor (Rhizopus nigricans); at left nearly mature sporangium with columella showing within; in the middle is ruptured sporangium with some of the gonidia clinging to the columella; at right two ruptured sporangia with everted columella.

Water Moulds (Saprolegnia)

The water moulds are very interesting plants to study because they are so easy to obtain, and it is so easy to observe a type of gonidium here to which we have referred in our studies of the algæ, the motile gonidium, or zoogonidium. (See appendix for directions for cultivating this mould.)

Appearance of the Saprolegnia

In the course of a few days we are quite certain to see in some of the cultures delicate whitish threads, radiating outward from the body of the fly in the water. A few threads should be examined from day to day to determine the stage of the fungus.

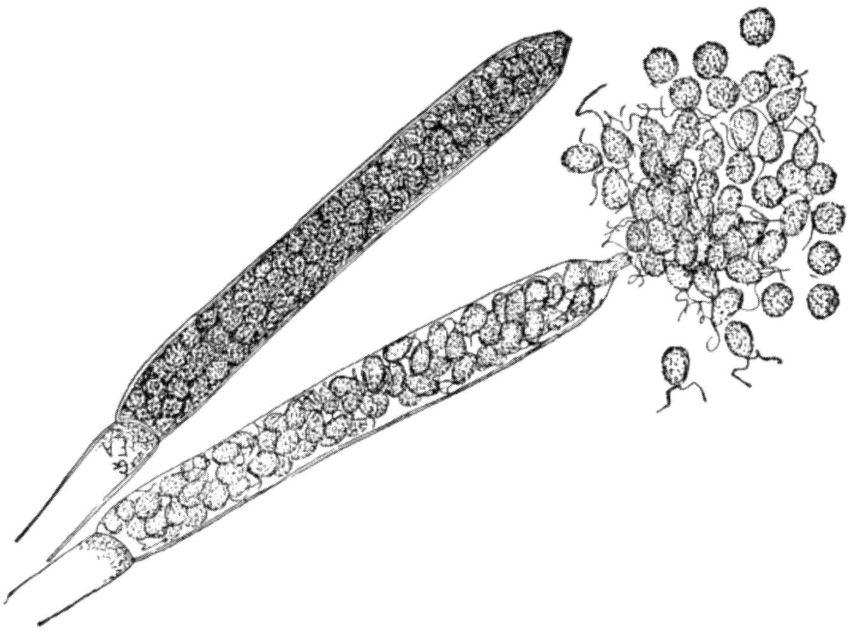

Figure 195. Sporangia of saprolegnia, one showing the escape of the zoogonidia.

Sporangia of Saprolegnia

The sporangia of saprolegnia can be easily detected because they are much stouter than the ordinary threads of the mycelium. Some of the threads should be mounted in fresh water. Search for some of those which show that the protoplasm is divided up into a great number of small areas, as shown in Figure 195. With the low power we should watch some of the older appearing ones, and if after a few minutes they do not open, other preparations should be made.

Figure 196. Branch of saprolegnia showing oogonia with oospores, eggs matured parthenogenetically.

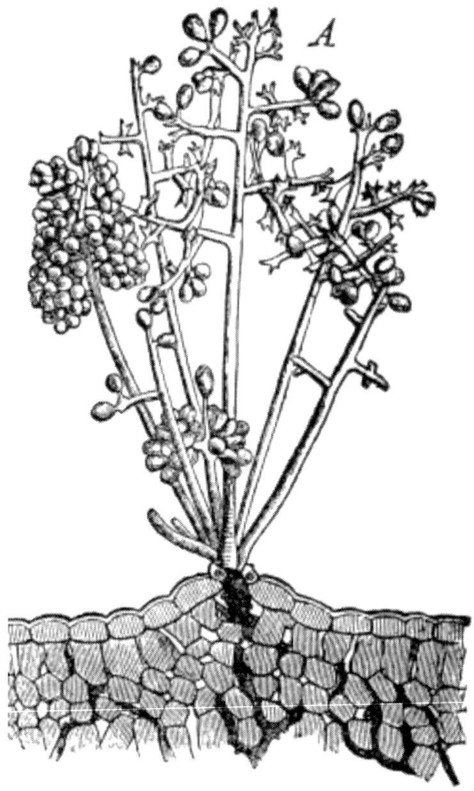

Figure 197. Downy mildew of grape (Plasmopora viticola), showing tuft of gonidiophores bearing gonidia, also intercellular mycelium. (After Millardet.)

Figure 198. Phytophthora infestans showing peculiar branches; gonidia below.

Zoogonidia of Saprolegnia

The sporangium opens at the end, and the zoogonidia swirl out and swim around for a short time, when they come to rest. With a good magnifying power the two cilia on the end may be seen, or we may make them more distinct by treatment with Schultz's solution, drawing some under the cover glass. The zoogonidium is oval and the cilia are at the pointed end. After they have been at rest for some time they often slip out of the thin wall, and swim again, this time with the two cilia on the side, and then the zoogonidium is this time more or less bean-shaped or reniform.

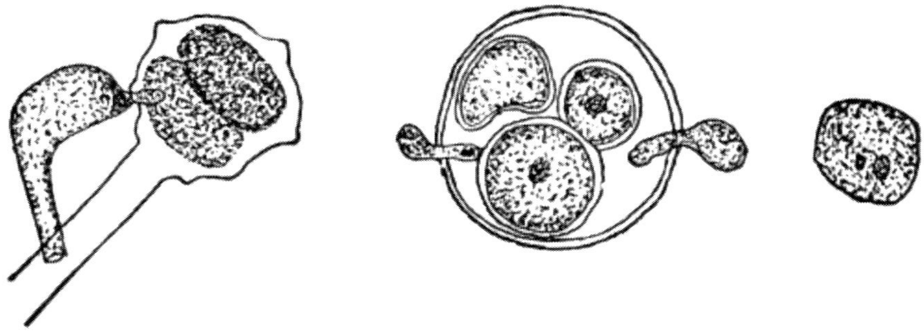

Figure 199. Fertilization in saprolegnia, tube of antheridium carrying in the nucleus of the sperm cell to the egg. In the right-hand Figure a smaller sperm nucleus is about to fuse with the nucleus of the egg. (After Humphrey and Trow.)

Figure 200. Branching hypha of Peronospora alsinearum.

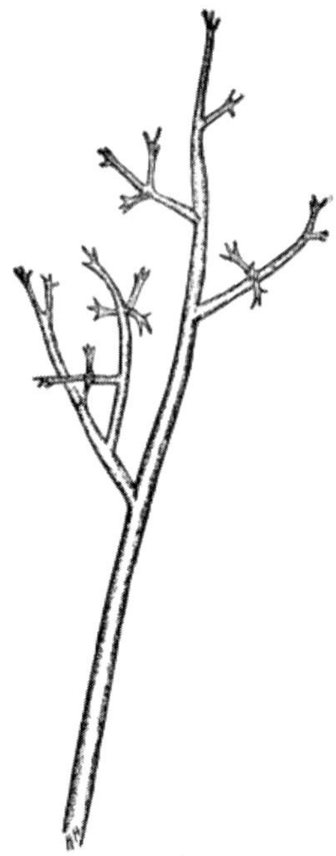

Figure 201. Branched hypha of downy mildew of grape showing peculiar branching (Plasmopara viticola).

Sexual Reproduction of Saprolegnia

When such cultures are older we often see large rounded bodies either at the end of a thread, or of a branch, which contain several smaller rounded bodies as shown in Figure 196.

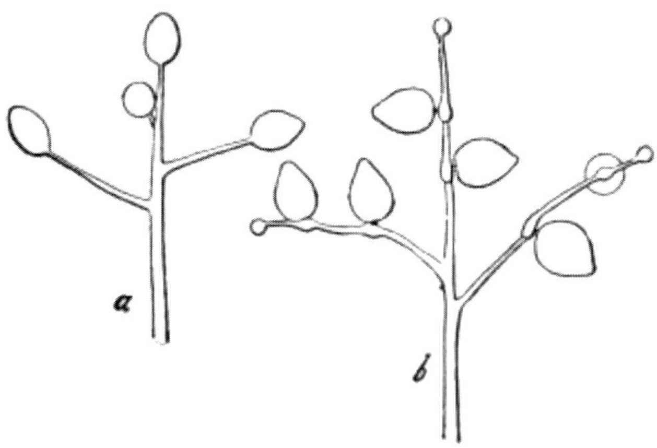

Figure 202. Gonidiophores and gonidia of potato blight (Phytophthora infestans). *b*, an older stage showing how the branch enlarges where it grows beyond the older gonidium.(After de Bary.)

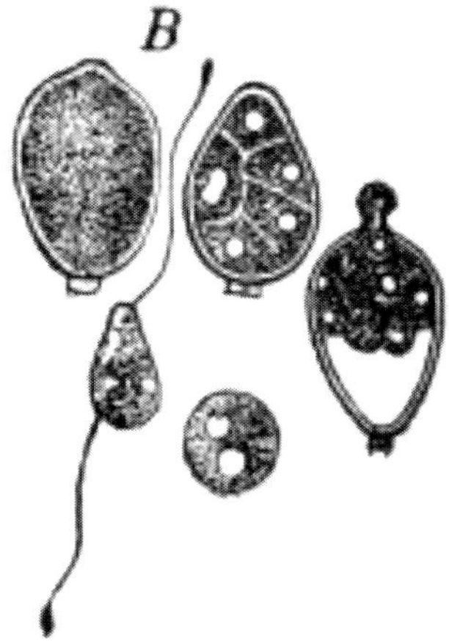

Figure 203. Gonidia of potato blight forming zoogonidia. (After de Bary.)

These are the oogonia (unless the plant is attacked by a parasite), and the round bodies inside are the egg-cells, if before fertilization, or the eggs, if after this process has taken place. Sometimes the slender antheridium can be seen coiled partly around the oogonium, and one end entering to come in contact with the egg-cell. But in some species the antheridium is not present, and that is the case with the species Figured at 196. In this case the eggs mature without fertilization. This maturity of the egg without fertilization is called *parthenogenesis*, which occurs in other plants also, but is a rather rare phenomenon.

In Figure 199 is shown the oogonium and an antheridium, and the antheridium is carrying in the male nucleus to the egg-cell. Spermatozoids are not developed here, but a nucleus in the antheridium reaches the egg-cell. It sinks in the protoplasm of the egg, comes in contact with the nucleus of the egg, and fuses with it. Thus fertilization is accomplished.

Downy Mildews

The downy mildews make up a group of plants which are closely related to the water moulds, but they are parasitic on land plants, and some species produce very serious diseases. The mycelium grows between the cells of the leaves, stems, etc., of their hosts, and sends haustoria into the cells to take up nutriment. Gonidia are formed on threads which grow through the stomates to the outside and branch as shown in Figures 198-201. The gonidia are borne on the tips of the branches. The kind of branching bears some relation to the different genera. Figure 200 is from Peronospora alsinearum on leaves of cerastium; Figures 197 and 199 are Plasmopara viticola, the grape mildew, while Figures 198 and 202 are from Phytophthora infestans which causes a disease known as potato blight. The gonidia of peronospora germinate by a germ tube, those of plasmopara first form zoogonidia, while in phytophthora the gonidium may either germinate forming a thread, or each gonidium may first form several zoogonidia, as shown in Figure 203.

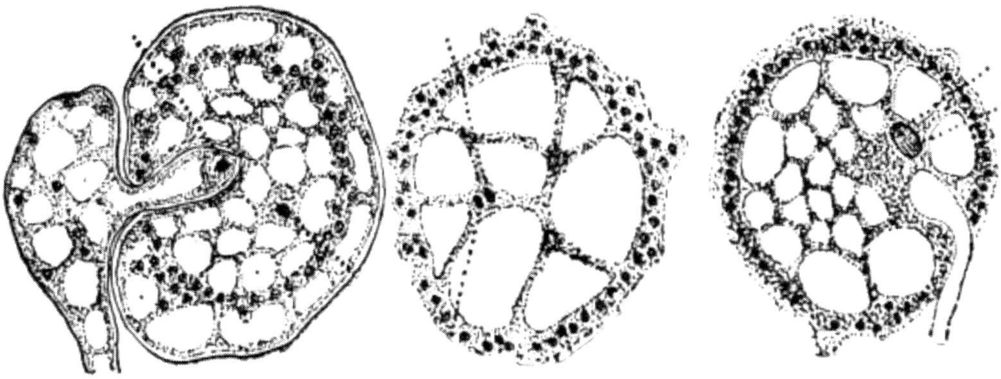

Figure 204. Fertilization in Peronospora alsinearum; tube from antheridium carrying in the sperm nucleus in Figure at the left, female nucleus near; fusion of the two nuclei shown in the two other Figures. (After Berlese.)

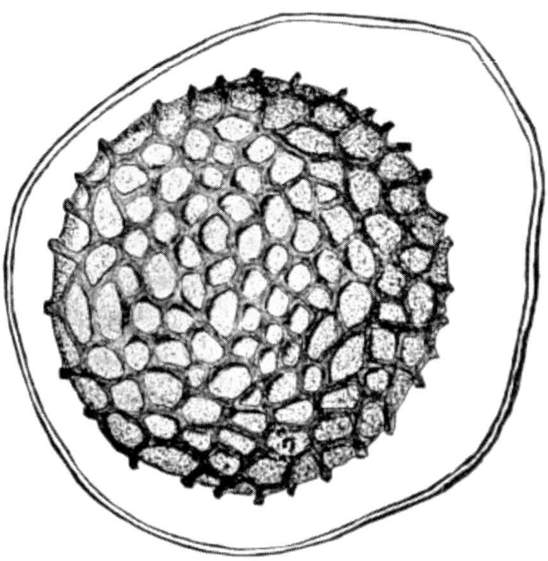

Figure 205. Ripe oospore of Peronospora alsinearum.

In sexual reproduction oogonia and antheridia are developed on the mycelium within the tissues. Figure 204 represents the antheridium entering the oogonium, and the male nucleus fusing with the female nucleus in fertilization. The sexual organs of Phytophthora infestans are not sufficiently known.

Mucor, saprolegnia, peronospora, and their relatives have few or no septa in the mycelium. In this respect they resemble certain of the algæ like vaucheria, but they lack chlorophyll. They are sometimes called the alga-like fungi and belong to a large group called *Phycomycetes*.

Chapter 20

Fungi Continued

"Rusts" (Uredineæ)

The fungi known as "rusts" are very important ones to study, since all the species are parasitic, and many produce serious injuries to crops.

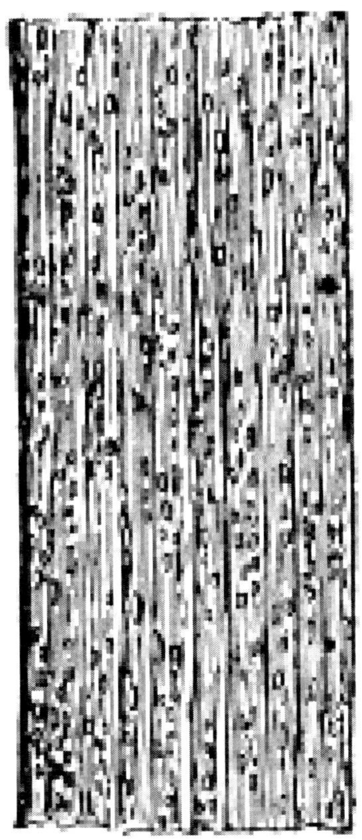

Figure 206. Wheat leaf with red-rust, natural size.

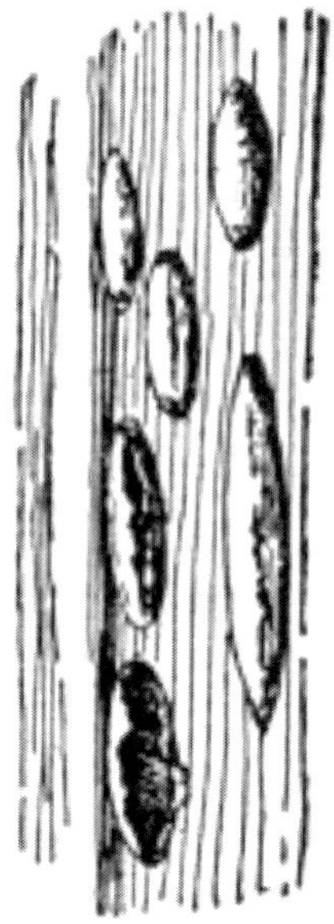

Figure 207. Portion of leaf enlarged to show sori.

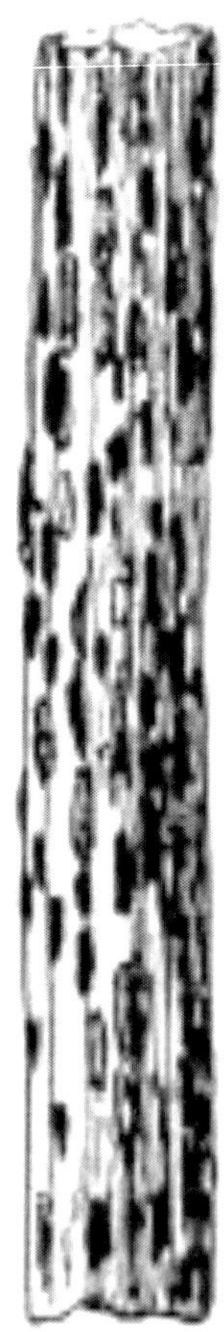

Figure 208. Natural size.

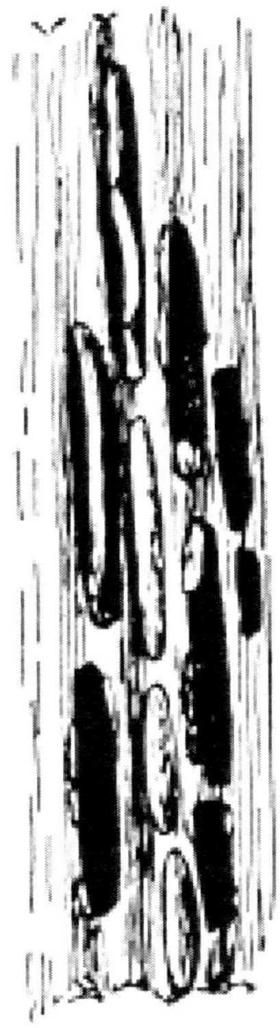

Figure 209. Enlarged.

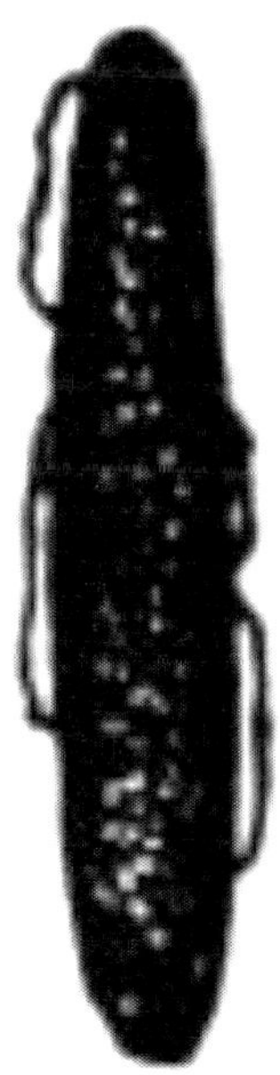

Figure 210. Single sorus.

Figures 206, 207.—Puccinia graminis, red-rust stage (uredo stage).

Figures 208-210.—Black rust of wheat, showing sori of teleutospores.

Wheat Rust (*Puccinia graminis*)

The wheat rust is one of the best known of these fungi, since a great deal of study has been given to it. One form of the plant occurs in long reddish-brown or reddish pustules, and is known as the "red-rust" (Figures 206, 207). Another form occurs in elongated black pustules, and this form is the one known as the "black rust" (Figures 208-211). These two forms occur on the stems, blades, etc., of the wheat, also on oats, rye, and some of the grasses.

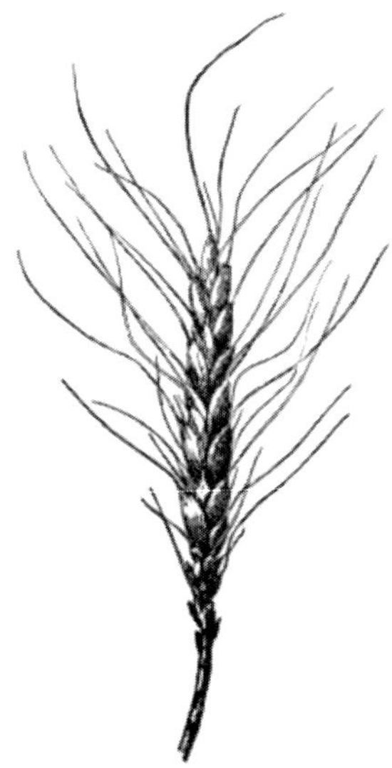

Figure 211. Head of wheat showing black rust spots on the chaff and awns.

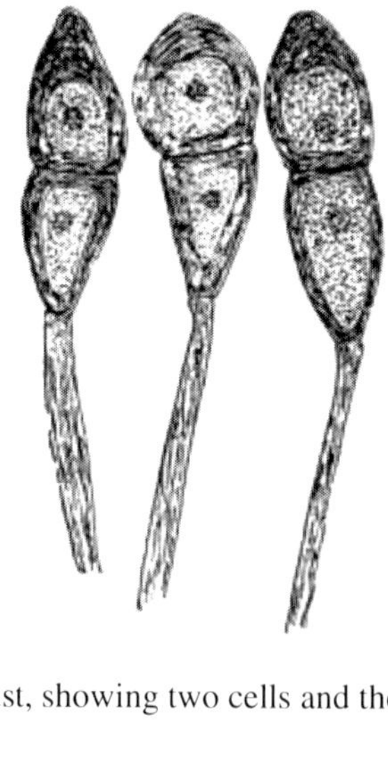

Figure 212. Teleutospores of wheat rust, showing two cells and the pedicel.

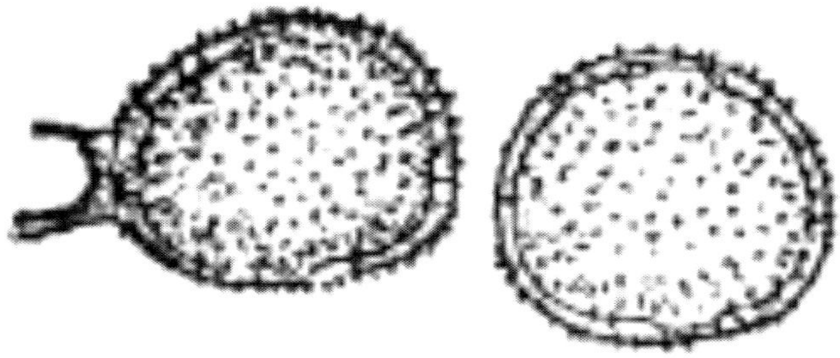

Figure 213. Uredospores of wheat rust, one showing remnants of the pedicel.

Teleutospores of the Black Rust Form

If we scrape off some portion of one of the black pustules (sori), tease it out in water on a slide, and examine with a microscope, we see numerous gonidia, composed of two cells, and having thick, brownish walls as shown in Figure 212. Usually there is a slender brownish stalk on one end. These gonidia are called *teleutospores*. They are somewhat oblong or elliptical, a little constricted where the septum separates the two cells, and the end cell varies from ovate to rounded. The mycelium of the fungus courses between the cells, just as is found in the case of the carnation rust, which belongs to the same family (see Parag. 186).

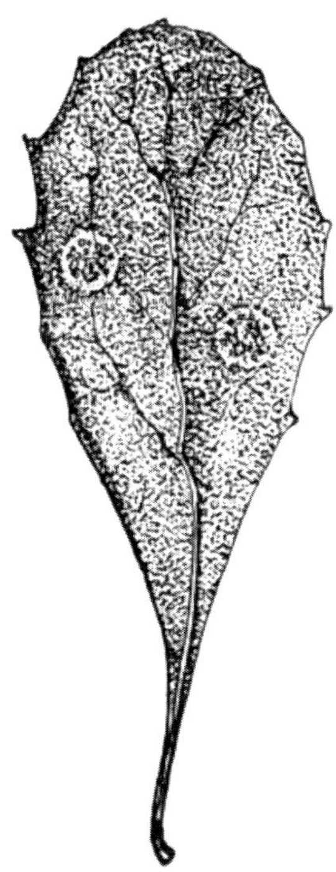

Figure 214. Barberry leaf with two diseased spots, natural size.

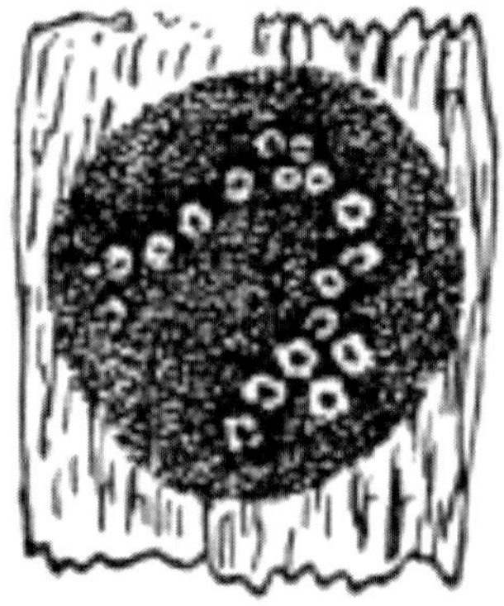

Figure 215. Single spot showing cluster-cups enlarged.

Figure 216. Two cluster-cups more enlarged, showing split margin.

Figures 214-216.—Cluster-cup stage of wheat rust.

Uredospores of the Red-Rust Form

If we make a similar preparation from the pustules of the red-rust form we see that instead of two-celled gonidia they are one-celled. The walls are thinner and not so dark in color, and they are covered with minute spines. They have also short stalks, but these fall away very easily. These one-celled gonidia of the red-rust form are called "uredospores." The uredospores and teleutospores are sometimes found in the same pustule.

It was once supposed that these two kinds of gonidia belonged to different plants, but now it is known that the one-celled form, the uredospores, is a form developed earlier in the season than the teleutospores.

Cluster-Cup Form on the Barberry

On the barberry is found still another form of the wheat rust, the "*cluster-cup*" stage. The pustules on the under side of the barberry leaf are cup-shaped, the cups being partly sunk in the tissue of the leaf, while the rim is more or less curved backward against the leaf, and split at several places. These cups occur in clusters on the affected spots of the barberry leaf as shown in Figure 215. Within the cups numbers of one-celled gonidia (orange in color, called æcidiospores) are borne in chains from short branches of the mycelium, which fill the base of the cup. In fact the wall of the cup (peridium) is formed of similar rows of cells, which, instead of separating into gonidia, remain united to form a wall. These cups are usually borne on the under side of the leaf.

Spermagonia

Upon the upper side of the leaves in the same spot occur small, orange-colored pustules which are flask-shaped. They bear inside, minute, rod-like bodies on the ends of slender threads, which ooze out on the surface of the leaf.

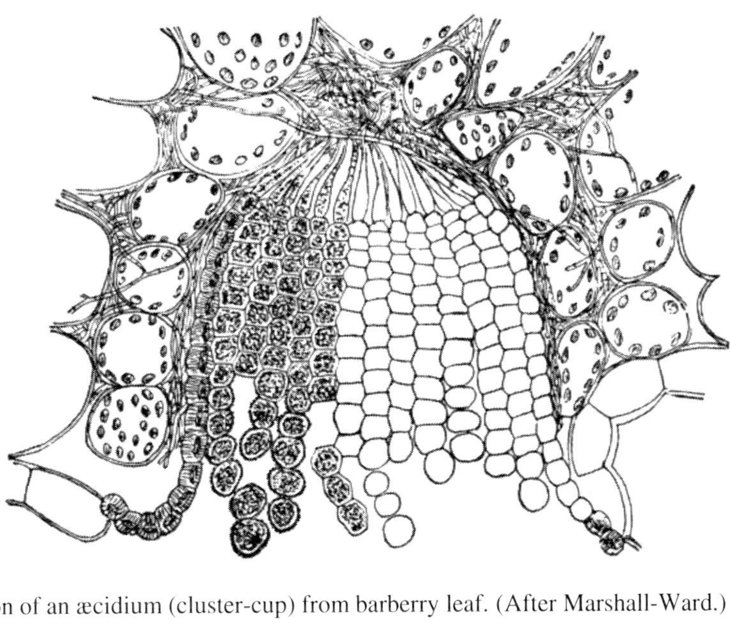

Figure 217. Section of an æcidium (cluster-cup) from barberry leaf. (After Marshall-Ward.)

These flask-shaped pustules are called *spermagonia*, and the minute bodies within them *spermatia*, since they were once supposed to be the male element of the fungus. Their function is not known. They appear in the spots at an earlier time than the cluster-cups.

How the Cluster-Cup Stage Was Found To Be a Part of the Wheat Rust

The cluster-cup stage of the wheat rust was once supposed also to be a different plant, and the genus was called *æcidium*. The occurrence of wheat rust in great abundance on the leeward side of affected barberry bushes in England suggested to the farmers that wheat rust was caused by barberry rust. It was later found that the æcidiospores of the barberry, when sown on wheat, germinate and the thread of mycelium enters the tissues of the wheat, forming mycelium between the cells. This mycelium then bears the uredospores, and later the teleutospores.

Uredospores Can Produce Successive Crops of Uredospores

The uredospores are carried by the wind to other wheat or grass plants, germinate, form mycelium in the tissues, and later the pustules with a second crop of uredospores. Several successive crops of uredospores may be developed in one season, so this is the form in which the fungus is greatly multiplied and widely distributed.

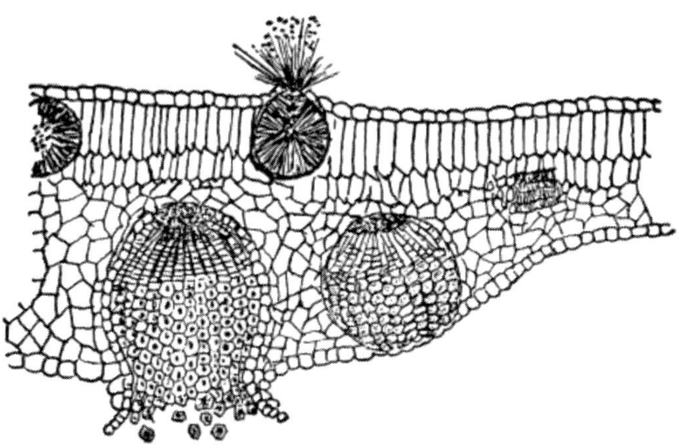

Figure 218. Section through leaf of barberry at point affected with the cluster-cup stage of the wheat rust; spermagonia above, æcidia below. (After Marshall-Ward.)

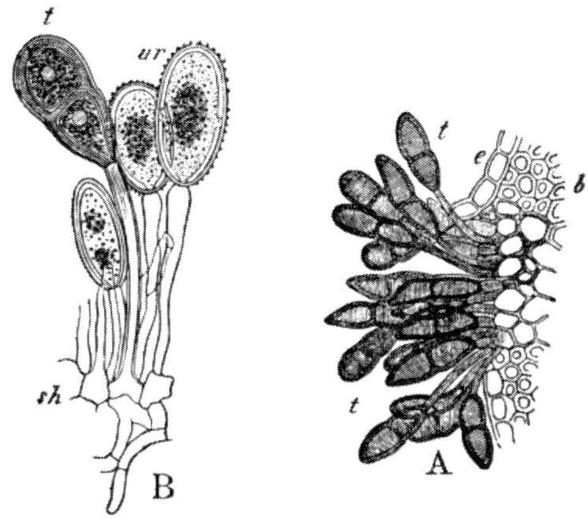

Figure 219. *A*, section through sorus of black rust of wheat, showing teleutospores. *B*, mycelium bearing both teleutospores and uredospores. (After de Bary.)

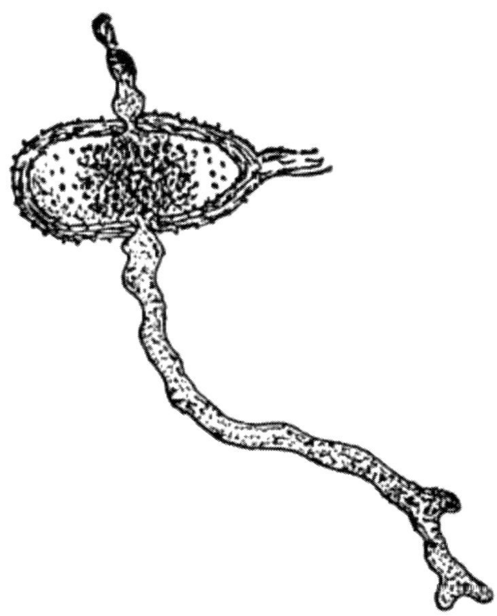

Figure 220. Germinating uredospore of wheat rust. (After Marshall-Ward.)

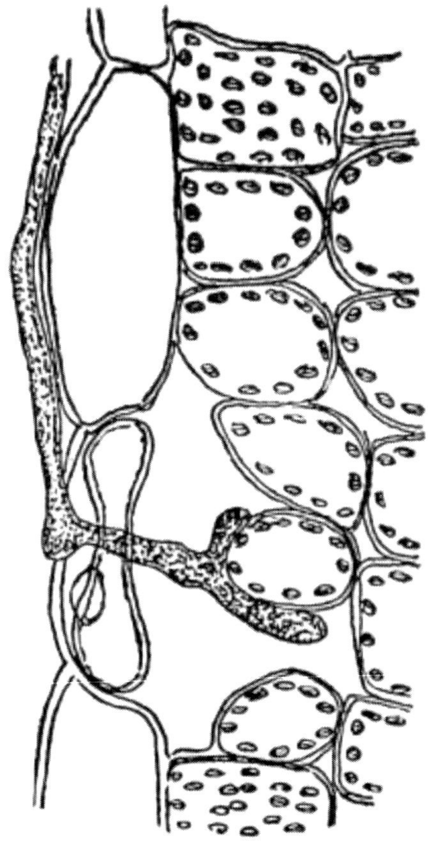

Figure 221. Germ tube entering the leaf through a stoma.

Teleutospores the Last Stage of the Fungus in the Season

The teleutospores are developed late in the season, or late in the development of the host plant (in this case the wheat is the host). They then rest during the winter. In the spring under favorable conditions each cell of the teleutospore germinates, producing a short mycelium called a *promycelium*, as shown in Figures 222, 223. This promycelium is usually divided into four cells. From each cell a short, pointed process is formed called a "*sterigma*." Through this the protoplasm moves and forms a small gonidium on the end, sometimes called a *sporidium*.

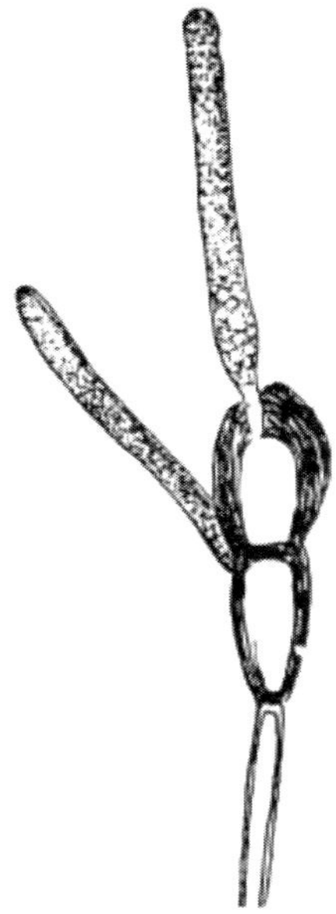

Figure 222. Teleutospore germinating, forming promycelium.

Figure 223. Promycelium of germinating teleutospore, forming sporidia.

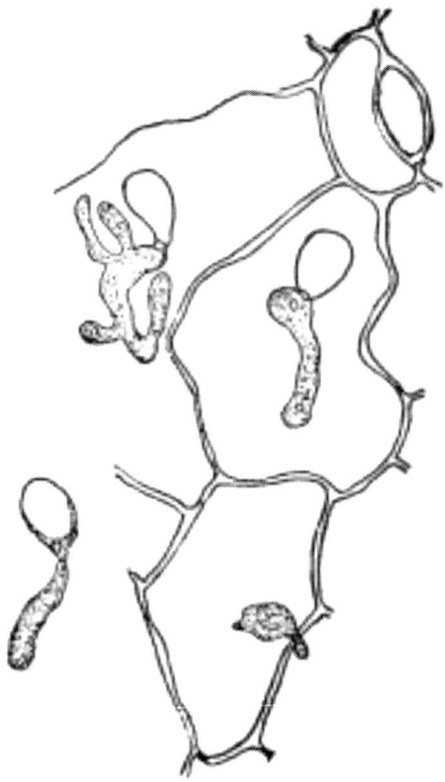

Figure 224. Germinating sporidia entering leaf of barberry by mycelium.

Figures 222-224.—*Puccinia graminis* (wheat rust). (After Marshall-Ward.)

How the Fungus Gets From the Wheat Back to the Barberry

If these sporidia from the teleutospores are carried by the wind so that they lodge on the leaves of the barberry, they germinate and produce the cluster-cup again. The plant has thus a very complex life history. Because of the presence of several different forms in the life cyle, it is called a *polymorphic* fungus. The presence of the barberry does not seem necessary in all cases for the development of the fungus from one year to another.

Synopsis of Life History of Wheat Rust

Cluster-Cup Stage on Leaf of Barberry

Mycelium between cells of leaf in affected spots.

Spermagonia (sing. spermagonium), small flask-shaped bodies sunk in upper side of leaf; contain "spermatia."

Æcidia (sing. æcidium), cup-shaped bodies in under side of leaf.

Wall or peridium, made up of outer layer of fungus threads which are divided into short cells but remain united.

At maturity bursts through epidermis of leaf; margin of cup curves outward and downward toward surface of leaf.

Central threads of the bundle are closely packed, but free. Threads divide into short angular cells which separate and become æcidiospores, with orange-colored content.

Æcidiospores carried by the wind to wheat, oats, grasses, etc. Here they germinate, mycelium enters at stomate, and forms mycelium between cells of the host.

Uredo Stage (Red-Rust) on Wheat, Oats, Grasses, etc.

Mycelium between cells of host.

Bears uredospores (1-celled) in masses under epidermis, which is later ruptured and uredospores set free.

Uredospores carried by wind to other individual hosts, and new crops of uredospores formed.

Teleutospore Stage (Black Rust), Also on Wheat, etc.

Mycelium between cells of host.

Bears teleutospores (2-celled) in masses (sori) under epidermis, which is later ruptured.

Teleutospores rest during winter. In spring each cell germinates and produces a promycelium, a short thread, divided into four cells.

Promycelium bears four sterigmata and four gonidia (or sporidia), which in favorable conditions pass back to the barberry, germinate, the tube enters between cells into the intercellular spaces of the host to produce the cluster-cup again, and thus the life cycle is completed.

Other Examples of the Rusts

Some of the rusts do great injury to fruit trees and also to forest trees. The "cedar apples" are abnormal growths on the leaves and twigs of the cedar stimulated by the presence of the mycelium of a rust known as Gymnosporangium macropus. The teleutospores are two-celled and are formed in the tissue of the "cedar apple" or gall. The teleutosori are situated at quite regular intervals over the surface of the gall at small circular depressions, and can be easily seen in late autumn and during the winter. A quantity of gelatine is developed along with the teleutospores. In early spring with the warm spring rains the gelatinous substance accompanying the teleutospores swells greatly, and causes the teleutospores to ooze out in long, dull, orange-colored strings, which taper gradually to a slender point and bristle all over the "cedar apple." Here the teleutospores germinate and produce the sporidia. The sporidia are carried to apple trees where they infect leaves and even the fruit, producing here the cluster-cups. There are no uredospores.

G. globosum is another species forming cedar apples, but the gelatinous strings of teleutospores are short and clavate, and the cluster-cups are formed on hawthorns. G. nidusavis forms "witches brooms" or "birds nests" in the branches of the cedar. The mycelium in the branches stimulates them to profuse branching so that numerous small branches are developed close together. The teleutosori form small pustules scattered over the branches. G. clavipes affects the branches of cedar only slightly deforming them or not at all, and the cluster-cups are formed on fruits, twigs, and leaves of the hawthorns or quinces, the cluster-cups being long, tubular, and orange in color.

Chapter 21

The Higher Fungi

The Series of the Higher Fungi

Of these there are two large series. One of these is represented by the sac fungi, and the other by the mushrooms, a good example of which is the common mushroom (Agaricus campestris).

Sac Fungi (Ascomycetes)

The sac fungi may be represented by the "powdery mildews"; examples, uncinula, microsphæra, podosphæra, etc. Figure 225 is from a photograph of two willow leaves affected by one of these mildews. The leaves are first partly covered with a whitish growth of mycelium, and numerous chains of colorless gonidia are borne on short erect threads. The masses of gonidia give the leaf a powdery appearance. The mycelium lives on the outer surface of the leaf, but sends short haustoria into the epidermal cells.

Fruit Bodies of the Willow Mildew

On this same mycelium there appear later numerous black specks scattered over the affected places of the leaf. These are the fruit bodies (*perithecia*). If we scrape some of these from the leaf, and mount them in water for microscopic examination, we shall be able to see their structure. Examining these first with a low power of the microscope, each one is seen to be a rounded body, from which radiate numerous filaments, the *appendages*. Each one of these appendages is coiled at the end into the form of a little hook. Because of these hooked appendages this genus is called *uncinula*. This rounded body is the *perithecium*.

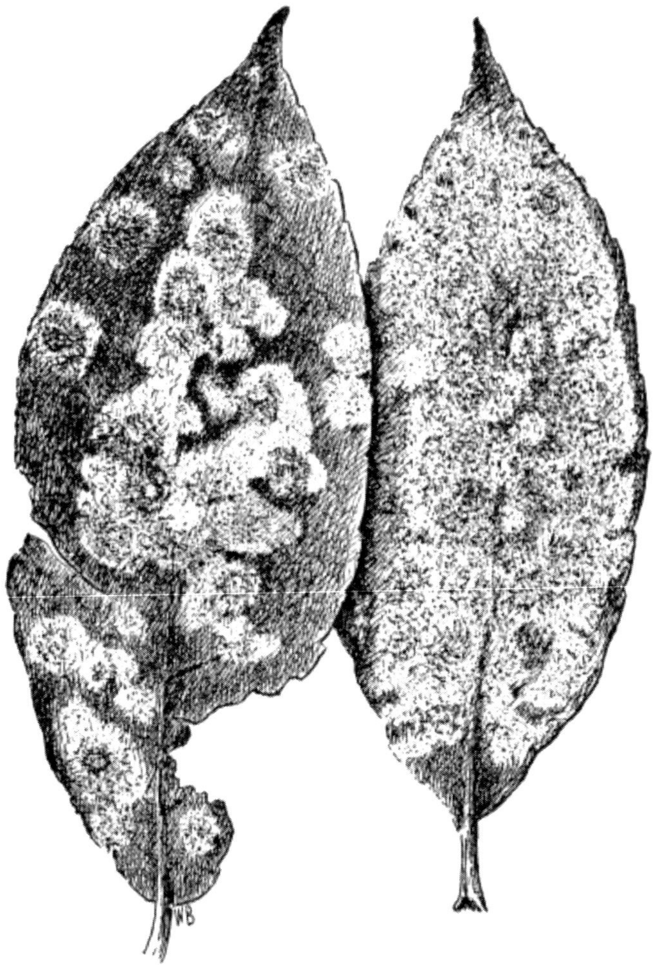

Figure 225. Leaves of willow showing willow mildew. The black dots are the fruit bodies (perithecia) seated on the white mycelium.

Asci and Ascospores

While we are looking at a few of these through the microscope with the low power, we should press on the cover glass with a needle until we see a few of the perithecia rupture. If this is done carefully we see several small ovate sacs issue, each containing a number of spores, as shown in Figure 227. Such a sac is an *ascus*, and the spores are *ascospores*.

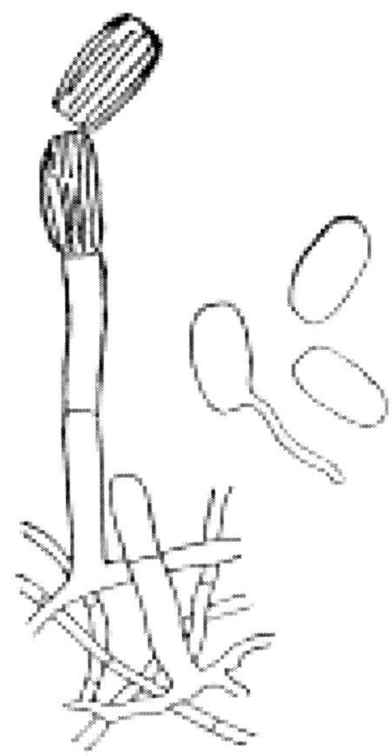

Figure 226. Willow mildew; bit of mycelium with erect conidiophores, bearing chain of gonidia; gonidium at left germinating.

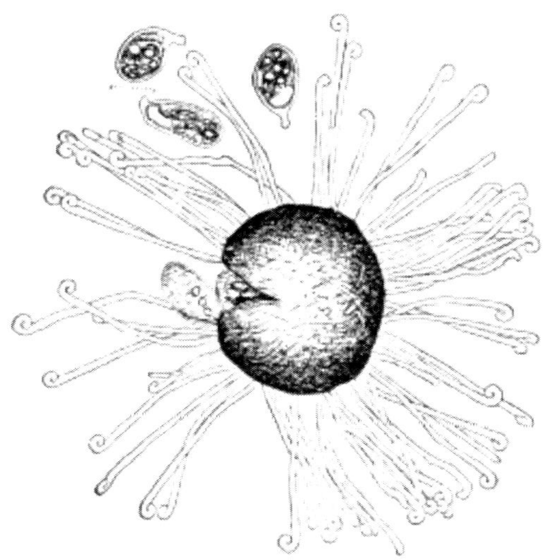

Figure 227. Fruit of willow mildew, showing hooked appendages. Genus uncinula.

Figures 227-228.—Perithecia (perithecium) of two powdery mildews, showing escape of asci containing the spores from the crushed fruit bodies.

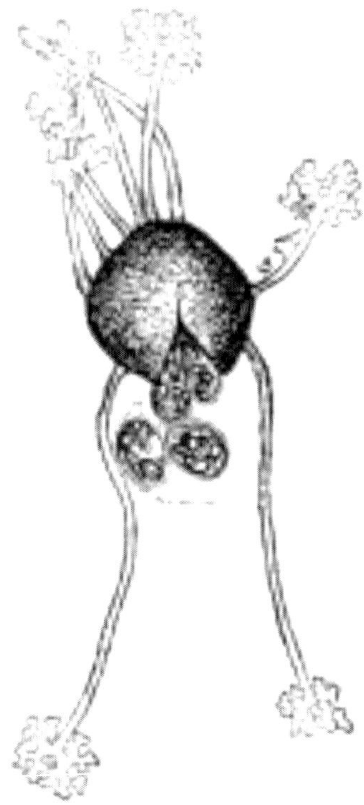

Figure 228. Fruit body of another mildew with dichotomous appendages. Genus microsphæra.

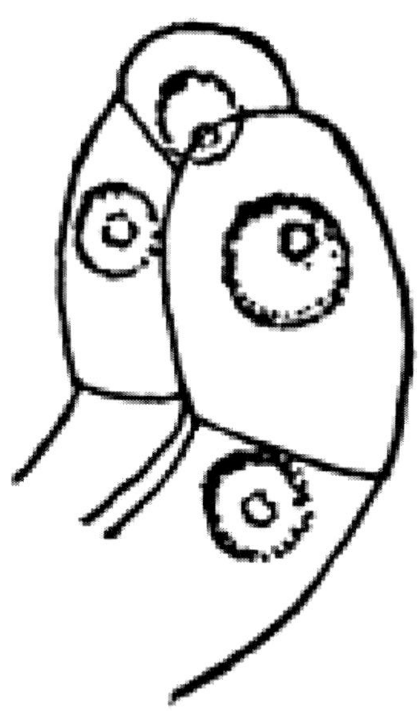

Figure 229. Contact of antheridium and carpogonium (carpogonium the larger cell); beginning of fertilization.

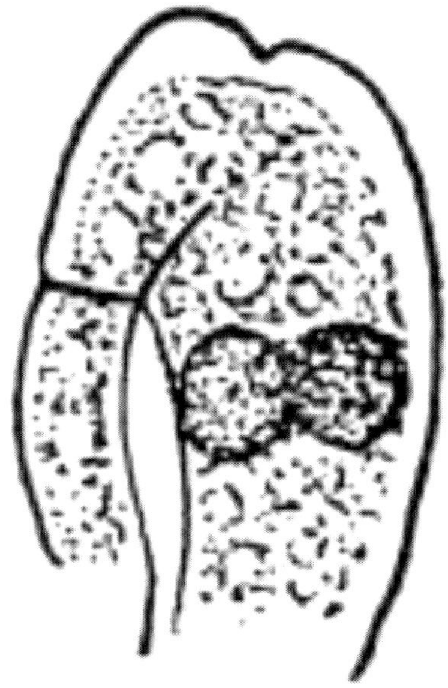

Figure 230. Disappearance of contact walls of antheridium and carpogonium, and fusion of the two nuclei.

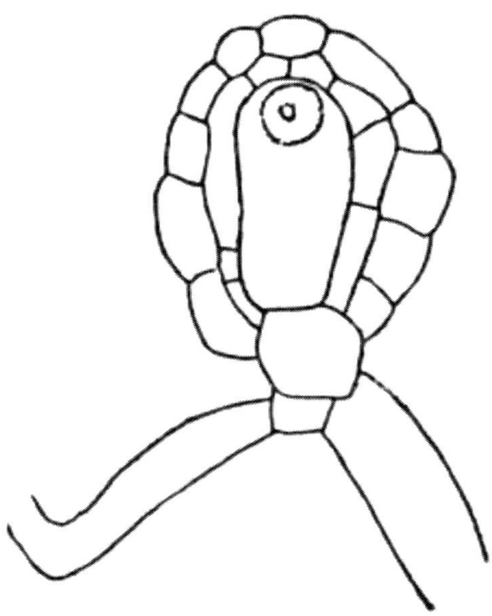

Figure 231. Fertilized egg surrounded by the enveloping threads which grow up around it.

Figures 229-231.—Fertilization in sphærotheca; one of the powdery mildews. (After Harper.)

Number of Spores in an Ascus

The ascus is the most important character showing the general relationship of the members of the sac fungi. While many of the powdery mildews have a variable number of spores in an ascus, a large majority of the ascomycetes have just 8 spores in an ascus, while some have 4, others 16, and some an indefinite number. The asci in a perithecium are more variable. In some ascomycetes there is no perithecium.

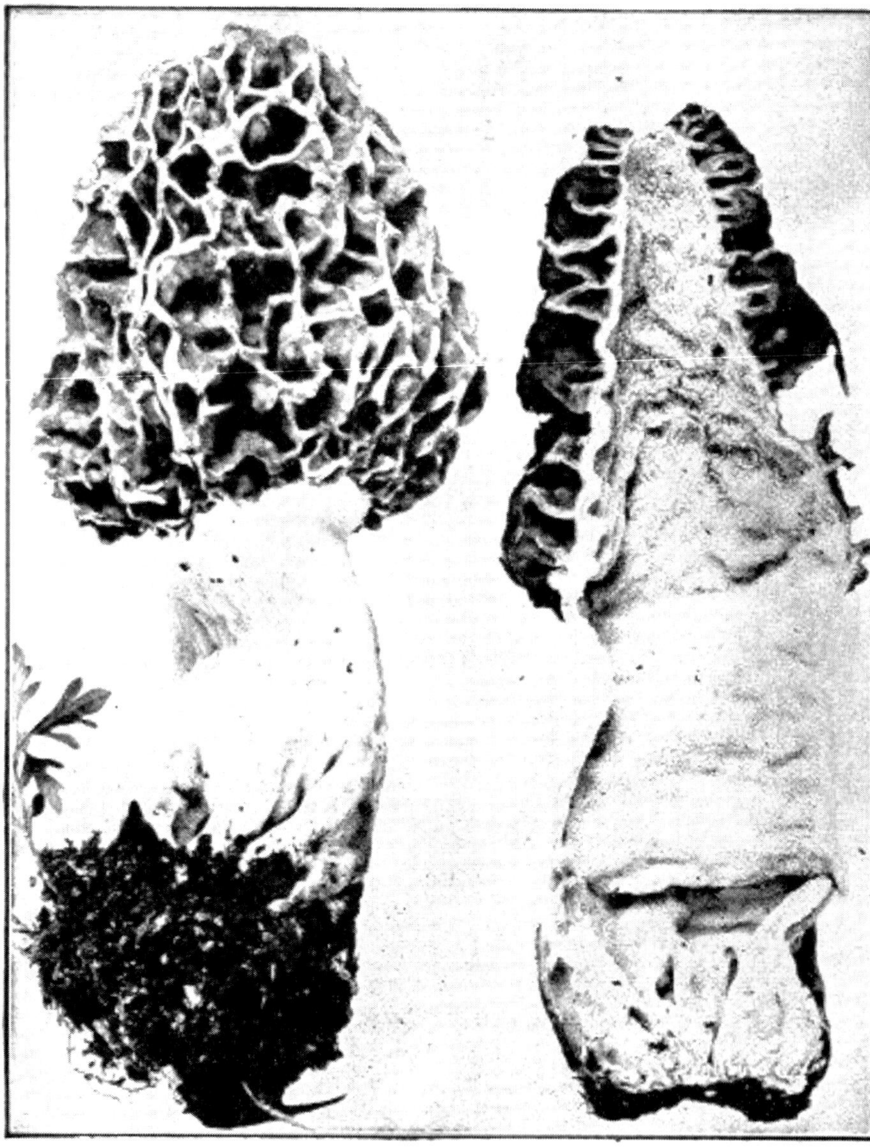

Figure 231*a*. Edible Morel. Morchella esculenta. The asci, forming hymenium, cover the pitted surface.

The Black Fungi

These are very common on dead logs, branches, leaves, etc., and may be collected in the woods at almost any season. The perithecia are often numerous, scattered or densely crowded as in Rosellinia. Sometimes they are united to form a crust which is partly formed from sterile elements as in Hypoxylon, or they form black clavate or branched bodies as in Xylaria. The black knot of the plum and cherry is also an example.

The lichens are mostly ascomycetes like the black fungi or cup fungi, while a few are basidiomycetes.

The Morels (Morchella)

There are several species of morels which are common in early spring on damp ground. Either one of the species is suitable for use if it is desired to include this in the study. Figure 231a illustrates the Morchella esculenta. The stem is cylindrical and stout. The fruiting portion forms the "head," and it is deeply pitted. The entire pitted surface is covered by the asci, which are cylindrical and eight spored. A thin section may be made of a portion for study, or a small piece may be crushed under the cover glass.

The Cup Fungi

These fungi are common on damp ground or on rotting logs in the summer. They may be preserved in 70 per cent alcohol for study. Many of them are shaped like broad open cups or saucers. The inner surface of the cup is the fruiting surface, and is covered with the cylindrical asci, which stand side by side. A bit of the cup may be sectioned or crushed under a cover glass for study.

Mushrooms (Basidiomycetes)

The large group of fungi to which the mushroom belongs is called the basidiomycetes because in all of them a structure resembling a club, or basidium, is present, and bears a limited number of spores, usually four, though in some genera the number is variable. Some place the rusts (Urodineæ) in the same series (basidium series), because of the short promycelium and four sporidia developed from each cell of the teleutospore.

The Gill-Bearing Fungi (Agaricaceæ)

A good example for this study is the common mushroom (*Agaricus campestris*).

This occurs from July to November in lawns and grassy fields. The plant is somewhat umbrella-shaped, as shown in Figure 232, and possesses a cylindrical stem attached to the under side of the convex cap or pileus. On the under side of the pileus are thin radiating plates, shaped somewhat like a knife blade. These are the gills, or lamellæ, and toward the stem they are rounded on the lower angle and are not attached to the stem. The longer ones extend from near the stem to the margin of the pileus, and the V-shaped spaces between them are occupied by successively shorter ones. Around the stem a little below the gills is a collar, termed the ring or annulus.

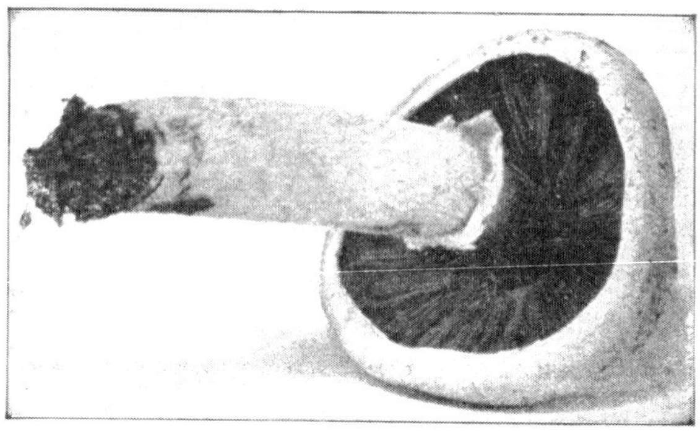

Figure 232. Agaricus campestris. View of under side showing stem, annulus, gills, and margin of pileus.

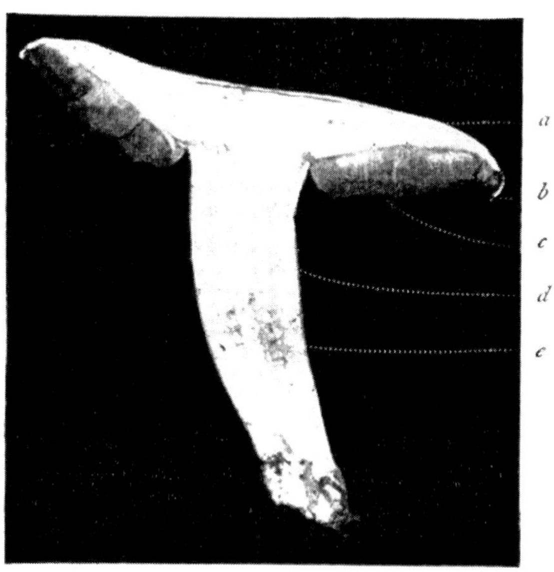

Figure 233. Agaricus campestris. Longitudinal section through stem and pileus. *a*, pileus; *b*, portion of veil on margin of pileus; *c*, gill; *d*, fragment of annulus; *e*, stipe.

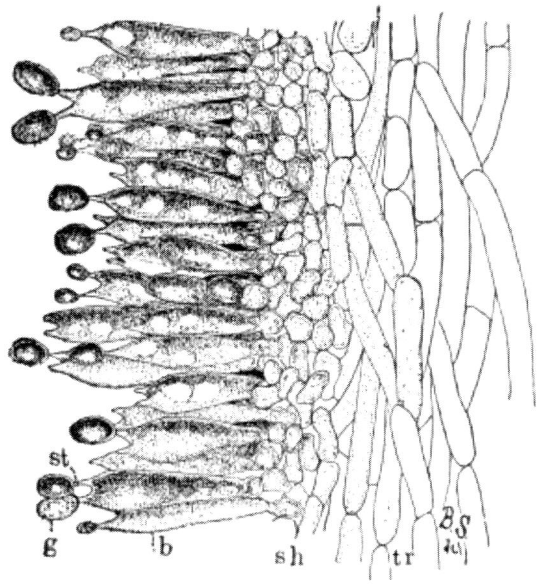

Figure 234. Portion of section of lamella of Agaricus campestris. *tr*, trama; *sh*, subhymenium; *b*, basidium; *st*, sterigma (*pl.* sterigmata); *g*, basidiospore.

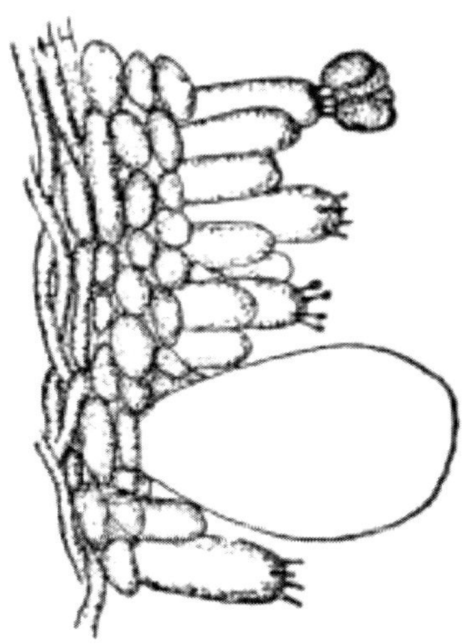

Figure 235. Portion of hymenium of Coprinus micaceus, showing large cystidium in the hymenium.

Fruiting Surface of the Mushroom

The surface of these gills is the fruiting surface of the mushroom, and bears the gonidia of the mushroom, which are dark purplish brown when mature, and thus the gills when old are dark in color. If we make a thin section across a few of the gills, we see that each side of the gill is covered with closely crowded club-shaped bodies, each one of which is a *basidium*. In Figure 234 a few of these are enlarged, so that the structure of the gill can be seen. Each basidium of the common mushroom has two spinous processes at the free end.

Each one is a *sterig'ma* (plural *sterig'mata*), and bears a gonidium. In a majority of the members of the mushroom family each basidium bears four spores. When mature these spores easily fall away, and a mass of them gives a purplish-black color to objects on which they fall, so that a print of the under surface of the cap showing the arrangement of the gills can be obtained by cutting off the stem, and placing the pileus on white paper for a time.

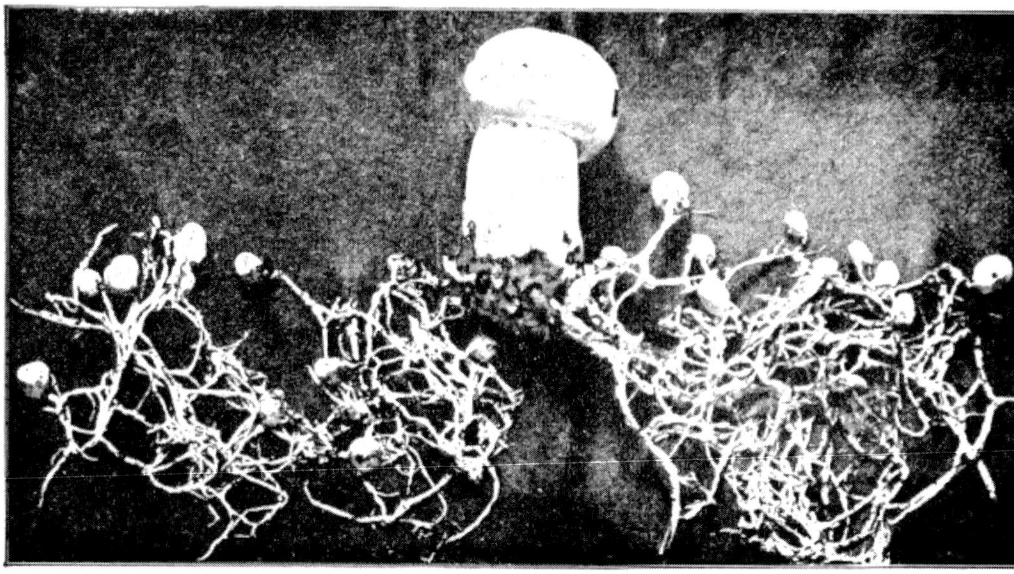

Figure 236. Agaricus campestris. Soil washed from "spawn" and "buttons," showing the minute young "buttons" attached to the strands of mycelium.

How the Mushroom is Formed

The mycelium of the mushroom lives in the ground, and grows here for several months or even years, and at the proper seasons develops the mature mushroom plant. The mycelium lives on decaying organic matter, and a large number of the threads grow closely together forming strands, or cords, of mycelium, which are quite prominent if they are uncovered by removing the soil, as shown in Figure 236.

From these strands the buttons arise by numerous threads growing side by side in a vertical direction, each thread growing independently at the end, but all lying very closely side by side. When the buttons are quite small the gills begin to form on the under margin of the knob. They are formed by certain of the threads growing downward in radiating ridges, just as many of these ridges being started as there are to be gills formed. At the same time, threads of the stem grow upward to meet those at the margin of the button in such a manner that they cover up the forming gills, and thus enclose the gills in a minute cavity. Sections of buttons at different ages will show this, as is seen in Figure 237. This curtain of mycelium which is thus stretched across the gill cavity is the veil. As the cap expands more and more this is stretched into a thin and delicate texture as shown in Figure 238. Finally, as shown in Figure 239, this veil is ruptured by the expansion of the pileus, and it either clings to the stem as a collar, or a portion of it remains clinging to the margin of the cap. When the buttons are very young the gills are white, but they soon become pink in color, and very soon after the veil breaks the spores mature, and then the gills are dark brown.

Figure 237. Agaricus campestris; sections of "buttons" of different sizes, showing formation of gills and veil covering them.

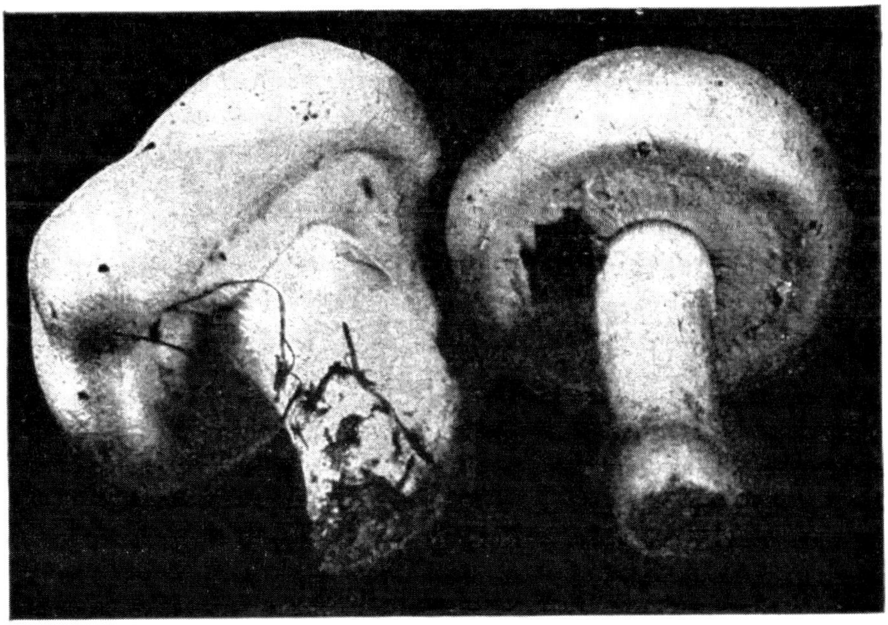

Figure 238. Agaricus campestris; nearly mature plants, showing veil still stretched across the gill cavity.

Figure 239. Agaricus campestris; under view of two plants just after rupture of veil, fragments of the latter clinging both to margin of pileus and to stem.

Figure 240. Agaricus campestris; plant in natural position just after rupture of veil, showing tendency to double annulus on the stem. Portions of the veil also dripping from margin of pileus.

Figure 241. Agaricus campestris; spore print.

Figure 242. "Fairy ring" formed by Agaricus arvensis (photograph by B. M. Duggar). The mycelium spreads centrifugally each year, consuming the available food, and thus the plants appear in a ring.

Figure 243.Amanita phalloides; white form, showing pileus, stipe, annulus, and volva.

Beware of the Poisonous Mushroom

The number of species of mushrooms, or toadstools as they are often called, is very great. Besides the common mushroom (Agaricus campestris) there are a large number of other edible species. But one should be very familiar with any species which is gathered for food, unless collected by one who certainly knows what the plant is, since carelessness in this respect sometimes results fatally from eating poisonous ones.

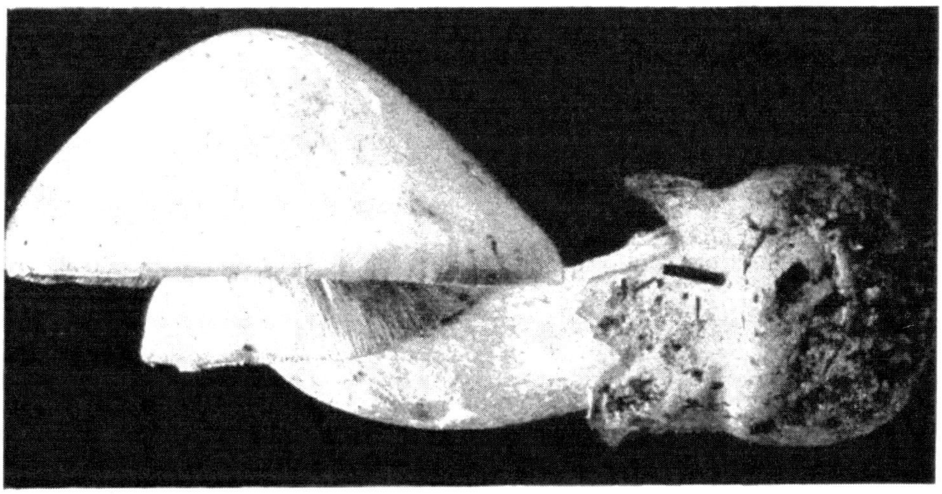

Figure 244. Amanita phalloides; plant turned to one side, after having been placed in a horizontal position, by the directive force of gravity.

A plant very similar in structure to the Agaricus campestris is the Lepiota naucina, but the spores are white, and thus the gills are white, except that in age they become a dirty pink. This plant occurs in grassy fields and lawns often along with the common mushroom. Great care should be exercised in collecting and noting the characters of these plants, for a very deadly poisonous species, the deadly amanita (Amanita phalloides) is perfectly white, has white spores, a ring, and grows usually in wooded places, but also sometimes occurs in the margins of lawns. In this plant the base of the stem is seated in a cup-shaped structure, the *volva*, shown in Figure 243. One should dig up the stem carefully so as not to tear off this volva if it is present, for with the absence of this structure the plant might easily be mistaken for the lepiota, and serious consequences would result.

Tube-Bearing Fungi (Polyporaceæ)

In the tube-bearing fungi, the fruiting surface, instead of lying over the surface of gills, lines the surface of tubes or pores on the under side of the cap. The fruit-bearing portion therefore is "honey-combed." The sulphur polyporus (Polyporus sulphureus) illustrates one form. The tube-bearing fungi are sometimes called "bracket" fungi, or "shelf" fungi, because the pileus is attached to the tree or stump like a shelf or bracket. One very common form in the woods is the plant so much sought by "artists," and often called Polyporus applanatus. It is hard and woody, reddish brown, brown or grayish on the upper side, according to age, and is marked by prominent and large concentric ridges. (This form is probably P. leucophæus.) The under side is white and honey-combed by numerous very minute pores. This plant is perennial, that is, it lives from year to year. Each year a new layer is added to the under side, and several new rings usually to the margin. If a plant two or three years old is cut in two, there will be seen several distinct tube layers or strata, each one representing a year's growth.

Figure 245. Edible Boletus. Boletus edulis. Fruiting surface honey-combed on under side of cap.

In some of these bracket fungi, each ring on the upper surface marks a year's growth as in the pine polyporus (P. pinicola). In the birch polyporus (P. fomentarius) the tubes are quite large. It also occurs on other trees. The beech polyporus (P. igniarius, also on other trees) often becomes very old. I have seen one specimen over eighty years old. Not all the tube-bearing fungi are bracket form. Some have a stem and cap (see Figure 245). Some are spread on the surface of logs.

Figure 246. Coral fungus. Hydnum coralloides, spines hanging down from branches.

Hedgehog Fungi (Hydnaceæ)

These plants are bracket in form or have a stem and cap, or are spread on the surface of wood; but the finest specimens resemble coral masses of fungus tissue (example, Hydnum, Figure 246). In most of them there are slender processes resembling teeth, spines or awls, which depend from the under surface (Figure 247). The fruiting surface covers these spines.

Coral Fungi or Fairy Clubs (Clavariaceæ)

These plants stand upright from the wood, leaves, or soil, on which they grow (example, Clavaria). The "coral" ones are branched, while the "fairy clubs" are simple. The fruiting surface covers the entire exposed surface of the plants (Figure 248).

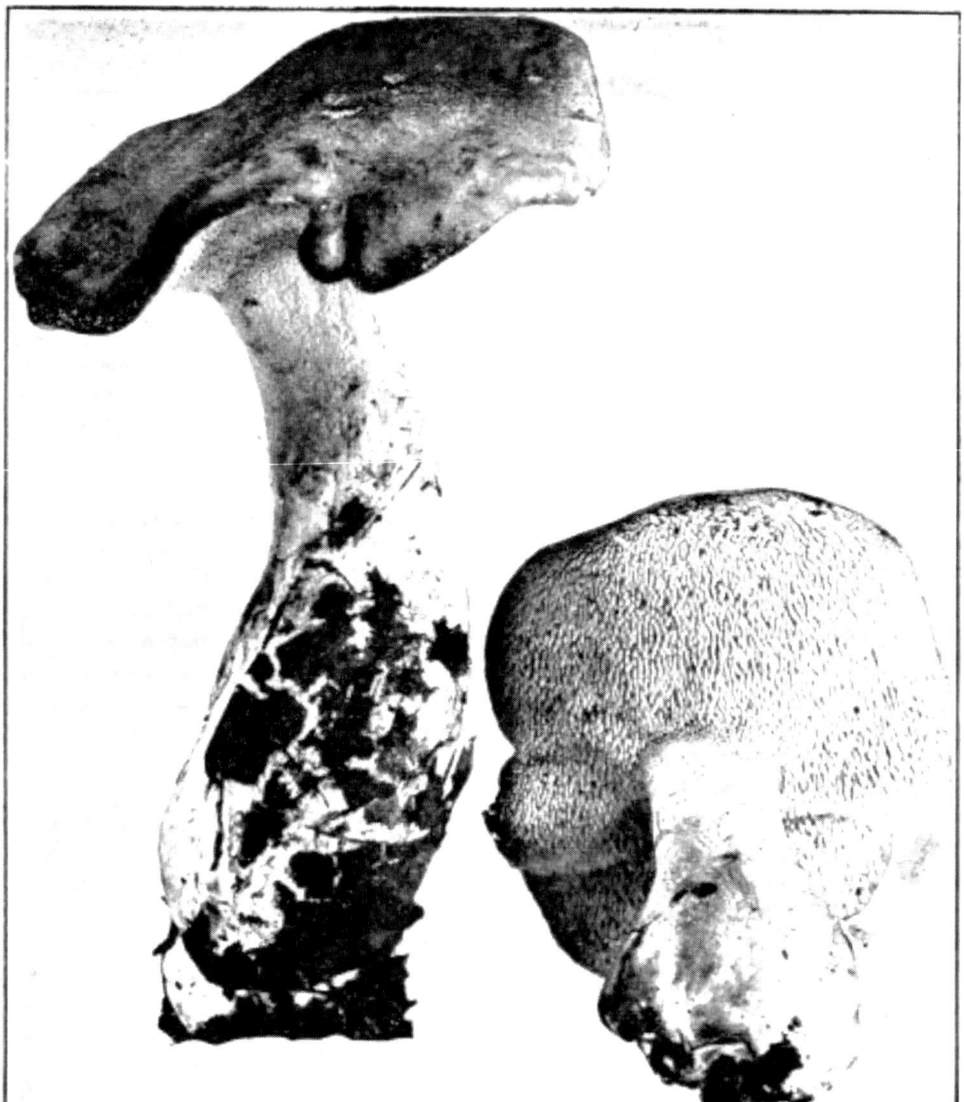

Figure 247. Hydnum repandum, spines hanging down from under side of cap.

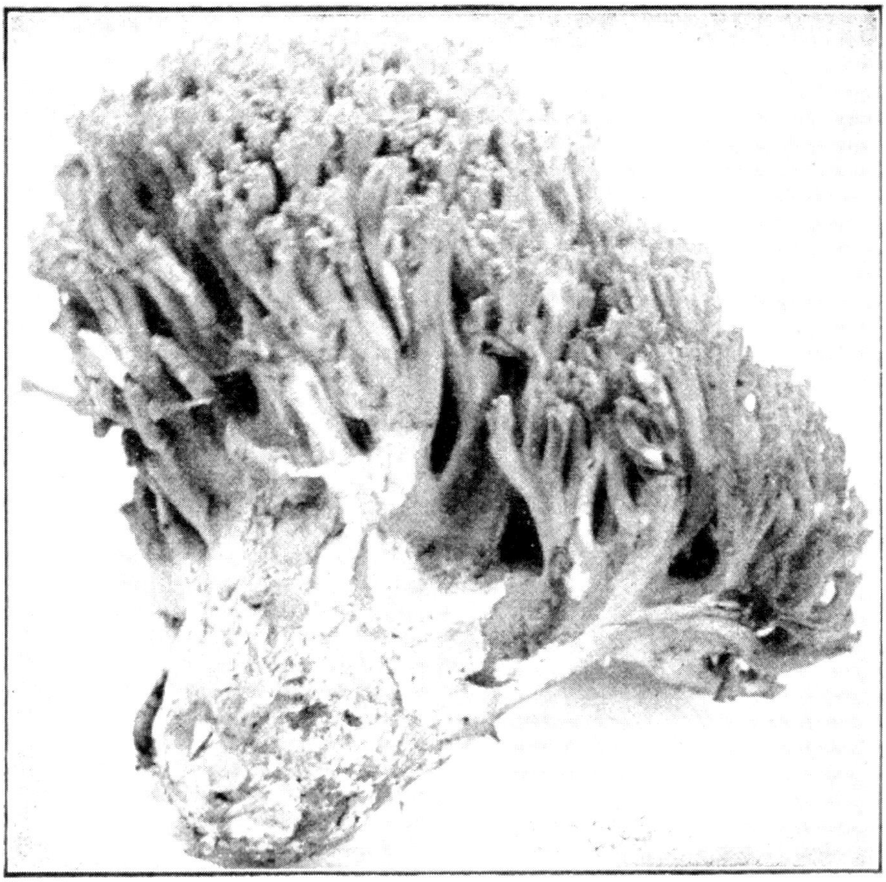

Figure 248. Clavaria botrytes.

Chapter 22

CLASSIFICATION OF THE FUNGI

CLASSIFICATION OF THE FUNGI

Those who believe that the fungi represent a natural group of plants arrange them in three large series related to each other somewhat as follows:

- The Gonidium Type or Series. The number of gonidia in the sporangium is indefinite and variable. It may be very large or very small, or even only one in a sporangium. To this series belong the lower fungi; examples: mucor, saprolegnia, peronospora, etc.
- The Basidium Type or Series. The number of gonidia on a basidium is limited and definite, and the basidium is a characteristic structure; examples: uredineæ (rusts), mushrooms, etc.
- The Ascus Type or Series. The number of spores in an ascus is limited and definite, and the ascus is a characteristic structure; examples: leaf curl of peach (exoascus), powdery mildews, black knot of plum, black rot of grapes, etc.

Others believe that the fungi do not represent a natural group, but that they have developed off from different groups of the algæ by becoming parasitic. As parasites they no longer needed chlorophyll, and consequently lost it.

According to this view the lower fungi have developed off from the lower algæ (saprolegnias, mucors, peronosporas, etc., being developed off from siphonaceous algæ like vaucheria), and the higher fungi being developed off from the higher algæ (the ascomycetes perhaps from the Rhodophyceæ).

A very general outline of classification,[19] according to the former of these views, might be presented here to show the general relationships of the fungi studied, with the addition of a few more in orders not represented above. It should be borne in mind that the author in presenting this view of classification does not necessarily commit himself to it. It is based on that presented in Engler & Prantl's Pflanzenfamilien. There are three classes.

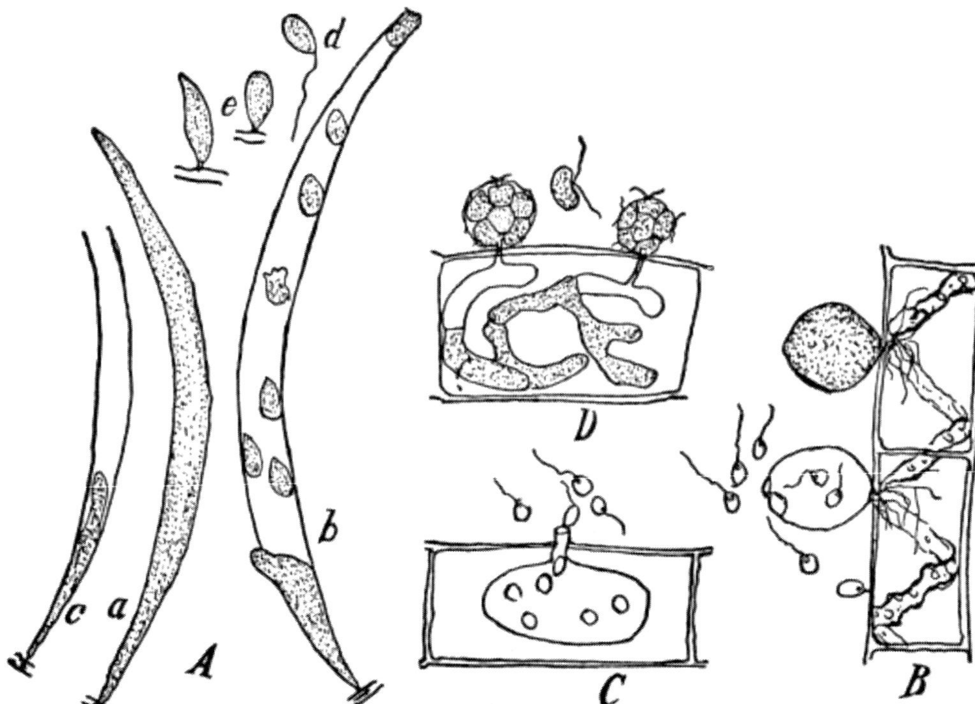

Figure 249. Chytrids. *A*, Harpochytrium hedenii, parasitic on spirogyra threads; *a*, sickle-form plant; *b*, the sporangium part with escaping zoospores; *c*, old plant proliferating by forming new sporangium in the old empty one; *d*, zoospore; *e*, two young plants just beginning to grow. *B*, Rhizophidium globosum parasitic on spirogyra. Globose sporangium with delicate threads inside of the host, zoospores escaping from one. *C*, Olpidium pendulum, parasitic in spirogyra cell. Elliptical sporangium with slender exit tube through which zoospores are escaping. *D*, Lagenidium rabenhorstii parasitic in spirogyra cell. Two slender sporangia with exit tubes through which protoplasm escapes forming a rounded mass at the end of tube, this protoplasm forming biciliate zoospores.

[19] Class Myxomycetes, or Mycetozoa.—To this class belong the "slime molds," low organisms consisting of masses of naked protoplasm which flows among decaying leaves and in decaying wood, coming to the surface to fruit. The fruit in many cases resembles miniature puff-balls, and these plants were formerly classed with the puff-balls. The spores germinate by forming swarm spores which unite to form a small plasmodium, which in turn grows to form a large plasmodium or protoplasmic mass. It is doubtful if they are any more plant than animal organisms. Examples: Trichia, Arcyria, Stemonitis, Physarum, Ceratiomyxa, etc., on rotten wood; Plasmodiophora brassicæ is a parasite causing club foot of cabbage, radishes, etc. It lives within the roots, causing large knots and swellings on the same.

Class Phycomycetes (Alga-like Fungi)

Subclass Oomycetes

These are the egg-spore fungi. They include the water mold (Saprolegnia), the downy mildew of the grape (Plasmopara), the potato blight (Phytophthora), the white rust of cruciferous plants (Cystopus = Albugo), the damping-off fungus (Pythium), and many parasites of the algæ known as chytrids, as Olpidium, Rhizophidium, Lagenidium, Chytridium, etc.

The two following orders are sometimes placed in a separate subclass, Archimycetes.

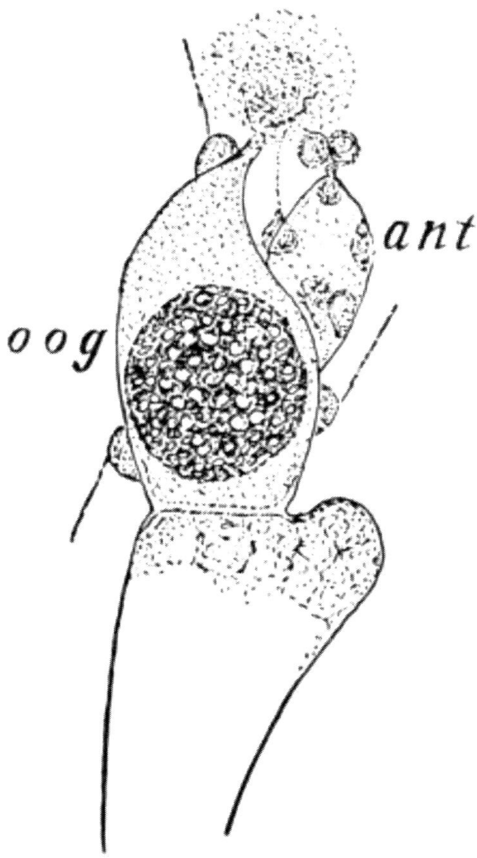

Figure 250. Monoblepharis insignis Thaxter. End of hypha bearing oogonium (*oog*) and antheridium (*ant*). Sperms escaping from antheridium and creeping up on the oogonium. (After Thaxter.)

Order Chytridiales (Chytridineæ)

These include the lowest fungi. Many of them are parasitic on algæ and lack mycelium, the swarm spore either with or without minute rhizoids, developing into a globose sporangium (Rhizophidium, Chytridium, Olpidium, etc., Figure 249), or the swarm spore attached to the wall of the host develops into a long sword-shaped body with a sterile base, which proliferates and forms a new sporangium in the old one (Harpochytrium), or with slight development of mycelium in aquatic plants (Cladochytrium). Some are parasitic in leaves and stems of land plants. Synchytrium decipiens is very common on the trailing legume, Amphicarpæa monoica.

Order Ancylistales (Ancylistineæ)

The members of this order have a slight development of mycelium and many are parasitic in algæ (Lagenidium, Figure 249).

Order Saprolegniales (Saprolegniineæ)

These include the water molds (Saprolegnia). See Chapter XIX.

Order Monoblepharidales (Monoblepharidineæ)

These are peculiar water molds, related to the Saprolegniales, but motile sperm cells are formed (Monoblepharis, etc., Figure 250).

Order Peronosporales (Peronosporineæ)

These include the downy mildews (Peronospora, Plasmopara, Phytopthora, etc.), and the white rust of crucifers and other plants (Cystopus = Albugo), Chapter XIX.

Subclass Zygomycetes

These Are the Conjugating Fungi

Order Mucorales (Mucorineæ).—This includes the black mold and its many relatives (Mucor, Rhizopus, etc.). Chapter XIX.

Order Entomophthorales (Entomophthorineæ)

This order includes the “fly fungus” (Empusa) and its many relatives parasitic on insects. In the autumn and winter dead flies are often found stuck to window-panes, with a white ring of the conidia around each fly.

Class Ascomycetes. (The Ascus Series)

Subclass Hemiascomycetes

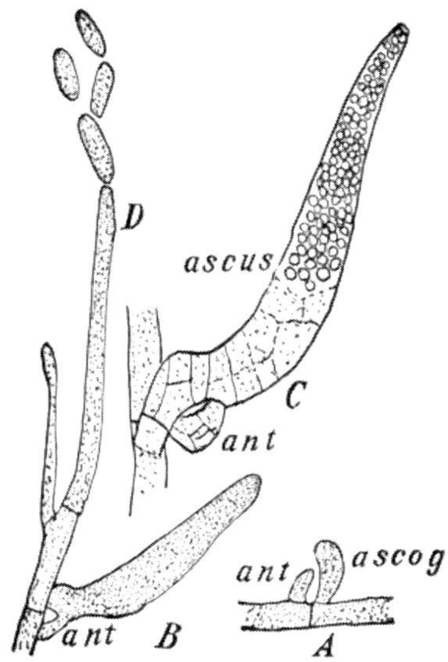

Figure 251. Dipodascus albidus. *A*, thread with sexual organs, ascogonium and antheridium; *B*, fertilized ascogonium developing ascus; *C*, ascus with spores; *D*, conidia. (After Lagerheim.)

Order Hemiascales (Hemiascineæ)

Fungi with a well developed, septate mycelium, but with a sporangium-like ascus, i.e., a large and indefinite number of spores in the ascus. Examples: Protomyces macrosporus in stems of Umbelliferæ, or P. polysporus in Ambrosia trifida. These two are by some placed in the Ustilagineæ. Dipodascus albidus grows in the exuding sap of Bromeliaceæ in Brazil and the sap of the beech in Sweden. The ascus is developed as the result of the fertilization of an ascogonium with an antheridium (see Figure 251).

Subclass Protoascomycetes

The asci are well-defined and usually with a limited and definite number of spores (usually 8, sometimes 1, 2, 4, 16, or more). Mycelium often well developed and septate. Asci scattered on the mycelium, not associated in definite fields or groups.

Order Protoascales (Protoascineæ).—The asci are separate cells, or are scattered irregularly in loose wefts of mycelium. No fruit body. (The yeast, Saccharomyces; and certain mold-like fungi, some of which are parasitic on mushrooms, as Endomyces, are examples.)

Subclass Euascomycetes

Asci associated in surfaces forming a hymenium, or in groups or intermingled in the elements of a fruit body. Fruit body usually present.

The following four or five orders comprise the Discomycetes, according to the usual classification.

Order Protodiscales (Protodiscineæ)

The asci are exposed and form large and indefinite groups, but there is no definite fruit body. Examples: leaf curl of peach, plum pocket, etc. (Exoascus).

Order Helvellales (Helvellineæ)

The asci form large fields over the upper portion of the fruit body. This order includes the morels (Figure 231*a*), helvellas, earth tongues (Geoglossum), etc.

Order Pezizales (Pezizineæ)

The asci form a definite field or fruiting surface surrounded on the sides and below by a wall of fungus tissue, forming a fruit body in the shape of a cup. These are known as the cup fungi (Peziza, Lachnea, etc.).

Order Phacidiales (Phacidiineæ)

Fungi mostly saprophytic, and fruit body similar to the cup fungi. Examples: Propolis in rotting wood, Rhytisma forming black crusts on leaves (maple for example), Urnula craterium, a large black beaker-shaped fungus on the ground.

Order Hysteriales (Hysteriineæ)

Fungi with a more or less elongated fruit body with an enclosing wall opening by a long slit. In some forms the fruit body has the appearance of a two-lipped body; in others it is shaped like a clam shell, the asci being inside. Example, Hysterographium common on dry, dead, decorticated sticks.

Order Tuberales (Tuberineæ)

The more or less rounded fruit bodies are usually subterranean. The most important fungi in this order are the truffles (Tuber). The mycelium of many species assists in the formation of mycorhiza on the roots of oaks, etc., and several species are partly cultivated, or protected, and collected for food. This is especially the case with Tuber brumale and its forms; more than a million francs worth of truffles are sold in France and Italy yearly. Dogs and pigs are employed in the collection of truffles from the ground.

Order Plectascales (Plectascineæ)

The fruit body of these plants is more or less globose, and contains the asci distributed irregularly through the mycelium of the interior. Some are subterranean (Elaphomyces), while others grow in decaying plants, or certain food substances (Eurotium, Sterigmatocystis, Penicillium). Penicillium in its conidial stage forms blue mold on fruit, bread, etc.

The following four orders comprise the Pyrenomycetes, according to the usual classification.

Order Perisporiales

The powdery mildews are good examples of this order (Uncinula, Microsphæra, etc., Chapter XXI).

Order Hypocreales[20]

The fruit bodies are colorless, or bright colored and entirely enclose the asci, sometimes opening by an apical pore. Nectria cinnabarina has clusters of minute orange oval fruit bodies, and is common on dead twigs. Cordyceps with a number of species is parasitic on insects, and on certain subterranean Ascomycetes, especially Elaphomyces (of the order *Plectascales* = *Plectascineæ*).

Order Dothidiales[21]

Fungi with black stroma formed of mycelium in which are cavities containing the asci. The cavities are usually shaped like a perithecium, but there is no wall distinct from the tissue of the stroma (Dothidea, Phyllachora, on grasses).

Order Sphæriales[22]

These contain the so-called black fungi, with separate or clustered, oval, fruit bodies, black in color. The black wall encloses the asci, and usually opens by an apical pore. Examples are found in the black knot of plum and cherry, black rot of grapes, and in Rosellinia, Hypoxylon, Xylaria, etc., on dead wood.

Order Laboulbeniales (Laboulbineæ)

These are peculiar fungi attached to the legs and bodies of insects by a short stalk, and provided with a sac-like fruit body which contains the asci. Example, Laboulbenia.

---

[20] As suborder in Engler and Prantl.
[21] As suborder in Engler and Prantl.
[22] As suborder in Engler and Prantl.

Class Basidiomycetes. (The Basidium Series)

Subclass Hemibasidiomycetes

Order Ustilaginales (Ustilagineæ)

This order includes the well-known smuts on corn, wheat, oats, etc. (Ustilago, Tilletia, etc.).

Subclass Æcidiomycetes

Order Uredinales[23] (Uredineæ)

This order includes the parasitic fungi known as rusts. Examples: wheat rust (Chapter XX), the cedar apple, etc.

The true Basidiomycetes include the following orders:

Subclass Protobasidiomycetes

Order Auriculariales[24]

This order includes trembling fungi in which the basidium is long and divided transversely into usually four cells (example, Auricularia), and similar forms. Pilacre petersii on dead wood represents an angiocarpous form.

Order Tremellales (Tremellineæ)

Trembling or gelatinous fungi with the globose basidium divided longitudinally into four cells (Tremella).

Subclass Eubasidiomycetes

Order Dacryomycetales (Dacryomycetineæ)

This order includes certain fungi of a gelatinous or waxy consistency, usually of bright colors. They resemble the Tremellales, but the basidia are slender and fork into two long sterigmata. (Example, Dacryomyces.) Gyrocephalus rufus is quite a large plant, 10-15 cm. high, growing on the ground in woods.

---

[23] The Uredinales and Auriculariales in Engler and Prantl are placed in order, Auriculariineæ.

[24] The Uredinales and Auriculariales in Engler and Prantl are placed in order, Auriculariineæ.

Order Exobasidiales (Exobasidiineæ)

The fungus causing azalea apples is an example (Exobasidium).

Order Hymeniales (Hymenomycetineæ)

In this order the basidia are usually club-shaped and undivided, and bear usually four spores on the end (sometimes two or six). There are several families.

Family Thelephoraceæ

The fruit bodies are more or less membranous and spread over wood or the ground, or somewhat leaf-like, growing on wood or the ground. The fruiting surface is nearly or quite even, and occupies the under side of the leaf-like bodies (Stereum, Thelephora) or the outside of the forms spread out on wood (Corticium, Coniophora).

Family Clavariaceæ

This order includes the fairy clubs, and some of the coral fungi. The larger number of species are in one genus (Clavaria, Figure 248).

Family Hydnaceæ

The fungi of this order are known as "hedgehog" fungi, because of the numerous awl-like teeth or spines over which the fruiting surface is spread, as in Hydnum (Figures 246, 247).

Family Polyporaceæ

The tube-bearing fungi (Polyporus, Boletus, etc., Figure 245).

Family Agaricaceæ

The gill-bearing fungi (Agaricus, Amanita, etc., see Chapter XXI).

The above five orders, according to the earlier classification (still used at the present time by some), made up the order Hymenomycetes, while the following five orders made up the Gasteromycetes. The Hymenomycetes, according to this system, included those plants in which the fruiting portion (hymenium) is either exposed from the first, or if covered by a veil or volva (as in Agaricus, Amanita, etc.) this ruptures and exposes the fruiting surface before, or at the time of, the ripening of the spores, while the Gasteromycetes included those in which the fruit body is closed until after the maturity of the spores.

Order Phallales (Phallineæ)

The "stink-horn" fungi, or "buzzard's nose." Usually foul-smelling fungi, the fruiting portion borne aloft on a stout stalk, and dissolving (Dictyophora, Ithyphallus, etc.).

Order Hymenogastrales (Hymenogastrineæ)

The basidia form a distinct hymenium which does not break down at maturity. Some of the plants resemble Boletus or Agaricus in the way the fruit bodies open (Secotium, etc.), while others open irregularly on the surface (Rhizopogon) or like an earth star (Sclerogaster), or portions of the surface become gelatinized (Phallogaster). The last-named one grows on very rotten wood, while most of the others grow on the ground.

Order Lycoperdales (Lycoperdineæ)

These include the "puff-balls," or "devil's snuff-box" (Lycoperdon), and the earth stars (Geaster). The basidia form a distinct hymenium, but at maturity the entire inner portion of the plant (except certain peculiar threads, the capillitium) disintegrates and with the spores forms a powdery mass.

Order Nidulariales (Nidulariineæ)

These are known as bird-nest fungi. The fruit body when mature is cup-shaped, or goblet-shaped, and contains minute flattened circular bodies (peridiola) containing the spores. The intermediate portions of the fruit body disintegrate and set the peridiola free, which then lie in the cup-shaped base like eggs in a nest.

Order Plectobasidiales (Plectobasidiineæ)

The basidia do not form a definite hymenium, but are interwoven with the threads inside, or are collected into knot-like groups. (Examples: Calostoma, Tulostoma, Astræus, Sphærobolus, etc.)

Lichens

The plant body of the lichens (see paragraphs 200, 201) consists of two component parts, the one a fungus, the other an alga. The fructification is that of the fungus. The fruit body shows the lichens to be related some to the Ascomycetes, others to the Hymenomycetes, and Gasteromycetes. They are usually classified as a distinct class or order from the fungi, but a natural arrangement would distribute them in several of the orders above. Their special relationship with these orders has not been satisfactorily worked out. For the present they are arranged as follows:

- Ascolichenes.
- *Pyrenocarpous lichens* (those with a fruit body like the Pyrenomycetes).
- *Gymnocarpous lichens* (those with a fruit body like the Discomycetes).
- Hymenolichenes (those with a fruit body like the Hymenomycetes).
- Gasterolichenes (those with a fruit body like the Gasteromycetes).

From a vegetative standpoint there are two types according to the distribution of the elements.

1st. Where the fungal and algal elements are evenly distributed in the plant body the lichen is said to be *homoiomerous*. There are two types of these:

a) *Filamentous lichens*, example, Ephebe pubescens.
b) *Gelatinous lichens*, example, Collema (with the alga nostoc), Physma (with the Chroococcaceæ).
c) 2d. Where the elements are stratified, as in Parmelia, etc., the lichen is said to be *heteromerous*. In these there are three types:
    a. *Crustaceous lichens*, the plant body is in the form of a thin incrustation on rocks, etc.
    b. *Foliaceous lichens*, the plant body is leaf-like and lobed and more or less loosely attached by rhizoids: Parmelia, Peltigera, etc.
    c. *Fruticose lichens*, the plant body is filamentous or band-like and branched, as in Usnea, Cladonia, etc.

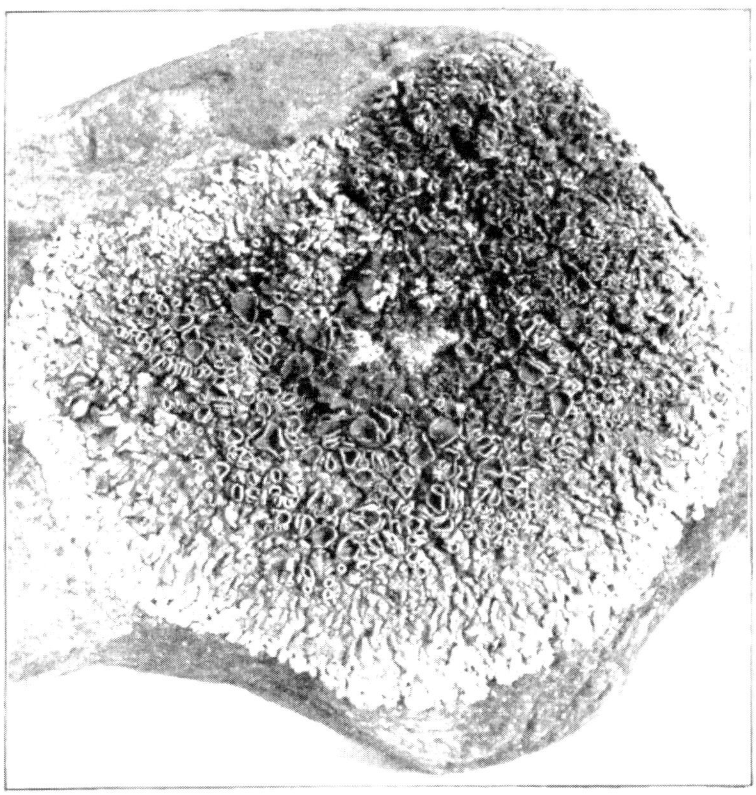

Figure 251*a*. Rock lichen (Parmelia contigua).

Chapter 23

LIVERWORTS (HEPATICÆ)

We come now to the study of representatives of another group of plants, a few of which we examined in studying the organs of assimilation and nutrition. I refer to what are called the liverworts. Two of these liverworts belonging to the genus riccia are illustrated in Figures 30, 252.

RICCIA

Form of the floating riccia (R. fluitans).—The general form of floating riccia is that of a narrow, irregular, flattened, ribbon-like object, which forks repeatedly, in a dichotomous manner, so that there are several lobes to a single plant. It receives its name from the fact that at certain seasons of the year it may be found floating on the water of pools or lakes. When the water lowers it comes to rest on the damp soil, and rhizoids are developed from the under side. Now the sexual organs, and later the fruit capsule, are developed.

Form of the Circular Riccia (*R. crystallina*)

The circular riccia is shown in Figure 252. The form of this one is quite different from the floating one, but the manner of growth is much the same. The branching is more compact and even, so that a circular plant is the result. This riccia inhabits muddy banks, lying flat on the wet surface, and deriving its soluble food by means of the little rootlets (rhizoids) which grow out from the under surface.

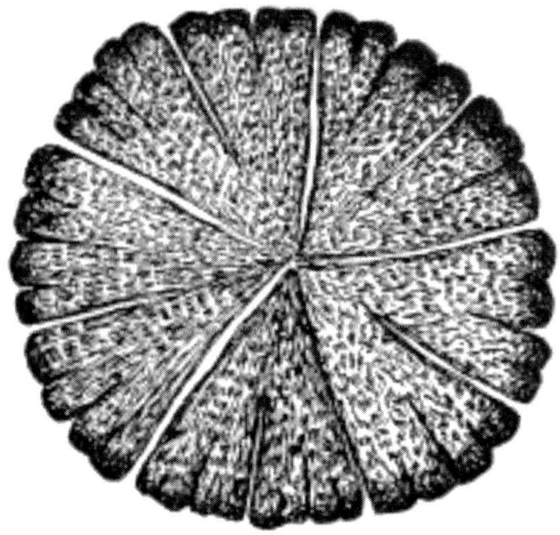

Figure 252. Thallus of Riccia crystallina.

Here and there on the margin are narrow slits, which extend nearly to the central point. They are not real slits, however, for they were formed there as the plant grew. Each one of these V-shaped portions of the thallus is a lobe, and they were formed in the young condition of the plant by a branching in a forked manner. Since growth took place in all directions radially the plant became circular in form. These large lobes we can see are forked once or twice again, as shown by the seeming shorter slits in the margin.

Sexual Organs

In order to study the sexual organs we must make thin sections through one of these lobes lengthwise and perpendicular to the thallus surface. These sections are mounted for examination with the microscope.

Archegonia

We are apt to find the organs in various stages of development, but we will select one of the flask-shaped structures shown in Figure 253 for study. This flask-shaped body we see is entirely sunk in the tissue of the thallus. This structure is the female organ, and is what we term in these plants the *archegonium*. It is more complicated in structure than the oogonium. The lower portion is enlarged and bellied out, and is the venter of the archegonium, while the narrow portion is the neck. We here see it in section. The wall is one cell layer in thickness. In the neck is a canal, and in the base of the venter we see a large rounded cell with a distinct and large nucleus. This cell is the *egg* cell.

Antheridia

The antheridia are also borne in cavities sunk in the tissue of the thallus. There is here no illustration of the antheridium of this riccia, but Figure 259 represents an antheridium of another liverwort, and there is not a great difference between the two kinds. Each one of those little rectangular sperm mother cells in the antheridium changes into a swiftly moving body like a little club with two long lashes attached to the smaller end. By the violent lashing of these organs the spermatozoid is moved through the water, or moisture which is on the surface of the thallus. It moves through the canal of the archegonium neck and into the egg, where it fuses with the nucleus of the egg, and thus fertilization is effected.

Embryo

In the plants which we have selected thus far for study, the egg, immediately after fecundation, we recollect, passed into a resting state, and was enclosed by a thick protecting wall. But in riccia, and in the other plants of the group which we are now studying, this is not the case. The egg, on the other hand, after acquiring a thin wall, swells up and fills the cavity of the venter.

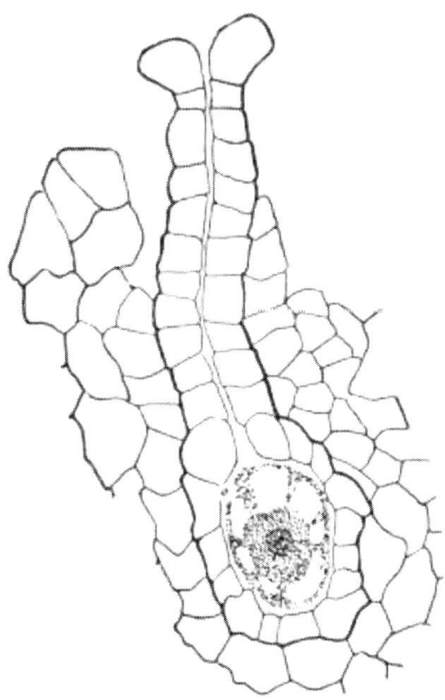

Figure 253. Archegonium of riccia, showing neck, venter, and the egg; archegonium is partly surrounded by the tissue of the thallus. (Riccia crystallina.)

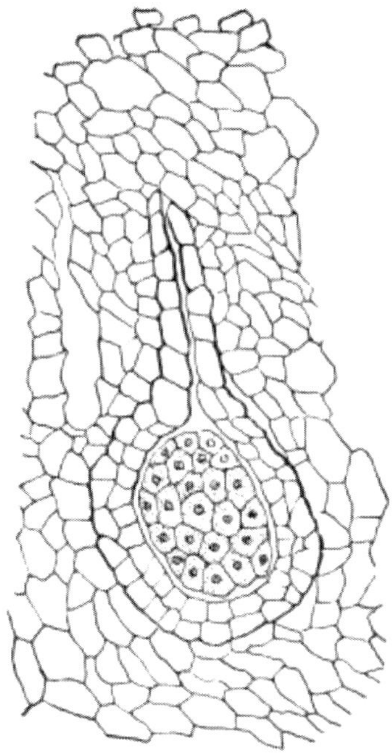

Figure 254. Young embryo (sporogonium) of riccia, within the venter of the archegonium; the latter has now two layers of cells. (*Riccia crystallina.*)

Then it divides by a cross wall into two cells. These two grow, and divide again, and so on until there is formed a quite large mass of cells rounded in form and still contained in the venter of the archegonium, which itself increases in size by the growth of the cells of the wall.

Sporogonium of Riccia

The fruit of riccia, which is developed from the fertilized egg in the archegonium, forms a rounded capsule still enclosed in the venter of the archegonium, which grows also to provide space for it. Therefore a section through the plant at this time, as described for the study of the archegonium, should show this capsule. The capsule then is a rounded mass of cells developed from the egg. A single outer layer of cells forms the wall, and therefore is sterile. All the inner cells, which are richer in protoplasm, divide into four cells each. Each of these cells becomes a spore with a thick wall, and is shaped like a triangular pyramid whose sides are of the same extent as the base (tetrahedral). These cells formed in fours are the *spores*. At this time the wall of the spore-case dissolves, the spores separate from each other and fill the now enlarged venter of the archegonium. When the thallus dies they are liberated, or escape between the loosely arranged cells of the upper surface.

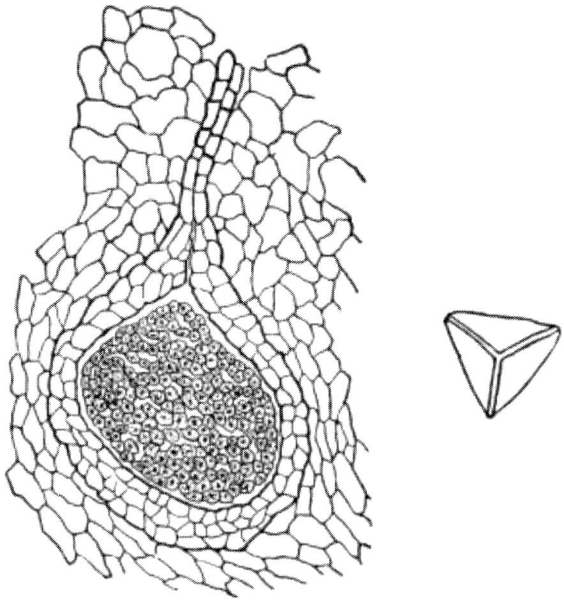

Figure 255. Nearly mature sporogonium of Riccia crystallina; mature spore at the right.

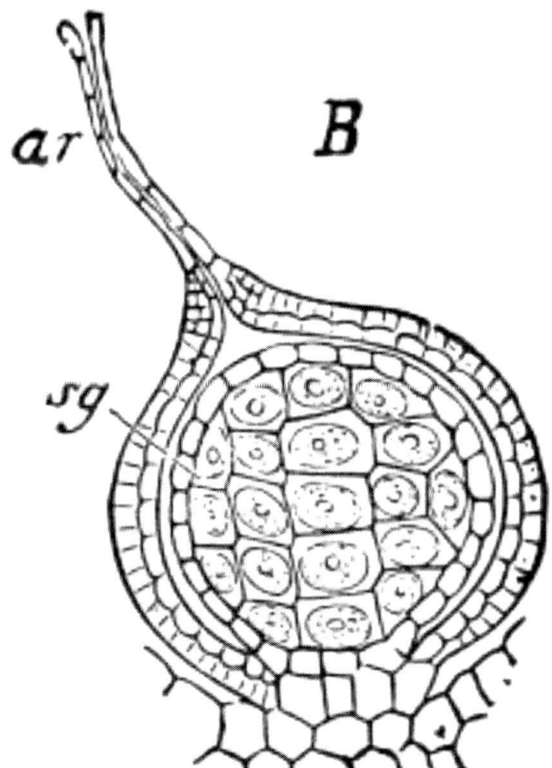

Figure 256. Riccia glauca; archegonium containing nearly mature sporogonium. *sg*, spore-producing cells surrounded by single layer of sterile cells, the wall of the sporogonium.

A New Phase in Plant Life

Thus we have here in the sporogonium of *riccia* a very interesting phase of plant life, in which the egg, after fertilization, instead of developing directly into the same phase of the plant on which it was formed, grows into a quite new phase, the sole function of which is the development of spores. Since the form of the plant on which the sexual organs are developed is called the *gametophyte*, this new phase in which the spores are developed is termed the *sporophyte*.

Now the spores, when they germinate, develop the *gametophyte*, or thallus, again. So we have this very interesting condition of things, the thallus (gametophyte) bears the sexual organs and the unfertilized egg. The fertilized egg, starting as it does from a single-celled stage, develops the sporogonium (sporophyte). Here the single-cell stage is again reached in the spore, which now develops the thallus.

Riccia Compared with Coleochæte, Œdogonium, Etc.

We have said that in the sporogonium of riccia we have formed a new phase in plant life. If we recur to our study of coleochæte we may see that there is here possibly a state of things which presages, as we say, this new phase which is so well formed in riccia. We recollect that after the fertilized egg passed the period of rest it formed a small rounded mass of cells, each of which now forms a zoospore. The zoospore in turn develops the normal thallus (gametophyte) of the coleochæte again. In coleochæte then we have two phases of the plant, each having its origin in a one-celled stage. Then if we go back to œdogonium, we remember that the fertilized egg, before it developed into the œdogonium plant again (which is the gametophyte), at first divides into *four* cells which become zoospores. These then develop the œdogonium plant.

Note. Too much importance should not be attached to this seeming homology of the sporophyte of œdogonium, coleochæte, and riccia, for the nuclear phenomena in the formation of the zoospores of œdogonium and coleochæte are not known. They form, however, a very suggestive series.

Marchantia

The marchantia (M. polymorpha) has been chosen for study because it is such a common and easily obtained plant, and also for the reason that with comparative ease all stages of development can be obtained. It illustrates also very well certain features of the structure of the liverworts.

Figure 257. Male plant of marchantia bearing antheridiophores.

The plants are of two kinds, male and female. The two different organs, then, are developed on different plants. In appearance, however, before the beginning of the structures which bear the sexual organs they are practically the same. The thallus is flattened like nearly all of the thalloid forms, and branches in a forked manner. The color is dark green, and through the middle line of the thallus the texture is different from that of the margins, so that it possesses what we term a midrib, as shown in Figures 257, 261. The growing point of the thallus is situated in the little depression at the free end. If we examine the upper surface with a hand lens we see diamond-shaped areas, and at the center of each of these areas are the openings known as the stomates.

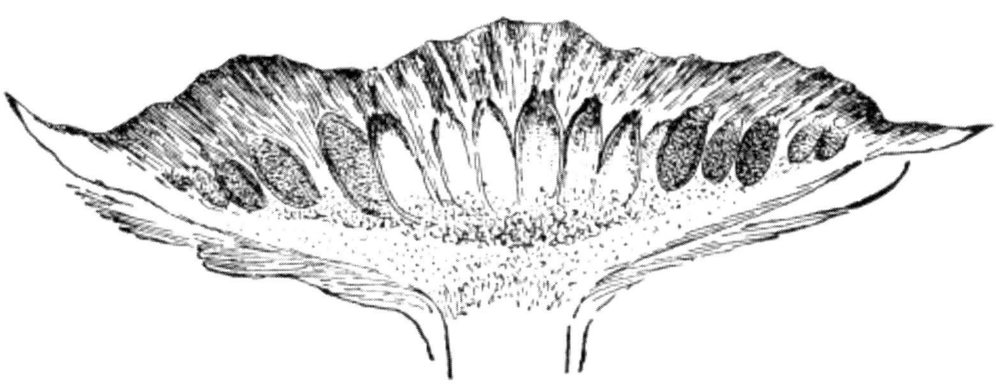

Figure 258. Section of antheridial receptacle from male plant of Marchantia polymorpha, showing cavities where the antheridia are borne.

Antheridial Plants

One of the male plants is Figured at 257. It bears curious structures, each held aloft by a short stalk. These are the antheridial receptacles (or male gametophores). Each one is circular, thick, and shaped somewhat like a biconvex lens. The upper surface is marked by radiating furrows, and the margin is crenate. Then we note, on careful examination of the upper surface, that there are numerous minute openings. If we make a thin section of this structure perpendicular to its surface we shall be able to unravel the mystery of its interior. Here we see, as shown in Figure 258, that each one of these little openings on the surface is an entrance to quite a large cavity. Within each cavity there is an oval or elliptical body, supported from the base of the cavity on a short stalk. This is an antheridium, and one of them is shown still more enlarged in Figure 259. This shows the structure of the antheridium, and that there are within several angular areas, which are divided by numerous straight cross-lines into countless tiny cuboidal cells, the *sperm mother cells*. Each of these, as stated in the former chapter, changes into a swiftly moving body resembling a serpent with two long lashes attached to its tail.

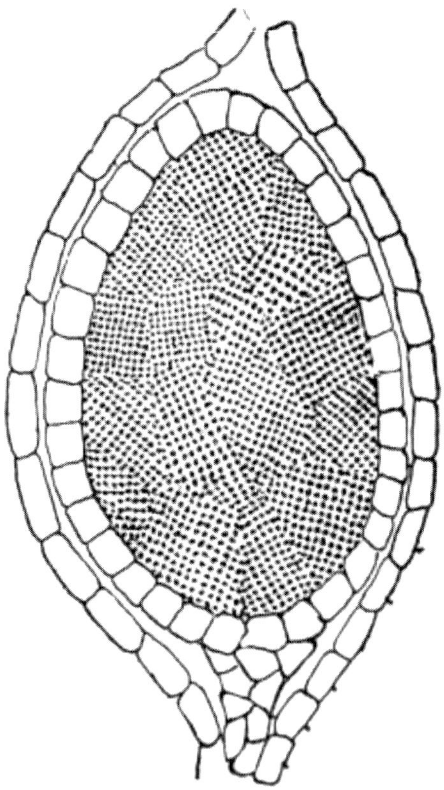

Figure 259. Section of antheridium of marchantia, showing the groups of sperm mother cells.

Figure 260. Spermatozoids of marchantia, uncoiling and one extended, showing the two cilia.

The way in which one of these sperm mother cells changes into this spermatozoid is very curious. We first note that a coiled spiral body is appearing within the thin wall of the cell, one end of the coil larger than the other. The other end terminates in a slender hair-like outgrowth with a delicate vesicle attached to its free end. This vesicle becomes more and more extended until it finally breaks and forms two long lashes which are clubbed at their free ends as shown in Figure 260.

Figure 261. Marchantia polymorpha, female plants bearing archegoniophores.

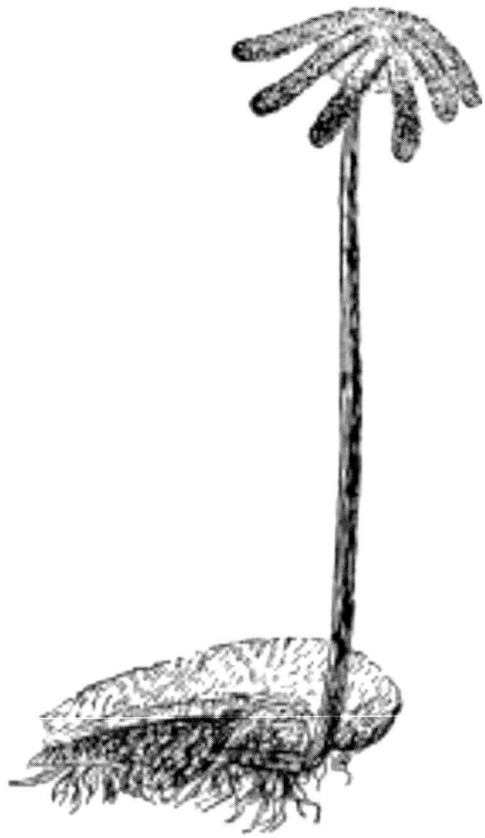

Figure 262. Marchantia polymorpha, showing origin of gametophore.

Archegonial Plants

In Figure 261 we see one of the female plants of marchantia. Upon this there are also very curious structures, which remind one of miniature umbrellas. The general plan of the archegonial receptacle (or female gametophore), for this is what these structures are, is similar to that of the antheridial receptacle, but the rays are more pronounced, and the details of structure are quite different, as we shall see. Underneath the arms there hang down delicate fringed curtains. If we make sections of this in the same direction as we did of the antheridial receptacle, we shall be able to find what is secreted behind these curtains. Such a section is Figured at 266. Here we find the archegonia, but instead of being sunk in cavities their bases are attached to the under surface, while the delicate, pendulous fringes afford them protection from drying. An archegonium we see is not essentially different in marchantia from what it is in riccia, and it will be interesting to learn whether the sporogonium is essentially different from what we find in riccia.

Homology of the Gametophore of Marchantia

To see the relation of the gametophore to the thallus of marchantia take portions of the thallus bearing the female receptacle. On the under side note that the prominent midrib continues beyond the thin lateral expansions and arches upward in the sinus or notch at the end, or at the side where the branch of the thallus has continued to grow beyond. The stalk of the gametophore is then a continuation of the midrib of the thallus. On the apex of this are organized several radial growing points which develop the digitate or ray-like receptacle. The gametophore is thus a specialized branch of the thallus. When young, or in many cases when nearly or quite mature, the gametophore, as one looks at the upper surface of the thallus, appears to arise from the upper surface, as in Figure 261. This is because the thin lateral expansions of the thallus project forward and overlap in advance of the stalk. It is sometimes necessary to tear these overlapping edges apart to see the real origin of the gametophore. But in quite old plants these expanded portions are farther apart and show clearly that the stalk arises from the midrib below and arches upward in the sinus, as in Figure 262.

Chapter 24

LIVERWORTS CONTINUED

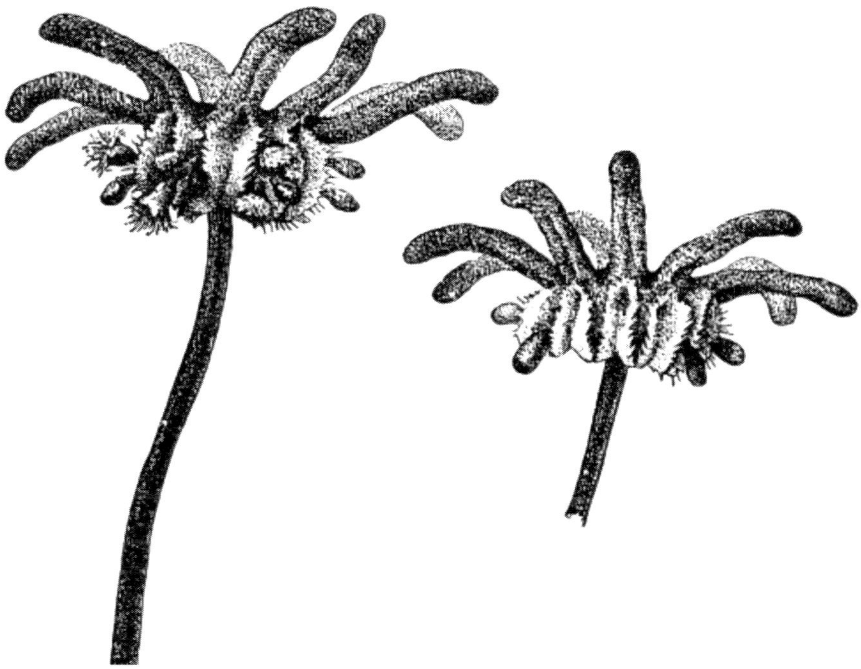

Figure 263. Archegonial receptacles of marchantia bearing ripe sporogonia. The capsule of the sporogonium projects outside, while the stalk is attached to the receptacle underneath the curtain. In the left Figure two of the capsules have burst and the elaters and spores are escaping.

SPOROGONIUM OF MARCHANTIA

If we examine the plant shown in Figure 181 we shall see oval bodies which stand out between the rays of the female receptacle, supported on short stalks. These are the sporogonia, or spore-cases. We judge at once that they are quite different from those which we have studied in riccia, since those were not stalked. We can see that some of the spore-cases have opened, the wall splitting down from the apex in several lines. This is caused by the drying of the wall. These tooth-like divisions of the wall now curl backward, and we can see the yellowish mass of the spores in slow motion, falling here and there. It appears also as if there were twisting threads which aided the spores in becoming freed from the capsule.

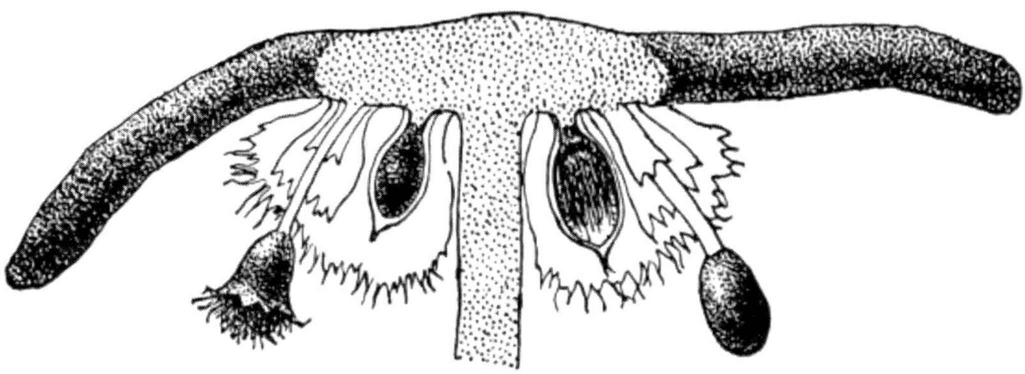

Figure 264. Section of archegonial receptacle of Marchantia polymorpha; ripe sporogonia. One is open, scattering spores and elaters; two are still enclosed in the wall of the archegonium. The junction of the stalk of the sporogonium with the receptacle is the point of attachment of the sporophyte of marchantia with the gametophyte.

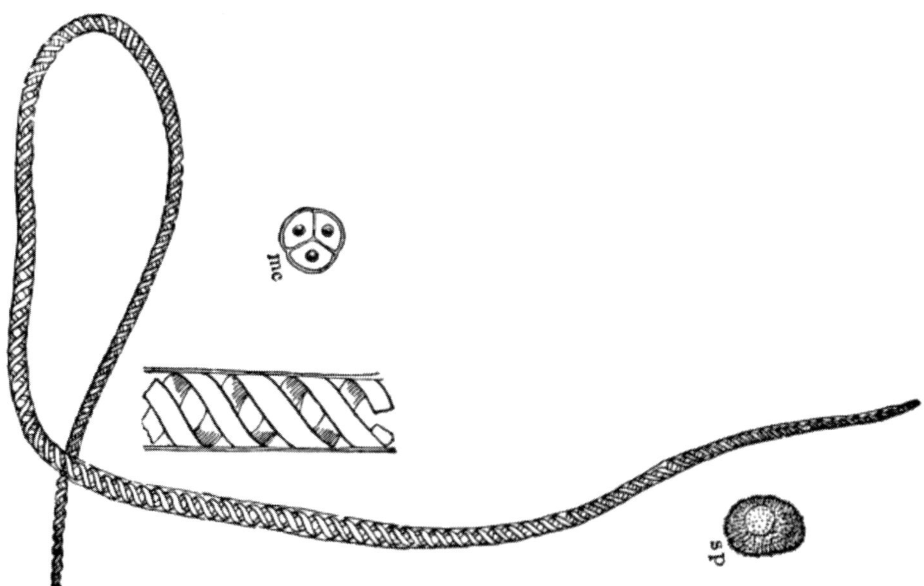

Figure 265. Elater and spore of marchantia. *sp*, spore; *mc*, mother cell of spores, showing partly formed spores.

SPORES AND ELATERS

If we take a bit of this mass of spores and mount it in water for examination with the microscope, we shall see that, besides the spores, there are very peculiar thread-like bodies, the markings of which remind one of a twisted rope. These are very long cells from the inner part of the spore-case, and their walls are marked by spiral thickenings. This causes them in drying, and also when they absorb moisture, to twist and curl in all sorts of ways. They thus aid in pushing the spores out of the capsule as it is drying.

SPOROPHYTE OF MARCHANTIA COMPARED WITH RICCIA

We must recollect that the sporogonium in marchantia is larger than in riccia, and that it is also not lying in the tissue of the thallus, but is only attached to it at one side by a slender stalk. This shows us an increase in the size and complex structure of this new phase of the plant, the *sporophyte*. This is one of the very interesting things which we have to note as we go on in the study of the higher plants.

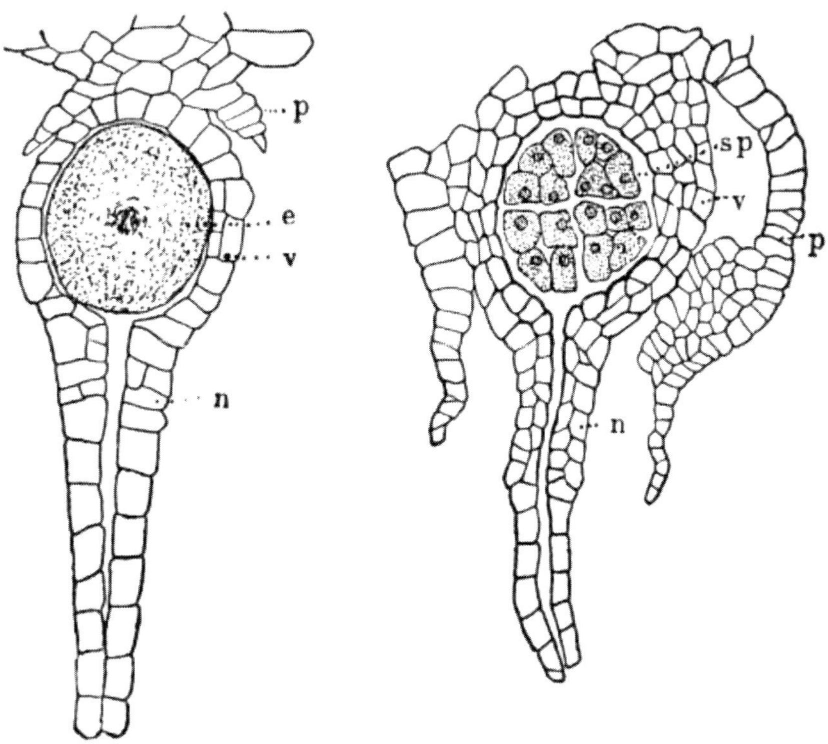

Figure 266. Marchantia polymorpha, archegonium at the left with egg; archegonium at the right with young sporogonium; *p*, curtain which hangs down around the archegonia; *e*, egg; *v*, venter of archegonium; *n*, neck of archegonium; *sp*, young sporogonium.

Sporophyte Dependent on the Gametophyte for Its Nutriment

We thus see that at no time during the development of the sporogonium is it independent from the gametophyte. This new phase of plants then, the sporophyte, has not yet become an independent plant, but must rely on the earlier phase for sustenance.

Development of the Sporogonium

It will be interesting to note briefly how the development of the marchantia sporogonium differs from that of riccia. The first division of the fertilized egg is the same as in riccia that is a wall which runs crosswise of the axis of the archegonium divides it into two cells.

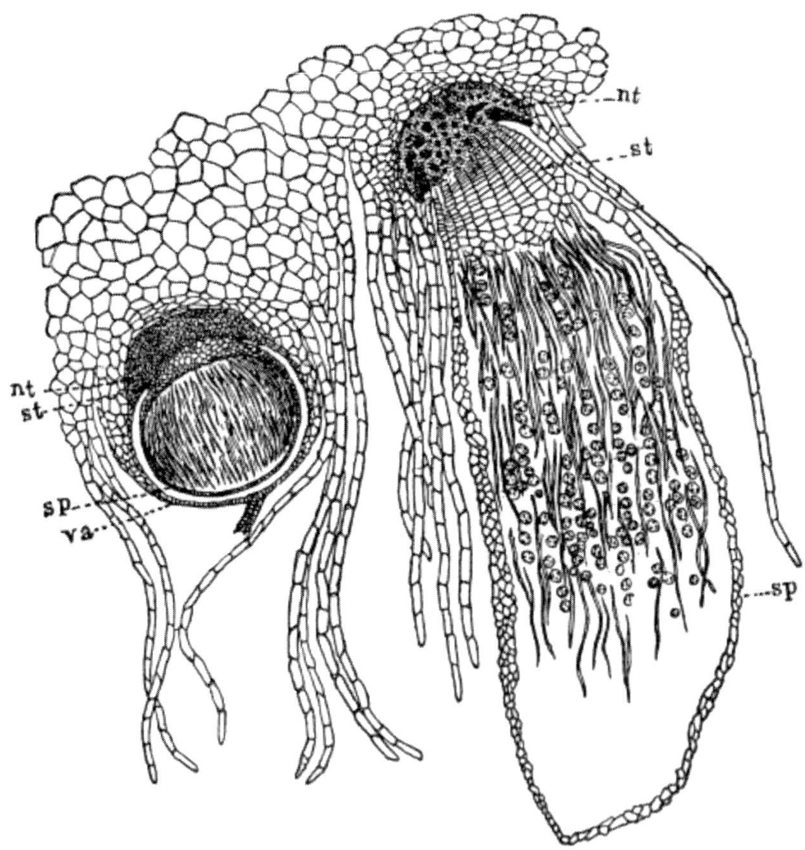

Figure 267. Section of developing sporogonia of marchantia; *nt*, nutritive tissue of gametophyte; *st*, sterile tissue of sporophyte; *sp*, fertile part of sporophyte; *va*, enlarged venter of archegonium.

In marchantia the cell at the base develops the stalk, so that here there is a radical difference. The outer cell forms the capsule. But here after the wall is formed the inner tissue does not all go to make spores, as is the case with riccia. But some of it forms the elaters. While in riccia only the outside layer of cells of the sporogonium remained sterile, in marchantia the basal half of the egg remains completely sterile and develops the stalk, and in the outer half the part which is formed from some of the inner tissue is also sterile.

EMBRYO

In the development of the embryo we can see all the way through this division line between the basal half, which is completely sterile, and the outer half, which is the fertile part. In Figure 267 we see a young embryo, and it is nearly circular in section although it is composed of numerous cells. The basal half is attached to the base of the inner surface of the archegonium, and at this time the archegonium still surrounds it. The archegonium continues to grow then as the embryo grows, and we can see the remains of the shrivelled neck. The portion of the embryo attached to the base of the archegonium is the sterile part and is called the "foot," and later develops the stalk. The sporogonium during all the stages of its development derives its nourishment from the gametophyte at this point of attachment at the base of the archegonium. Soon, as shown in Figure 267 at the right, the outer portion of the sporogonium begins to differentiate into the cells which form the elaters and those which form spores. These lie in radiating lines side by side, and form what is termed the *archesporium*. Each fertile cell forms four spores just as in riccia. They are thus called the mother cells of the spores, or spore mother cells.

HOW MARCHANTIA MULTIPLIES

New plants of marchantia are formed by the germination of the spores, and growth of the same to the thallus. The plants may also be multiplied by parts of the old ones breaking away by the action of strong currents of water, and when they lodge in suitable places grow into well-formed plants. As the thallus lives from year to year and continues to grow and branch the older portions die off, and thus separate plants may be formed from a former single one.

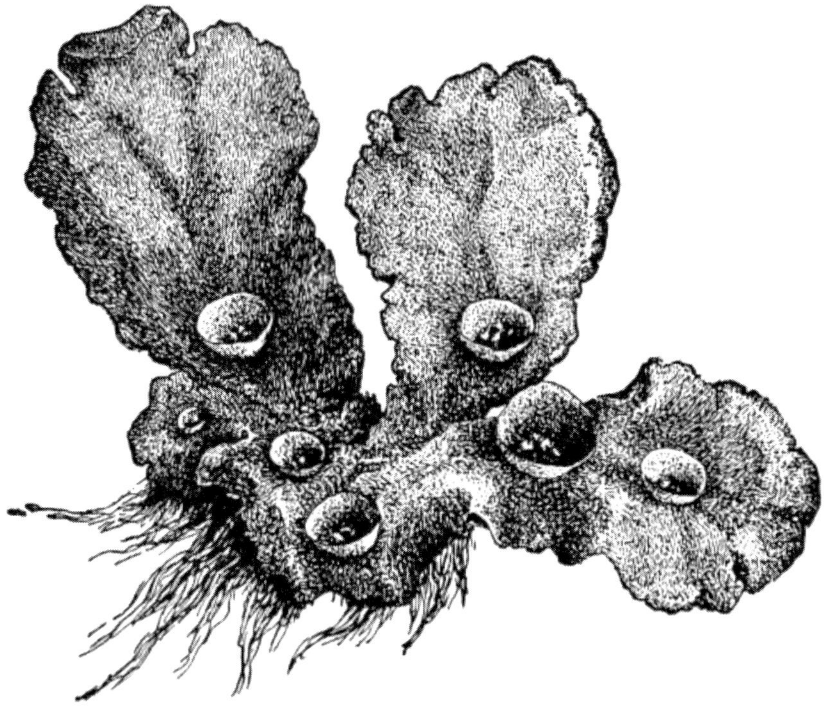

Figure 268. Marchantia plant with cupules and gemmæ; rhizoids below.

Buds, or Gemmæ, of Marchantia

But there is another way in which marchantia multiplies itself. If we examine the upper surface of such a plant as that shown in Figure 268, we shall see that there are minute cup-shaped or saucer-shaped vessels, and within them minute green bodies. If we examine a few of these minute bodies with the microscope we see that they are flattened, biconvex, and at two opposite points on the margin there is an indentation similar to that which appears at the growing end of the old marchantia thallus. These are the growing points of these little buds. When they free themselves from the cups they come to lie on one side. It does not matter on what side they lie, for whichever side it is, that will develop into the lower side of the thallus, and forms rhizoids, while the upper surface will develop the stomates.

Leafy-Stemmed Liverworts

We should now examine more carefully than we have done formerly a few of the leafy-stemmed liverworts (called foliose liverworts).

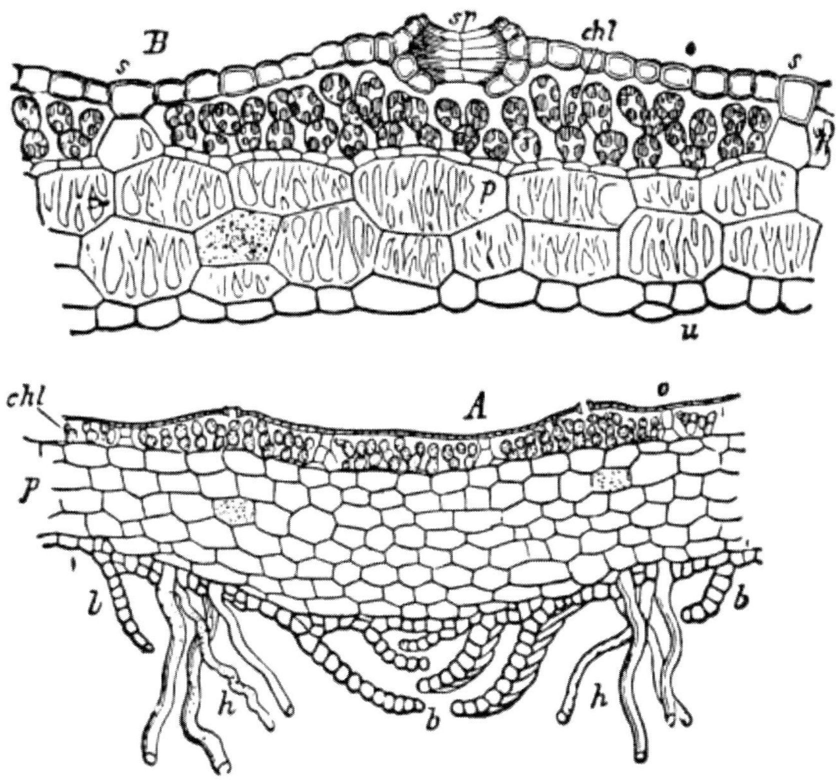

Figure 269. Section of thallus of marchantia. *A*, through the middle portion; *B*, through the marginal portion; *p*, colorless layer; *chl*, chlorophyll layer; *sp*, stomate; *h*, rhizoids; *b*, leaf-like outgrowths on underside (Goebel).

Frullania

(Figure 32).—This plant grows on the bark of logs, as well as on the bark of standing trees. It lives in quite dry situations. If we examine the leaves we will see how it is able to do this. We note that there are two rows of lateral leaves, which are very close together, so close in fact that they overlap like the shingles on a roof. Then, as the creeping stems lie very close to the bark of the tree, these overlapping leaves, which also hug close to the stem and bark, serve to retain moisture which trickles down the bark during rains. If we examine these leaves from the under side as shown in Figure 34, we see that the lower or basal part of each one is produced into a peculiar lobe which is more or less cup-shaped. This catches water and holds it during dry weather, and it also holds moisture which the plant absorbs during the night and in damp days. There is so much moisture in these little pockets of the under side of the leaf that minute animals have found them good places to live in, and one frequently discovers them in this retreat. There is here also a third row of poorly developed leaves on the under side of the stem.

Porella

Growing in similar situations is the plant known as porella. Sometimes there are a few plants in a group, and at other times large mats occur on the bark of a trunk. This plant, porella, also has closely overlapping leaves in rows on opposite sides of the stem, and the lower margin of each leaf is curved under somewhat as in frullania, though the pocket is not so well formed.

The larger plants are female that is thcy bear archegonia, while the male plants, those which bear antheridia, are smaller and the antheridia are borne on small lateral branches. The antheridia are borne in the axils of the leaves. Others of the leafy-stemmed liverworts live in damp situations. Some of these, as Cephalozia, grow on damp rotten logs. Cephalozia is much more delicate, and the leaves are farther apart. It could not live in such dry situations where the frullania is sometimes found. If possible the two plants should be compared in order to see the adaptation in the structure and form to their environment.

Sporogonium of a Foliose Liverwort

The sporogonium of the leafy-stemmed liverworts is well represented by that of several genera. We may take for this study the one illustrated in Figure 274, but another will serve the purpose just as well. We note here that it consists of a rounded capsule borne aloft on a long stalk, the stalk being much longer proportionately than in marchantia. At maturity the capsule splits down into four quadrants, the wall forming four valves, which spread apart from the unequal drying of the cells, so that the spores are set free, as shown in Figure 276. Some of the cells inside of the capsule develop elaters here also as well as spores. These are illustrated in Figure 278.

In this plant we see that the sporophyte remains attached to the gametophyte, and thus is dependent on it for sustenance. This is true of all the plants of this group. The sporophyte never becomes capable of an independent existence, and yet we see that it is becoming larger and more highly differentiated than in the simple riccia.

Figures 275-278.—Sporogonium of liverwort (jungermannia) opening by splitting into four parts, showing details of elaters and spores.

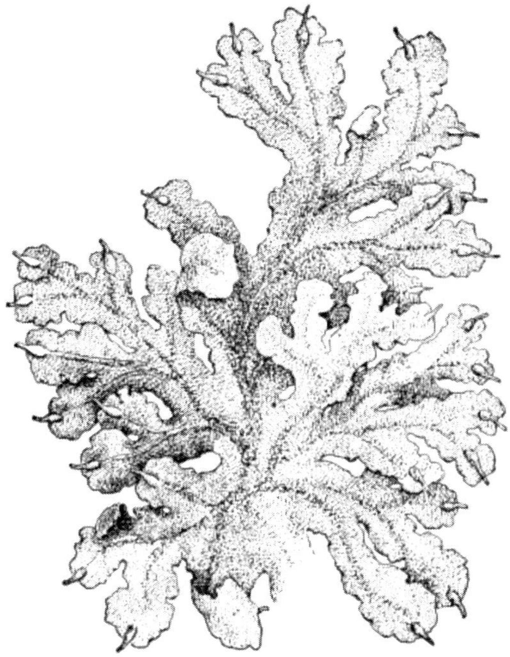

Figure 270. Thallus of a thalloid liverwort (blasia) showing lobed margin of the frond, intermediate between thalloid and foliose plant.

Figure 271. Foliose liverwort, male plant showing antheridia in axils of the leaves (a jungermannia).

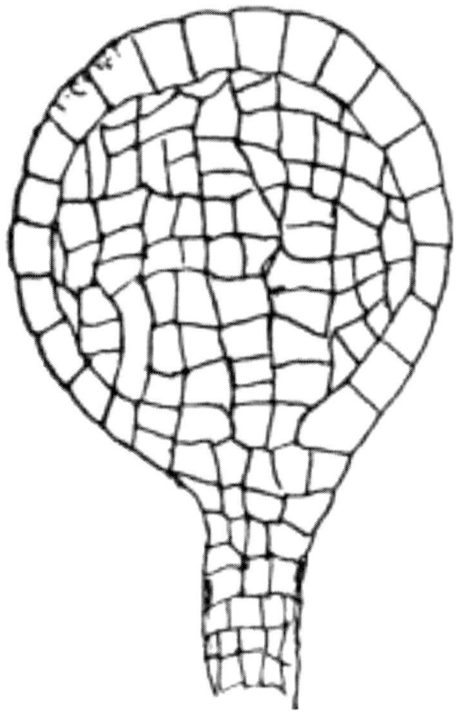

Figure 272. Antheridium of a foliose liverwort (jungermannia).

Figure 273. Foliose liverwort, female plant with rhizoids.

Figure 274. Fruiting plant of a foliose liverwort (jungermannia). Leafy part is the gametophyte; stalk and capsule is the sporophyte (sporogonium in the bryophytes).

Figure 275. Opening capsule showing escape of spores and elaters.

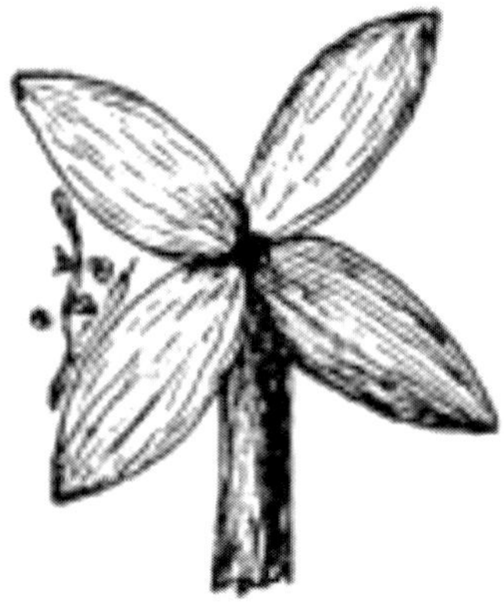

Figure 276. Capsule parted down to the stalk.

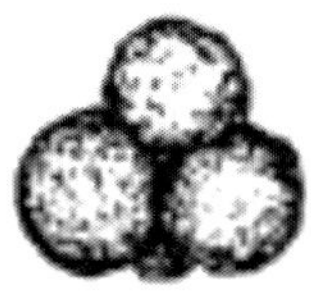

Figure 277. Four spores from mother cell held in a group.

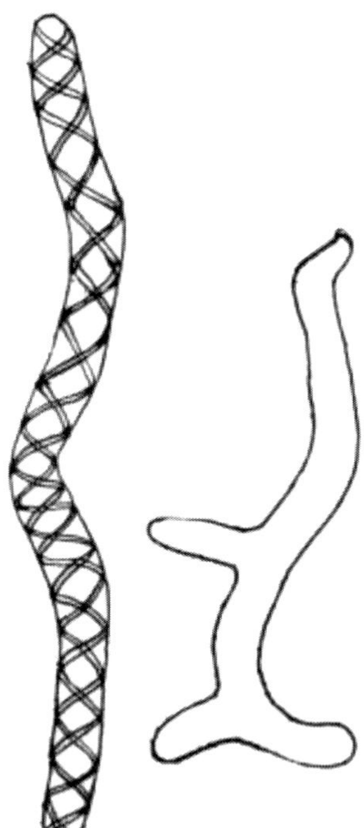

Figure 278. Elaters, at left showing the two spiral marks, at right a branched elater.

THE HORNED LIVERWORTS[25]

The horned liverworts take their name from the shape of the sporogonium. This is long, slender, cylindrical, pointed, and very slightly curved, suggesting the shape of a minute horn. Anthoceros is one of the most common and widely distributed species. The plant grows on damp soil or on mud.

Anthoceros

The gametophyte.—The gametophyte is thalloid. It is thin, flattened, green, irregularly ribbon-shaped and branched. It lies on the soil and is more or less crisped or wavy, or curled, the edges nearly plane, or somewhat irregular, and with minute lobes, or notches, especially near the growing end. The general form and branching can be seen in Figure 279. Where the plants are much crowded the thallus is more irregular, and often possesses numerous small lateral branches in addition to the main lobes. Upon the under side are the slender rhizoids, which attach to the soil. With a hand lens there can be seen also upon the under side small dark, rounded and thickened spots, where an alga (nostoc) is located.

Sexual Organs of Anthoceros

The sexual organs of anthoceros differ considerably from those of the other liverworts studied. In the first place they are immersed in the true tissue of the thallus, i.e., they do not project above the surface.

Antheridia

The antheridium arises from an internal cell of the thallus, a cell just below the upper surface. This cell develops usually a group of antheridia which lie in a cavity formed around this cell as the thallus continues to grow. They are situated along the middle line of the thallus, and can be seen by making a section in this direction. The antheridia are oval or rounded, have a wall of one layer of cells which contains the sperm cells, and each antheridium has a slender stalk. The sperms are like those of the true liverworts.

Archegonia

The archegonia are also borne along the middle line of the thallus. Each one arises at an early stage in the development of the tissue of the thallus from a superficial cell, but the archegonium does not project above the surface. The venter therefore which contains the

[25] May be used as an alternate study for marchantia.

egg is deep down in the thallus, the wall of the neck is formed from cells indistinguishable from the adjoining cells of the thallus and opens at the surface.

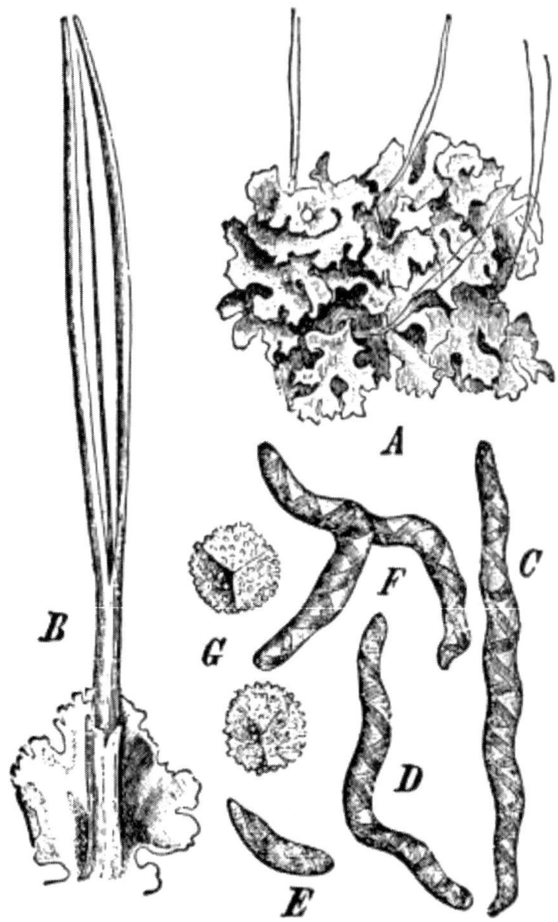

Figure 279. Anthoceros gracilis. *A*, several gametophytes, on which sporangia have developed; *B*, an enlarged sporogonium, showing its elongated character and dehiscence by two valves, leaving exposed the slender columella on the surface of which are the spores, *C*, *D*, *E*, *F*, elaters of various forms, *G*, spores. (After Schiffner.)

Sporophyte of Anthoceros

The Sporogonium

The sporogonium is developed from the fertilized egg, fertilization resulting of course from the fusion of one of the sperms with the nucleus of the egg. From the lower part of the embryo certain cells elongate and push out like rhizoids into the thallus (gametophyte), but never reach the outside so that the sporogonium derives its nutriment from the gametophyte in a parasitic manner like the true liverworts. It is surrounded at the base by a sheath, an outgrowth of the gametophyte.

Growing Point of the Sporogonium

A remarkable thing about the sporogonium of anthoceros, and its relatives, is that the growing point instead of being situated at the free end is located near the base, just above the nourishing foot. Thus the upper part of the sporogonium is older. In the old sporogonia there may be ripe spores near the free end, young ones near the middle, and undifferentiated growing tissue near the base. A longitudinal section of a sporogonium just as the spores are ripening will show this.

Structure of the Sporogonium

A longitudinal section of the sporogonium shows that the spore-bearing tissue occupies a comparatively small portion of the sporogonium. In the section there is a narrow layer (two cells thick) on either side and joined at the top. In the entire sporogonium this fertile tissue is in the shape of an inverted test tube situated inside of the sporogonium. The wall of the sporogonium is about four cells thick. The sterile tissue inside of the spore-bearing tube is the columella. The cells of the wall contain chlorophyll, and there are true stomata with guard cells in the epidermal layer.

Spores and Elaters

In the spore-bearing tissue there are two layers of cells (the archesporium). Each cell is a potential mother cell. The cells, however, of alternate tiers do not form spores. They elongate some what and are somewhat irregular and sometimes divide or branch. They are supposed to represent rudimentary *elaters*. The cells in the other tiers are actual mother cells, and each one forms four spores.

The sporophyte of anthoceros represents the highest type found in the liverworts. The spongy green parenchyma forming the wall, with the stomata in the epidermal layer, fits this tissue for the process of photosynthesis, so that this part of the sporophyte functions as the green leaf of the seed plants. It has been suggested by some that if the rhizoids on the nourishing foot could only extend outside and anchor in the soil, the sporophyte of anthoceros could live an independent existence. But we see that it stops short of that.

CLASSIFICATION OF THE LIVERWORTS

Class Hepaticæ

Order Marchantiales[26]

There are two families represented in the United States.

---

[26] As subclass in Engler and Prantl.

Family Ricciaceæ, including Riccia and Ricciocarpus

Family Marchantiaceæ, including Marchantia, Fegatella (= Conocephalus), Fimbriaria, Targionia, etc.

Order Jungermanniales[27]

There are two subdivisions of this order. *The Anacrogynæ* include chiefly thalloid forms with continued apical growth, the archegonia back of the apical cell. Examples: Blasia, Aneura, Pellia, etc.

The Acrogynæ include chiefly foliose forms, the archegonia arising from the apical cell and in such cases interrupting apical growth. Examples: Cephalozia, Frullania, Bazzania, Jungermannia, Ptilidium, Porella, etc.

Class Anthocerotes

The Anthocerotes have formerly been placed with the Hepaticæ as an order. But because of their wide divergence from the other liverworts in the development of the sexual organs, and especially in the structure of the sporophyte, they are now by some separated as a distinct class. There is one order.

Order Anthocerotales[28]

This includes one family (Anthocerotaceæ) with Anthoceros and Notothylas in Europe and North America, and Dendroceros in the tropics. The latter is epiphytic.

[27] As subclass in Engler and Prantl.
[28] As subclass in Engler and Prantl.

Chapter 25

Mosses (MUSCI)

We are now ready to take up the more careful study of the moss plant. There are a great many kinds of mosses, and they differ greatly from each other in the finer details of structure. Yet there are certain general resemblances which make it convenient to take for study almost any one of the common species in a neighborhood, which forms abundant fruit. Some, however, are more suited to a first study than others. (Polytrichum and funaria are good mosses to study.)

Mnium

We will select here the plant shown in Figure 280. This is known as a mnium (M. affine), and one or another of the species of mnium can be obtained without much difficulty. The mosses, as we have already learned, possess an axis (stem) and leaf-like expansions, so that they are leafy-stemmed plants also. Certain of the branches of the mnium stand upright, or nearly so, and the leaves are all of the same size at any given point on the stem, as seen in the Figure. There are three rows of these leaves, and this is true of most of the mosses.

The mnium plants usually form quite extensive and pretty mats of green in shady moist woods or ravines. Here and there among the erect stems are prostrate ones, with two rows of prominent leaves so arranged that it reminds one of some of the leafy-stemmed liverworts. If we examine some of the leaves of the mnium we see that the greater part of the leaf consists of a single layer of green cells, just as is the case in the leafy-stemmed liverworts. But along the middle line is a thicker layer, so that it forms a distinct midrib. This is characteristic of the leaves of mosses, and is one way in which they are separated from the leafy-stemmed liverworts, the latter never having a midrib.

The Fruiting Moss Plant

In Figure 280 is a moss plant "in fruit," as we say. Above the leafy stem a slender stalk bears the capsule, and in this capsule are borne the spores. The capsule then belongs to the *sporophyte phase* of the moss plant, and we should inquire whether the entire plant as we see it here is the sporophyte, or whether part of it is gametophyte. If a part of it is gametophyte and a part sporophyte, then where does the one end and the other begin? If we strip off the leaves at the end of the leafy stem, and make a longisection in the middle line, we should find that the stalk which bears the capsule is simply stuck into the end of the leafy stem, and is not organically connected with it. This is the dividing line, then, between the gametophyte and the sporophyte. We shall find that here the archegonium containing the egg is borne, which is a surer way of determining the limits of the two phases of the plant.

Figure 280. Portion of moss plant of Mnium affine, showing two sporogonia from one branch. Capsule at left has just shed the cap or operculum; capsule at right is shedding spores, and the teeth are bristling at the mouth. Next to the right is a young capsule with calyptra still attached; next are two spores enlarged.

THE MALE AND FEMALE MOSS PLANTS

The two plants of mnium shown in Figures 281, 282 are quite different, as one can easily see, and yet they belong to the same species. One is a female plant, while the other is a male plant. The sexual organs then in mnium, as in many others of the mosses, are borne on separate plants. The archegonia are borne at the end of the stem, and are protected by somewhat narrower leaves which closely overlap and are wrapped together. They are similar to the archegonia of the liverworts.

Figure 281. Female plant (gametophyte) of a moss (mnium), showing rhizoids below, and the tuft of leaves above which protect the archegonia.

The male plants of mnium are easily selected, since the leaves at the end of the stem form a broad rosette with the antheridia, and some sterile threads packed closely together in the center. The ends of the mass of antheridia can be seen with the naked eye, as shown in Figure 282. When the antheridia are ripe, if we make a section through a cluster, or if we merely tease out some from the end with a needle in a drop of water on the slide, then prepare for examination with the microscope, we can see the form of the antheridia. They are somewhat clavate or elliptical in outline, as seen in Figure 284. Between them there stand short threads composed of several cells containing chlorophyll grains. These are sterile threads (paraphyses).

Figure 282. Male plant (gametophyte) of a moss (mnium) showing rhizoids below and the antheridia at the center above surrounded by the rosette of leaves.

Sporogonium

In Figure 280 we see illustrated a sporogonium of mnium, which is of course developed from the fertilized egg-cell of the archegonium. There is a nearly cylindrical capsule, bent downward, and supported on a long slender stalk. Upon the capsule is a peculiar cap,[29] shaped like a ladle or spatula. This is the remnant of the old archegonium, which, for a time surrounded and protected the young embryo of the sporogonium, just as takes place in the liverworts. In most of the mosses this old remnant of the archegonium is borne aloft on the capsule as a cap, while in the liverworts it is thrown to one side as the sporogonium elongates.

[29] Called the calyptra.

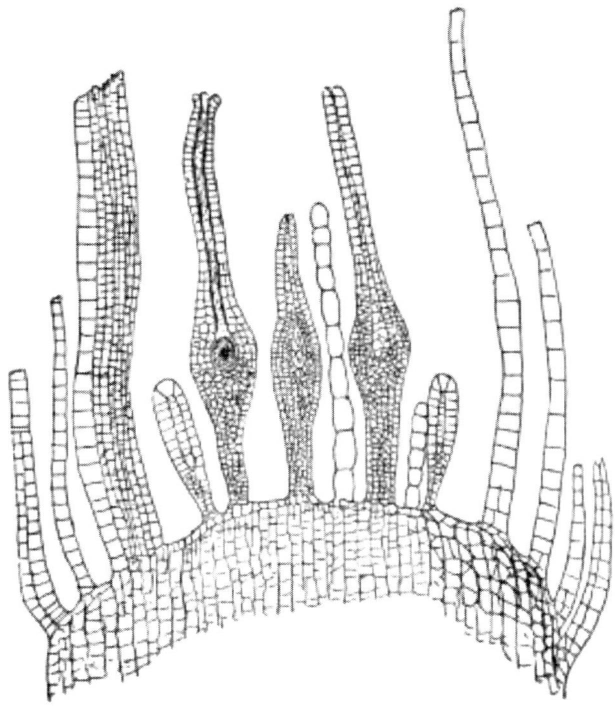

Figure 283. Section through end of stem of female plant of mnium, showing archegonia at the center. One archegonium shows the egg. On the sides are sections of the protecting leaves.

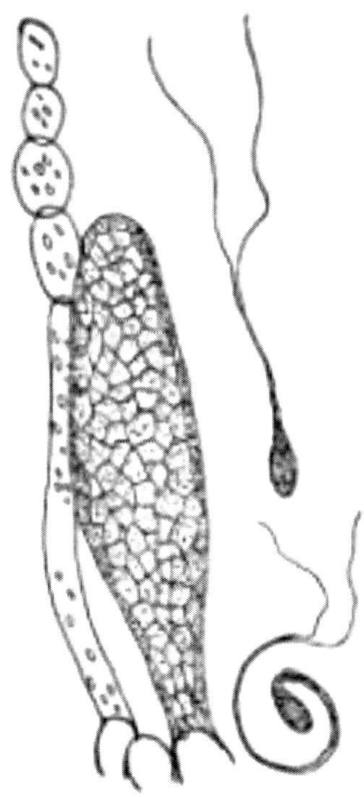

Figure 284. Antheridium of mnium with jointed paraphysis at the left; spermatozoids at the right.

Structure of the Moss Capsule

At the free end on the moss capsule as shown in the case of mnium in Figure 280, after the remnant of the archegonium falls away, there is seen a conical lid which fits closely over the end. When the capsule is ripe this lid easily falls away, and can be brushed off so that it is necessary to handle the plants with care if it is desired to preserve this for study.

When the lid is brushed away as the capsule dries more we see that the end of the capsule covered by the lid appears "frazzled." If we examine this end with the microscope we see that the tissue of the capsule here is torn with great regularity, so that there are two rows of narrow, sharp teeth which project outward in a ring around the opening. If we blow our "breath" upon these teeth they will be seen to move, and as the moisture disappears and reappears in the teeth, they close and open the mouth of the capsule, so sensitive are they to the changes in the humidity of the air. In this way all of the spores are prevented to some extent from escaping from the capsule at one time.

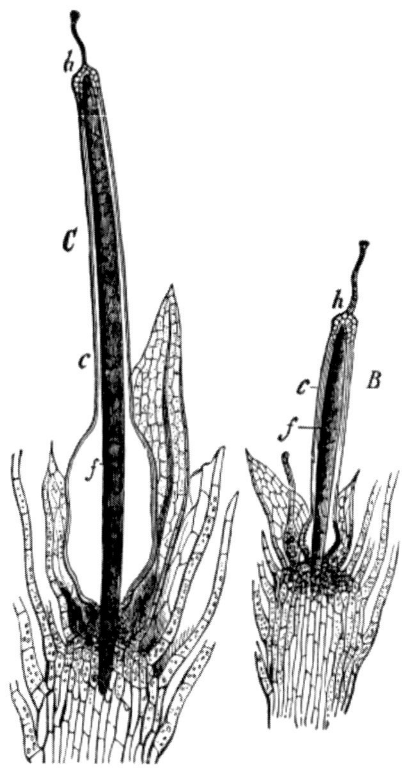

Figure 285. Two different stages of young sporogonium of a moss, still within the archegonium and wedging their way into the tissue of the end of the stem. *h*, neck of archegonium; *f*, young sporogonium. This shows well the connection of the sporophyte with the gametophyte.

Note. If we make a section longitudinal of the capsule of mnium, or some other moss, we find that the tissue which develops the spores is much more restricted than in the capsule of the liverworts which we have studied. The spore-bearing tissue is confined to a single layer which extends around the capsule some distance from the outside of the wall, so that a central cylinder is left of sterile tissue. This is the columella, and is present in nearly all the mosses. Each of the cells of the fertile layer divides into four spores.

Development of the Sporogonium

The egg-cell after fertilization divides by a wall crosswise to the axis of the archegonium. Each of these cells continues to divide for a time, so that a cylinder pointed at both ends is formed. The lower end of this cylinder of tissue wedges its way down through the base of the archegonium into the tissue of the end of the moss stem as shown in Figure 285. This forms the foot through which the nutrient materials are passed from the gametophyte to the sporogonium. The upper part continues to grow, and finally the upper end differentiates into the mature capsule.

Protonema of the Moss

When the spores of a moss germinate they form a thread-like body, with chlorophyll. This thread becomes branched, and sometimes quite extended tangles of these threads are formed. This is called the protonema, that is *first thread*. The older threads become finally brown, while the later ones are green. From this protonema at certain points buds appear which divide by close oblique walls. From these buds the leafy stem of the moss plant grows. Threads similar to these protonemal threads now grow out from the leafy stem, to form the rhizoids. These supply the moss plant with nutriment, and now the protonema usually dies, though in some few species it persists for long periods.

Classification of the Mosses

Class Muscineæ (MUSCI)

Order Sphagnales[30]

This order includes the peat mosses. There is but one family (Sphagnaceæ) and but a single genus (Sphagnum). The peat mosses are widely distributed over the globe, chiefly occurring in moors, or "bogs," usually low ground around the shores of lakes, ponds, or along streams, but they often occur on wet dripping rocks in cool shady places. Small ponds are sometimes filled in by their growth. As the sphagnum growing in such an abundance

[30] As subclass in Engler and Prantl.

of water only partially decays, "ground" is built up rather rapidly, and the sphagnum remains are known as "peat." This "ground"-building peculiarity of sphagnum sometimes enables the plant (often in conjunction with others) to fill in ponds completely. (See Atoll Moor, Chapter LV.)

The gametophyte of sphagnum, like that of all the mosses, is dimorphic, but the first part (or protonema) which develops from the spores is thalloid, and therefore more like the thallose liverworts. The leafy axis (or gametophore) which develops from the thalloid form is very characteristic (see Chapter LV).

The archegonia are borne on the free end of the main axis, while the antheridia are borne on short branches which are brightly colored, red, yellow, etc. The sporophyte (sporogonium) is globose and possesses a broad foot anchored in the end of a naked prolongation of the end of the leafy gametophore. This naked prolongation of the gametophore looks like the stalk of the sporogonium, but a study of its connection with the sporogonium shows that it is part of the gametophyte, which is only developed after the fertilization of the egg in the archegonium. In the sporogonium there is a short columella, and the archesporium is in the form of an inverted cup.

Order Andreæales[31]

This order includes the single genus Andreæa. The plants are small but form extensive mats, growing on rocks in arctic or alpine regions usually. They are sometimes found in great abundance on bare, rather dry rocks on mountains. The protonema is somewhat thalloid. The sporogonium opens by splitting longitudinally into four valves. An elongated columella is present so that the archesporium is shaped like an inverted test tube.

Order Archidiales[32]

This order contains the single genus Archidium, and by some is placed as an aberrant genus in the Bryales. There is no columella in the simple sporogonium. The archesporium occupies all the internal part of the sporogonium, some cells being fertile and others sterile.

Order Bryales[33]

These include the higher mosses, and a very large number of genera and species. The protonema is filamentous and branched except in a few forms where it is partly thalloid as in Tetraphis (= Georgia). (Tetraphis pellucida is a common moss on very rotten logs. The capsule has four prominent teeth.) In a few of the lower genera (Phascum, Pleuridium, etc.) the capsule opens irregularly, but in the larger number the capsule opens by a lid (operculum). A cylindrical columella is present, and the archesporium is in the form of a tube open at both ends. (Examples: Polytrichum, Bryum, Mnium, Hypnum, etc.)

---

[31] As subclass in Engler and Prantl.
[32] As subclass in Engler and Prantl.
[33] As subclass in Engler and Prantl.

Table Showing Relation of Gametophyte and Sporophyte in the Liverworts and Mosses

	GAMETOPHYTE. (Prominent part of the plant. Leads an independent existence.)				SPOROPHYTE (Attached to gametophyte and dependent on it for nourishment.)			Beginning of Gametophyte.
	Vegetative Phase	Vegetative Mulitipl-ication.	Sexual Organs.		Beginning of Sporophyte.	Sterile Part.	Fertile Part.	
Riccia.	Thallus flattened, ribbon-like, forked, or nearly circular.	Sometimes by branching and dying away of older parts.	Immersed by surrounding, upward growth of thallus.		Fertilized egg. (Develops sporogonium.)	Wall of sporogonium, of one-layer cells.	Central mass (archesporium) develops	Spores.
			Antheridia, with spermatozoids.	Archegonia, with egg in each.				
Marchantia.	Thallus flattened, ribbon-like, forked, male and female plants bear gametophores.	By dying away of older parts, and by gemmæ.	Borne on special receptacles on different plants.		Fertilized egg. (Develops sporogonium.)	Sterile part of stalked is stalk, wall of capsule of several layers, elaters.	Central part of capsule (archesporium) develops ..... and elaters.	Spores.
			Antheridia, with spermatozoids, borne on antheridiphores, or male gametophores.	Archegonia, borne on female gametophore (or archegoniophore), each with an egg.				
Jungermannia (or Cephalozia, Porella, etc.)	A plant with apparent leaves and stem; margins of thallus have become cut into lobes. Male and female plants.	By dying away of older parts.	On different plants.		Fertilized egg. (Develops sporogonium.)	Sterile part of stalked capsule is stalk, wall of capsule of several layers, elaters.	Central part of capsule (archesporium) develops ...and elaters.	Spores.
			Antheridia, with spermatozoids, in axils of leaves of male plant.	Archegonia, each with egg, on female plant.				

	GAMETOPHYTE. (Prominent part of the plant. Leads an independent existence.)				SPOROPHYTE (Attached to gametophyte and dependent on it for nourishment.)			Beginning of Gametophyte.
	Vegetative Phase	Vegetative Mulitipl-ication.	Sexual Organs.		Beginning of Sporophyte.	Sterile Part.	Fertile Part.	
Mosses. Mnium, Polytrichum etc.	Plant with apparent leafy axis, 3 rows of leaves (similar to jungermannia), borne on an earlier protonemal stage. Male and female plants.	By branching, by growth of protonema from axis, leaves, or even sporogonium. (In some genera by gemmæ.)	On different plants. Antheridia, with spermatozoids, at end of stem of male plant.	Archegonia each with egg, on female plant. (Calyptra found on sporogonium is remnant of archegonium.)	Fertilized egg. (Develops sporogonium.)	Sterile part of stalked capsule is stalk, wall of capsule of several layers, columella, lid, teeth etc., of the highly specialized capsule.	Cylindrical layer of cells around columella is the archesporium; it develops	Spores.

Chapter 26

FERNS

In taking up the study of the ferns we find plants which are very beautiful objects of nature and thus have always attracted the interest of those who love the beauties of nature. But they are also very interesting to the student, because of certain remarkable peculiarities of the structure of the fruit bodies, and especially because of the intermediate position which they occupy within the plant kingdom, representing in the two phases of their development the primitive type of plant life on the one hand, and on the other the modern type. We will begin our study of the ferns by taking that form which is the more prominent, the fern plant itself.

THE CHRISTMAS FERN

One of the ferns which is very common in the Northern States, and occurs in rocky banks and woods, is the well-known Christmas fern (*Aspidium acrostichoides*) shown in Figure 286. The leaves are the most prominent part of the plant, as is the case with most if not all our native ferns. The stem is very short and for the most part under the surface of the ground, while the leaves arise very close together, and thus form a rosette as they rise and gracefully bend outward. The leaf is elongate and reminds one somewhat of a plume with the pinnæ extending in two rows on opposite sides of the midrib. These pinnæ alternate with one another, and at the base of each pinna is a little spur which projects upward from the upper edge. Such a leaf is said to be pinnate. While all the leaves have the same general outline, we notice that certain ones, especially those toward the center of the rosette, are much narrower from the middle portion toward the end. This is because of the shorter pinnæ here.

Figure 286. Christmas fern (Aspidium acrostichoides).

FRUIT "DOTS" (SORUS, INDUSIUM)

If we examine the under side of such short pinnæ of the Christmas fern we see that there are two rows of small circular dots, one row on either side of the pinna. These are called the "fruit dots," or sori (a single one is a sorus). If we examine it with a low power of the microscope, or with a pocket lens, we see that there is a circular disk which covers more or less completely very minute objects, usually the ends of the latter projecting just beyond the edge if they are mature. This circular disk is what is called the *indusium*, and it is a special outgrowth of the epidermis of the leaf here for the protection of the spore-cases. These minute objects underneath are the fruit bodies, which in the case of the ferns and their allies are called *sporangia*. This indusium in the case of the Christmas fern, and also in some others, is attached to the leaf by means of a short slender stalk which is fastened to the middle of the under side of this shield, as seen in cross-section in Figure 292.

SPORANGIA

If we section through the leaf at one of the fruit dots, or if we tease off some of the sporangia so that the stalks are still attached, and examine them with the microscope, we can see the form and structure of these peculiar bodies. Different views of a sporangium are shown in Figure 293. The slender portion is the stalk, and the larger part is the spore-case proper. We should examine the structure of this spore-case quite carefully, since it will help us to understand better than we otherwise could the remarkable operations which it performs in scattering the spores.

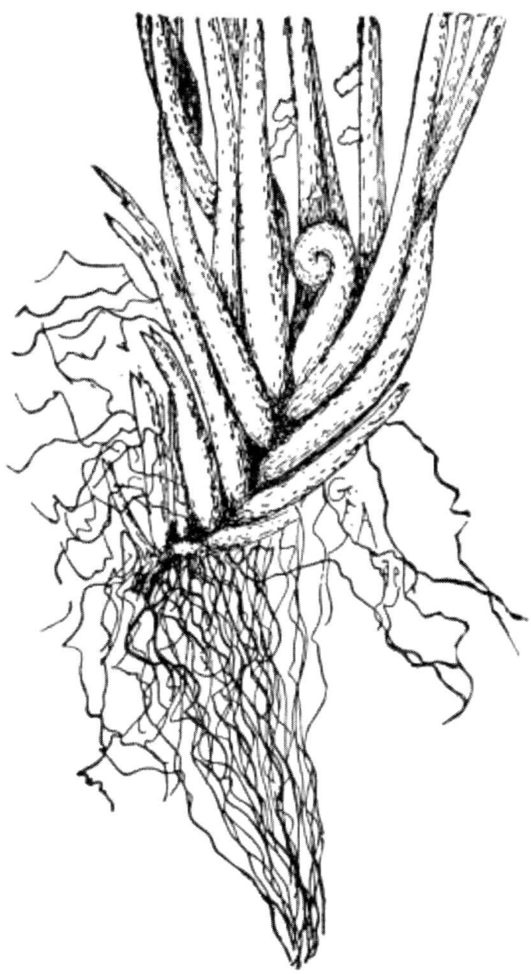

Figure 287. Rhizome with bases of leaves, and roots of the Christmas fern.

Structure of a Sporangium

If we examine one of the sporangia in side view as shown in Figure 293, we note a prominent row of cells which extend around the margin of the dorsal edge from near the attachment of the stalk to the upper front angle. The cells are prominent because of the thick inner walls, and the thick radial walls which are perpendicular to the inner walls. The walls on the back of this row and on its sides are very thin and membranous. We should make this out carefully, for the structure of these cells is especially adapted to a special function which they perform. This row of cells is termed the *annulus*, which means a little ring. While this is not a complete ring, in some other ferns the ring is nearly complete.

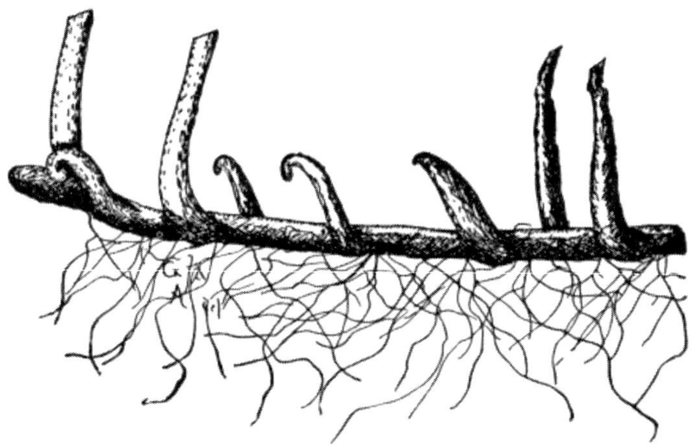

Figure 288. Rhizome of sensitive fern (*Onoclea sensibilis*).

Figure 289. Under side of pinna of *Aspidium spinulosum* showing fruit dots (sori).

In the front of the sporangium is another peculiar group of cells. Two of the longer ones resemble the lips of some creature, and since the sporangium opens between them they are sometimes termed the lip cells. These lip cells are connected with the upper end of the annulus on one side and with the upper end of the stalk on the other side by thin-walled cells, which may be termed connective cells, since they hold each lip cell to its part of the opening sporangium. The cells on the side of the sporangium are also thin-walled. If we now examine a sporangium from the back, or dorsal edge as we say, it will appear as in the left-hand Figure. Here we can see how very prominent the annulus is. It projects beyond the surface of the other cells of the sporangium. The spores are contained inside this case.

Opening of the Sporangium and Dispersion of the Spores

If we take some fresh fruiting leaves of the Christmas fern, or of any one of many of the species of the true ferns just at the ripening of the spores, and place a portion of it on a piece of white paper in a dry room, in a very short time we shall see that the paper is being dusted with minute brown objects which fly out from the leaf. Now if we take a portion of the same leaf and place it under the low power of the microscope, so that the full rounded sporangia can be seen, in a short time we note that the sporangium opens, the upper half curls backward as shown in Figure 294, and soon it snaps quickly, to near its former position, and the spores are at the same time thrown for a considerable distance. This movement can sometimes be seen with the aid of a good hand lens.

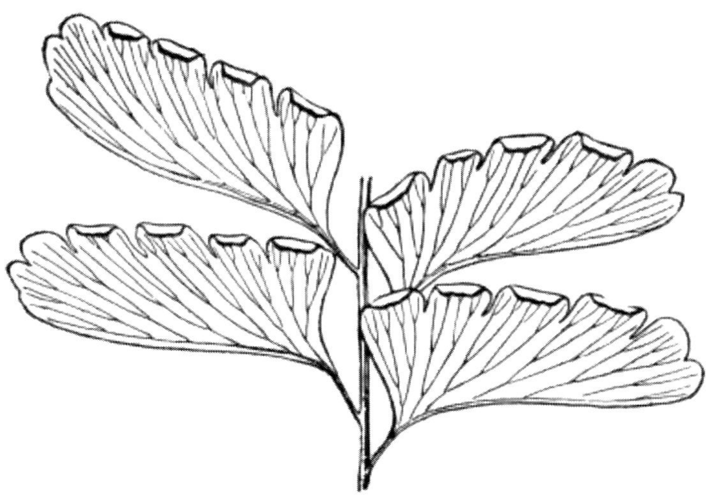

Figure 290. Four pinnæ of adiantum, showing recurved margins which cover the sporangia.

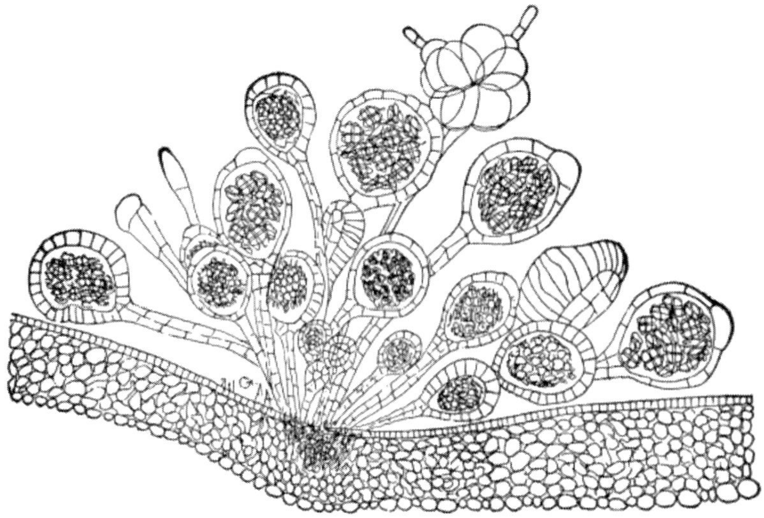

Figure 291. Section through sorus of Polypodium vulgare showing different stages of sporangium, and one multicellular capitate hair.

How Does This Opening and Snapping of the Sporangium Take Place?

We are now more curious than ever to see just how this opening and snapping of the sporangium takes place. We should now mount some of the fresh sporangia in water and cover with a cover glass for microscopic examination. A drop of glycerine should be placed at one side of the cover glass on the slip so that the edge of the glycerine will come in touch with the water. Now as one looks through the microscope to watch the sporangia, the water should be drawn from under the cover glass with the aid of some bibulous paper, like filter paper, placed at the edge of the cover glass on the opposite side from the glycerine. As the glycerine takes the place of the water around the sporangia it draws the water out of the cells of the annulus, just as it took the water out of the cells of the spirogyra as we learned some time ago. As the water is drawn out of these cells there is produced a pressure from without, the atmospheric pressure upon the glycerine. This causes the walls of these cells of the annulus to bend inward, because, as we have already learned, the glycerine does not pass through the walls nearly so fast as the water comes out.

Now the structure of the cells of this annulus, as we have seen, is such that the inner walls and the perpendicular walls are stout, and consequently they do not bend or collapse when this pressure is brought to bear on the outside of the cells. The thin membranous walls on the back (dorsal walls) and on the sides of the annulus, however, yield readily to the pressure and bend inward. This, as we can readily see, pulls on the ends of each of the perpendicular walls drawing them closer together.

This shortens the outer surface of the annulus and causes it to first assume a nearly straight position, then curve backward until it quite or nearly becomes doubled on itself. The sporangium opens between the lip cells on the front and the lateral walls of the sporangium are torn directly across. The greater mass of spores are thus held in the upper end of the open sporangium, and when the annulus has nearly doubled on itself it suddenly snaps back again in position.

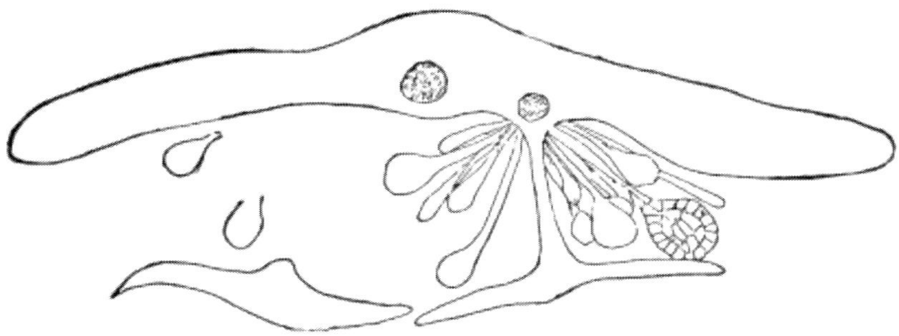

Figure 292. Section through sorus and shield-shaped indusium of aspidium.

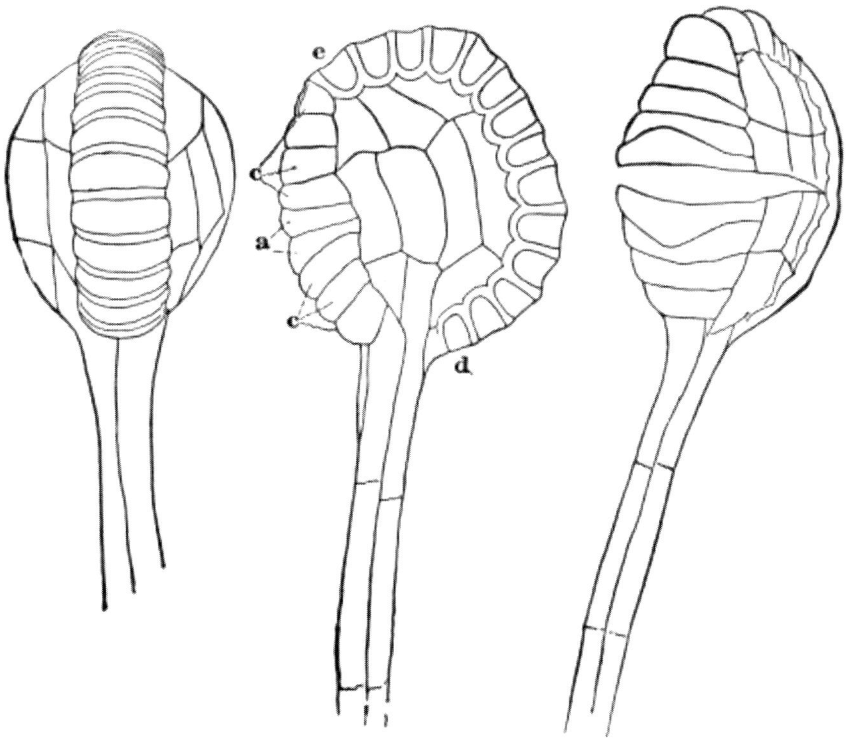

Figure 293. Rear, side, and front views of fern sporangium. *d*, *e*, annulus; *a*, lip cells.

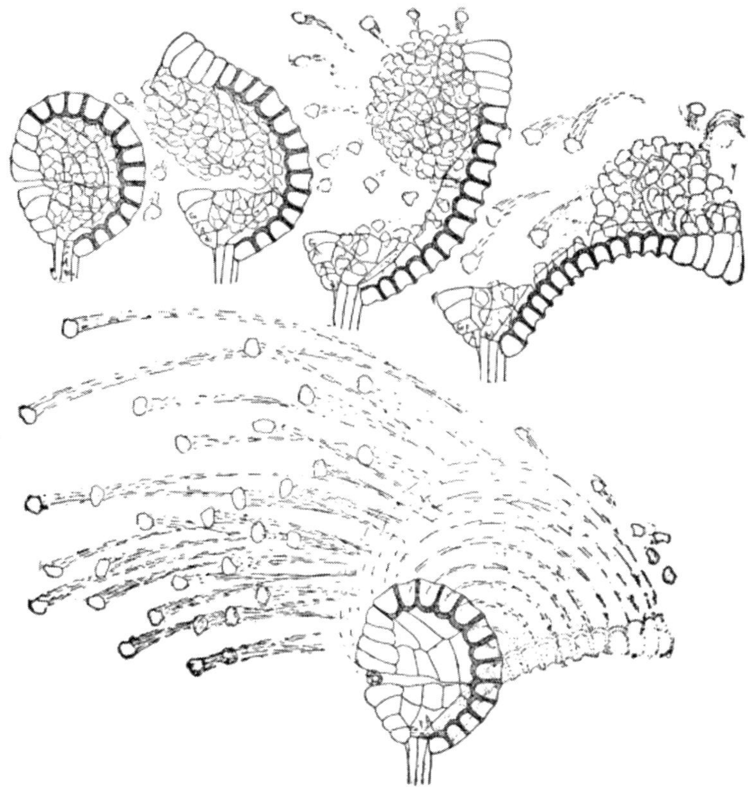

Figure 294. Dispersion of spores from sporangium of Aspidium acrostichoides, showing different stages in the opening and snapping of the annulus.

While treating with the glycerine we can see all this movement take place. Each cell of the annulus acts independently, but often they all act in concert. When they do not all act in concert, some of them snap sooner than others, and this causes the annulus to snap in segments.

The Movements of the Sporangium Can Take Place in Old and Dried Material

If we have no fresh material to study the sporangium with, we can use dried material, for the movements of the sporangia can be well seen in dried material, provided it was collected at about the time the sporangia are mature, that is at maturity, or soon afterward. We take some of the dry sporangia (or we may wash the glycerine off those which we have just studied) and mount them in water, and quickly examine them with a microscope. We notice that in each cell of the annulus there is a small sphere of some gas. The water which bathes the walls of the annulus is absorbed by some substance inside these cells. This we can see because of the fact that this sphere of gas becomes smaller and smaller until it is only a mere dot, when it disappears in a twinkling. The water has been taken in under such pressure that it has absorbed all the gas, and the farther pressure in most cases closes the partly opened sporangium more completely.

Now we should add glycerine again and draw out the water, watching the sporangia at the same time. We see that the sporangia which have opened and snapped once will do it again. And so they may be made to go through this operation several times in succession. We should now note carefully the annulus, that is after the sporangia have opened by the use of glycerine. So soon as they have snapped in the glycerine we can see those minute spheres of gas again, and since there was no air on the outside of the sporangia, but only glycerine, this gas must, it is reasoned, have been given up by the water before it was all drawn out of the cells.

The Common Polypody

We may now take up a few other ferns for study. Another common fern is the polypody, one or more species of which have a very wide distribution. The stem of this fern is also not usually seen, but is covered with the leaves, except in the case of those species which grow on the surface of rocks. The stem is slender and prostrate, and is covered with numerous brown scales. The leaves are pinnate in this fern also, but we find no difference between the fertile and sterile leaves (except in some rare cases). The fruit dots occupy much the same positions on the under side of the leaf that they do in the Christmas fern, but we cannot find any indusium. In the place of an indusium are club-shaped hairs as shown in Figure 291. The enlarged ends of these clubs reaching beyond the sporangia give some protection to them when they are young.

Other Ferns

We might examine a series of ferns to see how different they are in respect to the position which the fruit dots occupy on the leaf. The common brake, which sometimes covers extensive areas and becomes a troublesome weed, has a stout and smooth underground stem (rhizome) which is often 12 to 20 *cm* beneath the surface of the soil. There is a long leaf stalk, which bears the lamina, the latter being several times pinnate. The margins of the fertile pinnæ are inrolled, and the sporangia are found protected underneath in this long sorus along the margin of the pinna.

The beautiful maidenhair fern and its relatives have obovate pinnæ, and the sori are situated in the same positions as in the brake. In other ferns, as the walking fern, the sori are borne along by the side of the veins of the leaf.

Opening of the Leaves of Ferns

The leaves of ferns open in a peculiar manner. The tip of the leaf is the last portion developed, and the growing leaf appears as if it was rolled up as in Figure 287 of the Christmas fern. As the leaf elongates this portion unrolls.

Longevity of Ferns

Most ferns live from year to year, by growth adding to the advance of the stem, while by decay of the older parts the stem shortens up behind. The leaves are short-lived, usually dying down each year, and a new set arising from the growing end of the stem. Often one can see just back or below the new leaves the old dead ones of the past season, and farther back the remains of the petioles of still older leaves.

Budding of Ferns

A few ferns produce what are called bulbils or bulblets on the leaves. One of these, which is found throughout the greater part of the eastern United States, is the bladder fern (Cystopteris bulbifera), which grows in shady rocky places. The long graceful delicate leaves form in the axils of the pinnæ, especially near the end of the leaf, small oval bulbs as shown in Figure 295. If we examine one of these bladder-like bulbs we see that the bulk of it is made up of short thick fleshy leaves, smaller ones appearing between the outer ones at the smaller end of the bulb. This bulb contains a stem, young root, and several pairs of these fleshy leaves. They easily fall to the ground or rocks, where, with the abundant moisture usually present in localities where the fern is found, the bulb grows until the roots attach the plant to the soil or in the crevices of the rocks. A young plant growing from one of these bulbils is shown in Figure 295.

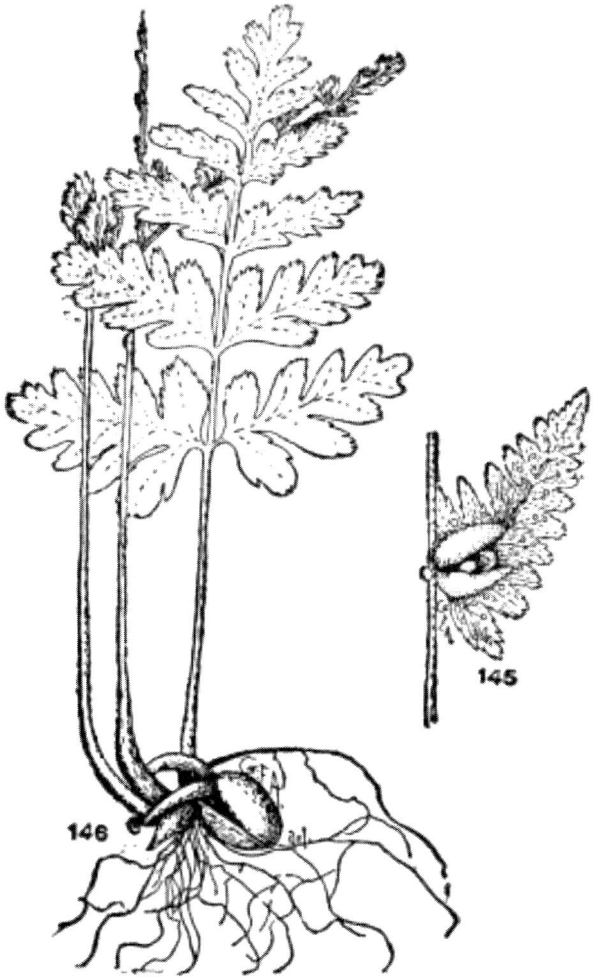

Figure 295. *Cystopteris bulbifera*, young plant growing from bulb. At right is young bulb in axil of pinna of leaf.

GREENHOUSE FERNS

Some of the ferns grown in conservatories have similar bulblets. Figure 296 represents one of these which is found abundantly on the leaves of Asplenium bulbiferum. These bulbils have leaves which are very similar to the ordinary leaf except that they are smaller. The bulbs are also much more firmly attached to the leaf, so that they do not readily fall away.

Plant conservatories usually furnish a number of very interesting ferns, and one should attempt to make the acquaintance of some of them, for here one has an opportunity during the winter season not only to observe these interesting plants, but also to obtain material for study. In the tree ferns which often are seen growing in such places we see examples of the massive trunks and leaves of some of the tropical species.

Figure 296. Bulbil growing from leaf of asplenium (*A*, bulbiferum).

THE FERN PLANT IS A SPOROPHYTE

We have now studied the fern plant, as we call it, and we have found it to represent the spore-bearing phase of the plant, that is the *sporophyte* (corresponding to the sporogonium of the liverworts and mosses).

Is There a Gametophyte Phase in Ferns?

But in the sporophyte of the fern, which we should not forget is the fern plant, we have a striking advance upon the sporophyte of the liverworts and mosses. In the latter plants the sporophyte remained attached to the gametophyte, and derived its nourishment from it. In the ferns, as we see, the sporophyte has a root of its own, and is attached to the soil. Through the aid of root hairs of its own it takes up mineral solutions. It possesses also a true stem, and true leaves in which carbon conversion takes place. It is able to live independently, then. Does a gametophyte phase exist among the ferns? Or has it been lost? If it does exist, what is it like, and where does it grow? From what we have already learned we should expect to find the gametophyte begin with the germination of the spores which are developed on the sporophyte, that is on the fern plant itself. We should investigate this and see.

Chapter 27

Ferns Continued

Gametophyte of Ferns

Sexual Stage of Ferns

We now wish to see what the sexual stage of the ferns is like. Judging from what we have found to take place in the liverworts and mosses we should infer that the form of the plant which bears the sexual organs is developed from the spores. This is true, and if we should examine old decaying logs, or decaying wood in damp places in the near vicinity of ferns, we should probably find tiny, green, thin, heart-shaped growths, lying close to the substratum. These are also found quite frequently on the soil of pots in plant conservatories where ferns are grown. Gardeners also in conservatories usually sow fern spores to raise new fern plants, and usually one can find these heart-shaped growths on the surface of the soil where they have sown the spores. We may call the gardener to our aid in finding them in conservatories, or even in growing them for us if we cannot find them outside. In some cases they may be grown in an ordinary room by keeping the surfaces where they are growing moist, and the air also moist, by placing a glass bell jar over them.

In Figure 297 is shown one of these growths enlarged. Upon the under side we see numerous thread-like outgrowths, the rhizoids, which attach the plant to the substratum, and which act as organs for the absorption of nourishment. The sexual organs are borne on the under side also, and we will study them later. This heart-shaped, flattened, thin, green plant is the *prothallium* of ferns, and we should now give it more careful study, beginning with the germination of the spores.

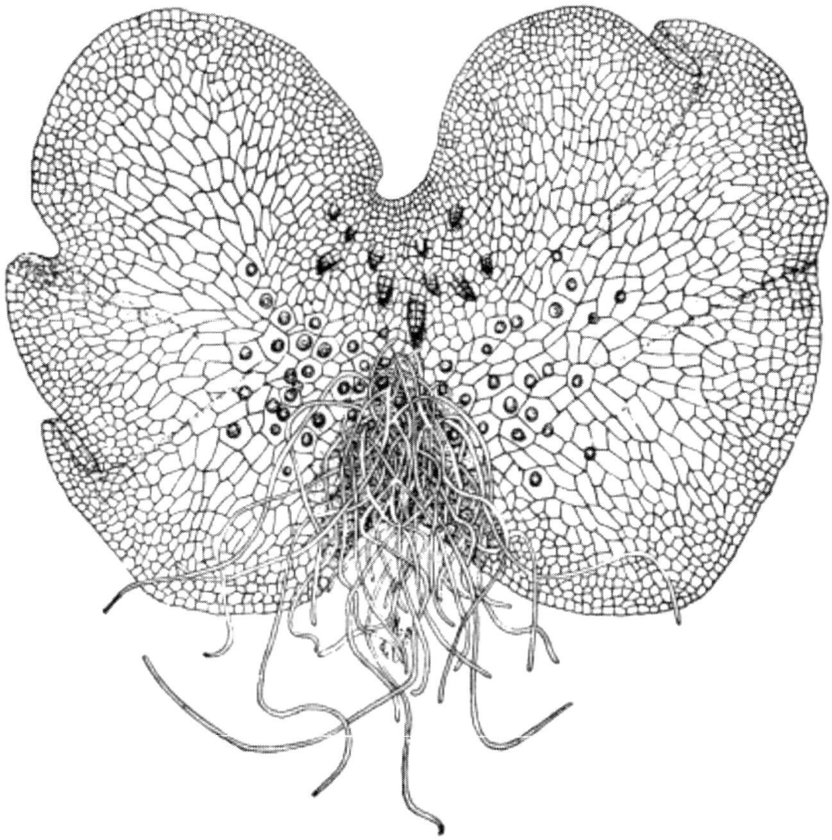

Figure 297. Prothallium of fern, under side, showing rhizoids, antheridia scattered among and near them, and the archegonia near the sinus.

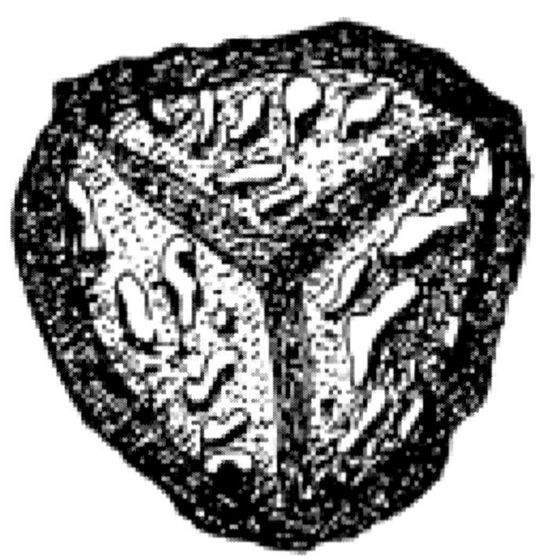

Figure 298.Spore of Pteris serrulatashowing the three-rayed elevation along the side of which the spore wall cracks during germination.

Figure 299. Spore of Aspidium acrostichoides with winged exospore.

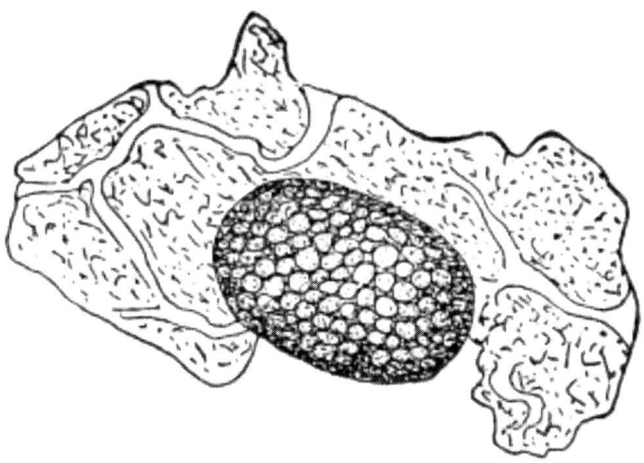

Figure 300. Spore crushed to remove exospore and show endospore.

Spores

We can easily obtain material for the study of the spores of ferns. The spores vary in shape to some extent. Many of them are shaped like a three-sided pyramid. One of these is shown in Figure 298. The outer wall is roughened, and on one end are three elevated ridges which radiate from a given point. A spore of the Christmas fern is shown in Figure 299. The outer wall here is more or less winged. At Figure 300 is a spore of the same species from which the outer wall has been crushed, showing that there is an inner wall also. If possible we should study the germination of the spores of some fern.

Germination of the Spores

After the spores have been sown for about one week to ten days we should mount a few in water for examination with the microscope in order to study the early stages. If germination has begun, we find that here and there are short slender green threads, in many cases attached to brownish bits, the old walls of the spores. Often one will sow the sporangia along with the spores, and in such cases there may be found a number of spores still within the old sporangium wall that are germinating, when they will appear as in Figure 302.

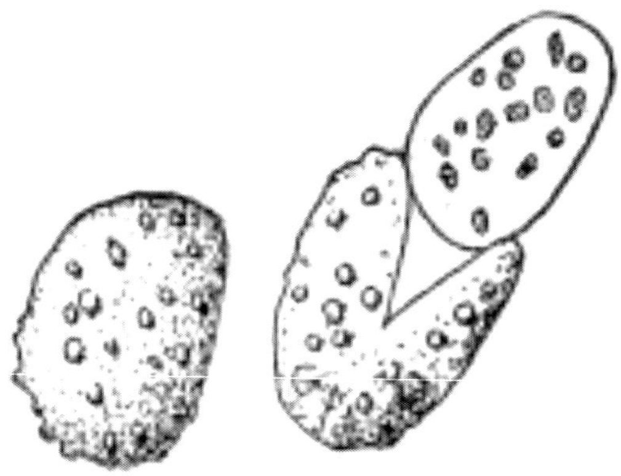

Figure 301. Spores of asplenium; exospore removed from the one at the right.

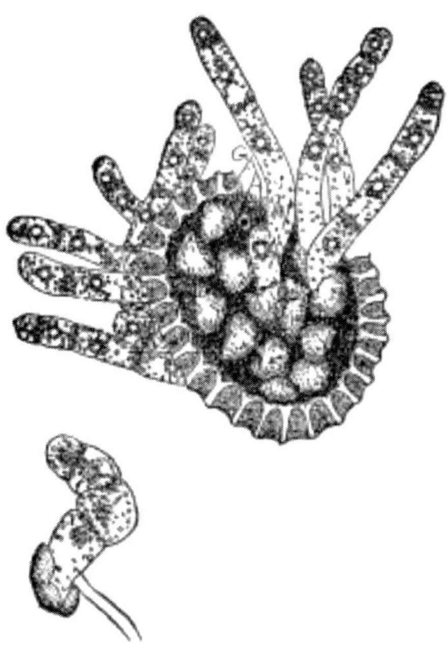

Figure 302. Germinating spores of Pteris aquiline still in the sporangium.

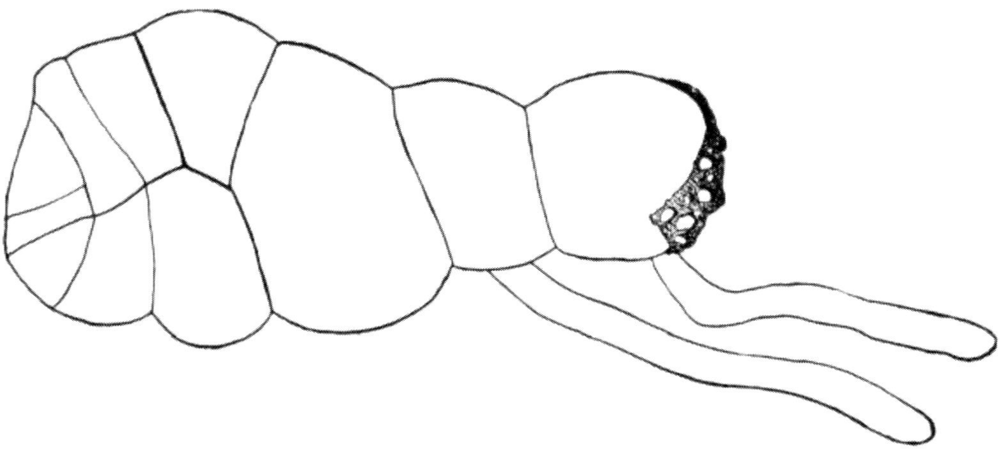

Figure 303. Young prothallium of a fern (niphobolus).

Protonema

These short green threads are called *protonemal* threads, or *protonema*, which means a *first thread*, and it here signifies that this short thread only precedes a larger growth of the same object. In Figures 302, 303 are shown several stages of germination of different spores. Soon after the short germ tube emerges from the crack in the spore wall, it divides by the formation of a cross wall, and as it increases in length other cross walls are formed. But very early in its growth we see that a slender outgrowth takes place from the cell nearest the old spore wall. This slender thread is colorless, and is not divided into cells. It is the first rhizoid, and serves both as an organ of attachment for the thread, and for taking up nutriment.

Prothallium

Very soon, if the sowing has not been so crowded as to prevent the young plants from obtaining nutriment sufficient, we will see that the end of this protonema is broadening, as shown in Figure 303. This is done by the formation of the cell walls in different directions. It now continues to grow in this way, the end becoming broader and broader, and new rhizoids are formed from the under surface of the cells. The growing point remains at the middle of the advancing margin, and the cells which are cut off from either side, as they become old, widen out. In this way the "wings," or margins of the little, green, flattened body, are in advance of the growing point, and the object is more or less heart-shaped, as shown in Figure 297. Thus we see how the prothallium of ferns is formed.

Sexual Organs of Ferns

If we take one of the prothallia of ferns which have grown from the sowings of fern spores, or one of those which may be often found growing on the soil of pots in conservatories, mount it in water on a slip, with the under side uppermost, we can then examine it for the sexual organs, for these are borne in most cases on the under side.

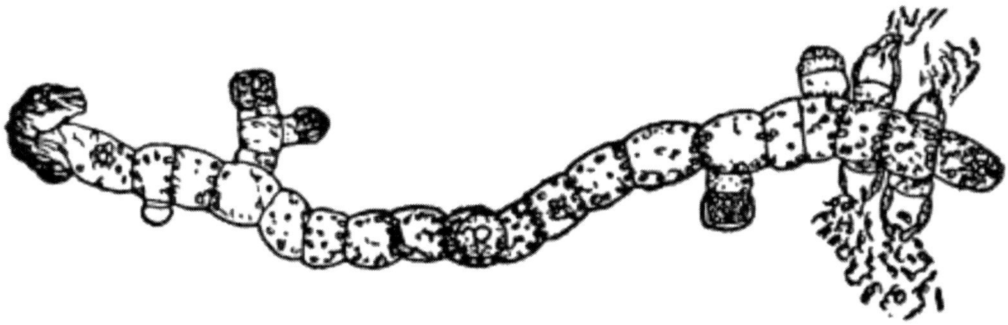

Figure 304. Male prothallium of a fern (niphobolus), in form of an alga or protonema. Spermatozoids escaping from antheridia.

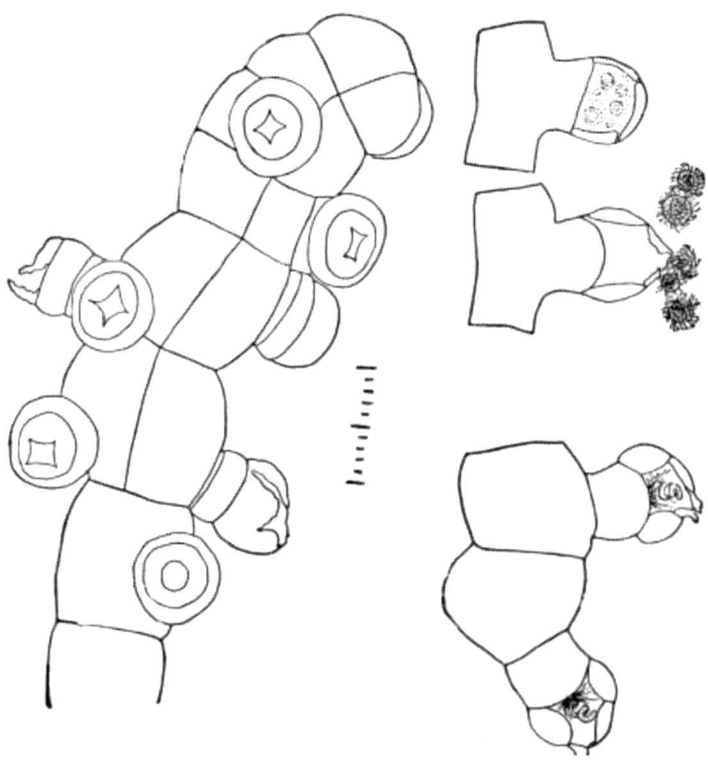

Figure 305. Male prothallium of fern (niphobolus), showing opened and unopened antheridia; section of unopened antheridium; spermatozoids escaping; spermatozoids which did not escape from the antheridium.

Antheridia

If we search among the rhizoids we see small rounded elevations as shown in Figure 297 or 305 scattered over this portion of the prothallium. These are the antheridia. If the prothallia have not been watered for a day or so, we may have an opportunity of seeing the spermatozoids coming out of the antheridium, for when the prothallia are freshly placed in water the cells of the antheridium absorb water. This presses on the contents of the antheridium and bursts the cap cell if the antheridium is ripe, and all the spermatozoids are shot out. We can see here that each one is shaped like a screw, with the coils at first close. But as the spermatozoid begins to move this coil opens somewhat and by the vibration of the long cilia which are on the smaller end it whirls away. In such preparations one may often see them spinning around for a long while, and it is only when they gradually come to rest that one can make out their form.

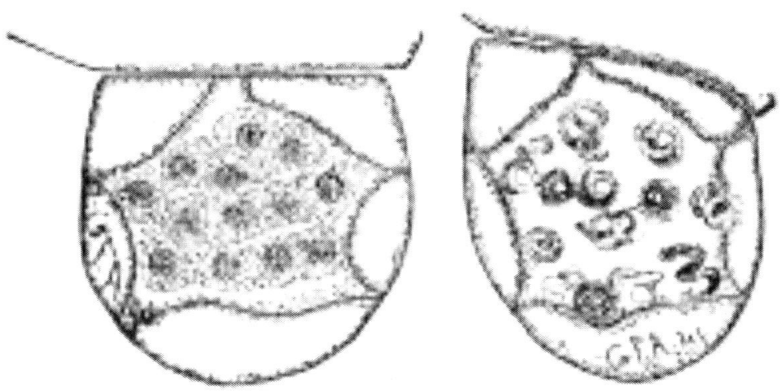

Figure 306. Section of antheridia showing sperm cells, and spermatozoids in the one at the right.

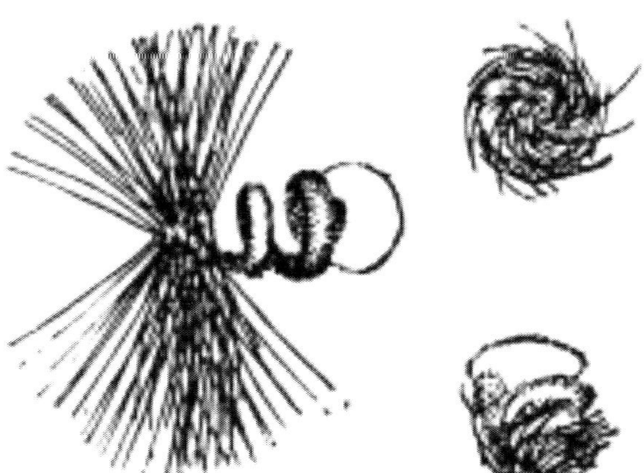

Figure 307. Different views of spermatozoids; in a quiet condition; in motion (Adiantum concinnum).

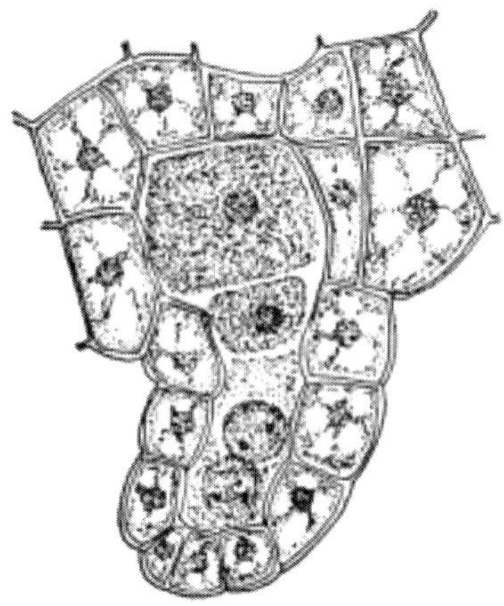

Figure 308. Archegonium of fern. Large cell in the center is the egg, next is the ventral canal cell, and in the canal of the neck are two nuclei of the canal cell.

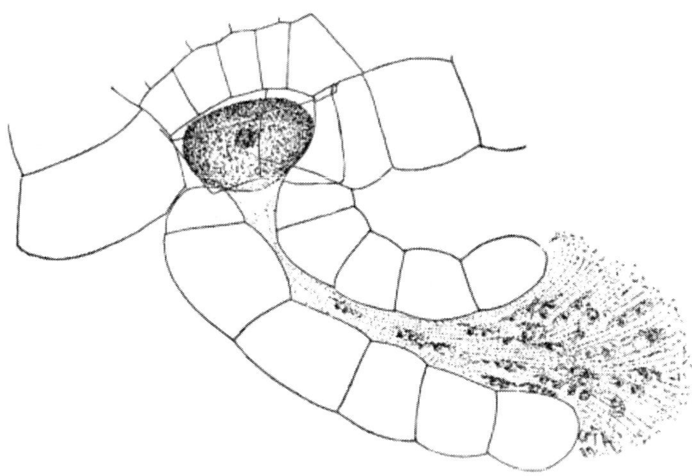

Figure 309. Mature and open archegonium of fern (Adiantum cuneatum) with spermatozoids making their way down through the slime to the egg.

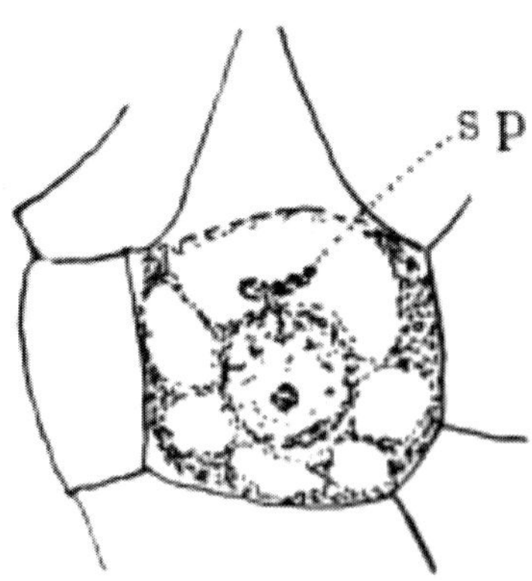

Figure 310. Fertilization in a fern (Marattia). *sp*, spermatozoid fusing with the nucleus of the egg. (After Campbell.)

Archegonia

If we now examine closely on the thicker part of the under surface of the prothallium, just back of the "sinus," we may see longer stout projections from the surface of the prothallium. These are shown in Figure 297. They are the archegonia. One of them in longisection is shown in Figure 308. It is flask-shaped, and the broader portion is sunk in the tissue of the prothallium. The egg is in the larger part. The spermatozoids when they are swimming around over the under surface of the prothallium come near the neck, and here they are caught in the viscid substance which has oozed out of the canal of the archegonium. From here they slowly swim down the canal, and finally one sinks into the egg, fuses with the nucleus of the latter, and the egg is then fertilized. It is now ready to grow and develop into the fern plant. This brings us back to the sporophyte, which begins with the fertilized egg.

SPOROPHYTE

Embryo

The egg first divides into two cells as shown in Figure 228, then into four. Now from each one of these quadrants of the embryo a definite part of the plant develops, from one the first leaf, from one the stem, from one the root, and from the other the organ which is called the foot, and which attaches the embryo to the prothallium, and transports nourishment for the embryo until it can become attached to the soil and lead an independent existence.

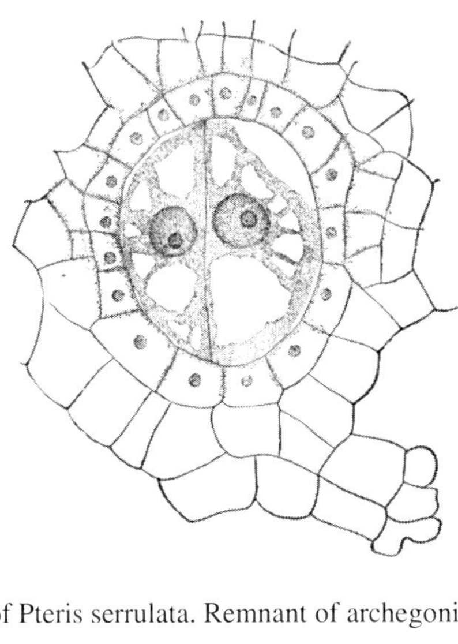

Figure 311. Two-celled embryo of Pteris serrulata. Remnant of archegonium neck below.

During this time the wall of the archegonium grows somewhat to accommodate the increase in size of the embryo, as shown in Figures 312, 313. But soon the wall of the archegonium is ruptured and the embryo emerges, the root attaches itself to the soil, and soon the prothallium dies.

The embryo is first on the under side of the prothallium, and the first leaf and the stem curves upward between the lobes of the heart-shaped body, and then grows upright as shown in Figure 314. Usually only one embryo is formed on a single prothallium, but in one case I found a prothallium with two well-formed embryos, which are Figured in 315.

Comparison of Ferns with Liverworts and Mosses

In the ferns then we have reached a remarkable condition of things as compared with that which we found in the mosses and liverworts. In the mosses and liverworts the sexual phase of the plant (gametophyte) was the prominent one, and consisted of either a thallus or a leafy axis, but in either case it bore the sexual organs and led an independent existence; that is it was capable of obtaining its nourishment from the soil or water by means of organs of absorption belonging to itself, and it also performed the office of photosynthesis.

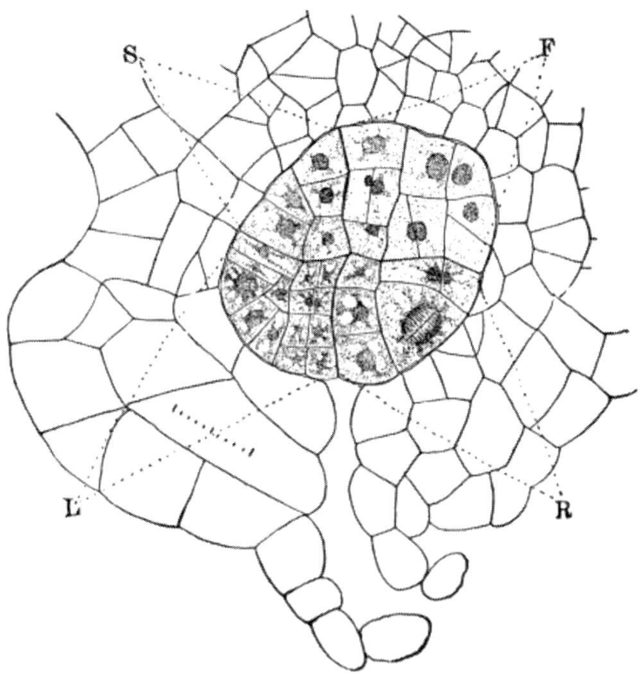

Figure 312. Young embryo of fern (Adiantum concinnum) in enlarged venter of the archegonium. *S*, stem; *L*, first leaf or cotyledon; *R*, root; *F*, foot.

The spore-bearing phase (sporophyte) of the liverworts and mosses, on the other hand, is quite small as compared with the sexual stage, and it is completely dependent on the sexual stage for its nourishment, remaining attached permanently throughout all its development, by means of the organ called a foot, and it dies after the spores are mature.

Now in the ferns we see several striking differences. In the first place, as we have already observed, the spore-bearing phase (sporophyte) of the plant is the prominent one, and that which characterizes the plant. It also leads an independent existence, and, with the exception of a few cases, does not die after the development of the spores, but lives from year to year and develops successive crops of spores. There is a *distinct advance* here in the *size*, *complexity*, and *permanency* of this phase of the plant.

On the other hand the sexual phase of the ferns (gametophyte), while it still is capable of leading an independent existence, is short-lived (with very few exceptions). It is also much smaller than most of the liverworts and mosses, especially as compared with the size of the spore-bearing phase. The gametophyte phase or stage of the plants, then, is decreasing in size and durance as the sporophyte stage is increasing. We shall be interested to see if this holds good of the fern allies, that is of the plants which belong to the same group as the ferns. And as we come later to take up the study of the higher plants we must bear in mind to carry on this comparison, and see if this progression on the one hand of the sporophyte continues, and if the retrogression of the gametophyte continues also.

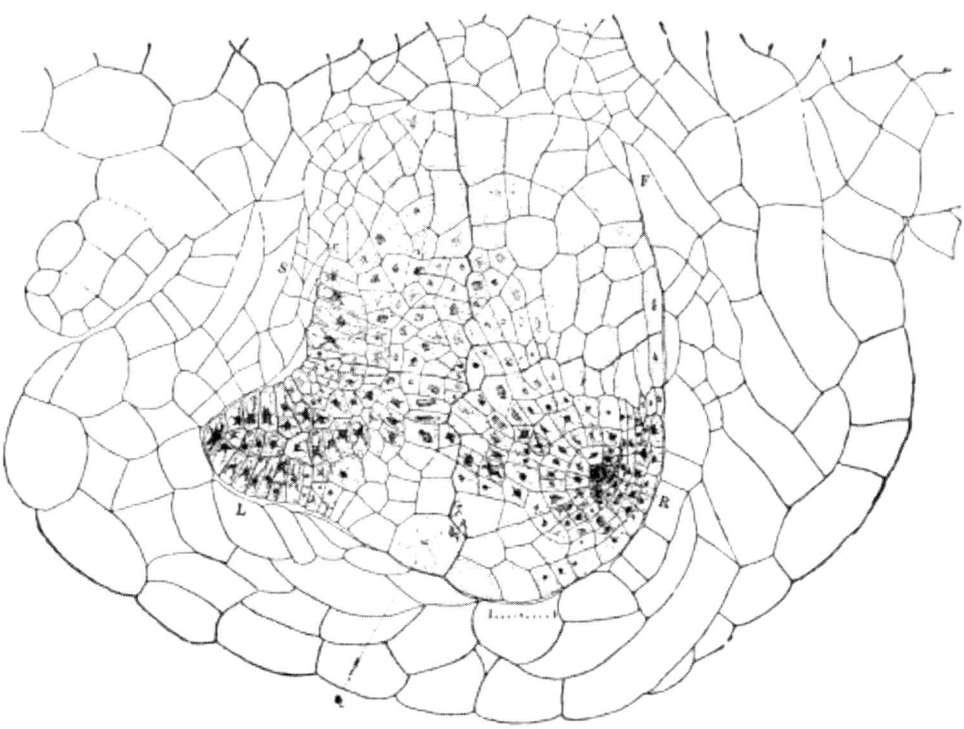

Figure 313. Embryo of fern (Adiantum concinnum) still surrounded by the archegonium, which has grown in size, forming the"calyptra." *L*, leaf; *S*, stem; *R*, root; *F*, foot.

Figure 314. Young plant of Pteris serrulata still attached to prothallium.

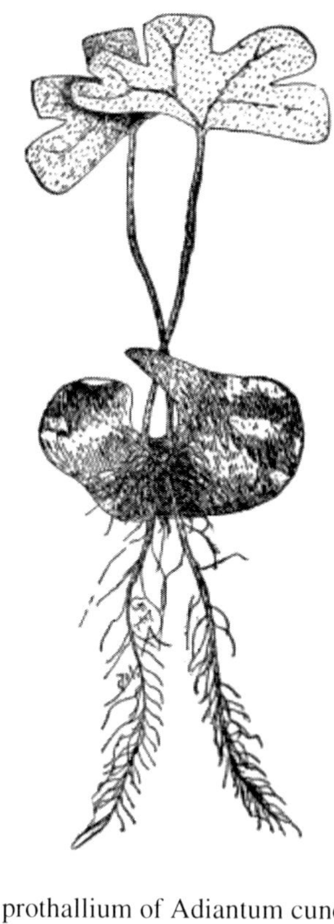

Figure 315. Two embryos from one prothallium of Adiantum cuneatum.

Chapter 28

Dimorphism of Ferns

In comparing the different members of the leaf series there are often striking illustrations of the transition from one form to another, as we have noted in the case of the trillium flower. This occurs in many other flowers, and in some, as in the water-lily, these transformations are always present, here showing a transition from the petals to the stamens. In the bud-scales of many plants, as in the butternut, walnut, currant, etc., there are striking gradations between the form of the simple bud-scales and the form of the leaf. Some of the most interesting of these transformations are found in the dimorphic ferns.

Dimorphism in the Leaves of Ferns

In the common polypody fern, the maidenhair, and in many other ferns, all the leaves are of the same form. That is, there is no difference between the fertile leaf and the sterile leaf. On the other hand, in the case of the Christmas fern we have seen that the fertile leaves are slightly different from the sterile leaves, the former having shorter pinnæ on the upper half of the leaf. The fertile pinnæ are here the shorter ones, and perform but little of the function of carbon conversion. This function is chiefly performed by the sterile leaves and by the sterile portions of the fertile leaves. This is a short step toward the division of labor between the two kinds of leaves, one performing chiefly the labor of carbon conversion, the other chiefly the labor of bearing the fruit.

The Sensitive Fern

This division of labor is carried to an extreme extent in the case of some ferns. Some of our native ferns are examples of this interesting relation between the leaves like the common sensitive fern (Onoclea sensibilis) and the ostrich fern (O. struthiopteris) and the cinnamon-fern (Osmunda cinnamomea). The sensitive fern is here shown in Figure 316. The sterile leaves are large, broadly expanded, and pinnate, the pinnæ being quite large. The fertile leaves are shown also in the Figure, and at first one would not take them for leaves at all. But if we examine them carefully we see that the general plan of the leaf is the same: the two rows of pinnæ which are here much shorter than in the sterile leaf, and the pinnules, or smaller divisions of the pinnæ, are inrolled into little spherical masses which lie close on the side of the pinnæ. If we unroll one of these pinnules we find that there are several fruit dots within protected by this roll. In fact when the spores are mature these pinnules open somewhat, so that the spores may be disseminated.

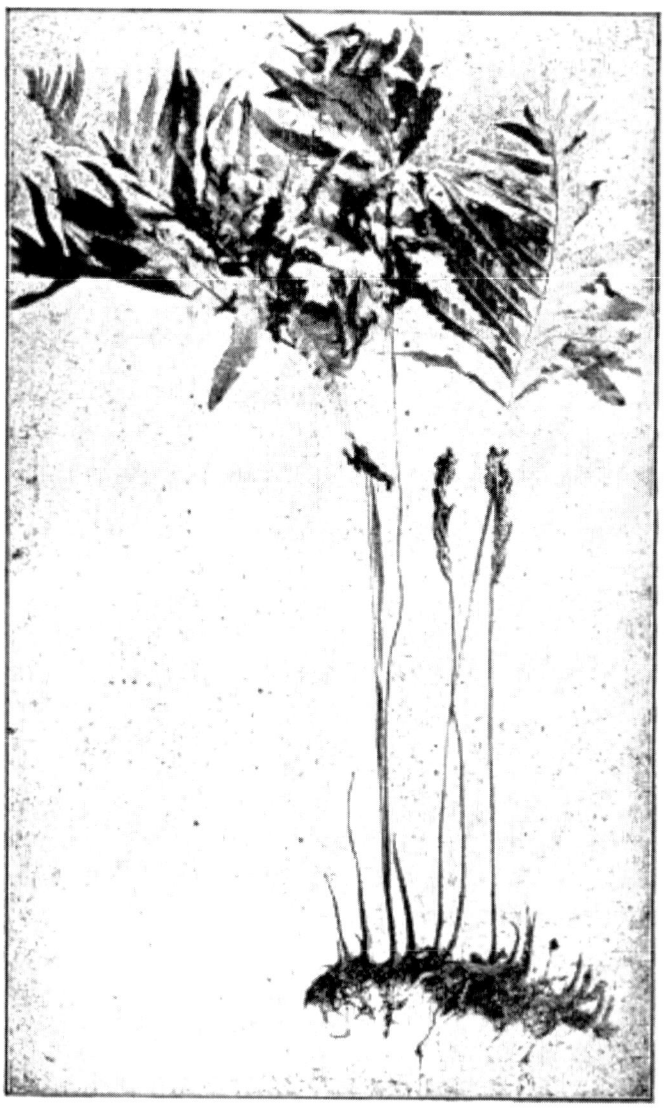

Figure 316. Sensitive fern; normal condition of vegetative leaves and sporophylls.

Figure 317. Sensitive fern; one fertile leaf nearly changed to vegetative leaf.

There is very little green color in these fertile leaves, and what green surface there is is very small compared with that of the broad expanse of the sterile leaves. So here there is practically a complete division of labor between these two kinds of leaves, the general plan of which is the same, and we recognize each as being a leaf.

Figure 318. Sensitive fern, showing one vegetative leaf and two sporophylls completely transformed.

Transformation of the Fertile Leaves of Onoclea to Sterile Ones

It is not a very rare thing to find plants of the sensitive fern which show intermediate conditions of the sterile and the fertile leaf. A number of years ago it was thought by some that this represented a different species, but now it is known that these intermediate forms are partly transformed fertile leaves. It is a very easy matter in the case of the sensitive fern to produce these transformations by experiment. If one in the spring, when the sterile leaves attain a height of 12 to 16 *cm* (8-10 inches), cuts them away, and again when they have a second time reached the same height, some of the fruiting leaves which develop later will be transformed. A few years ago I cut off the sterile leaves from quite a large patch of the sensitive fern, once in May, and again in June. In July, when the fertile leaves were appearing above the ground, many of them were changed partly or completely into sterile leaves. In all some thirty plants showed these transformations, so that every conceivable gradation was obtained between the two kinds of leaves.

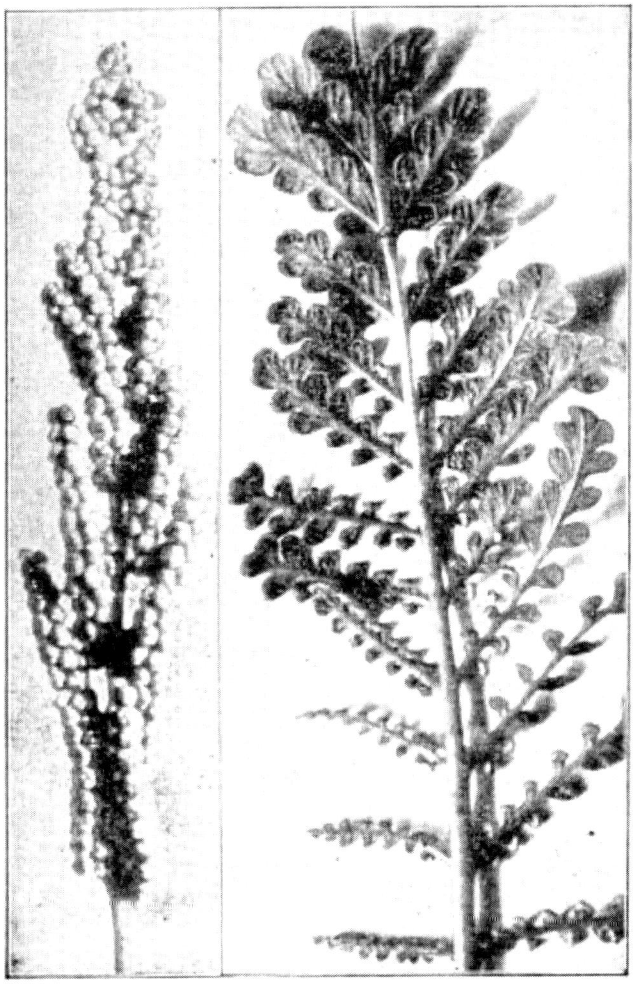

Figure 319. Normal and transformed sporophyll of sensitive fern.

It is quite interesting to note the form of these changed leaves carefully, to see how this change has affected the pinnæ and the sporangia. We note that the tip of the leaf as well as the tips of all the pinnæ are more expanded than the basal portions of the same. This is due to the fact that the tip of the leaf develops later than the basal portions. At the time the stimulus to the change in the development of the fertile leaves reached them they were partly formed, that is the basal parts of the fertile leaves were more or less developed and fixed and could not change.

Those portions of the leaf, however, which were not yet completely formed, under this stimulus, or through correlation of growth, are incited to vegetative growth, and expand more or less completely into vegetative leaves.

The Sporangia Decrease As the Fertile Leaf Expands

If wc now examine the sporangia on the successive pinnæ of a partly transformed leaf we find that in case the lower pinnæ are not changed at all, the sporangia are normal. But as we pass to the pinnæ which show increasing changes, that is those which are more and more expanded, we see that the number of sporangia decrease, and many of them are sterile, that is they bear no spores. Farther up there are only rudiments of sporangia, until on the more expanded pinnæ sporangia are no longer formed, but one may still see traces of the indusium. On some of the changed leaves the only evidences that the leaf began once to form a fertile leaf are the traces of these indusia. In some of these cases the transformed leaf was even larger than the sterile leaf.

The Ostrich Fern

Similar changes were also produced in the case of the ostrich fern, and in Figure 319 is shown at the left a normal fertile leaf, then one partly changed, and at the right one completely transformed.

Dimorphism in Tropical Ferns

Very interesting forms of dimorphism are seen in some of the tropical ferns. One of these is often seen growing in plant conservatories, and is known as the staghorn fern (Platycerium alcicorne). This in nature grows attached to the trunks of quite large trees at considerable elevations on the tree, sometimes surrounding the tree with a massive growth. One kind of leaf, which may be either fertile or sterile, is narrow, and branched in a peculiar manner, so that it resembles somewhat the branching of the horn of a stag. Below these are other leaves which are different in form and sterile. These leaves are broad and hug closely around the roots and bases of the other leaves. Here they serve to catch and retain moisture, and they also catch leaves and other vegetable matter which falls from the trees. In this position the leaves decay and then serve as food for the fern.

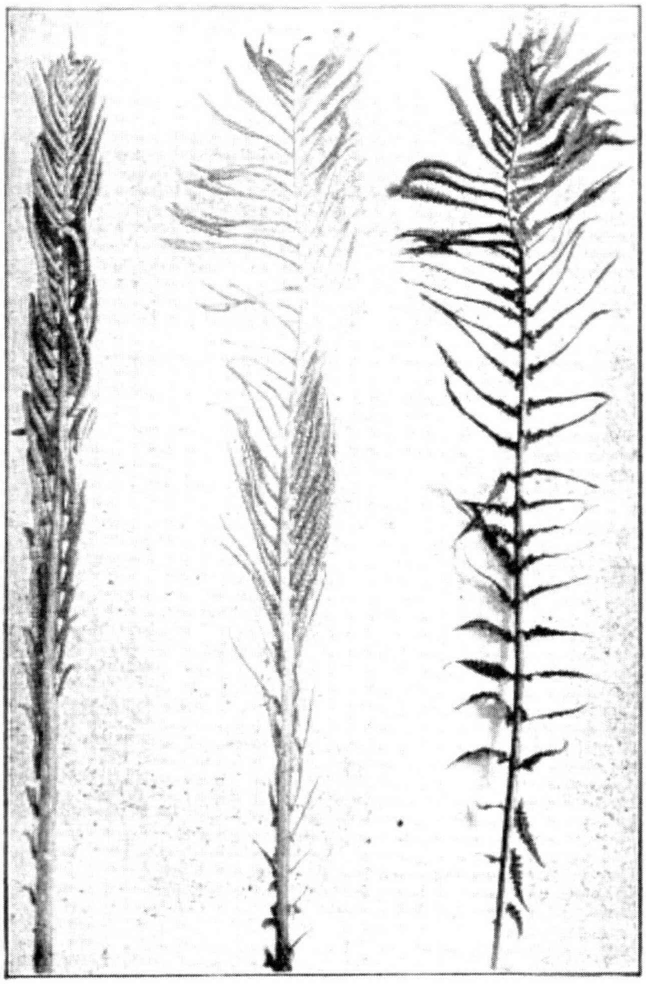

Figure 320. Ostrich fern, showing one normal sporophyll, one partly transformed, and one completely transformed.

Chapter 29

HORSETAILS

Among the relatives of the ferns are the horsetails, so called because of the supposed resemblance of the branched stems of some of the species to a horse's tail, as one might infer from the plant shown in Figure 325. They do not bear the least resemblance to the ferns which we have been studying. But then relationship in plants does not depend on mere resemblance of outward form, or of the prominent part of the plant.

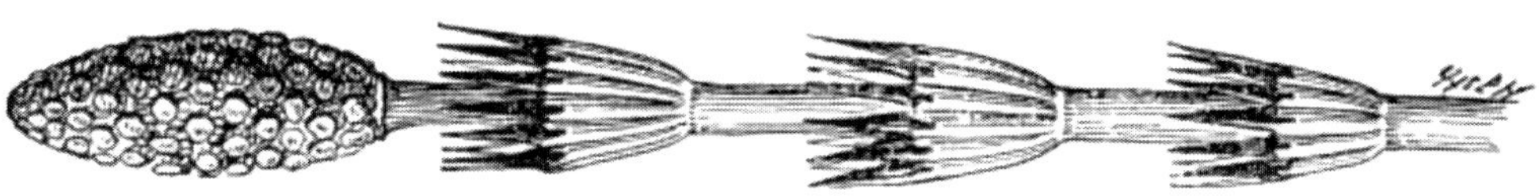

Figure 321. Portion of fertile plant of Equisetum arvense showing whorls of leaves and the fruiting spike.

THE FIELD EQUISETUM. FERTILE SHOOTS

Figure 321 represents the common horsetail (Equisetum arvense). It grows in moist sandy or gravelly places, and the fruiting portion of the plant (for this species is dimorphic), that is the portion which bears the spores, appears above the ground early in the spring. It is one of the first things to peep out of the recently frozen ground. This fertile shoot of the plant does not form its growth this early in the spring. Its development takes place under the ground in the autumn, so that with the advent of spring it pushes up without delay. This shoot is from 10 to 20 *cm*. high, and at quite regular intervals there are slight enlargements, the nodes of the stem. The cylindrical portions between the nodes are the internodes. If we examine the region of the internodes carefully we note that there are thin membranous scales, more or less triangular in outline, and connected at their bases into a ring around the stem. Curious as it may seem, these are the leaves of the horsetail. The stem, if we examine it farther, will be seen to possess numerous ridges which extend lengthwise and which alternate with furrows. Farther, the ridges of one node alternate with those of the internode both above and below. Likewise the leaves of one node alternate with those of the nodes both above and below.

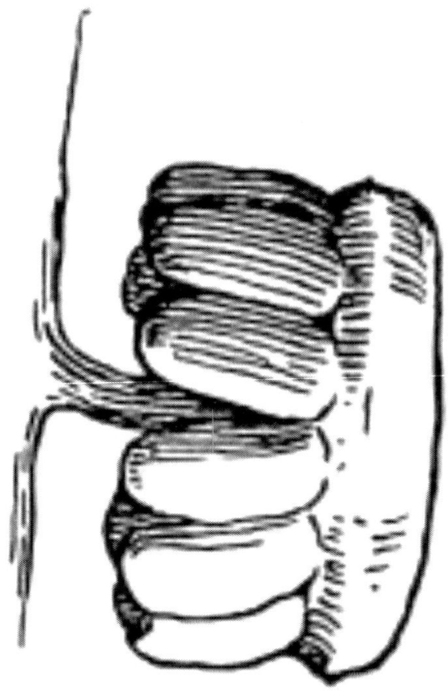

Figure 322. Peltate sporophyll of equisetum (side view) showing sporangia on under side.

SPORANGIA

The end of this fertile shoot we see possesses a cylindrical to conic enlargement. This is the *fertile spike*, and we note that its surface is marked off into regular areas if the spores have not yet been disseminated. If we dissect off a few of these portions of the fertile spike, and examine one of them with a low magnifying power, it will appear like the Figure 322. We see here that the angular area is a disk-shaped body, with a stalk attached to its inner surface, and with several long sacs projecting from its inner face parallel with the stalk and surrounding the same. These elongated sacs are the *sporangia*, and the disk which bears them, together with the stalk which attaches it to the stem axis, is the *sporophyll*, and thus belongs to the leaf series. These sporophylls are borne in close whorls on the axis.

Spores

When the spores are ripe the tissue of the sporangium becomes dry, and it cracks open and the spores fall out. If we look at Figure 323 we see that the spore is covered with a very singular coil which lies close to the wall. When the spore dries this uncoils and thus rolls the spore about. Merely breathing upon these spores is sufficient to make them perform very curious evolutions by the twisting of these four coils which are attached to one place of the wall. They are formed by the splitting up of an outer wall of the spore.

Sterile Shoot of the Common Horsetail

When the spores are ripe they are soon scattered, and then the fertile shoot dies down. Soon afterward, or even while some of the fertile shoots are still in good condition, sterile shoots of the plant begin to appear above the ground. One of these is shown in Figure 325. This has a much more slender stem and is provided with numerous branches. If we examine the stem of this shoot, and of the branches, we see that the same kind of leaves are present and that the markings on the stem are similar. Since the leaves of the horsetail are membranous and not green, the stem is green in color, and this performs the function of photosynthesis. These green shoots live for a great part of the season, building up material which is carried down into the underground stems, where it goes to supply the forming fertile shoots in the fall. On digging up some of these plants we see that the underground stems are often of great extent, and that both fertile and sterile shoots are attached to one and the same.

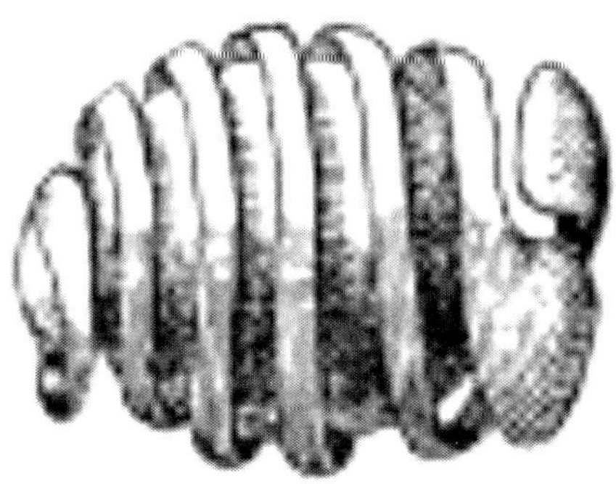

Figure 323. Spore of equisetum with elaters coiled up.

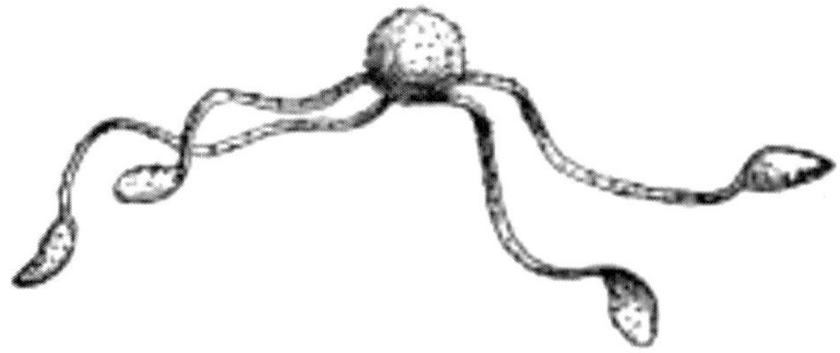

Figure 324. Spore of equisetum with elaters uncoiled.

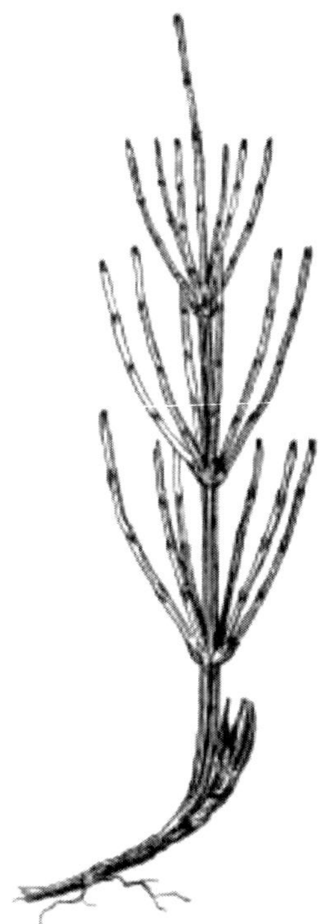

Figure 325. Sterile plant of horsetail (*Equisetum arvensis*).

THE SCOURING RUSH, OR SHAVE GRASS

Another common species of horsetail in the Northern States grows on wet banks, or in sandy soil which contains moisture along railroad embankments. It is the scouring rush (E. hyemale), so called because it was once used for polishing purposes. This plant like all the species of the horsetails has underground stems. But unlike the common horsetail, there is but one kind of aerial shoot, which is green in color and fertile. The shoots range as high as one meter or more, and are quite stout. The new shoots which come up for the year are unbranched, and bear the fertile spike at the apex. When the spores are ripe the apex of the shoot dies, and the next season small branches may form from a number of the nodes.

GAMETOPHYTE OF EQUISETUM

The spores of equisetum have chlorophyll when they are mature, and they are capable of germinating as soon as mature. The spores are all of the same kind as regards size, just as we found in the case of the ferns. But they develop prothallia of different sizes, according to the amount of nutriment which they obtain. Those which obtain but little nutriment are smaller and develop only antheridia, while those which obtain more nutriment become larger, more or less branched, and develop archegonia. This character of an independent prothallium (gametophyte) with the characteristic sexual organs, and the also independent sporophyte, with spores, shows the relationship of the horsetails with the ferns. We thus see that these characters of the reproductive organs, and the phases and fruiting of the plant, are more essential in determining relationships of plants than the mere outward appearances.

Chapter 30

CLUB MOSSES

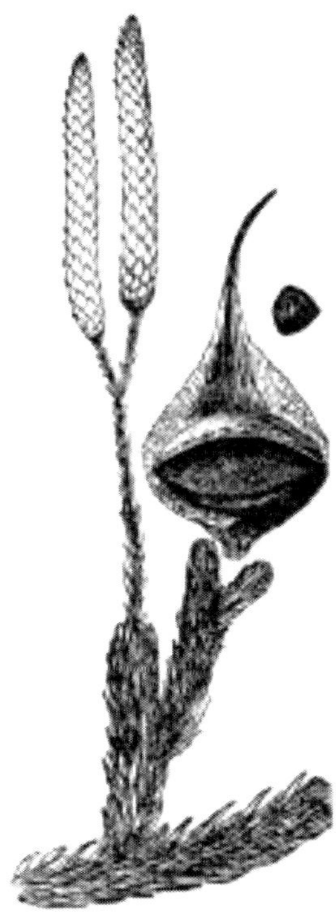

Figure 326. Lycopodium clavatum, branch bearing two fruiting spikes; at right sporophyll with open sporangium; single spore near it.

What are called the "club mosses" make up another group of interesting plants which rank as allies of the ferns. They are not of course true mosses, but the general habit of some of the smaller species, and especially the form and size of the leaves, suggest a resemblance to the larger of the moss plants.

The Clavate Lycopodium

Here is one of the club mosses (Figure 326) which has a wide distribution and which is well entitled to hold the name of club because of the form of the upright club-shaped branches. As will be seen from the illustration, it has a prostrate stem. This stem runs for considerable distances on the surface of the ground, often partly buried in the leaves, and sometimes even buried beneath the soil. The leaves are quite small, are flattened-awl-shaped, and stand thickly over the stem, arranged in a spiral manner, which is the usual arrangement of the leaves of the club mosses. Here and there are upright branches which are forked several times. The end of one or more of these branches becomes produced into a slender upright stem which is nearly leafless, the leaves being reduced to mere scales. The end of this leafless branch then terminates in one or several cylindrical heads which form the club.

Fruiting Spike of Lycopodium Clavatum

This club is the fruiting spike or head (sometimes termed a *strobilus*). Here the leaves are larger again and broader, but still not so large as the leaves on the creeping shoots, and they are paler. If we bend down some of the leaves, or tear off a few, we see that in the axil of the leaf, where it joins the stem, there is a somewhat rounded, kidney-shaped body. This is the spore-case or sporangium, as we can see by an examination of its contents. There is but a single spore-case for each of the fertile leaves (sporophyll). When it is mature, it opens by a crosswise slit as seen in Figure 326. When we consider the number of spore-cases in one of these club-shaped fruit bodies we see that the number of spores developed in a large plant is immense. In mass the spores make a very fine, soft powder, which is used for some kinds of pyrotechnic material, and for various toilet purposes.

Lycopodium Lucidulum

Another common species is Figured at 327. This is Lycopodium lucidulum. The habit of the plant is quite different. It grows in damp ravines, woods, and moors. The older parts of the stem are prostrate, while the branches are more or less ascending. It branches in a forked manner. The leaves are larger than in the former species, and they are all of the same size, there being no appreciable difference between the sterile and fertile ones. The characteristic club is not present here, but the spore-cases occupy certain regions of the stem, as shown at 327. In a single season one region of the stem may bear spore-cases, and then a sterile portion of the same stem is developed, which later bears another series of spore-cases higher up.

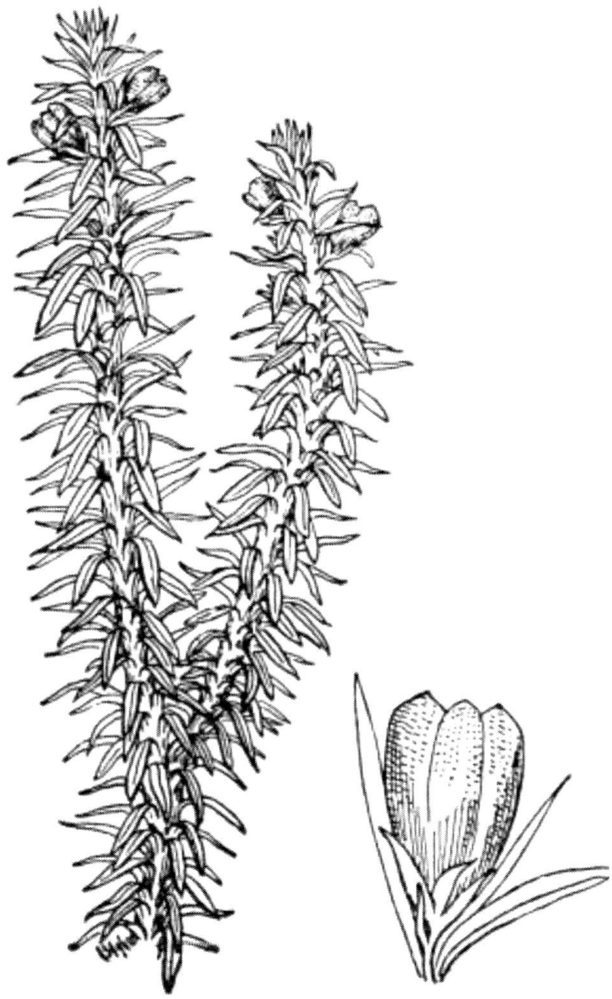

Figure 327. Lycopodium lucidulum, bulbils in axils of leaves near the top, sporangia in axils of leaves below them. At right is a bulbil enlarged.

Bulbils on Lycopodium Lucidulum

There is one curious way in which this club moss multiplies. One may see frequently among the upper leaves small wedge-shaped or heart-shaped green bodies but little larger than the ordinary leaves. These are little buds which contain rudimentary shoot and root and several thick green leaves. When they fall to the ground they grow into new lycopodium plants, just as the bulbils of cystopteris do which were described in the chapter on ferns.

Note.—The prothallia of the species of lycopodium which have been studied are singular objects. In *L. cernuum* a cylindrical body sunk in the earth is formed, and from the upper surface there are green lobes. In L. phlegmaria and some others slender branched, colorless bodies are formed which according to Treub grow as a saphrophyte in decayed bark of trees. Many of the prothallia examined have a fungus growing in their tissue which is supposed to play some part in the nutrition of the prothallium.

The Little Club Mosses (Selaginella)

Closely related to the club mosses are the selaginellas. These plants resemble closely the general habit of the club mosses, but are generally smaller and the leaves more delicate. Some species are grown in conservatories for ornament, the leaves of such usually having a beautiful metallic lustre. The leaves of some are arranged as in lycopodium, but many species have the leaves in four to six rows. Figure 328 represents a part of a selaginella plant (*S. apus*). The fruiting spike possesses similar leaves, but they are shorter, and their arrangement gives to the spike a four-sided appearance.

Figure 328. Selaginella with three fruiting spikes. (Selaginella apus.)

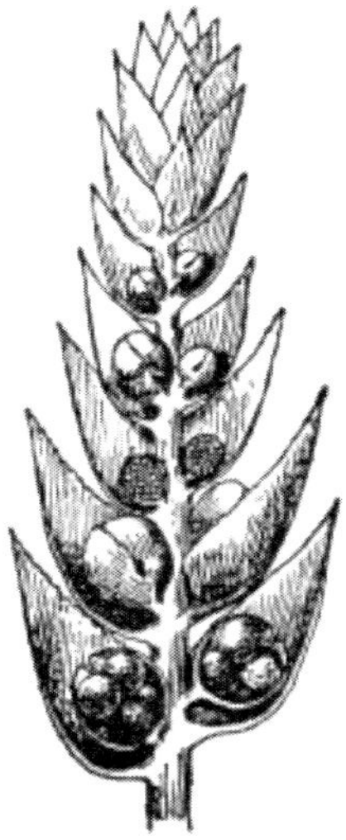

Figure 329. Fruiting spike showing large and small sporangia.

Figure 330. Large sporangium.

Figure 331. Small sporangium.

Sporangia

On examining the fruiting spike, we find as in lycopodium that there is but a single sporangium in the axil of a fertile leaf. But we see that they are of two different kinds, small ones in the axils of the upper leaves, and large ones in the axils of a few of the lower leaves of the spike. The *microspores* are borne in the smaller spore-cases and the *macrospores* in the larger ones. Figures 329-331 give the details. Thcre are many microspores in a single small spore-case, but 3-4 macrospores in a large spore-case.

Male Prothallia

The prothallia of selaginella are much reduced structures. The microspores when mature are already divided into two cells. When they grow into the mature prothallium a few more cells are formed, and some of the inner ones form the spermatozoids, as seen in Figure 332. Here we see that the antheridium itself is larger than the prothallia. Only antheridia are developed on the prothallia formed from the microspores, and for this reason the prothallia are called *male prothallia*. In fact a male prothallium of selaginella is nearly all antheridium, so reduced has the gametophyte become here.

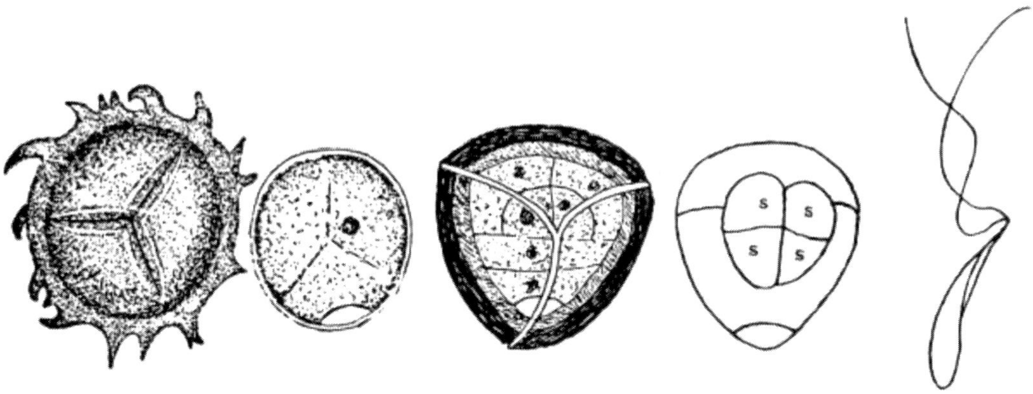

Figure 332. Details of microspore and male prothallium of selaginella; 1st, microspore; 2d, wall removed to show small prothallial cell below; 3d, mature male prothallium still within the wall; 4th, small cell below is the prothallial cell, the remainder is antheridium with wall and four sperm cells within; 5th spermatozoid. After Beliaieff and Pfeffer.

Female Prothallia

The female prothallia are developed from the macrospores. The macrospores when mature have a rough, thick, hard wall. The female prothallium begins to develop inside of the macrospore before it leaves the sporangium. The protoplasm is richer near the wall of the spore and at the upper end. Here the nucleus divides a great many times, and finally cell walls are formed, so that a tissue of considerable extent is formed inside the wall of the spore, which is very different from what takes place in the ferns we have studied. As the prothallium matures the spore is cracked at the point where the three angles meet, as shown in Figure 334. The archegonia are developed in this exposed surface, and several can be seen in the illustration.

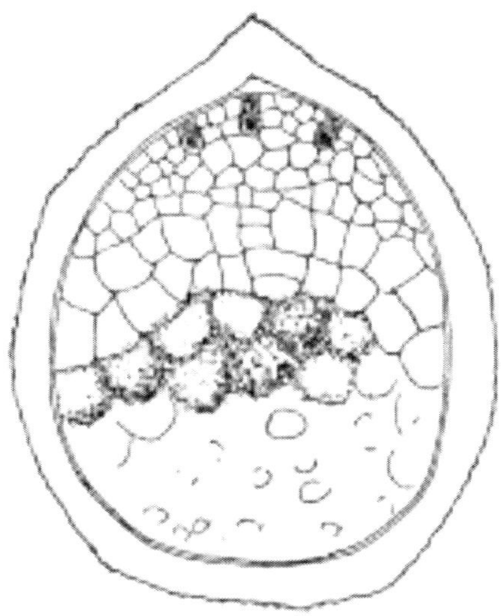

Figure 333. Section of mature macrospore of selaginella, showing female prothallium and archegonia. After Pfeffer.

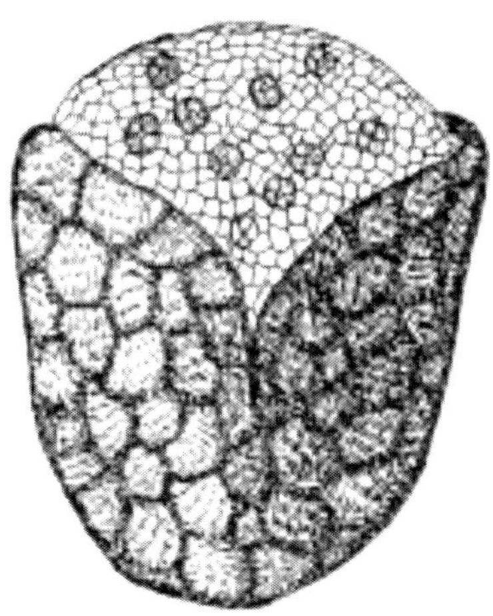

Figure 334. Mature female prothallium of selaginella, just bursting open the wall of macrospore, exposing archegonia. After Pfeffer.

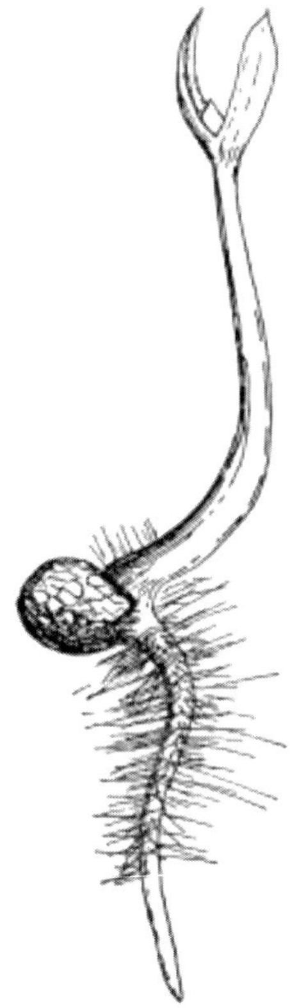

Figure 335. Seedling of selaginella still attached to the macrospore. After Campbell.

EMBRYO

After fertilization the egg divides in such a way that a long cell called a suspensor is cut off from the upper side, which elongates and pushes the developing embryo down into the center of the spore, or what is now the female prothallium. Here it derives nourishment from the tissues of the prothallium, and eventually the root and stem emerge, while a process called the "foot" is still attached to the prothallium. When the root takes hold on the soil the embryo becomes free.

Chapter 31

QUILLWORTS (ISOETES)

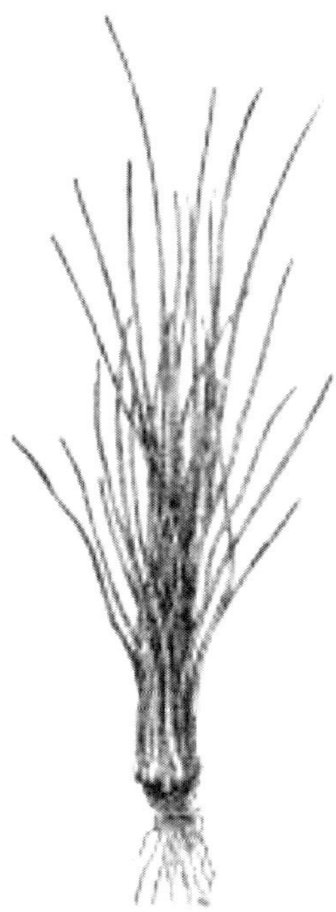

Figure 336. Isoetes, mature plant, sporophyte stage.

The quillworts, as they are popularly called, are very curious plants. They grow in wet marshy places. They receive their name from the supposed resemblance of the leaf to a quill. Figure 336 represents one of these quillworts (Isoetes engelmannii). The leaves are the prominent part of the plant, and they are about all that can be seen except the roots, without removing the leaves. Each leaf, it will be seen, is long and needle-like, except the basal part, which is expanded, not very unlike, in outline, a scale of an onion. These expanded basal portions of the leaves closely overlap each other, and the very short stem is completely covered at all times. Figure 338 is from a longitudinal section of a quillwort. It shows the form of the leaves from this view (side view), and also the general outline of the short stem, which is triangular. The stem is therefore a very short object.

SPORANGIA OF ISOETES

If we pull off some of the leaves of the plant we see that they are somewhat spoon-shaped as in Figure 337. In the inner surface of the expanded base we note a circular depression which seems to be of a different texture from the other portions of the leaf. This is a *sporangium*. Beside the spores on the inside of the sporangium, there are strands of sterile tissue which extend across the cavity. This is peculiar to isoetes of all the members of the class of plants to which the ferns belong, but it will be remembered that sterile strands of tissue are found in some of the liverworts in the form of elaters.

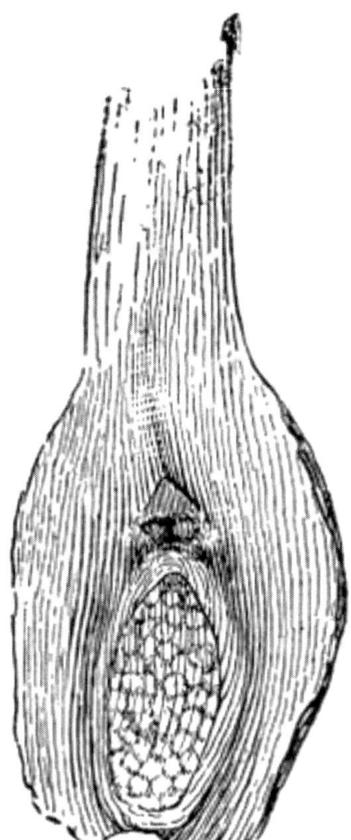

Figure 337. Base of leaf of isoetes, showing sporangium with macrospores. (Isoetes engelmannii.)

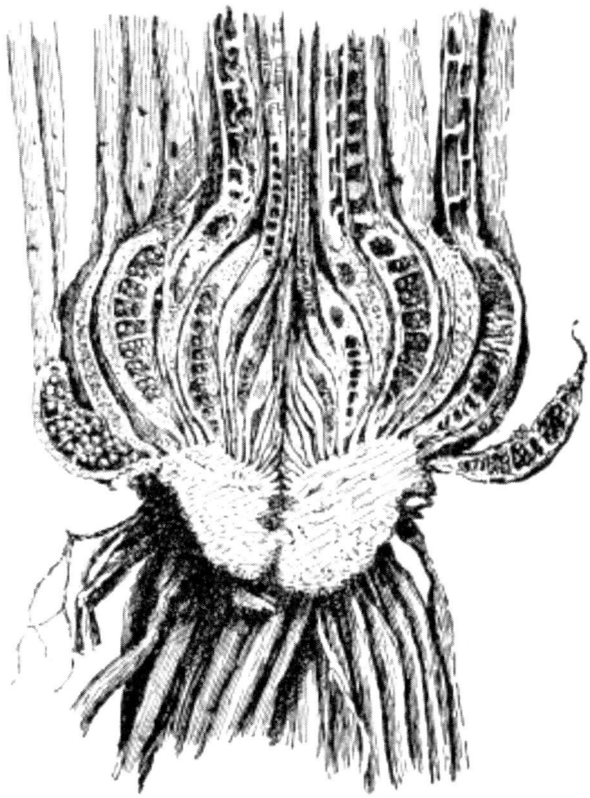

Figure 338. Section of plant of Isoetes engelmannii, showing cup-shaped stem, and longitudinal sections of the sporangia in the thickened bases of the leaves.

The spores of isoetes are of two kinds, small ones (microspores) and large ones (macrospores), so that in this respect it agrees with selaginella, though it is so very different in other respects. When one kind of spore is borne in a sporangium usually all in that sporangium are of the same kind, so that certain sporangia bear microspores, and others bear macrospores. But it is not uncommon to find both kinds in the same sporangium. When a sporangium bears only microspores the number is much greater than when one bears only macrospores.

If we examine some of the microspores of isoetes we see that they are shaped like the quarters of an apple, that is they are of the bilateral type as seen in some of the ferns (asplenium).

MALE PROTHALLIA

In isoetes, as in selaginella, the microspores develop only male prothallia, and these are very rudimentary, one division of the spore having taken place before the spore is mature, just as in selaginella.

Female Prothallia

These are developed from the macrospores. The latter are of the tetrahedral type. The development of the female prothallium takes place in much the same way as in selaginella, the entire prothallium being enclosed in the macrospore, though the cell divisions take place after it has left the sporangium. When the archegonia begin to develop the macrospore cracks at the three angles and the surfacc bearing the archegonia projects slightly as in selaginella. Absorbing organs in the form of rhizoids are very rarely formed.

Embryo

The embryo lies well immersed in the tissue of the prothallium, though there is no suspensor developed as in selaginella.

Chapter 32

COMPARISON OF FERNS AND THEIR RELATIVES

COMPARISON OF SELAGINELLA AND ISOETES WITH THE FERNS

On comparing selaginella and isoetes with the ferns, we see that the sporophyte is, as in the ferns, the prominent part of the plant. It possesses root, stem, and leaves. While these plants are not so large in size as some of the ferns, still we see that there has been a great advance in the sporophyte of selaginella and isoetes upon what exists in the ferns. There is a division of labor between the sporophylls, in which some of them bear microsporangia with microspores, and some bear macrosporangia with only macrospores. In the ferns and horsetails there is only one kind of sporophyll, sporangium, and spore in a species. By this division of labor, or differentiation, between the sporophylls, one kind of spore, the microspore, is compelled to form a male prothallium, while the other kind of spore, the macrospore, is compelled to form a female prothallium. This represents a progression of the sporophyte of a very important nature.

On comparing the gametophyte of selaginella and isoetes with that of the ferns, we see that there has been a still farther retrogression in size from that which we found in the independent and large gametophyte of the liverworts and mosses. In the ferns, while it is reduced, it still forms rhizoids, and leads an independent life, absorbing its own nutrient materials, and assimilating carbon. In selaginella and isoetes the gametophyte does not escape from the spore, nor does it form absorbing organs, nor develop assimilative tissue. The reduced prothallium develops at the expense of food stored by the sporophyte while the spore is developing. Thus, while the gametophyte is separate from the sporophyte in selaginella and isoetes, it is really dependent on it for support or nourishment.

The important general characters possessed by the ferns and their so-called allies, as we have found, are as follows: The spore-bearing part, which is the fern plant, leads an independent existence from the prothallium, and forms root, stem, and leaves. The spores are borne in sporangia on the leaves. The prothallium also leads an independent existence, though in isoetes and selaginella it has become almost entirely dependent on the sporophyte. The prothallium bears also well-developed antheridia and archegonia. The root, stem, and leaves of the sporophyte possess vascular tissue. All the ferns and their allies agree in the possession of these characters. The mosses and liverworts have well-developed antheridia and archegonia, and the higher plants have vascular tissue. But no plant of either of these groups possesses the combined characters which we find in the ferns and their relatives. The latter are, therefore, the fern-like plants, or *pteridophyta*. The living forms of the pteridophyta are classified as follows into families or orders. (See page 295.)

Classification of the Pteridophytes

Of the living pteridophytes four classes may be recognized.

Class Filicineæ[34]

This class includes the ferns. Four orders may be recognized.

Order Ophioglossales

(One Family, Ophioglossaceæ).—This order includes the grapeferns (Botrychium), so called because of the large botryoid cluster of sporangia, resembling roughly a cluster of grapes; and the adder-tongue (Ophioglossum), the sporangia being embedded in a long tongue-like outgrowth from the green leaf. Botrychium and Ophioglossum are widely distributed. The roots are fleshy, nearly destitute of root hairs, and contain an endophytic fungus, so that the roots are mycorhiza. The gametophyte is subterranean, and devoid of chlorophyll. In Botrychium virginianum, an endophytic fungus has been found in the prothallium. Another genus (Helminthostachys) with one species is limited to the East Indies.

Order Marattiales

(One Family, Marattiaceæ).—These are tropical ferns, with only four or five living genera (Marattia, Danæa, etc.). They resemble the typical ferns, but the sporangia are usually united, several forming a compound sporangium, or *synangium*.

The Ophioglossales and Marattiales are known as eusporangiate ferns, while the following order includes the leptosporangiate ferns.

[34] As class Filicales in Engler and Prantl.

Table Showing Relation of Gametophyte and Sporophyte in the Pteridophytes

<table>
<tr><td colspan="5">GAMETOPHYTE.
(Becoming smaller, mostly independent.
In selaginella and isoetes becoming
dependent on the sporophyte.)</td><td colspan="3">SPOROPHYTE
(Largest part of the plant. The fern plant. Independent of, and more hardy than, the gametophyte. Usually perennial.)</td><td rowspan="2">Beginning of Gametophyte.</td></tr>
<tr><td></td><td colspan="2">Vegetative Part.</td><td colspan="2">Sexual Organs.</td><td>Beginning of Sporophyte.</td><td>Vegetative Part.</td><td>Fruiting Part.</td></tr>
<tr><td rowspan="2">Ferns.
(Polypodiaceæ.)</td><td colspan="2" rowspan="2">A green, thin, expanded, heart-shaped growth, with rhizoids.</td><td colspan="2">Usually both kinds on the same prothallium.</td><td rowspan="2">Fertilized egg. (Develops into fern plant.)</td><td rowspan="2">Root, stem, leaf.</td><td rowspan="2">Sporangia on leaf. All of one kind. Sporangium contains ....</td><td rowspan="2">Spores.</td></tr>
<tr><td>Antheridia with spermatozoids.</td><td>Archegonia, each with egg.</td></tr>
<tr><td rowspan="2">Equisetum.</td><td colspan="2" rowspan="2">A green, thin, expanded, lobed growth, with rhizoids.</td><td colspan="2">Usually the two kinds cn different prothallia.</td><td rowspan="2">Fertilized egg. (Develops into equisetum plant.)</td><td rowspan="2">Root, stem, leaf.</td><td rowspan="2">Sporangia on sporophylls. All of one kind. Sporangium contains ....</td><td rowspan="2">Spores.</td></tr>
<tr><td>Antheridia, on small male prothallia, with spermatozoids.</td><td>Archegonia on larger female prothallia, each with an egg.</td></tr>
<tr><td rowspan="2">Isoetes.</td><td colspan="2">Colorless, rounded mass of cells, inside of spore wall, usually no rhizoids, or but few. Two kinds.</td><td colspan="2">On different prothallia.</td><td rowspan="2">Fertilized egg. (Develops into isoetes plant.)</td><td rowspan="2">Root, stem, leaf. Stem very short. Leaves bear sporangia in cavities at base; outer leaves usually bear macrosporangia, inner ones microsporangia.</td><td>Sporangia of two kinds. Small ones contain ....</td><td>Microspores.</td></tr>
<tr><td>Small ones, male. Developed into small prothallial cell, and antherid cell while still in sporangium.</td><td>Large ones, female. Developed from nutriment stored in macrospore from sporophyte.</td><td>One antheridium, much larger than the single prothallial cell. Antheridium with spermatozoids.</td><td>Few archegonia, in apex of oval colorless, female prothallium, each with egg.</td><td>Large ones contain ....</td><td>Macrospores.</td></tr>
</table>

Table Showing Relation of Gametophyte and Sporophyte in the Pteridophytes (Continued)

<table>
<tr><td colspan="4">GAMETOPHYTE.
(Becoming smaller, mostly independent.
In selaginella and isoetes becoming
dependent on the sporophyte.)</td><td colspan="6">SPOROPHYTE
(Largest part of the plant. The fern plant. Independent of, and more hardy than, the gametophyte. Usually perennial.)</td><td rowspan="2">Beginning
of
Gametophyte.</td></tr>
<tr><td></td><td>Vegetative Part.</td><td colspan="2">Sexual Organs.</td><td>Beginning
of
Sporophyte.</td><td colspan="3">Vegetative Part.</td><td colspan="2">Fruiting Part.</td></tr>
<tr><td rowspan="2">Selaginella.</td><td colspan="3">Colorless, rounded mass of cells inside of spore wall, no rhizoids, or but few. Two kinds.</td><td colspan="2">On different prothallia.</td><td rowspan="2">Fertilized egg. (Develops into selaginella plant.)</td><td colspan="2" rowspan="2">Root, stem, leaf. Spore-bearing leaves grouped on the end of stem in a spike. Lower ones bear macrosporangia, upper ones bear microsporangia.</td><td>Sporangia of two kinds. Small ones contain ....</td><td>Microspores.</td></tr>
<tr><td colspan="2">Small ones, male. Developed into small prothallial cell, and antherid cell while in sporangium.</td><td>Large ones, female. Developed while still in sporangium and dependent on sporophyte.</td><td>One antheridium, much larger than the single prothallial cell. Antheridium with spermatozoids.</td><td>Few archegonia, in apex of oval, colorless, female prothallium, each with egg.</td><td>Large ones contain ....</td><td>Macrospores.</td></tr>
</table>

Order Filicales

This order includes the typical ferns. Eight families are recognized.

Family Osmundaceæ

Three genera are known in this family. Osmunda has a number of species, three of which are found in the Eastern United States; the cinnamon-fern (O. cinnamomea), the royal fern (O. regalis), and Clayton's fern (O. claytoniana). No species of this family are found on the Pacific coast.

Family Gleicheniaceæ

These ferns are found chiefly in the tropics, and in the mountain regions of the temperate zones of South America. There are two genera, Gleichenia containing all but one of the known species.

Family Matoniaceæ

One genus, Matonia, in the Malayan region.

Family Schizæceæ

These are chiefly tropical, but two species are found in eastern North America, Schizæa pusilla and Lygodium palmatum, the latter a climbing fern.

Family Hymenophyllaceæ

These are known as the filmy ferns because of their thin, delicate leaves. They grow only in damp or wet regions, mostly in the tropics, but a few species occur in the southern United States.

Family Cyatheaceæ

These are known as the tree ferns, because of the large size which many of them attain. They occur chiefly in tropical mountainous regions, many of them palm-like and imposing because of the large trunks and leaves. Dicksonia, Cyathea, Cibotium, Alsophila, are some of the most conspicuous genera.

Family Parkeriaceæ

There is a single species in this family (Ceratopteris thalictroides), abundant in the tropics and extending into Florida. It is aquatic.

Family Polypodiaceæ

This family includes the larger number of living ferns and many genera and species are found in North America. Examples, Polypodium, Pteridium (= Pteris), Adiantum, etc.

Order Hydropterales (or Salviniales)

The members of this order are peculiar, aquatic ferns, some floating on the water (Azolla, Salvinia), while others are anchored to the soil by roots (Marsilia, Pilularia). They are known as water ferns. The sporangia are of two kinds, one containing large spores (macrospores) and the other small spores (microspores). They are therefore heterosporous ferns.

Family Salviniaceæ

There are two genera, Salvinia and Azolla.

Family Marsiliaceæ

Two genera, Marsilia and Pilularia. In this family the sporangia are enclosed in a sporocarp, which forms a pod-like structure.

Class Equisetineæ[35]

Order Equisetales

The single order contains a single family, Equisetaceæ, among the living forms, and but a single genus, Equisetum. There are about twenty-four species, with fourteen in the United States (see Chapter 29).

Class Lycopodiineæ[36]

Order Lycopodiales

The first two families of this order include the homosporous Lycopodiineæ, while the Selaginellaceæ are heterosporous.

Family Lycopodiaceæ

There are two genera. Lycopodium (club moss) includes many species, most of them tropical, but a number in temperate and subarctic regions. The gametophyte of many species is tuberous, lacks chlorophyll, and in some there lives an endophytic fungus. Phylloglossum with one species is found in Australia.

---

[35] As class Equisetales in Engler and Prantl.

[36] As class Lycopodiales in Engler and Prantl.

Family Psilotaceæ

There are two genera. Psilotum chiefly in the tropics has one species (P. triquetrum) in the region of Florida.

Family Selaginellaceæ

These include the little club mosses, with one genus, Selaginella (see Chapter 30).

Class Isoetineæ

Order Isoetales

With one family Isoetaceæ and one genus Isoetes (see Chapter 31). There are about fifty species, with about sixteen in the United States.

Chapter 33

GYMNOSPERMS

THE WHITE PINE

General Aspect of the White Pine

The white pine (Pinus strobus) is found in the Eastern United States. In favorable situations in the forest it reaches a height of about 50 meters (about 160 feet), and the trunk a diameter of over 1 meter. In well-formed trees the trunk is straight and towering; the branches where the sunlight has access and the trees are not crowded, or are young, reaching out in graceful arms, form a pyramidal outline to the tree. In old and dense forests the lower branches, because of lack of sunlight, have died away, leaving tall, bare trunks for a considerable height.

The Long Shoots of the Pine

The branches are of two kinds. Those which we readily recognize are the long branches, so called because the growth in length each year is considerable. The terminal bud of the long branches, as well as of the main stem, continues each year the growth of the main branch or shoot; while the lateral long branches arise each year from buds which are crowded close together around the base of the terminal bud. The lateral long branches of each year thus appear to be in a whorl. The distance between each false whorl of branches, then, represents one year's growth in length of the main stem or long branch.

The Dwarf Shoots of the Pine

The dwarf branches are all lateral on the long branches, or shoots. They are scattered over the year's growth, and each bears a cluster of five long, needle-shaped, green leaves, which remain on the tree for several years. At the base of the green leaves are a number of chaff-like scales, the previous bud scales. While the dwarf branches thus bear green leaves, and scales, the long branches bear only thin scale-like leaves which are not green.

Spore-Bearing Leaves of the Pine

The two kinds of spore-bearing leaves of the pine, and their close relatives, are so different from anything which we have yet studied, and are so unlike the green leaves of the pine, that we would scarcely recognize them as belonging to this category. Indeed there is great uncertainty regarding their origin.

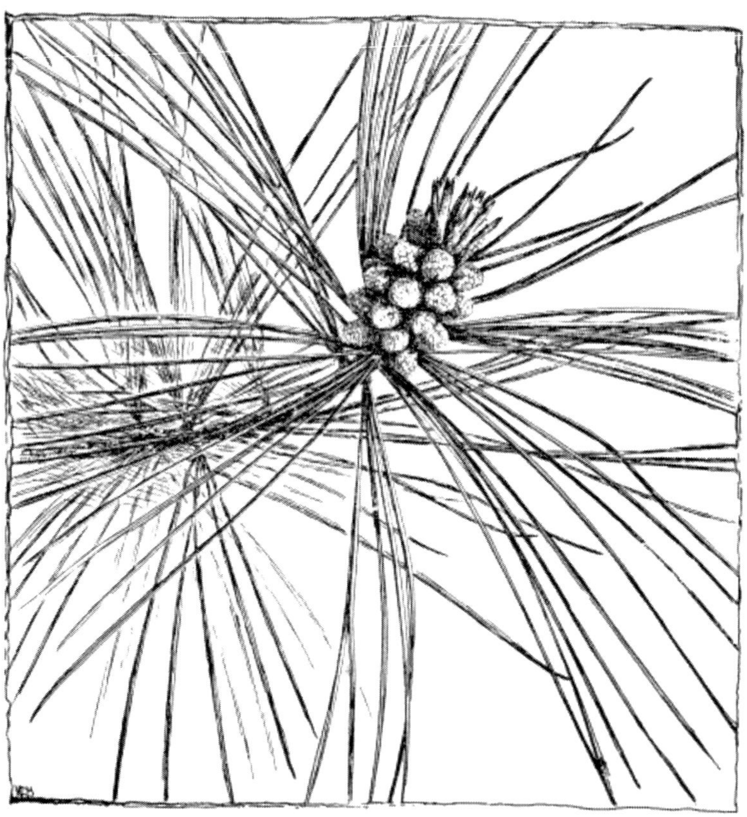

Figure 339. Spray of white pine showing cluster of male cones just before the scattering of the pollen.

Male Cones, or Male Flowers

The male cones are borne in clusters as shown in Figure 339. Each compact, nearly cylindrical, or conical mass is termed a cone, or flower, and each arises in place of a long lateral branch. One of these cones is shown considerably enlarged in Figure 340. The central axis of each cone is a lateral branch, and belongs to the stem series. The stem axis of the cone can be seen in Figure 341. It is completely covered by stout, thick, scale-like outgrowths. These scales are obovate in outline, and at the inner angle of the upper end there are several rough, short spines. They are attached by their inner lower angle, which forms a short stalk or petiole, and continues through the inner face of the scale as a "midrib." What corresponds to the lamina of the scale-like leaf bulges out on each side below and makes the bulk of the scale. These prominences on the under side are the sporangia (microsporangia). There are thus two sporangia on a sporophyll (microsporophyll). When the spores (microspores), which here are usually called pollen grains, are mature, each sporangium, or anther locule, splits down the middle as shown in Figure 342, and the spores are set free.

Figure 340. Staminate cone of white pine, with bud scales removed on one side.

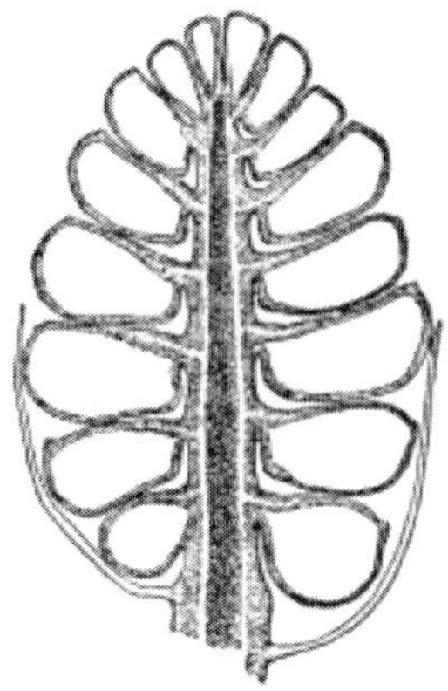

Figure 341. Section of staminate cone, showing sporangia.

Figure 342. Two sporophylls removed, showing opening of sporangia.

Figure 343. Pollen grain of white pine.

Microspores of the Pine, or Pollen Grains

A mature pollen grain of the pine is shown in Figure 343. It is a queer-looking object, possessing on two sides an air sac, formed by the upheaval of the outer coat of the spore at these two points. When the pollen is mature, the moisture dries out of the scale (or stamen, as it is often called here) while it ripens. When a limb, bearing a cluster of male cones, is jarred by the hand, or by currents of air, the split suddenly opens, and a cloud of pollen bursts out from the numerous anther locules. The pollen is thus borne on the wind and some of it falls on the female flowers.

Figure 344. White pine, branch with cluster of mature cones shedding the seed. A few young cones four months old are shown on branch at the left. Drawn from photograph.

Form of the Mature Female Cone

A cluster of the white pine cones is shown in Figure 344. These are mature, and the scales have spread as they do when mature and becoming dry, in order that the seeds may be set at liberty. The general outline of the cone is lanceolate, or long oval, and somewhat curved. It measures about 10-15*cm* long. If we remove one of the scales, just as they are beginning to spread, or before the seeds have scattered, we shall find the seeds attached to the upper surface at the lower end. There are two seeds on each scale, one at each lower angle. They are ovate in outline, and shaped somewhat like a biconvex lens. At this time the seeds easily fall away, and may be freed by jarring the cone. As the seed is detached from the scale a strip of tissue from the latter is peeled off. This forms a "wing" for the seed. It is attached to one end and is shaped something like a knife blade. On the back of the scale is a small appendage known as the cover scale.

Figure 345. Mature cone of white pine at time of scattering of the seed, nearly natural size.

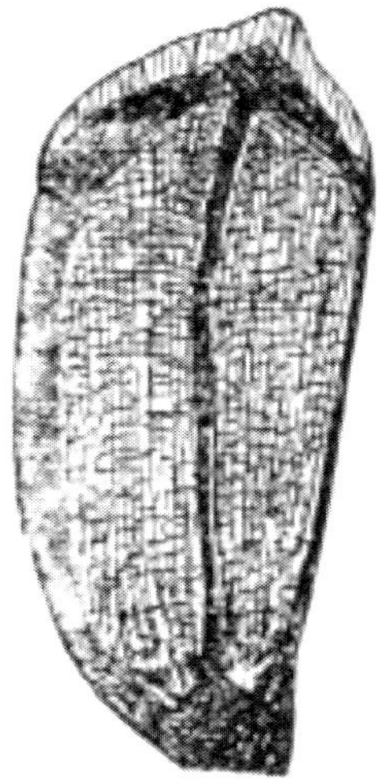

Figure 346. Sterile scale. Seeds undeveloped.

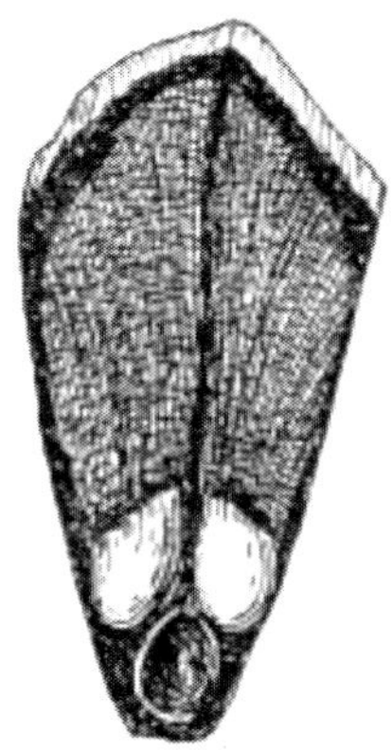

Figure 347. Scale with well-developed seeds.

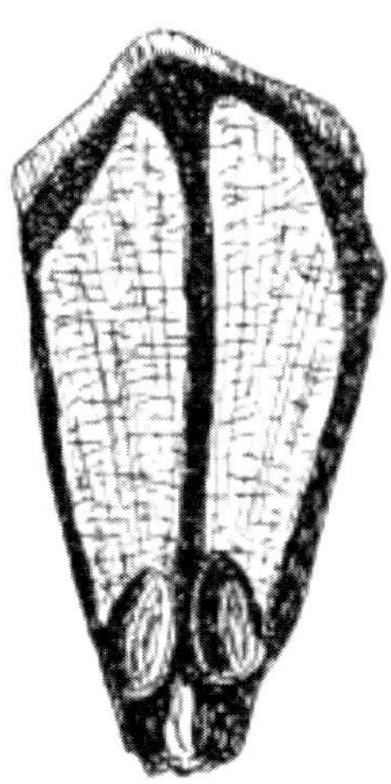

Figure 348. Seeds have split off from scale.

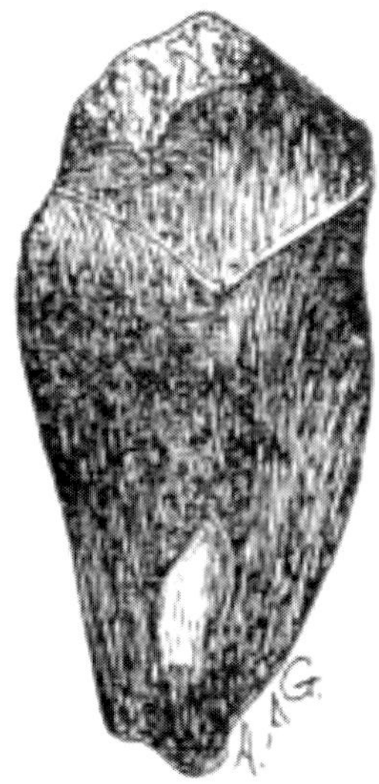

Figure 349. Back of scale with small cover scale.

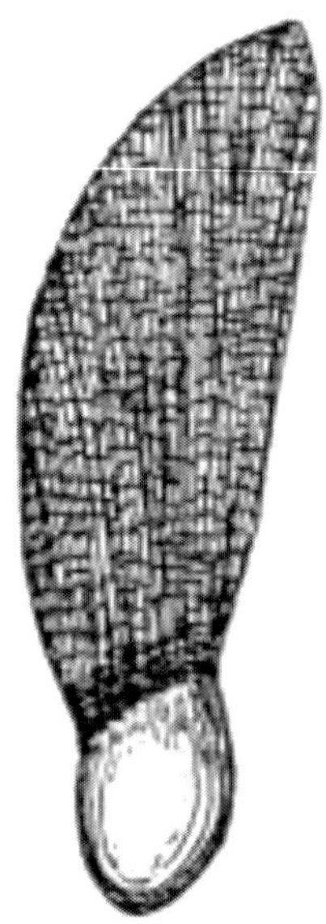

Figure 350. Winged seed free from scale.

Figures 346-350.—White pine showing details of mature scales and seed.

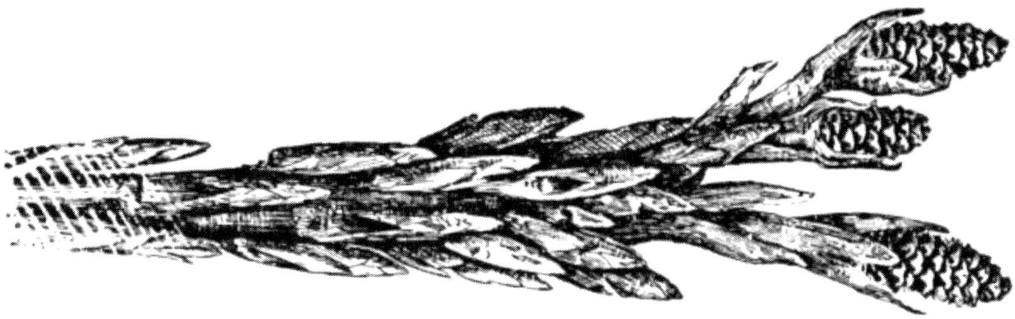

Figure 351. Female cones of the pine at time of pollination, about natural size.

Formation of the Female Pine Cone

The female flowers begin their development rather late in the spring of the year. They are formed from terminal buds of the higher branches of the tree. In this way the cone may terminate the main shoot of a branch, or of the lateral shoots in a whorl. After growth has proceeded for some time in the spring, the terminal portion begins to assume the appearance of a young female cone or flower. These young female cones, at about the time that the pollen is escaping from the anthers, are long ovate, measuring about 6-10 *mm* long. They stand upright as shown in Figure 351.

Form of a "Scale" of the Female Flower

If we remove one of the scales from the cone at this stage we can better study it in detail. It is flattened, and oval in outline, with a stout "rib," if it may be so called, running through the middle line and terminating in a point. The scale is in two parts as shown in Figure 354, which is a view of the under side. The small "outgrowth" which appears as an appendage is the cover scale, for while it is smaller in the pine than the other portion, in some of the relatives of the pine it is larger than its mate, and being on the outside, covers it. (The inner scale is sometimes called the ovuliferous scale, because it bears the ovules.)

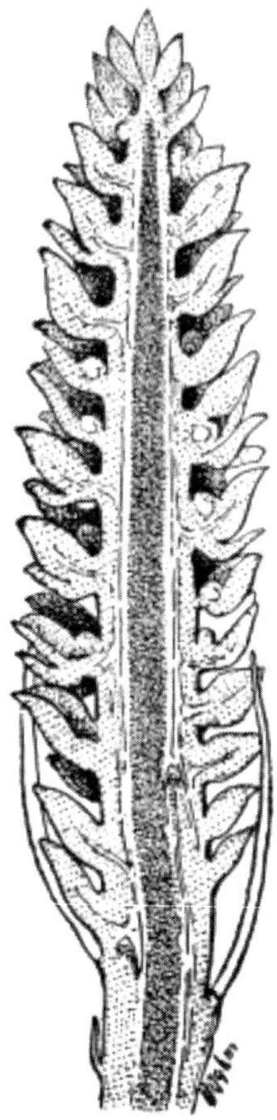

Figure 352. Section of female cone of white pine, showing young ovules (macrosporangia) at base of the ovuliferous scales.

Figure 353. Scale of white pine with the two ovules at base of ovuliferous scale.

Figure 354. Scale of white pine seen from the outside, showing the cover scale.

Ovules, or Macrosporangia, of the Pine

At each of the lower angles of the scale is a curious oval body with two curved, forceps-like processes at the lower and smaller end.

Figure 355. Branch of white pine showing young female cones at time of pollination on the ends of the branches, and one-year-old cones below, near the time of fertilization.

These are the macrosporangia, or, as they are called in the higher plants, the ovules. These ovules, as we see, are in the positions of the seeds on the mature cones. In fact the wall of the ovule forms the outer coat of the seed, as we will later see.

Pollination

At the time when the pollen is mature the female cones are still erect on the branches, and the scales, which during the earlier stages of growth were closely pressed against one another around the axis, are now spread apart. As the clouds of pollen burst from the clusters of the male cones, some of it is wafted by the wind to the female cones. It is here caught in the open scales, and rolls down to their bases, where some of it falls between these forceps-like processes at the lower end of the ovule. At this time the ovule has exuded a drop of a sticky fluid in this depression between the curved processes at its lower end. The pollen sticks to this, and later, as this viscid substance dries up, it pulls the pollen close up in the depression against the lower end of the ovule. This depression is thus known as the *pollen chamber*.

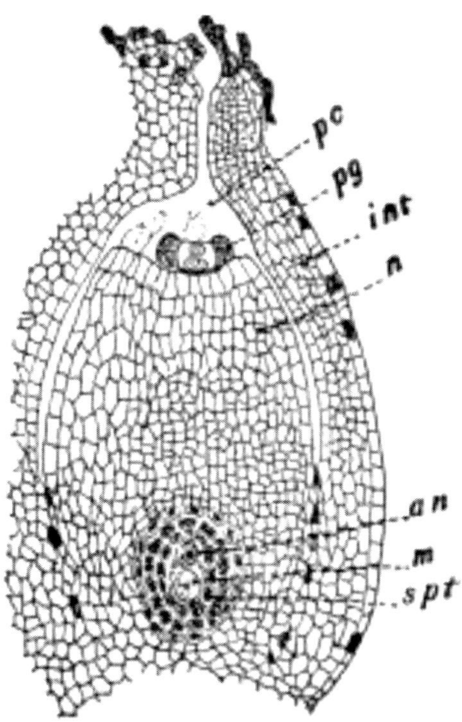

Figure 356. Macrosporangium of pine (ovule). *int*, integument; *n*, nucellus; *m*, macrospore; *pc*, pollen chamber; *pg*, pollen grain; *an*, axile row; *spt*, spongy tissue. (After Ferguson.)

Now the open scales on the young female cone close up again so tightly that water from rains is excluded. What is also very curious, the cones, which up to this time have been standing erect, so that the open scale could catch the pollen, now turn so that they hang downward. This more certainly excludes the rains, since the overlapping of the scales forms a shingled surface. Quantities of resin are also formed in the scales, which exudes and makes the cone practically impervious to water.

The female cone now slowly grows during the summer and autumn, increasing but little in size during this time. During the winter it rests, that is, ceases to grow. With the coming of spring, growth commences again and at an accelerated rate. The increase in size is more rapid. The cone reaches maturity in September. We thus see that nearly eighteen months elapse from the beginning of the female flower to the maturity of the cone, and about fifteen months from the time that pollination takes place.

Female Prothallium of the Pine

To study this we must make careful longitudinal sections through the ovule (better made with the aid of a microtome). Such a section is shown in Figure 358. The outer layer of tissue, which at the upper end (point where the scale is attached to the axis of the cone) stands free, is the ovular coat, or *integument*. Within this integument, near the upper end, there is a cone-shaped mass of tissue. This mass of tissue is the *nucellus*, or the *macrosporangium* proper. In the lower part of the nucellus in Figure 356 can be seen a rounded mass of "spongy tissue" (*spt*), which is a special nourishing tissue of the nucellus, or sporangium, around the macrospore. Within this can be seen an axile row of three cells (*an: m*). The lowest one, which is larger than the other two, is the *macrospore*. Sometimes there are four of these cells in the axile row. This axile row of three or four cells is formed by the two successive divisions of a mother cell in the nucellus. So it would appear that these three or four cells are all spores.

Only one of them, however, the lower one, develops; the others are disorganized and disappear. The nucleus of the macrospore now divides several times to form several free nuclei in the now enlarging cavity, much as the nucleus of the macrospore in Selaginella and Isoetes divides within the spore. The development thus far takes place during the first summer, and now with the approach of winter the very young female prothallium goes into rest about the stage shown in Figure 358. The conical portion of the nucellus which lies above is the nucellar cap.

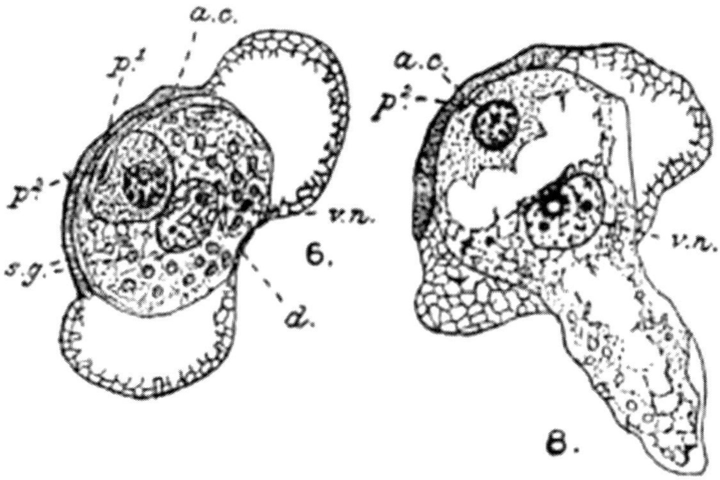

Figure 357. Pollen grains of pine. One of them germinating. p^1 and p^2, the two disintegrated prothallial cells, = sterile part of male gametophyte; *a.c.*, central cell of antheridium; *v.n.*, vegetative nucleus or tube nucleus of the single-wall cell of antheridium; *s.g.*, starch grains. (After Ferguson.)

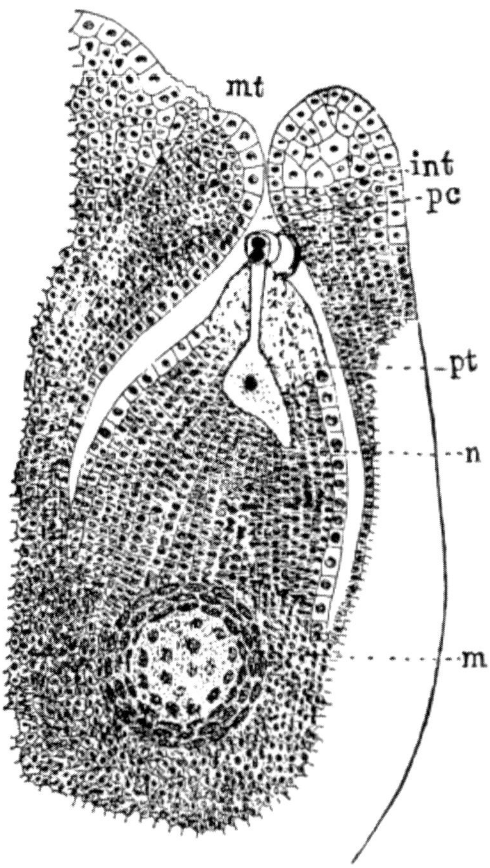

Figure 358. Section of ovule of white pine. *int*, integument; *pc*, pollen chamber; *pt*, pollen tube; *n*, nucleus; *m*, macrospore cavity.

Male Prothallia

By the time the pollen is mature the male prothallium is already partly formed. In Figure 343 we can see two well-formed cells. Two other cells are formed earlier, but they become so flattened that it is difficult to make them out when the pollen grain is mature. These are shown in Figure 357, p^1 and p^2, and they are the only sterile cells of the male prothallium in the pines. The large cell is the antheridium wall, its nucleus *v.n.* in Figure 357. The smaller cell, *a.c.*, is the central cell of the antheridium. During the summer and autumn the male prothallium makes some farther growth, but this is slow. The larger cell, called the vegetative cell or tube cell, which is in reality the wall of the antheridium, elongates by the formation of a tube, forming a sac, known as the pollen tube. It is either simple or branched. It grows down into the tissue of the nucellus, and at a stage represented in Figure 358, winter overtakes it and it rests. At this time the central cell has divided into two cells, and the vegetative nucleus is in the pollen tube.

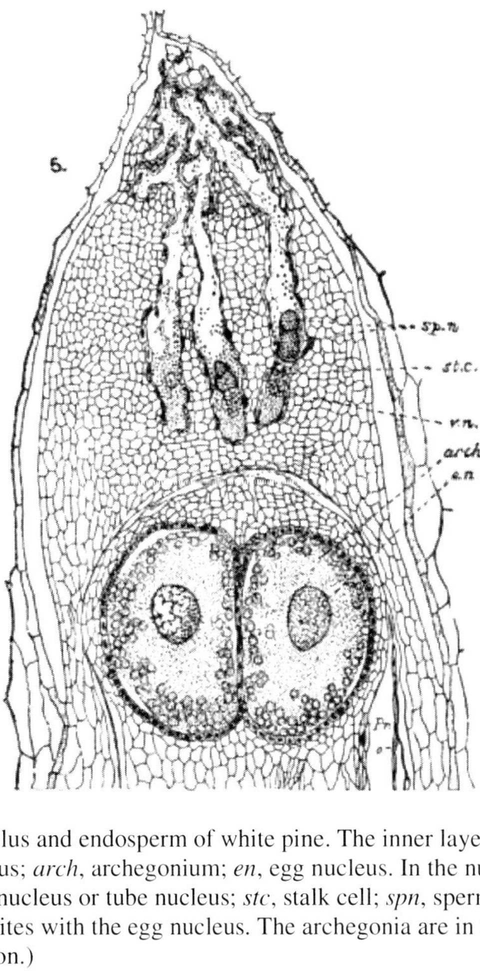

Figure 359. Section of nucellus and endosperm of white pine. The inner layer of cells of the integument shown just outside of nucellus; *arch*, archegonium; *en*, egg nucleus. In the nucellar cap are shown three pollen tubes. *vn*, vegetative nucleus or tube nucleus; *stc*, stalk cell; *spn*, sperm nuclei, the larger one in advance is the one which unites with the egg nucleus. The archegonia are in the endosperm or female gametophyte. (After Ferguson.)

The Endosperm

In the following spring growth of all these parts continues. The nuclei in the macrospore divide to form more, and eventually cell walls are formed between them making a distinct tissue, known as the *endosperm*. This endosperm continues to grow until a large part of the nucellus is consumed for food.

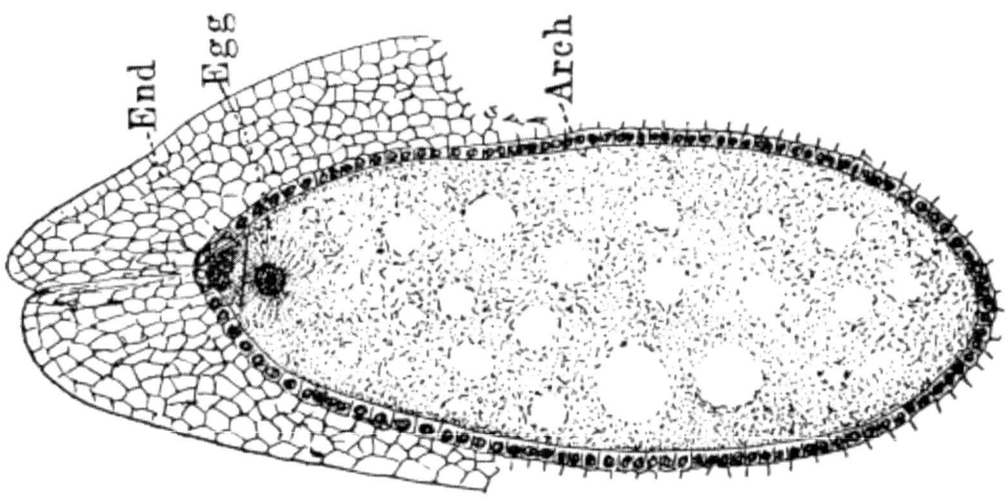

Figure 360. Last division of the egg in the white pine cutting off the ventral canal cell at the apex of the archegonium. *End*, endosperm; *Arch*, archegonium.

Female Prothallium and Archegonia

The endosperm is the female prothallium. This is very evident from the fact that several archegonia are developed in it usually on the side toward the pollen chamber. The archegonia are sexual organs, and since the sexual organs are developed on the gametophyte, therefore, the endosperm is the female gametophyte, or prothallium. In Figure 359 are represented two archegonia in the endosperm and the pollen tubes are growing down through the nucellus. The archegonia are quite large, the wall is a sheath or jacket of cells which encloses the very large egg which has a large nucleus in the center.

Pollen Tube and Sperm Cells

While the endosperm (female prothallium) and archegonia are developing the pollen tube continues its growth down through the nucellar cap, as shown in Figure 359. At the same time the two cells which were formed in the pollen grain (antheridium) from the central cell move down into the tube. One of these is the "generative" cell, or "body" cell, and the other is called the stalk cell, though it is more properly a sterile half of the central cell. The nucleus of the generative cell, about the time the archegonium is mature, divides to form two nuclei, which are the sperm nuclei, and the one in advance is the larger, though it is much smaller than the egg nucleus.

Fertilization

Very soon after the archegonia are mature (early in June in the northern United States) the pollen tube grows through into the archegonium and empties the two sperm nuclei, the vegetative nucleus and the stalk cell, into the protoplasm of the large egg. The larger of the two sperm nuclei at once comes in contact with the very large egg nucleus and sinks down into a depression of the same, as shown in Figure 361.

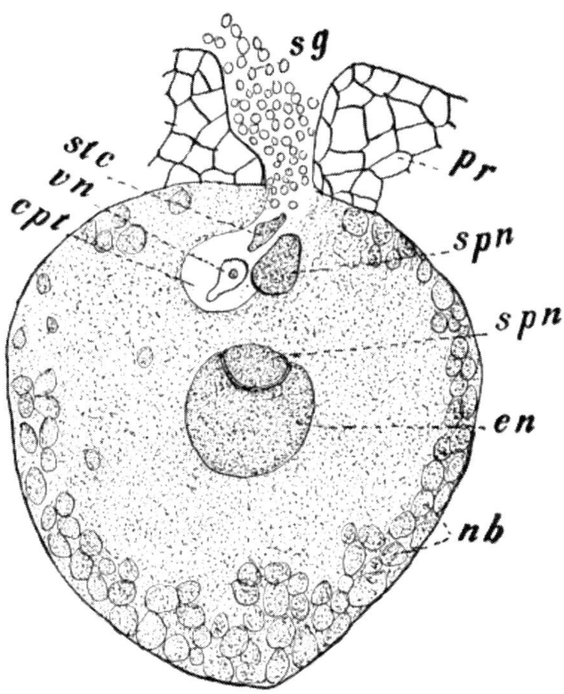

Figure 361. Archegonium of white pine at stage of fertilization, *en*, egg nucleus; *spn*, sperm nucleus in conjugation with it; *nb*, nutritive bodies in cytoplasm of large egg; *cpt*, cavity of pollen tube; *vn*, vegetative nucleus or tube nucleus; *stc*, stalk cell; *spn*, second sperm nucleus: *pr*, portion of prothallium or endosperm; *sg*, starch grains in pollen tube. The sheath of jacket cells of the archegonium is not shown. (After Ferguson.)

These two nuclei, in the pines, do not fuse into a resting nucleus, but at once organize the nuclear Figure for the first division of the embryo. Two nuclei are thus formed, and these divide to form four nuclei which sink to the bottom of the archegonium and there organize the embryo which pushes its way into the endosperm from which it derives its food (Figure 362).

Homology of the Parts of the Female Cone

Opinions are divided as to the homology of the parts of the female cone of the pine. Some consider the entire cone to be homologous with a flower of the angiosperms. The entire scale according to this view is a carpel, or sporophyll, which is divided into the cover scale and the ovuliferous scale. This division of the sporophyll is considered similar to that which we have in isoetes, where the sporophyll has a ligule above the sporangium, or as in ophioglossum, where the leaf is divided into a fertile and a sterile portion.

Others believe that the ovuliferous scale is composed of two leaves situated laterally and consolidated representing a shoot in the axis of the bract. There is some support for this in the fact that in certain abnormal cones which show proliferation a short axis appears in the axil of the bract and bears lateral leaves, and in some cases all gradations are present between these lateral leaves on the axis and their consolidation into an ovuliferous scale. In the normal condition of the ovuliferous scale the axis has disappeared and the shoot is represented only by the consolidated leaves, which would represent then the macrosporophylls (or carpels) each bearing one macrosporangium (ovule).

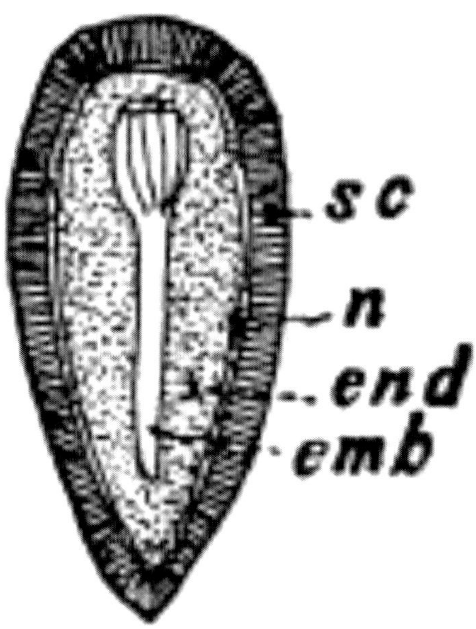

Figure 362. Pine seed, section of, *sc*, seed coat; *n*, remains of nucellus; *end*, endosperm (= female gametophyte); *emb*, embryo = young sporophyte. Seed coat and nucellus = remains of old sporophyte.

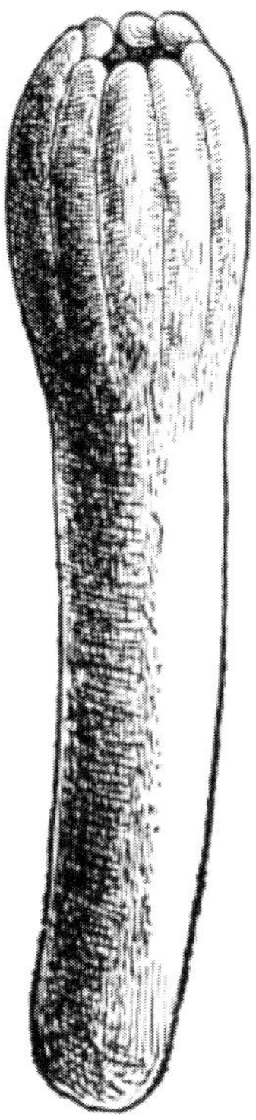

Figure 363. Embryo of white pine removed from seed, showing several cotyledons.

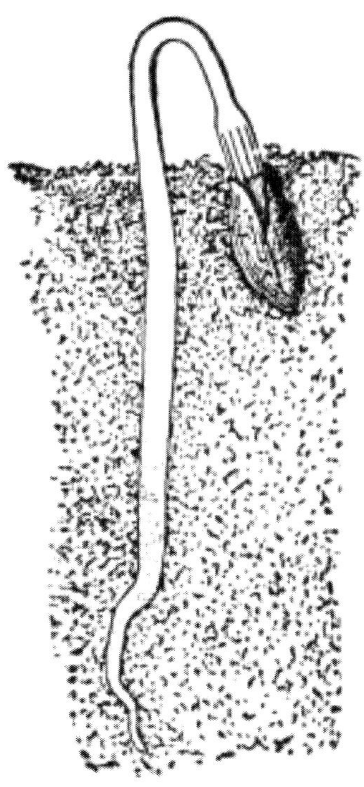

Figure 364. Pine seedling just emerging from the ground.

One of the most interesting and plausible views is that of Celakovsky. He believes that the axial shoot is reduced to two ovules, that the ovules have two integuments, but the outer integument of each has become proliferated into scales which are consolidated. In this proliferation of the outer integument it is thrown off from the ovule so that it only remains attached to one side and the larger part of the ovule is thus left with only one integument. This view is supported by the fact that in gingko, for example (another gymnosperm), the outer integument (the "collar") sometimes proliferates into a leaf. Celakovsky's view is, therefore, not very different from the second one mentioned above.

Figure 365. White pine seedling casting seed coats.

Chapter 34

Further Studies on Gymnosperms

Cycas

Figure 366. Macrosporophyll of Cycas revoluta.

In such gymnosperms as cycas, illustrated in the frontispiece, there is a close resemblance to the members of the fern group, especially the ferns themselves. This is at once suggested by the form of the leaves. The stem is short and thick. The leaves have a stout midrib and numerous narrow pinnæ. In the center of this rosette of leaves are numerous smaller leaves, closely overlapping like bud scales. If we remove one of these at the time the fruit is forming we see that in general it conforms to the plan of the large leaves. There are a midrib and a number of narrow pinnæ near the free end, the entire leaf being covered with woolly hairs. But at the lower end, in place of the pinnæ, we see oval bodies. These are the macrosporangia (ovules) of cycas, and correspond to the macrosporangia of selaginella, and the leaf is the macrosporophyll.

Female Prothallium of Cycas

In Figures 367, 368, are shown mature ovules, or macrosporangia, of cycas. In 368, which is a roentgen-ray photograph of 367, the oval prothallium can be seen. So in cycas, as in selaginella, the female prothallium is developed entirely inside of the macrosporangium, and derives the nutriment for its growth from the cycas plant, which is the sporophyte. Archegonia are developed in this internal mass of cells. This aids us in determining that it is the prothallium. In cycas it is also called endosperm, just as in the pines.

Figure 367. Macrosporangium of Cycas revoluta.

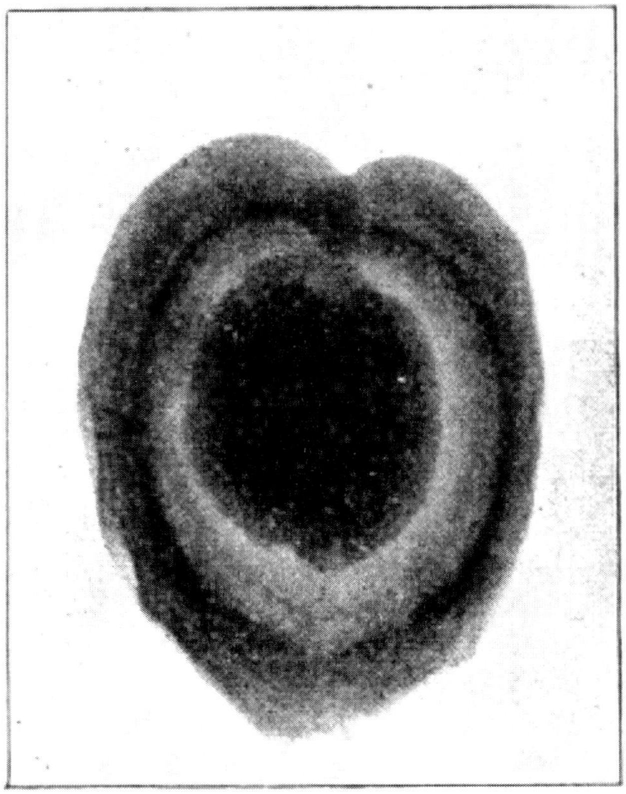

Figure 368. Roentgen photograph of same, showing female prothallium.

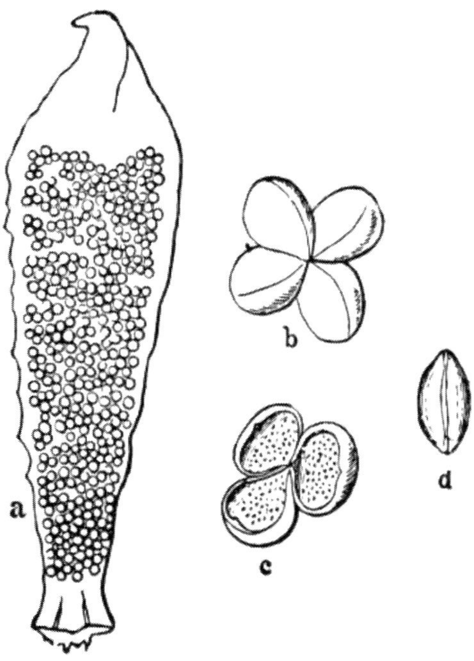

Figure 369. A sporophyll (stamen) of cycas; sporangia in groups on the under side. *b*, group of sporangia; *c*, open sporangia. (From Warming.)

If we cut open one of the mature ovules, we can see the endosperm (prothallium) as a whitish mass of tissue. Immediately surrounding it at maturity is a thin, papery tissue, the remains of the nucellus (macrosporangium), and outside of this are the coats of the ovule, an outer fleshy one and an inner stony one.

Microspores, or Pollen, of Cycas

The cycas plant illustrated in the frontispiece is a female plant. Male plants also exist which have small leaves in the center that bear only microsporangia. These leaves, while they resemble the ordinary leaves, are smaller and correspond to the stamens. Upon the under side, as shown in Figure 369, the microsporangia are borne in groups of three or four, and these contain the microspores, or pollen grains. The arrangement of these microsporangia on the under side of the cycas leaves bears a strong resemblance to the arrangement of the sporangia on the under side of the leaves of some ferns.

The gingko tree is another very interesting plant belonging to this same group. It is a relic of a genus which flourished in the remote past, and it is interesting also because of the resemblance of the leaves to some of the ferns like adiantum, which suggests that this form of the leaf in gingko has been inherited from some fern-like ancestor.

Figure 370. Zamia integrifolia, showing thick stem, fern-like leaves, and cone of male flowers.

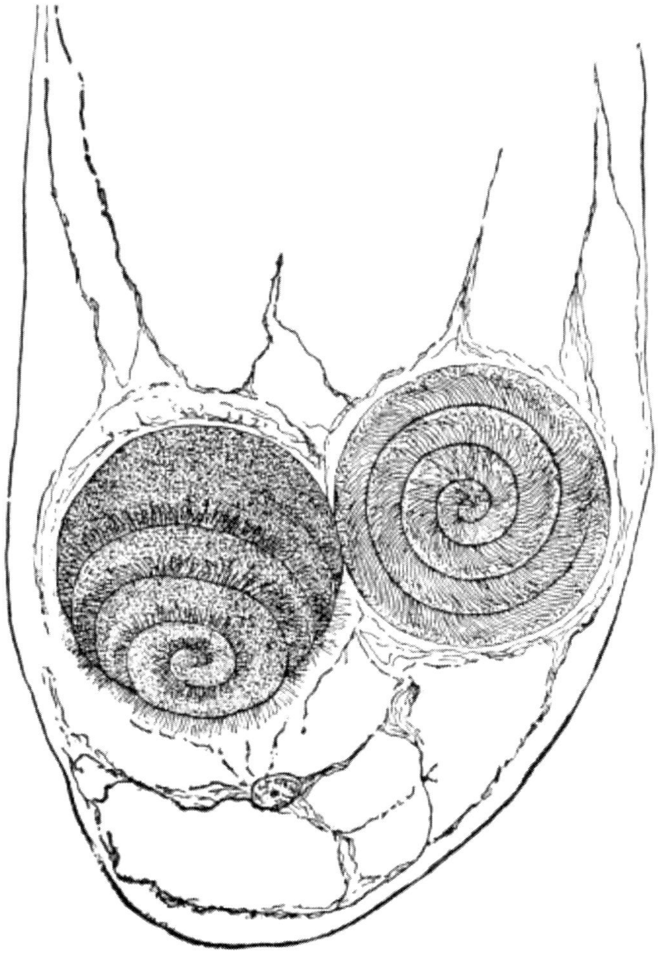

Figure 371. Two spermatozoids in end of pollen tube of cycas. (After drawing by Hirase and Ikeno.)

While the resemblance of the leaves of some of the gymnosperms to those of the ferns suggests fern-like ancestors for the members of this group, there is stronger evidence of such ancestry in the fact that a prothallium can well be determined in the ovules. The endosperm with its well-formed archegonia is to be considered a prothallium.

Spermatozoids in Some Gymnosperms

But within the past two years it has been discovered in gingko, cycas, and zamia, all belonging to this group, that the sperm cells are well-formed spermatozoids. In zamia each one is shaped somewhat like the half of a biconvex lens, and around the convex surface are several coils of cilia. After the pollen tube has grown down through the nucellus, and has reached a depression at the end of the prothallium (endosperm) where the archegonia are formed, the spermatozoids are set free from the pollen tube, swim around in a liquid in this depression, and later fuse with the egg. In gingko and cycas these spermatozoids were first discovered by Ikeno and Hirase in Japan, and later in zamia by Webber in this country. In Figures 371-374 the details of the male prothallia and of fertilization are shown.

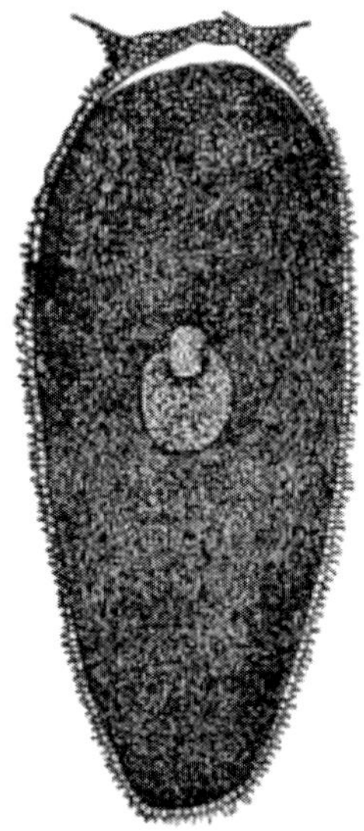

Figure 372. Fertilization in cycas, small spermatozoid fusing with the larger female nucleus of the egg. The egg protoplasm fills the archegonium. (From drawings by Hirase and Ikeno.)

Figure 373. Spermatozoid of gingko. Some abnormal forms have a tail. (After Ikeno and Hirase.)

The Sporophyte in the Gymnosperms

In the pollen grains of the gymnosperms we easily recognize the characters belonging to the spores in the ferns and their allies, as well as in the liverworts and mosses. They belong to the same series of organs, are borne on the same phase or generation of the plant, and are practically formed in the same general way, the variations between the different groups not being greater than those within a single group. These spores we have recognized as being the product of the sporophyte. We are able then to identify the sporophyte as that phase or generation of the plant formed from the fertilized egg and bearing ultimately the spores. We see from this that the sporophyte in the gymnosperms is the prominent part of the plant, just as we found it to be in the ferns. The pine tree, then, as well as the gingko, cycas, yew, hemlock-spruce, black spruce, the giant redwood of California, etc., are sporophytes.

While the sporangia (anther sacs) of the male flowers open and permit the spores (pollen) to be scattered, the sporangia of the female flowers of the gymnosperms rarely open. The macrospore is developed within sporangium (nucellus) to form the female prothallium (endosperm).

The Gametophyte Has Become Dependent on the Sporophyte

In this respect the gymnosperms differ widely from the pteridophytes, though we see suggestions of this condition of things in Isoetes and Selaginella, where the female prothallium is developed within the macrospore, and even in Selaginella begins, and nearly completes, its development while still in the sporangium.

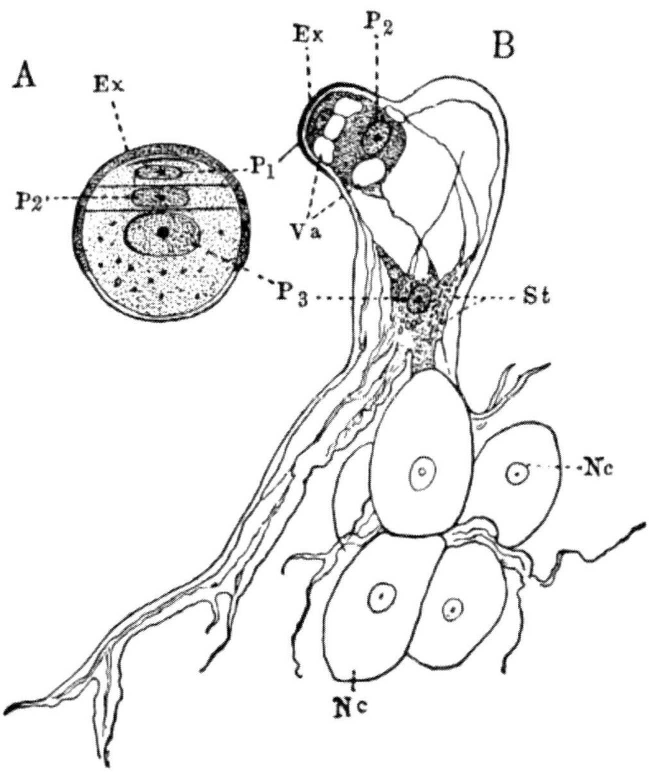

Figure 374. Gingko biloba. *A*, mature pollen grain; *B*, germinating pollen grain, the branched tube entering among the cells of the nucellus; *Ex*, exine (outer wall of spore); P_1, prothallial cell; P_2, antheridial cell (divides later to form stalk cell and generative cell); P_3, vegetative cell; *Va*, vacuoles; *Nc*, nucellus. (After drawings by Hirase and Ikeno.)

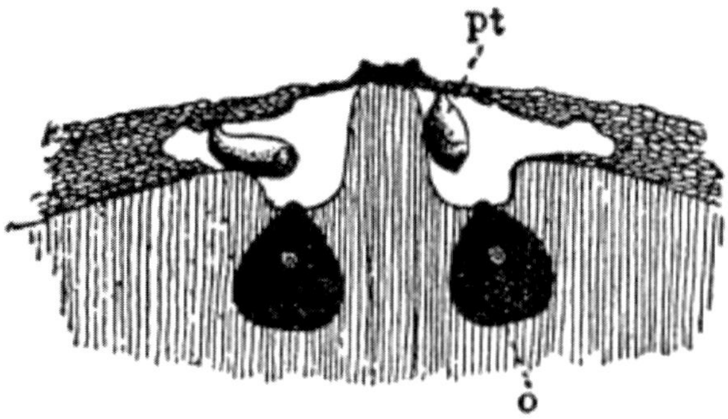

Figure 375. Gingko biloba, diagrammatic representation of the relation of pollen tube to the archegonium in the end of the nucellus. *pt*, pollen tube; *o*, archegonium. (After drawing by Hirase and Ikeno.)

In comparing the female prothallium of the gymnosperms with that of the fern group we see a remarkable change has taken place. The female prothallium of the gymnosperms is very much reduced in size. Especially, it no longer leads an independent existence from the sporophyte, as is the case with nearly all the fern group. It remains enclosed within the macrosporangium (in cycas if not fertilized it sometimes grows outside of the macrosporangium and becomes green), and derives its nourishment through it from the sporophyte, to which the latter remains organically connected. This condition of the female prothallium of the gymnosperms necessitated a special adaptation of the male prothallium in order that the sperm cells may reach and fertilize the egg-cell.

Gymnosperms are Naked Seed Plants

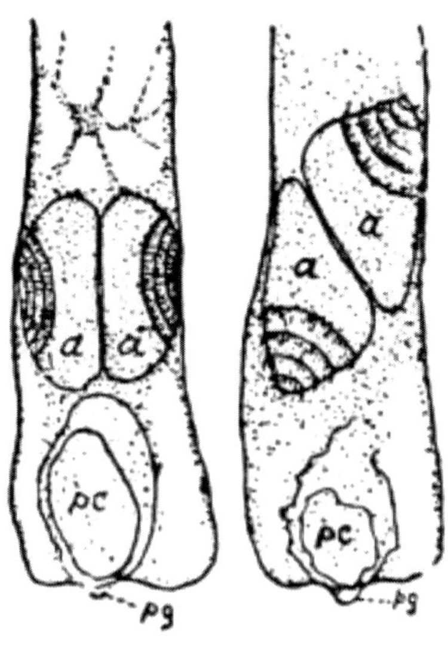

Figure 376. Spermatozoids of zamia in pollen tube; *pg*, pollen grain; *a*, *a*, spermatozoids. (After Webber.)

Figure 377. Spermatozoid of zamia showing spiral row of cilia. (After Webber.)

The pine, as we have seen, has naked seeds. That is, the seeds are not enclosed within the carpel, but are exposed on the outer surface. All the plants of the great group to which the pine belongs have naked seeds. For this reason the name "*gymnosperms*" has been given to this great group.

Classification of Gymnosperms

The gingko tree has until recently been placed with the pines, yew, etc., in the order *Pinales*, but the discovery of the spermatozoids in the pollen tube suggests that it is not closely allied with the Pinales, and that it represents an order coordinate with them. Engler arranges the living gymnosperms somewhat as follows:

Class Gymnospermæ

Order 1.	Cycadales;	family Cycadaceæ.	Cycas, Zamia, etc.
Order 2.	Gingkoales;	family Gingkoaceæ.	Gingko.
Order 3.	Pinales (or Coniferæ);	family 1. Taxaceæ.	Taxus, the common yew in the eastern United States, and Torreya, in the western United States, are examples.
		family 2. Pinaceæ.	Sequoia (redwood of California), firs, spruces, pines, cedars, cypress, etc.
Order 4.	Gnetales.	Welwitschia mirabilis, deserts of southwest Africa; Ephedra, deserts of the Mediterranean and of West Asia. Gnetum, climbers (Lianas), from tropical Asia and America.	

Table Showing Homologies of Sporophyte and Gametophyte in the Pine

	Terms Corresponding to those used in Pteridophytes.		Common Terms.	
	Sporophyte	=	Pine tree.	
	Spore-bearing part	=	Male and female cones.	
Sporophyte	Microsporophyll	=	Stamen.	
	Microsporangium	=	Pollen sac.	
Male gametophyte	Microspore	=	Pollen grain.	
	Mature microspore is rudimentary male prothallium	=	Mature pollen grain.	
	with rudimentary antheridium			
	Large cell (part of antheridium wall?)	=	Vegetative cell of pollen grain.	
	Antheridium cell	=	Small cell of pollen	
	Antheridium cell divides to form stalk cell and	=	Generative cell.	
	central cell of antheridium (male sexual organ)			
	Central cell of antheridium divides to form	=	Paternal cells, or	
	two sperm cells		generative cells.	
Sporophyte	Macrosporophyll	=	Ovuliferous scale (cover scale	
			and carpellary outgrowth); or	
			three carpels united into	
			ovuliferous scale, the central one	
			sterile (in axil of cover scale).	
	Macrosporangium covered by integument	=	Nucellus covered by	
	by integument		integument = ovule.	
Female gametophyte	Macrospore (remains in sporangium)	=	Large cell in center of nucellus	
			which develops embryo sac and	
			endosperm (remains in nucellus).	
	Female prothallium (in sporangium)	=	Endosperm, in nucellus.	
	Archegonia (female sexual organs)	=	Corpuscula, in endosperm.	
	Egg	=	Maternal cell, or germ cell.	
Young sporophyte	Egg (fertilized)	=	Germ cell.	
	Young sporophyte	=	Pine embryo in nucellus	
			and integument.	
	Young sporophyte	=	Embryo	Seed.
	In remains of gametophyte	=	Endosperm	
	And sporangium	=	Nucellus	
	Surrounded by new growth of old sporophyte	=	Integument	

Chapter 35

MORPHOLOGY OF THE ANGIOSPERMS: TRILLIUM; DENTARIA

TRILLIUM

General Appearance

As one of the plants to illustrate this group we may take the wake-robin, as it is sometimes called, or trillium. There are several species of this genus in the United States; the commonest one in the eastern part is the "white wake-robin" (Trillium grandiflorum). This occurs in or near the woods. A picture of the plant is shown in Figure 378. There is a thick, fleshy, underground stem, or rhizome as it is usually called. This rhizome is perennial, and is marked by ridges and scars. The roots are quite stout and possess coarse wrinkles. From the growing end of the rhizome each year the leafy, flowering stem arises. This is 20-30*cm* (8-12 inches) in height. Near the upper end is a whorl of three ovate leaves, and from the center of this rosette rises the flower stalk, bearing the flower at its summit.

Parts of the Flower. Calyx

Now if we examine the flower we see that there are several leaf-like structures. These are arranged also in threes just as are the leaves. First there is a whorl of three, pointed, lanceolate, green, leaf-like members, which make up the *calyx* in the higher plants, and the parts of the calyx are *sepals*, that is, each leaf-like member is a *sepal*. But while the sepals are part of the flower, so called, we easily recognize them as belonging to the *leaf series*.

Corolla

Next above the calyx is a whorl of white or pinkish members, in Trillium grandiflorum, which are also leaf-like in form, and broader than the sepals, being usually somewhat broader at the free end. These make up what is the *corolla* in the higher plants, and each member of the corolla is a *petal*. But while they are parts of the flower, and are not green, their form and position would suggest that they also belong to the leaf series.

Figure 378. Trillium grandiflorum.

Andrœcium

Within and above the insertion of the corolla is found another tier, or whorl, of members which do not at first sight resemble leaves in form. They are known in the higher plants as *stamens*. As seen in Figure 379 each stamen possesses a stalk (= filament), and extending along on either side for the greater part of the length are four ridges, two on each side.

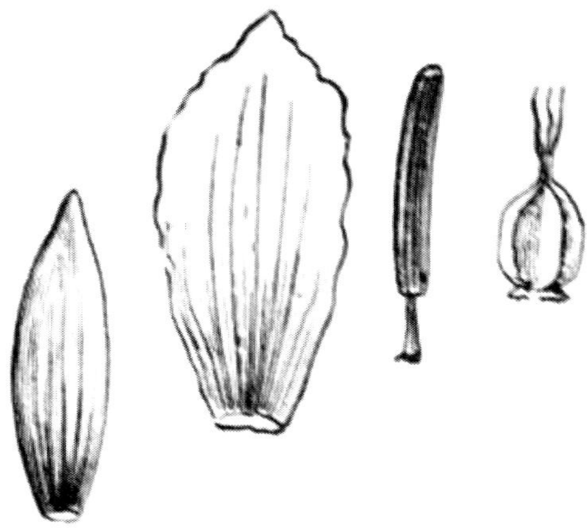

Figure 379. Sepal, petal, stamen, and pistil of Trillium grandiflorum.

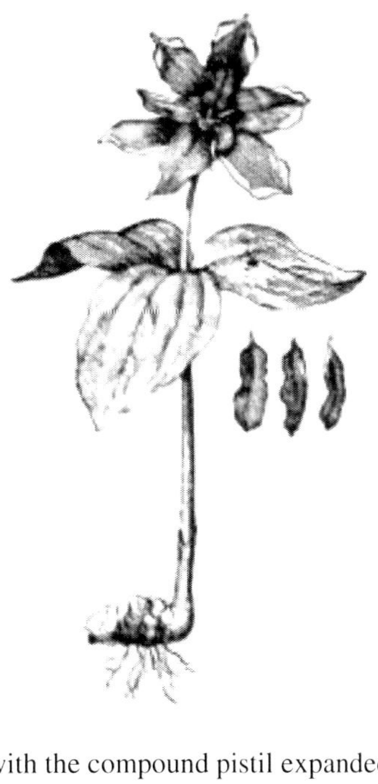

Figure 380. Trillium grandiflorum, with the compound pistil expanded into three leaf-like members. At the right these three are shown in detail.

This part of the stamen is the *anther*, and the ridges form the anther sacs, or lobes. Soon after the flower is opened, these anther sacs open also by a split in the wall along the edge of the ridge. At this time we see quantities of yellowish powder or dust escaping from the ruptured anther locules. If we place some of this under the microscope we see that it is made up of minute bodies which resemble spores; they are rounded in form, and the outer wall is spiny. They are in fact spores, the microspores of the trillium, and here, as in the gymnosperms, are better known as *pollen*.

The Stamen A Sporophyll

Since these pollen grains are the spores, we would infer, from what we have learned of the ferns and gymnosperms, that this member of the flower which bears them is a sporophyll; and this is the case. It is in fact what is called the *microsporophyll*. Then we see also that the anther sacs, since they enclose the spores, would be the sporangia (microsporangia). From this it is now quite clear that the stamens belong also to the leaf series. They are just six in number, twice the number found in a whorl of leaves, or sepals, or corolla. It is believed, therefore, that there are two whorls of stamens in the flower of trillium.

Gynœcium

Next above the stamens and at the center of the flower is a stout, angular, ovate body which terminates in three long, slender, curved points. This is the pistil, and at present the only suggestion which it gives of belonging to the leaf series is the fact that the end is divided into three parts, the number of parts in each successive whorl of members of the flower. If we cut across the body of this pistil and examine it with a low power we see that there are three chambers or cavities, and at the junction of each the walls suggest to us that this body may have been formed by the infolding of the margins of three leaf-like members, the places of contact having then become grown together. We see also that from the incurved margins of each division of the pistil there stand out in the cavity oval bodies. These are the *ovules*. Now the ovules we have learned from our study of the gymnosperms are the *sporangia* (here the macrosporangia). It is now more evident that this curious body, the pistil, is made up of three leaf-like members which have fused together, each member being the equivalent of a sporophyll (here the macrosporophyll). This must be a fascinating observation, that plants of such widely different groups and of such different grades of complexity should have members formed on the same plan and belonging to the same series of members, devoted to similar functions, and yet carried out with such great modifications that at first we do not see this common meeting ground which a comparative study brings out so clearly.

Transformations of the Flower of Trillium

If anything more were needed to make it clear that the parts of the flower of trillium belong to the leaf series we could obtain evidence from the transformations which the flower of trillium sometimes presents. In Figure 381 is a sketch of a flower of trillium, made from a photograph.

Figure 381. Abnormal trillium. The nine parts of the perianth are green, and the outer whorls of stamens are expanded into petal-like members.

Figure 382. Transformed stamen of trillium showing anther locules on the margin.

One set of the stamens has expanded into petal-like organs, with the anther sacs on the margin. In Figure 380 is shown a plant of Trillium grandiflorum in which the pistil has separated into three distinct and expanded leaf-like structures, all green except portions of the margin.

DENTARIA

General Appearance

For another study we may take a plant which belongs to another division of the higher plants, the common "pepper root," or "toothwort" (Dentaria diphylla) as it is sometimes called. This plant occurs in moist woods during the month of May, and is well distributed in the northeastern United States. A plant is shown in Figure 383. It has a creeping underground rhizome, whitish in color, fleshy, and with a few scales. Each spring the annual flower-bearing stem rises from one of the buds of the rhizome, and after the ripening of the seeds, dies down.

The leaves are situated a little above the middle point of the stem. They are opposite and the number is two, each one being divided into three dentate lobes, making what is called a compound leaf.

Figure 383. Toothwort (*Dentaria diphylla*).

Figure 384. Flower of the toothwort (*Dentaria diphylla*).

Parts of the Flower

The flowers are several, and they are borne on quite long stalks (pedicels) scattered over the terminal portion of the stem. We should now examine the parts of the flower beginning with the calyx. This we can see, looking at the under side of some of the flowers, possesses four scale-like sepals, which easily fall away after the opening of the flower. They do not resemble leaves so much as the sepals of trillium, but they belong to the leaf series, and there are two pairs in the set of four. The corolla also possesses four petals, which are more expanded than the sepals and are whitish in color. The stamens are six in number, one pair lower than the others, and also shorter. The filament is long in proportion to the anther, the latter consisting of two lobes or sacs, instead of four as in trillium. The pistil is composed of two carpels, or leaves fused together. So we find in the case of the pepper root that the parts of the flower are in twos, or multiples of two. Thus they agree in this respect with the leaves; and while we do not see such a strong resemblance between the parts of the flower here and the leaves, yet from the presence of the pollen (microspores) in the anther sacs (microsporangia) and of ovules (macrosporangia) on the margins of each half of the pistil, we are, from our previous studies, able to recognize here that all the members of the flower belong to the leaf series.

In trillium and in the pepper root we have seen that the parts of the flower in each apparent whorl are either of the same number as the leaves in a whorl, or some multiple of that number. This is true of a large number of other plants, but it is not true of all. A glance at the spring-beauty (Claytonia virginiana), and at the anemone (or Isopyrum biternatum, Figure 563) will serve to show that the number of the different members of the flower may vary. The trillium and the dentaria were selected as being good examples to study first, to make it very clear that the members of the flower are fundamentally leaf structures, or rather that they belong to the same series of members as do the leaves of the plant.

Synopsis of Members of the Sporophyte in Angiosperms

Higher plant.								
Sporophyte phase	{	Root.			{	Foliage leaves.		
(or modern phase)		Shoot.	{	Stem.		Perianth leaves.	}	
				Leaf.		Spore-bearing leaves		
						with sporangia.		Flower.
						(Sporangia sometimes		
						on shoot.)		

Chapter 36

GAMETOPHYTE AND SPOROPHYTE OF ANGIOSPERMS

MALE PROTHALLIUM OF ANGIOSPERMS

The first division which takes place in the nucleus of the pollen grain occurs, in the case of trillium and many others of the angiosperms, before the pollen grain is mature. In the case of some specimens of T. grandiflorum in which the pollen was formed during the month of October of the year before flowering, the division of the nucleus into two nuclei took place soon after the formation of the four cells from the mother cell. The nucleus divided in the young pollen grain is shown in Figure 385. After this takes place the wall of the pollen grain becomes stouter, and minute spiny projections are formed.

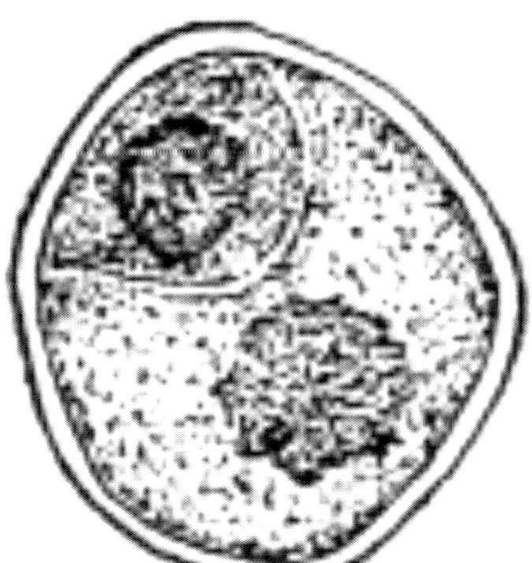

Figure 385. Nearly mature pollen grain of trillium. The smaller cell is the generative cell.

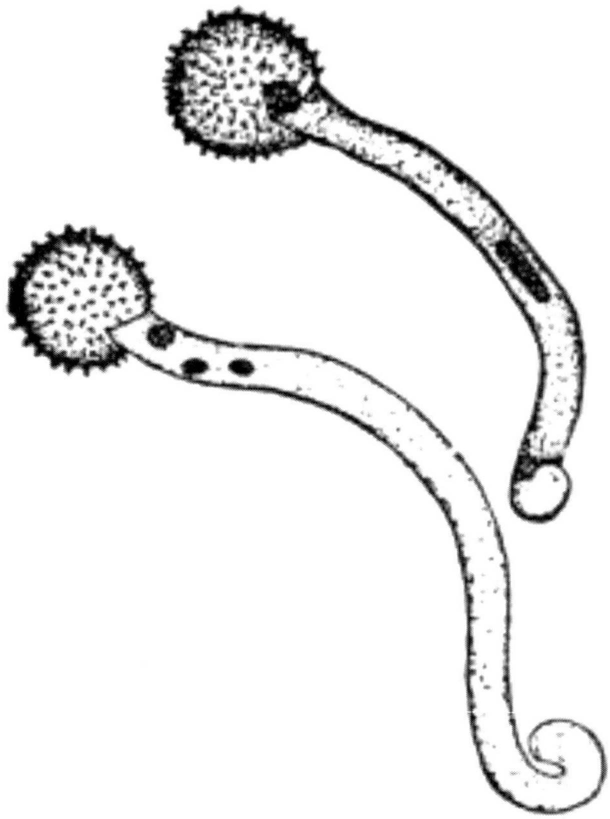

Figure 386. Germinating spores (pollen grains) of peltandra; generative nucleus in one undivided, in other divided to form the two sperm nuclei; vegetative nucleus in each near the pollen grain.

The larger cell is the vegetative cell of the prothallium, while the smaller one, since it later forms the sperm cells, is the generative cell. This generative cell then corresponds to the central cell of the antheridium, and the vegetative cell perhaps corresponds to a wall cell of the antheridium. If this is so, then the male prothallium of angiosperms has become reduced to a very simple antheridium. The farther growth takes place after fertilization. In some plants the generative cell divides into the two sperm cells at the maturity of the pollen grain. In other cases the generative cell divides in the pollen tube after the germination of the pollen grain. For study of the pollen tube the pollen may be germinated in a weak solution of sugar, or on the cut surface of pear fruit, the latter being kept in a moist chamber to prevent drying the surface.

In the spring after flowering the pollen escapes from the anther sacs, and as a result of pollination is brought to rest on the stigma of the pistil. Here it germinates, as we say, that is, it develops a long tube which makes its way down through the style, and in through the micropyle to the embryo sac, where, in accordance with what takes place in other plants examined, one of the sperm cells unites with the egg, and fertilization of the egg is the result.

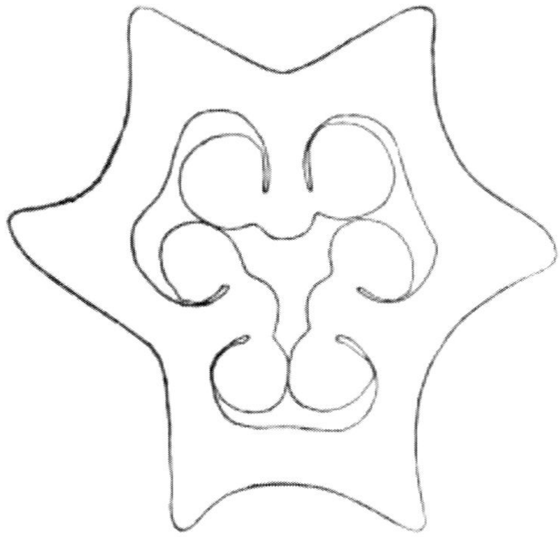

Figure 387. Section of pistil of trillium, showing position of ovules (macrosporangia).

Figure 388. Mandrake (Podophyllum peltatum).

Macrospore and Embryo Sac

In trillium the three carpels are united into one, and in dentaria the two carpels are also united into one compound pistil. Simple pistils are found in many plants, for example in the ranunculaceæ, the buttercups, columbine, etc. These simple pistils bear a greater resemblance to a leaf, the margins of which are folded around so that they meet and enclose the ovules or sporangia.

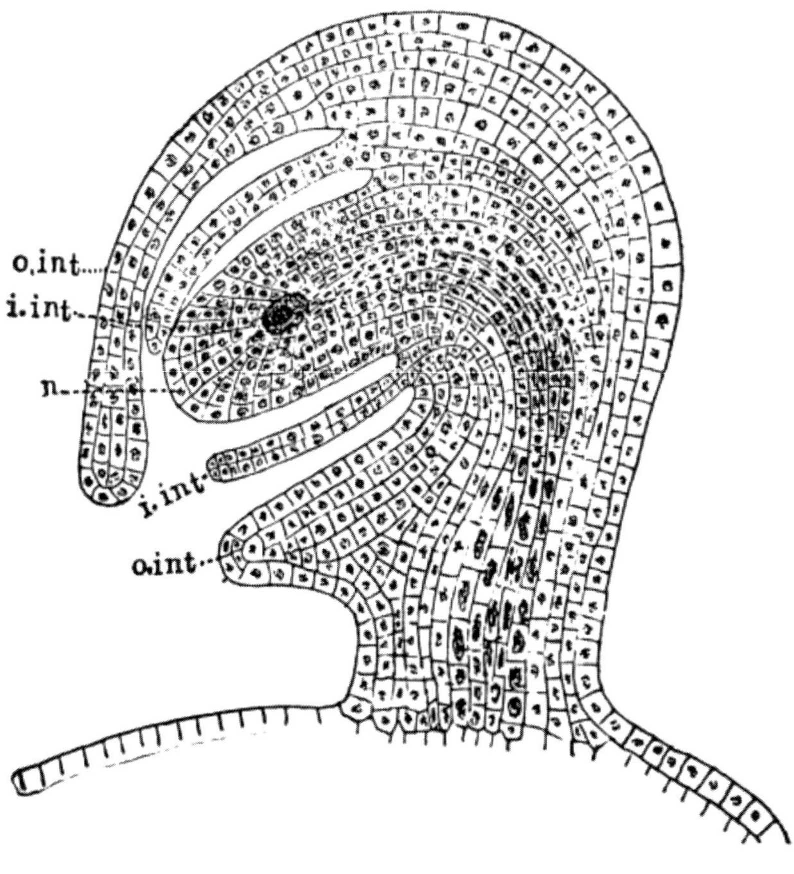

Figure 389. Young ovule (macrosporangium) of podophyllum. *n*, nucellus containing the one-celled stage of the macrospore; *i.int*, inner integument; *o.int*, outer integument.

If we cut across the compound pistil of trillium we find that the infoldings of the three pistils meet to form three partial partitions which extend nearly to the center, dividing off three spaces. In these spaces are the ovules which are attached to the infolded margins. If we make cross-sections of a pistil of the May-apple (podophyllum) and through the ovules when they are quite young, we shall find that the ovule has a structure like that shown in Figure 389. At *m* is a cell much larger than the surrounding ones. This is called the macrospore. The tissue surrounding it is called here the nucellus, but because it contains the macrospore it must be the macrosporangium. The two coats or integuments of the ovule are yet short and have not grown out over the end of the nucellus. This macrospore increases in size, forming first a cavity or sac in the nucellus, the *embryo sac*. The nucleus divides several times until eight are formed, four in the micropylar end of the embryo sac and four in the opposite end. In some plants it has been found that one nucleus from each group of four moves toward the middle of the embryo sac. Here they fuse together to form one nucleus, the *endosperm nucleus* or *definitive nucleus* shown in Figure 390. One of the nuclei at the micropylar end is the egg, while the two smaller ones nearer the end are the *synergids*. The egg-cell is all that remains of the archegonium in this reduced prothallium. The three nuclei at the lower end are the *antipodal* cells.

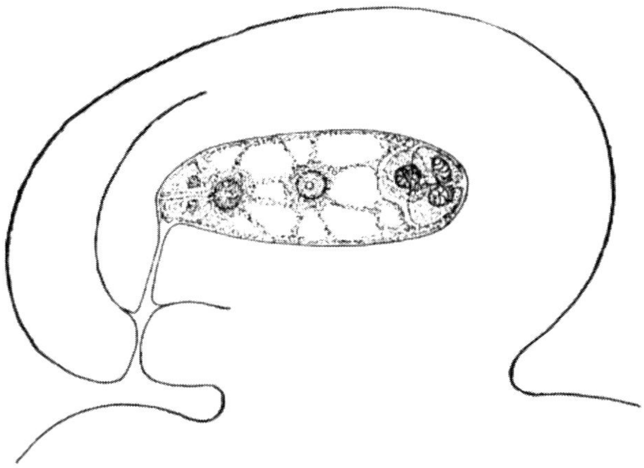

Figure 390. Podophyllum peltatum, ovule containing mature embryo sac; two synergids, and eggs at left, endosperm nucleus in center, three antipodal cells at right.

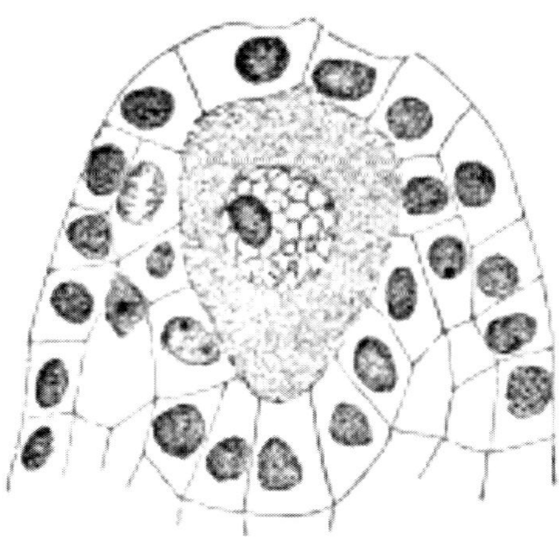

Figure 391. Macrospore (one-celled stage) of lilium.

Embryo Sac is the Young Female Prothallium

In Figures 391-393 are shown the different stages in the development of the embryo sac in lilium. The embryo sac at this stage is the young female prothallium, and the egg is the only remnant of the female sexual organ, the archegonium, in this reduced gametophyte.

Fertilization

When the pollen tube has reached the embryo sac (paragraph 652) it opens and the two sperm cells are emptied near the egg. The first sperm nucleus enters the protoplasm surrounding the egg nucleus and uniting with the latter brings about fertilization. The second sperm nucleus often unites with the endosperm nucleus (or with one or both of the "polar nuclei"), bringing about what some call a second fertilization. Where this takes place in addition to the union of the first sperm nucleus with the egg nucleus it is called *double fertilization*. The sperm nucleus is usually smaller than the egg nucleus, but often grows to near or quite the size of the egg nucleus before union. See Figures 394 and 395.

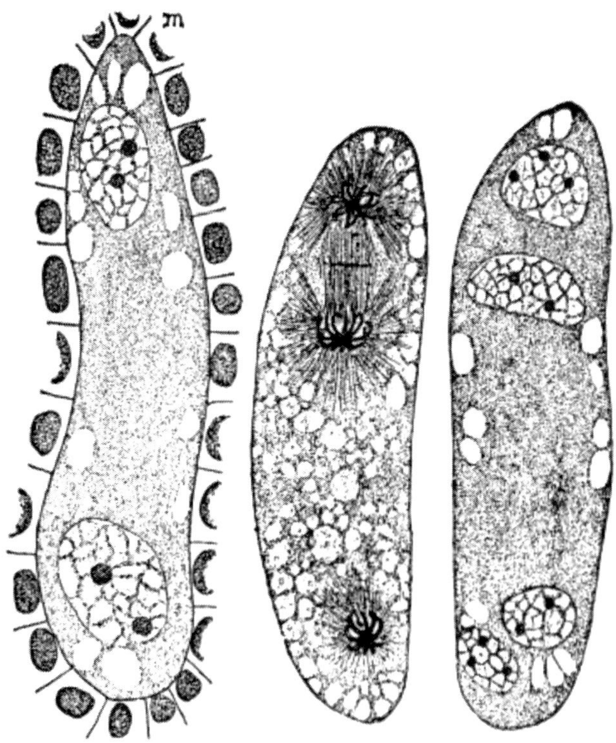

Figure 392. Two-and four-celled stage of embryo sac of lilium. The middle one shows division of nuclei to form the four-celled stage. (Easter lily.)

Fertilization in Plants is Fundamentally the Same as in Animals

In all the great groups of plants as represented by spirogyra, œdogonium, vaucheria, peronospora, ferns, gymnosperms, and in the angiosperms, fertilization, as we have seen, consists in the fusion of a male nucleus with a female nucleus. Fertilization, then, in plants is identical with that which takes place in animals.

Embryo

After fertilization the egg develops into a short row of cells, the *suspensor* of the embryo. At the free end the embryo develops. In Figures 397 and 398 is a young embryo of trillium.

Endosperm, the Mature Female Prothallium

During the development of the embryo the endosperm nucleus divides into a great many nuclei in a mass of protoplasm, and cell walls are formed separating them into cells.

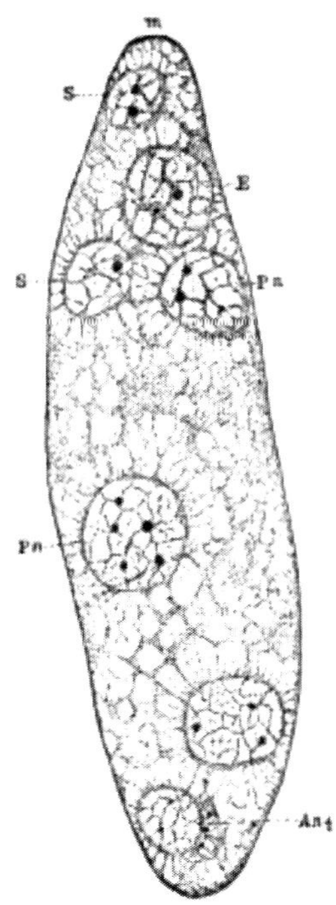

Figure 393. Mature embryo sac (young prothallium) of lilium. *m*, micropylar end; *S*, synergids; *E*, egg; *Pn*, polar nuclei; *Ant*, antipodals. (Easter lily.)

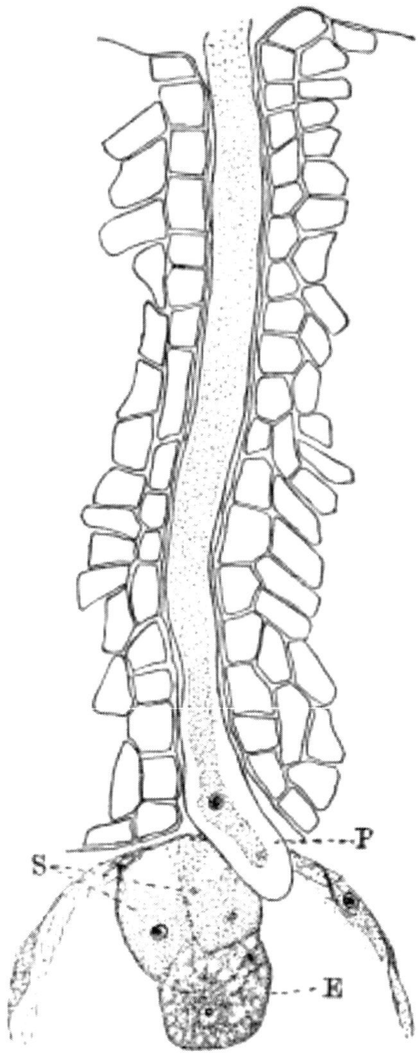

Figure 394. Section through nucellus and upper part of embryo sac of cotton at time of entrance of pollen tube. *E*, egg; *S*, synergids; *P*, pollen tube with sperm cell in the end. (Duggar.)

This mass of cells is the *endosperm*, and it surrounds the embryo. It is the *mature female prothallium*, belated in its growth in the angiosperms, usually developing only when fertilization takes place, and its use has been assured.

Seed

As the embryo is developing it derives its nourishment from the endosperm (or in some cases perhaps from the nucellus). At the same time the integuments increase in extent and harden as the seed is formed.

Perisperm

In most plants the nucellus is all consumed in the development of the endosperm, so that only minute fragments of disorganized cell walls remain next the inner integument. In some plants, however, (the water-lily family, the pepper family, etc.,) a portion of the nucellus remains intact in the mature seed. In such seeds the remaining portion of the nucellus is the *perisperm.*

Presence or Absence of Endosperm in the Seed

In many of the angiosperms all of the endosperm is consumed by the embryo during its growth in the formation of the seed. This is the case in the rose family, crucifers, composites, willows, oaks, legumes, etc., as in the acorn, the bean, pea and others. In some, as in the bean, a large part of the nutrient substance passing from the endosperm into the embryo is stored in the cotyledons for use during germination. In other plants the endosperm is not all consumed by the time the seed is mature. Examples of this kind are found in the buttercup family, the violet, lily, palm, jack-in-the-pulpit, etc. Here the remaining endosperm in the seed is used as food by the embryo during germination.

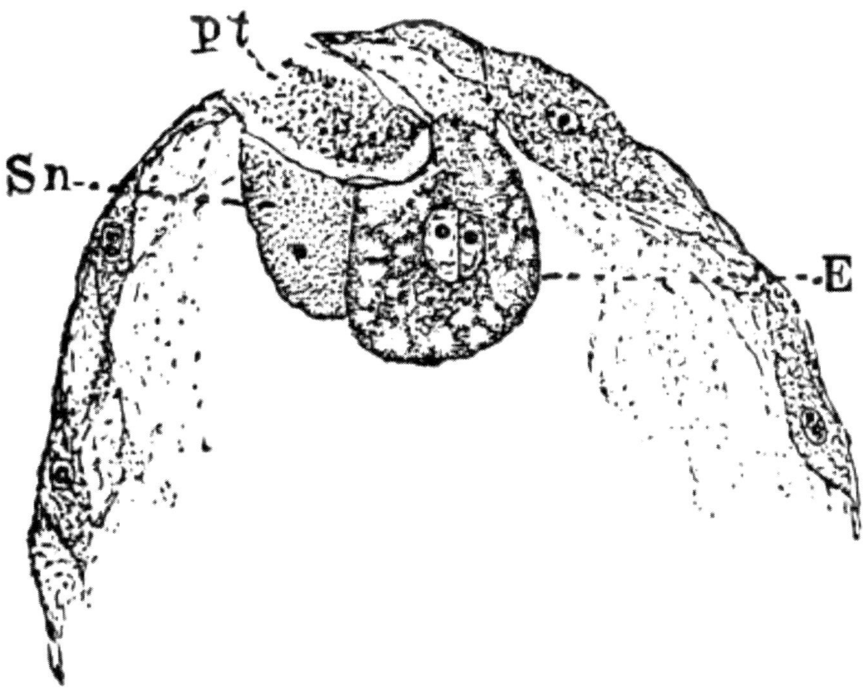

Figure 395. Fertilization of cotton. *pt*, pollen tube; *Sn*, synergids; *E*, egg, with male and female nucleus fusing. (Duggar.)

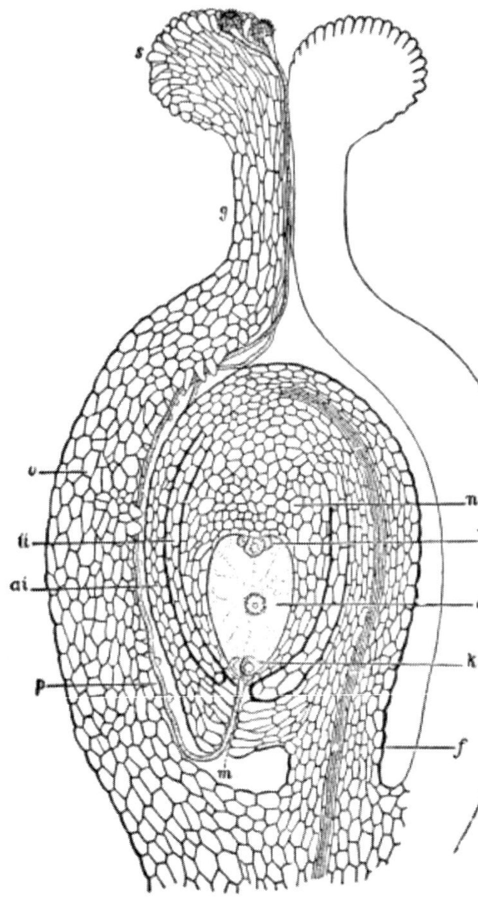

Figure 396. Diagrammatic section of ovary and ovule at time of fertilization in angiosperm. *f*, funicle of ovule; *n*, nucellus; *m*, micropyle; *b*, antipodal cells of embryo sac; *e*, endosperm nucleus; *k*, egg-cell and synergids; *ai*, outer integument of ovule; *ii*, inner integument. The track of the pollen tube is shown down through the style, walls of the ovary to the micropylar end of the embryo sac.

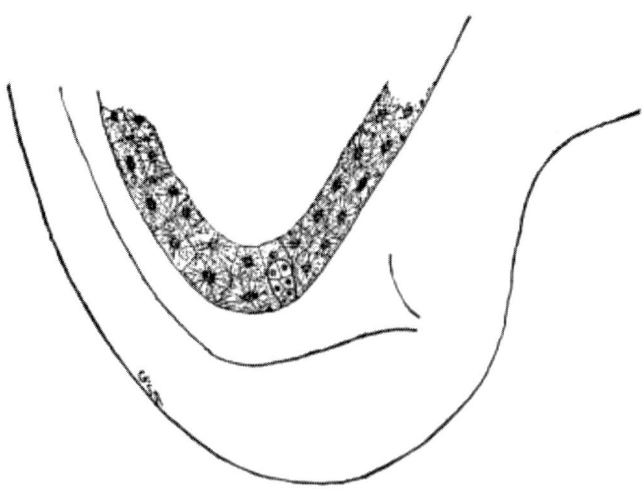

Figure 397. Section of one end of ovule of trillium, showing young embryo in endosperm.

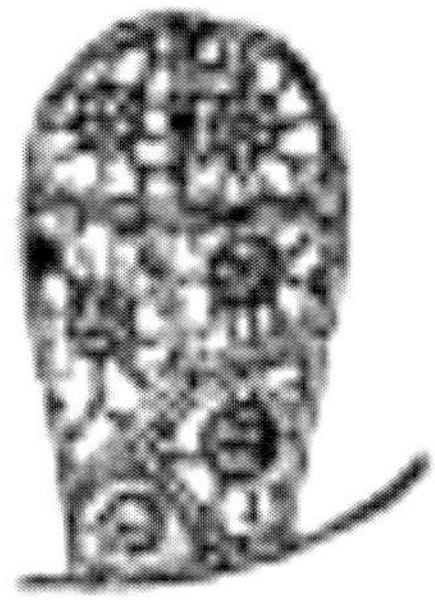

Figure 398. Embryo enlarged.

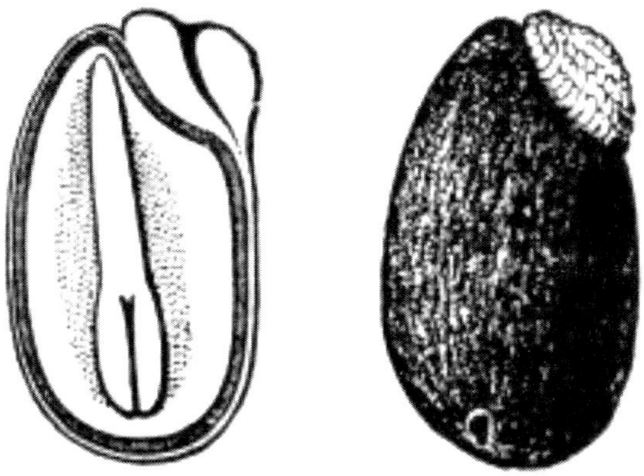

Figure 399. Seed of violet, external view, and section. The section shows the embryo lying in the endosperm.

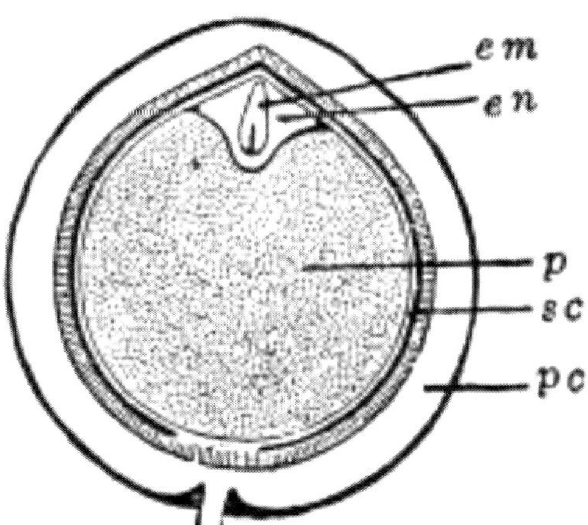

Figure 400. Section of fruit of pepper (Piper nigrum), showing small embryo lying in a small quantity of whitish endosperm at one end, the perisperm occupying the larger part of the interior, surrounded by pericarp.

Outer Parts of the Seed

While the embryo is forming within the ovule and the growth of the endosperm is taking place, where this is formed, other correlated changes occur in the outer parts of the ovule, and often in adjacent parts of the flower. These unite in making the "seed," or the "fruit." Especially in connection with the formation of the seed a new growth of the outer coat, or integument, of the ovule occurs, forming the outer coat of the seed, known as the *testa*, while the inner integument is absorbed. In some cases the inner integument of the ovule also forms a new growth, making an inner coat of the seed (rosaceæ). In still other cases neither of the integuments develops into a testa, and the embryo sac lies in contact with the wall of the ovary. Again an additional envelope grows up around the seed; an example of this is found in the case of the red berries of the "yew" (taxus), the red outer coat being an extra growth, called an *aril.*

In the willow and the milkweed an aril is developed in the form of a tuft of hairs. (In the willow it is an outgrowth of the funicle, = stalk of the ovule, and is called a funicular aril; while in the milkweed it is an outgrowth of the micropyle, = the open end of the ovule, and is called a micropylar aril.)

Increase in Size During Seed Formation

Synopsis of the seed

		Aril, rarely present.
		Ovular coats (one or two usually present), the testa.
		Funicle (stalk of ovule), raphe (portion of funicle when bent on to the side of ovule), micropyle, hilum (scar where seed was attached to ovary).
	Ripened ovule.	Remnant of the nucellus (central part of ovule); sometimes nucellus remains as Perisperm in some albuminous seeds.
The seed.	Endosperm, present in albuminous seeds.	
	Embryo within surrounded by endosperm when this is present, or by the remnant of nucellus, and by the ovular coats which make the testa. In many seeds (example, bean) the endospermis transferred to the cotyledons which become fleshy (exalbuminous seeds).	

Accompanying this extra growth of the different parts of the ovule in the formation of the seed is an increase in the size, so that the seed is often much greater in size than the ovule at the time of fertilization. At the same time parts of the ovary, and in many plants, the adherent parts of the floral envelopes, as in the apple; or of the receptacle, as in the strawberry; or in the involucre, as in the acorn; are also stimulated to additional growth, and assist in making the fruit.

PARTS OF THE OVULE

In Figure 401 are represented three different kinds of ovules, which depend on the position of the ovule with reference to its stalk. The funicle is the stalk of the ovule, the hilum is the point of attachment of the ovule with the ovary, the raphe is the part of the funicle in contact with the ovule in inverted ovules, the chalaza is the portion of the ovule where the nucellus and the integuments merge at the base of the ovule, and the micropyle is the opening at the apex of the ovule where the coats do not meet.

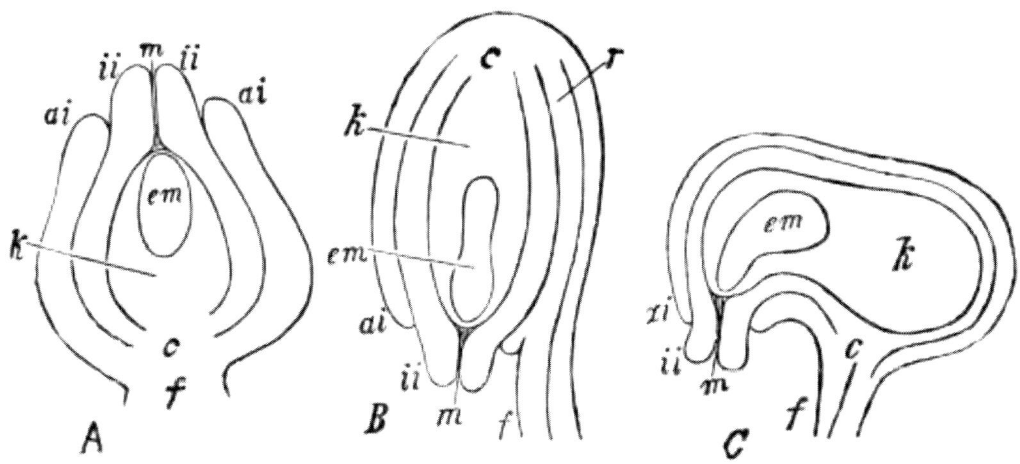

Figure 401. *A*, represents a straight (orthotropous) ovule of polygonum; *B*, the inverted (anatropous) ovule of the lily; and *C*, the right-angled (campylotropous) ovule of the bean. *f*, funicle; *c*, chalaza; *k*, nucellus; *ai*, outer integument; *ii*, inner integument; *m*, micropyle; *em*, embryo sac.

COMPARISON OF ORGAN AND MEMBER

The Stamens and Pistils Are Not the Sexual Organs

Before the sexual organs and sexual processes in plants were properly understood it was customary for botanists to speak of the stamens and pistils of flowering plants as the sexual organs. Some of the early botanists, a century ago, found that in many plants the seed would not form unless first the pollen from the stamens came to be deposited on the stigma of the pistil. A little further study showed that the pollen germinated on the stigma and formed a tube which made its way down through the pistil and into the ovule.

This process, including the deposition of the pollen on the stigma, was supposed to be fertilization, the stamen was looked on as the male sexual organ, and the pistil as the female sexual organ. We have found out, however, by further study, and especially by a comparison of the flowering plants and the lower plants, that the stamens and pistils are not the sexual organs of the flower.

The Stamens and Pistils Are Spore-Bearing Leaves

The stamen is the spore-bearing leaf, and the pollen grains are not unlike spores; in fact they are the small spores of the angiosperms. The pistil is also a spore-bearing leaf, the ovule the sporangium, which contains the large spore called an *embryo sac*. In the ferns we know that the spore germinates and produces the green heart-shaped prothallium. The prothallium bears the sexual organs. Now the fern leaf bears the spores and the spore forms the prothallium. So it is in the flowering plants. The stamen bears the small spores—pollen grains—and the pollen grain forms the prothallium. The prothallium in turn forms the sexual organs. The process is in general the same as it is in the ferns, but with this special difference: the prothallium and the sexual organ of the flowering plants are very much reduced.

Difference between Organ and Member

While it is not strictly correct then to say that the stamen is a sexual organ, or male organ, we might regard it as a *male member* of the flower, and we should distinguish between *organ* and *member*. It is an *organ* when we consider *pollen production*, but it is not a sexual organ. When we consider *fertilization* it is *not a sexual organ, but a male member* of the flower which bears the small spore.

The following table will serve to indicate these relations.

Stamen	= spore-bearing leaf = male member of flower.
Anther locule	= sporangium.
Pollen grain	= small spore = reduced male prothallium and sexual organ.

So the pistil is not a sexual organ, but might be regarded as the female member of the flower.

Pistil	=	spore-bearing leaf = female member of flower.
Ovule	=	sporangium.
Embryo sac	=	large spore = female prothallium containing the egg.
The egg	=	a reduced archegonium = the female sexual organ.

Progression and Retrogression in Sporophyte and Gametophyte

Sporophyte is Prominent and Highly Developed

In the angiosperms then, as we have seen from the plants already studied, the trillium, dentaria, etc., are sporophytes, that is they represent the spore-bearing, or sporophytic, stage. Just as we found in the case of the gymnosperms and ferns, this stage is the prominent one, and the one by which we characterize and recognize the plant. We see also that the plants of this group are still more highly specialized and complex than the gymnosperms, just as they were more specialized and complex than the members of the fern group. From the very simple condition in which we possibly find the sporophyte in some of the algæ like spirogyra, vaucheria, and coleochæte, there has been a gradual increase in size, specialization of parts, and complexity of structure through the bryophytes, pteridophytes, and gymnosperms, up to the highest types of plant structure found in the angiosperms. Not only do we find that these changes have taken place, but we see that, from a condition of complete dependence of the spore-bearing stage on the sexual stage (gametophyte), as we find it in the liverworts and mosses, it first becomes free from the gametophyte in the members of the fern group, and is here able to lead an independent existence. The sporophyte, then, might be regarded as the modern phase of plant life, since it is that which has become and remains the prominent one in later times.

The Gametophyte Once Prominent Has Become Degenerate

On the other hand we can see that just as remarkable changes have come upon the other phase of plant life, the sexual stage, or gametophyte. There is reason to believe that the gametophyte was the stage of plant life which in early times existed almost to the exclusion of the sporophyte, since the characteristic thallus of the algæ is better adapted to an aquatic life than is the spore-bearing state of plants. At least, we now find in the plants of this group as well as in the liverworts, that the gametophyte is the prominent stage. When we reach the members of the fern group, and the sporophyte becomes independent, we find that the gametophyte is decreasing in size, in the higher members of the pteridophytes, the male prothallium consisting of only a few cells, while the female prothallium completes its development still within the spore wall. And in selaginella it is entirely dependent on the sporophyte for nourishment.

As we pass through the gymnosperms we find that the condition of things which existed in the bryophytes has been reversed, and the gametophyte is now entirely dependent on the sporophyte for its nourishment, the female prothallium not even becoming free from the sporangium, which remains attached to the sporophyte, while the remnant of a male prothallium, during the stage of its growth, receives nourishment from the tissues of the nucellus through which it bores its way to the egg-cell.

In the angiosperms this gradual degradation of the male and female prothallia has reached a climax in a one-celled male prothallium with two sperm cells, and in the embryo sac with no clearly recognizable traces of an archegonium to identify it as a female prothallium. The development of the endosperm subsequent, in most cases, to fertilization, providing nourishment for the sporophytic embryo at one stage or another, is believed to be the last remnant of the female prothallium in plants.

The Seed

The seed is the only important character possessed by the higher plants (the gymnosperms and angiosperms) which is not possessed by one or another of the lower great groups. With the gradual evolution of the higher plants from the lower there has been developed at certain periods organs or structural characters which were not present in some of the lower groups. Thus the thallus is the plant body of the algæ and fungi, so that these two groups of plants are sometimcs called *Thallophytes*. In the Bryophytes (liverworts and mosses) the thallus is still present, but there is added the highly organized archegonium in place of the simple female gamete or oogonium, or carpogonium of the algæ and fungi, and the sporophyte has become a distinct though still dependent structure. In the Pteridophytes the thallus is still present as the prothallium, archegoina are also present, and while both of these structures are retrograding the sporophyte has become independent and has organized for the first time a true vascular system for conduction of water and food. In the gymnosperms and angiosperms the thallus is present in the endosperm; distinct, though reduced, archegonia are present in most gymnosperms and represented only by the egg in the angiosperms; the vascular system is still more highly developed while the seed for the first time is organized, and characterizes these plants so that they are called seed plants, or *Spermatophytes*.

VARIATION, HYBRIDIZATION, MUTATION

Variation

It is a well-known fact that plants as well as animals are subject to variation. Under certain conditions, some of which are partly understood and others are unknown, the progeny of plants differ in one or more characters from their parents. Some of these variations are believed to be due to the influence of environment (see Parts III and IV). Others are the result of the crossing of individuals which show greater or lesser differences in one or more characters, or the crossing of different species (*hybridization*). The most profound variations are those which spring suddenly into existence (*mutation*).

Hybridization

Two different species are "crossed" where the egg-cell of one species is fertilized by the sperm of another species. The progeny resulting from such a cross is a *hybrid*. Hybrids sometimes resemble one parent, sometimes another, sometimes both. Where the parents differ only in respect to one character of an organ or structure, there is a regular law in respect to the progeny if they are self-fertilized. In the first generation all the individuals are alike and resemble one of the parents, and the special differential character of that parent is called the *dominant* character. In the second generation 75% possess the dominant character, while 25% resemble the other original parent, and its differential character is called *recessive*. These are *pure* recessives, since successive generations, if self-fertilized, are always recessive. Of the 75% which show the dominant character in the second generation, one-third (or 25% of the whole number) are pure dominants if self-fertilization is continued, while 50% are really "cross breds" like the first generation, and if self-fertilized split up again into approximately 25 dominants, 50 cross breds, and 25 recessives. This is what is called Mendel's law. Where the original parents differ in respect to more than one character, the result is more complicated (see Mendel's Principles of Heredity; also de Vries, Das Spaltungsgesetz der Bastarde, Ber. d. deutsch. bot. Gesell., 18, 83, 1900).

Mutation

This term is applied to those variations which appear so suddenly that some of the progeny of two like individuals differ from all the others to a marked degree. Some of these mutations are so different as to be regarded as new species. Some of the primroses show mutations, and Œnothera gigas is a mutation from Œnothera lamarkiana (see de Vries, Die Mutationstheorie, Leipzig).

Table Showing Homologies of Sporophyte and Gametophyte in Angiosperms (Terms Corresponding to those used in Pteridophytes. Common Terms)

	Terms Corresponding to those used in Pteridophytes.		Common Terms.	
	Sporophyte	=	Higher plant.	
	Spore-bearing part	=	Stamens and carpels.	
Sporophyte	Microsporophyll	=	Stamen	Anther
				Filament.
	Microsporangium	=	Pollen sac, usually two or four.	
Male gametophyte	Microspore at maturity usually of 2 or 3	=	Pollen grain.	
	cells {young male prothallium}			
	1. Large cell (part of antheridium wall?), with			
	its nucleus surrounded by wall of spore	=	Vegetative cell.	
	2. Small cell with nucleus, no wall, floating			
	in protoplasm of large cell is the central			
	cell of antheridium (male sexual organ)	=	Generative cell.	
	Mature male prothallium	=	Pollen grain with tube.	
	Antheridium cell divided, 2 sperm cells	=	Paternal cells, or generative cells.	
Sporophyte	Macrosporophyll	=	Carpel or simple pistil	Stigma. Style. Ovary.

<table>
<tr><td></td><td>Terms Corresponding to those used in Pteridophytes.</td><td></td><td>Common Terms.</td></tr>
<tr><td></td><td>Macrosporangium, covered by 1 or 2 coats</td><td>=</td><td>Nucellus, covered by 1 or 2 coats = ovule.</td></tr>
<tr><td rowspan="6">Female gametophyte</td><td>Macrospore, cell in end of macrosporangium, does not become free, cavity enlarges</td><td>=</td><td>Uninuclear state of embryo sac.</td></tr>
<tr><td>Macrospore divides into 8 cells to form young female prothallium</td><td>=</td><td>Embryo sac.</td></tr>
<tr><td>Remnant of archegonium, egg (female sexual organ)</td><td>=</td><td>Maternal cell, or germ cell.</td></tr>
<tr><td>Growing part of prothallium</td><td>=</td><td>Two polar nuclei fused, making endosperm nucleus.</td></tr>
<tr><td>Mature female prothallium</td><td>=</td><td>Endosperm, developed by many divisions of endosperm nucleus.</td></tr>
<tr><td>Young sporophyte surrounded by parts of the gametophyte and new growth of old sporophyte</td><td>After fecundation of egg, egg divides to form embryo. Embryo in endosperm (sometimes latter nearly or quite absent) surrounded by coats</td><td>=</td><td>Seed.</td></tr>
<tr><td colspan="3">Young sporophyte surrounded by remnants of gametophyte and new parts of old sporophyte (remains of endosperm and of nucellus, and ovular coat) = the seed.</td><td></td></tr>
</table>

Chapter 37

Morphology of the Nucleus and Significance of Gametophyte and Sporophyte

In the development of the spores of the liverworts, mosses, ferns, and their allies, as well as in the development of the microspores of the gymnosperms and angiosperms, we have observed that four spores are formed from a single mother cell. These mother cells are formed as a last division of the fertile tissue (archesporium) of the sporangium. In ordinary cell division the nucleus always divides prior to the division of the cell. In many cases it is directly connected with the laying down of the dividing cell wall.

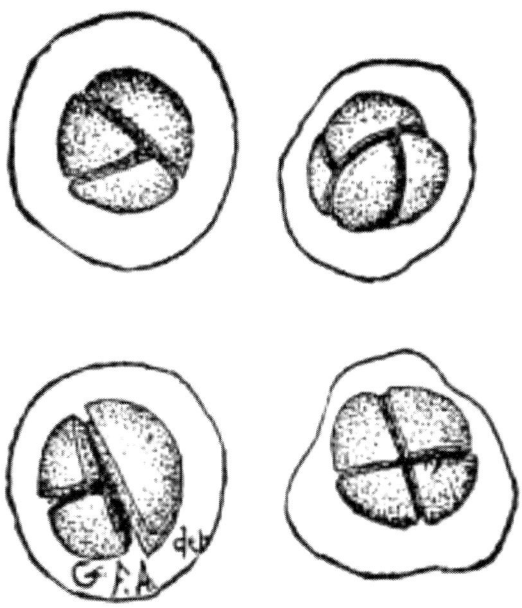

Figure 402. Forming spores in mother cells (Polypodium vulgare).

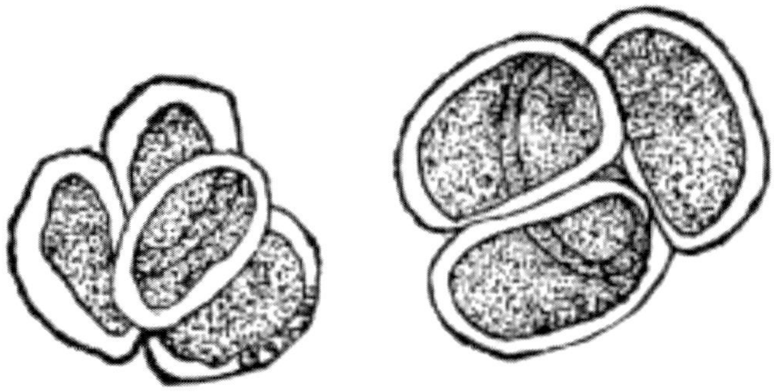

Figure 403. Spores just mature and wall of mother cell broken (Asplenium bulbiferum).

Direct Division of the Nucleus

The nucleus divides in two different ways. On the one hand the process is very simple. The nucleus simply fragments, or cuts itself in two. This is direct division.

Indirect Division of the Nucleus

On the other hand very complicated phenomena precede and attend the division of the nucleus, giving rise to a succession of nuclear Figures presented by a definite but variable series of evolutions on the part of the nuclear substance. This is *indirect division* of the nucleus, or *karyokinesis*. Indirect division of the nucleus is the usual method, and it occurs in the normal growth and division of the cell. The nuclear Figures which are formed in the division of the mother cell into the four spores are somewhat different from those occurring in vegetative division, but their study will serve to show the general character of the process.

Chromatin and Linin of the Nucleus

In Figure 404 is represented a pollen mother cell of the May-apple (podophyllum). The nucleus is in the resting stage. There is a network consisting of very delicate threads, the *linin* network. Upon this network are numerous small granules, and at the junction of the threads are distinct knots. The nucleolus is quite large and prominent. The numerous small granules upon the linin stain very deeply when treated with certain dyes used in differentiating the nuclear structure. This deeply staining substance is the *chromatin* of the nucleus.

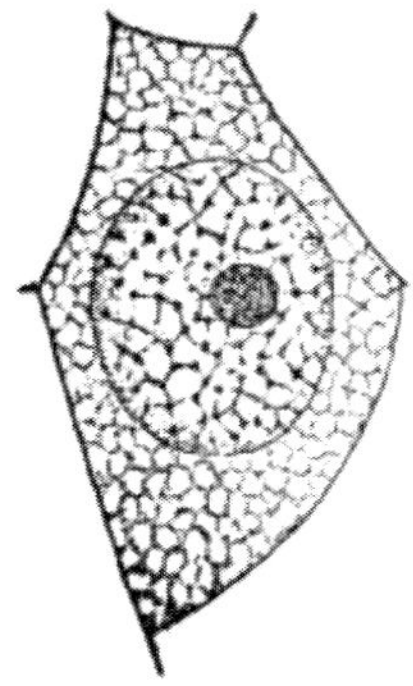

Figure 404. Pollen mother cell of podophyllum, resting nucleus. Chromatin forming a network.

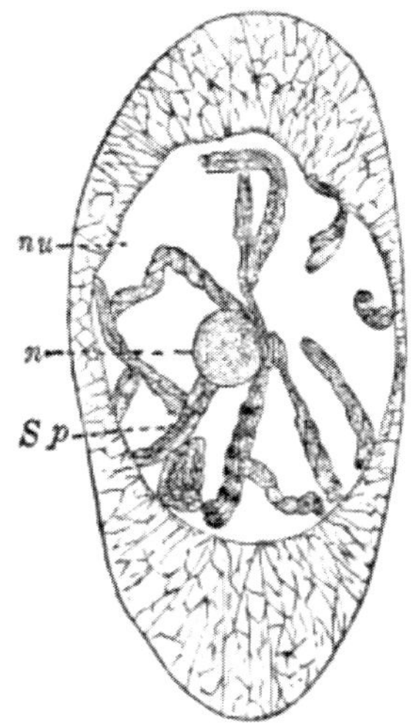

Figure 405. Spirem stage of nucleus. *nu*, nuclear cavity; *n*, nucleolus; *Sp*, spirem.

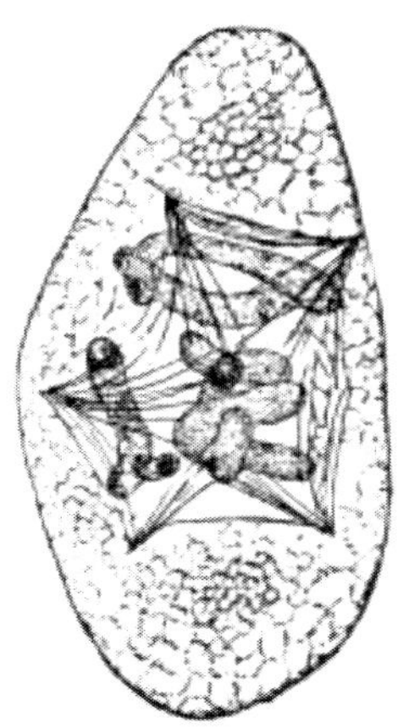

Figure 406. Forming spindle, threads from protoplasm with several poles, roping the chromosomes up to nuclear plate.

(Figures 404-406 after Mottier.)

THE CHROMATIN SKEIN

One of the first nuclear Figures in the preparatory stages of division is the chromatin *skein* or *spirem*. The chromatin substance unites to form this. The spirem is in the form of a narrow continuous ribbon, or band, woven into an irregular skein, or gnarl, as shown in Figure 405. This band splits longitudinally into two narrow ones, and then each divides into a definite number of segments, about eight in the case of podophyllum. Sometimes the longitudinal splitting of the band appears to take place after the separation into the chromatin segments. The segments remain in pairs until they separate at the nuclear plate.

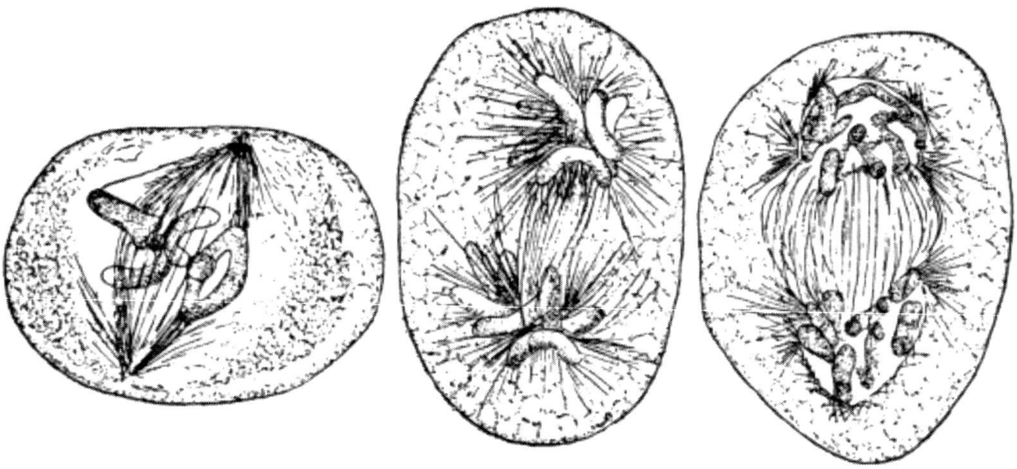

Figure 407. Karyokinesis in pollen mother cells of podophyllum. At the left the spindle with the chromosomes separating at the nuclear plate; in the middle Figure the chromosomes have reached the poles of the spindle, and at the right the chromosomes are forming the daughter nuclei. (After Mottier.)

CHROMOSOMES, NUCLEAR PLATE, AND NUCLEAR SPINDLE

Each one of these rod-like chromatin segments is a *chromosome*. The pairs of chromosomes arrange themselves in a median plane of the nucleus, radiating somewhat in a stellate fashion, forming the *nuclear plate*, or *monaster*. At the same time threads of the protoplasm (kinoplasm) become arranged in the form of a spindle, the axis of which is perpendicular to the nuclear plate of chromosomes, as shown in Figure 407, at left. Each pair of chromosomes now separate in the line of the division of the original spirem, one chromosome of each pair going to one pole of the spindle, while the other chromosome of each pair goes to the opposite pole. The chromosomes here unite to form the daughter nuclei. Each of these nuclei now divide as shown in Figure 409 (whether the chromosomes in this second division in the mother cell split longitudinally or divide transversely has not been definitely settled), and four nuclei are formed in the pollen mother cell. The protoplasm about each one of these four nuclei now surrounds itself with a wall and the spores are formed.

Figure 408.Different stages in the separation of divided U-shaped chromosomes at the nuclear plate. (After Mottier.) In podophyllum.

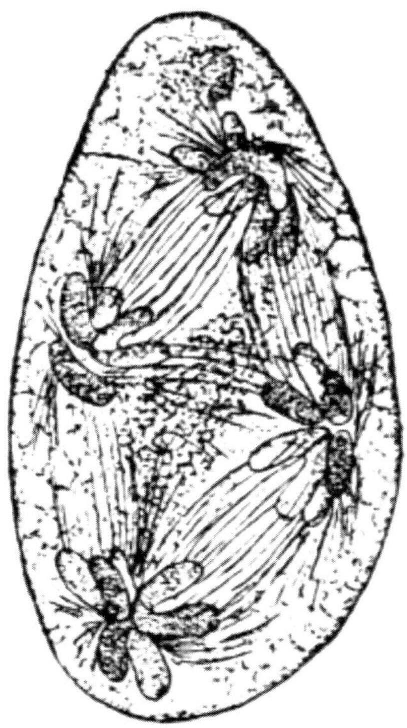

Figure 409. Second division of nuclei in pollen mother cell of podophyllum, chromosomes at poles.

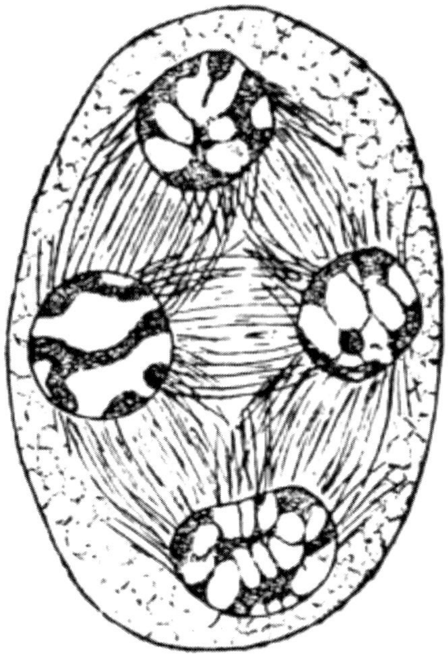

Figure 410. Chromosomes uniting at poles to form the nuclei of the four spores. (After Mottier.)

The Number of Chromosomes Usually the Same in a Given Species Throughout One Phase of the Plant

In those plants which have been carefully studied, the number of chromosomes in the dividing nucleus has been found to be fairly constant in a given species, through all the divisions in that stage or phase of the plant, especially in the embryonic, or young growing parts. For example, in the prothallium, or gametophyte, of certain ferns, as osmunda, the number of chromosomes in the dividing nucleus is always twelve. So in the development of the pollen of lilium from the mother cells, and in the divisions of the antherid cell to form the generative cells or sperm cells, there are always twelve chromosomes so far as has been found. In the development of the egg of lilium from the macrospore there are also twelve chromosomes.

When Fertilization Takes Place the Number of Chromosomes is Doubled in the Embryo

In the spermatozoid of osmunda then, as well as in the egg, since these are developed on the gametophyte, there are twelve chromosomes each. The same is true in the sperm cell (generative cell) of lilium, and also in the egg-cell. When these nuclei unite, as they do in fertilization, the paternal nucleus with the maternal nucleus, the number of chromosomes in the fertilized egg, if we take lilium as an example, is twenty-four instead of twelve; the number is doubled. The fertilized egg is the beginning of the sporophyte, as we have seen. Curiously throughout all the divisions of the nucleus in the embryonic tissues of the sporophyte, so far as has been determined, up to the formation of the mother cells of the spores, the number of chromosomes is usually the same.

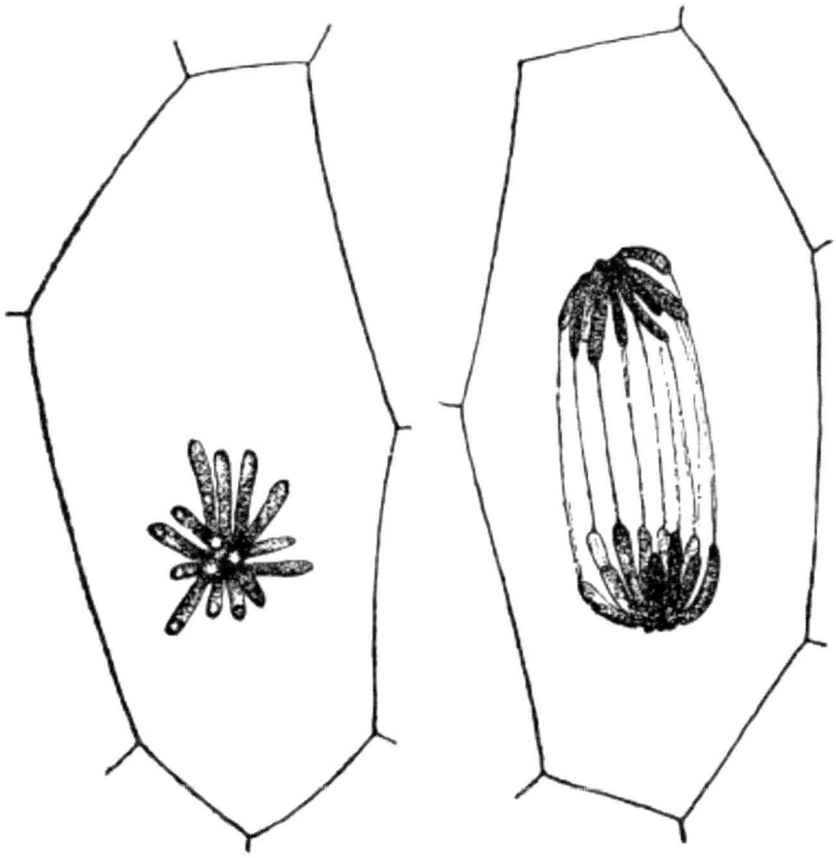

Figure 411. Karyokinesis in sporophyte cells of podophyllum (twice the number of chromosomes here that are found in the dividing spore mother cells).

Reduction of the Number of Chromosomes in the Nucleus

If there were no reduction in the number of chromosomes at any point in the life cycle of plants, the number would thus become infinitely large. A reduction, however, does take place. This usually occurs, either in the mother cell of the spores or in the divisions of its nucleus, at the time the spores are formed. In the mother cells a sort of pseudo-reduction is effected by the chromatin band separating into one half the usual number of nuclear segments. So that in lilium during the first division of the nucleus of the mother cell the chromatin band divides into twelve segments, instead of twenty-four as it has done throughout the sporophyte stage. So in podophyllum during the first division in the mother cell it separates into eight instead of into sixteen. Whether a qualitative reduction by transverse division of the spirem band, unaccompanied by a longitudinal splitting, takes place during the first or second karyokinesis is still in doubt. Qualitative reduction does take place in some plants according to Beliaieff and others. Recently the author has found that it takes place in Trillium grandiflorum during the second karyokinesis, and in Arisæma triphyllum the chromosomes divide both transversely and longitudinally during the first karyokinesis forming four chromosomes, and a qualitative reduction takes place here.

Significance of Karyokinesis and Reduction

The precision with which the chromatin substance of the nucleus is divided, when in the spirem stage, and later the halves of the chromosomes are distributed to the daughter nuclei, has led to the belief that this substance bears the hereditary qualities of the organism, and that these qualities are thus transmitted with certainty to the offspring. In reduction not only is the original number of chromosomes restored, it is believed by some that there is also a qualitative reduction of the chromatin, i.e. that each of the four spores possesses different qualitative elements of the chromatin as a result of the reducing division of the nucleus during their formation.

The increase in number of chromosomes in the nucleus occurs with the beginning of the sporophyte, and the numerical reduction occurs at the beginning of the gametophyte stage. The full import of karyokinesis and reduction is perhaps not yet known, but there is little doubt that a profound significance is to be attached to these interesting phenomena in plant life.

The Gametophyte May Develop Directly from the Tissue of the Sporophyte

If portions of the sporophyte of certain of the mosses, as sections of a growing seta, or of the growing capsule, be placed on a moist substratum, under favorable conditions some of the external cells will grow directly into protonemal threads. In some of the ferns, as in the sensitive fern (onoclea), when the fertile leaves are expanding into the sterile ones, protonemal outgrowths occur among the aborted sporangia on the leaves of the sporophyte. Similar rudimentary protonemal growths sometimes occur on the leaves of the common brake (pteris) among the sporangia, and some of the rudimentary sporangia become changed into the protonema. In some other ferns, as in asplenium (A. filix-fœmina, var. clarissima), prothallia are borne among the aborted sporangia, which bear antheridia and archegonia. In these cases the gametophyte develops from the tissue of the sporophyte without the intervention or necessity of the spores. This is *apospory*.

Figure 412. Apogamy in Pteris cretica.

The Sporophyte May Develop Directly from the Tissue of the Gametophyte

In some of the ferns, Pteris cretica for example, the embryo fern sporophyte arises directly from the tissue of the prothallium, without the intervention of sexual organs, and in some cases no sexual organs are developed on such prothallia. Sexual organs, then, and the fusion of the spermatozoid and egg nucleus are not here necessary for the development of the sporophyte. This is *apogamy*. Apogamy occurs in some other species of ferns, and in other groups of plants as well, though it is in general a rare occurrence except in certain species, where it may be the general rule.

Types of Nuclear Division

The nuclear Figures in the vegetative cells are usually different from those in the spore mother cells. In the spore mother cells there are two types of nuclear division. (1) The first division in the mother cell is called *heterotypic*. The early stages of this division usually extend over a longer period than the second, and the Figures are more complex. Before the chromosomes arrive at the nuclear plate they are often in the form of rings, or tetrads, or in the form of X, V, or Y, and the number is usually one half the number in the preceding cells of the sporophyte. (2) The *homotypic* division immediately follows the heterotypic and the Figures are simpler, often the chromosomes being of a hook form, or sometimes much stouter than in the heterotypic division. In the vegetative cells (sometimes called somatic cells, or body cells in contrast with reproductive cells) there is another type, called by some the *vegetative type*. The chromosomes here are often in the form of the letter U, and the Figures are much simpler than in the heterotypic division. In the somatic cells of the sporophyte, as stated above, the number of chromosomes is double that found in the heterotypic and homotypic divisions of the mother cells and in the somatic cells of the gametophyte, Figure 411 represents a late stage in the division of somatic cells in the sporophyte of podophyllum. The root-tips of various plants as the onion, lily, etc., are excellent places in which to study nuclear division in the somatic cells of the sporophyte.

Comparison with Animals

In animals there does not seem to be anything which corresponds with the gametophyte of plants unless the sperm cells and eggs themselves represent it. Heterotypic and homotypic division with the accompanying reduction of the number of the chromosomes takes place in animals usually in the mother cells of the sperms and eggs. At the time of fertilization the number of chromosomes is doubled, so that all the somatic cells (except in rare instances) from the fertilized egg to the mother cells of sperms and eggs have the doubled number of chromosomes. Reduction, therefore, takes place in animals just prior to the formation of the gametes, while in plants it takes place just prior to the formation of the gametophytes.

Perhaps There is not A Fundamental Difference between Gametophyte and Sporophyte

This development of sporophyte, or leafy-stemmed plant of the fern (parag. 685), from the tissue of the gametophyte is taken by some to indicate that there is not such a great difference between the gametophyte and sporophyte of plants as others contend. In accordance with this view it has been suggested that the leafy-stemmed moss plant, as well as the leafy stem of the liverworts, is homologous with the sporophyte or leafy stem of the fern plant; that it arises by budding from the protonema; and that the sexual organs are borne then on the sporophyte.

PART III. PLANT MEMBERS IN RELATION TO ENVIRONMENT

Chapter 38

The Organization of the Plant

Organization of Plant Members[37]

It is now generally conceded that the earliest plants to appear in the world were very simple in form and structure. Perhaps the earliest were mere bits of naked protoplasm, not essentially different from early animal life. The simplest ones which are clearly recognized as plants are found among the lower algæ and fungi. These are single cells of very minute size, roundish, oval, or oblong, existing during their growing period in water or in a very moist substratum or atmosphere. Examples are found in the red snow plant (*Sphærella nivalis*), the Pleurococcus, the bacteria; and among small colonies of these simple organisms (Pandorina) or the thread-like forms (Spirogyra, Œdogonium, etc.). It is evident that some of the life relations of such very simple organisms are very easily obtained—that is, the adjustment to environment is not difficult. All of the living substance is very closely surrounded by food material in solution. These food solutions are easily absorbed. Because of the minute size of the protoplasts and of the plant body, they do not have to solve problems of transport of food to distant parts of the body. When we pass to more bulky organisms consisting of large numbers of protoplasts closely compacted together, the problem of relation to environment and of food transport become felt; the larger the

[37] Suggestions to the teacher.—In the study of the flowering plants in the secondary school and in elementary courses three general topics are suggested. 1st, the study of the form and members of the plant and their arrangement, as in Chapters XXXVIII-XLV. 2d, the study of a few plants representative of the more important families, in order that the members of the plant, as studied under the first topic, may be seen in correlation with the plant as a whole in a number of different types. 3d, the study of plants in their relation to environment, as in Chapter XLVI. The first and second topics can be conducted consecutively in the classroom and laboratory. The third topic can be studied at opportune times during the progress of topics 1 and 2. For example, while studying topic 1 excursions can be made to study winter conditions of buds, shoots, etc., if in winter period, or the relations of leaves, etc., to environment, if in the growing period. While studying topic 2 excursions can be made to study flower relations, and also vegetation relations to environment (see Chapters XLVI-LVII of the author's "College Text-book of Botany"). It is believed that a study of these three general topics is of much more value to the beginning student than the ordinary plant analysis and determination of species.

organism usually the greater are these problems. A point is soon reached at which there is a gain by a differentiation in the work of different protoplasts, some for absorption, some for conduction, some for the light relation, some for reproduction, and so on. There is also a gain in splitting the form of the plant body up into parts so that a larger surface is exposed to environment with an economy in the amount of building material required. In this differentiation of the plant body into parts, there are two general problems to be solved, and the plant to be successful in its struggle for existence must control its development in such a way as to preserve the balance between them. (1) A ready display of a large surface to environment for the purpose of acquiring food and the disposition of waste. (2) The protection of the plant from injuries incident to an austere environment.

It is evident with the great variety of conditions met with in different parts of the same locality or region, and in different parts of the globe, that the plant has had very complex problems to meet and in the solution of them it has developed into a great variety of forms. It is also likely that different plants would in many cases meet these difficulties in different ways, sometimes with equal success, at other times with varied success. Just as different persons, given some one piece of work to do, are likely to employ different methods and reach results that are varied as to their value. While we cannot attribute consciousness or choice to plants in the sense in which we understand these qualities in higher animals, still there is something in their "constitution" or "character" whereby they respond in a different manner to the same influences of environment. This is, perhaps, imperceptible to us in the different individuals of the same species, but it is more marked in different species. Because of our ignorance of this occult power in the plant, we often speak of it as an "inherent" quality.

Perhaps the most striking examples one might use to illustrate the different line of organization among plants in two regions where the environment is very different are to be found in the adaptation of the cactus or the yucca to desert regions, and the oak or the cucurbits to the land conditions of our climate. The cactus with stem and leaf function combined in a massive trunk, or the yucca with bulky leaves expose little surface in comparison to the mass of substance, to the dry air. They have tissue for water storage and through their thick epidermis dole it out slowly since there is but little water to obtain from dry soil.

The cucurbits and the oak in their foliage leaves expose a very large surface in proportion to the mass of their substance, to an atmosphere not so severely dry as that of the desert, while the roots are able to obtain an abundant supply of water from the moist soil. The cactus and the yucca have differentiated their parts in a very different way from the oak or the cucurbits, in order to adapt themselves to the peculiar conditions of the environment.

When we say that certain plants have the power to adapt themselves to certain conditions of environment, we do not mean to say that if the cucurbits were transferred to the desert they would take on the form of the cactus or the yucca. They could do neither.

They would perish, since the change would be too great for their organization. Nor do we mean, that, if the cactus or yucca were transferred from the desert to our climate, they would change into forms with thin foliage leaves. They could not. The fact is that they are enabled to live in our climate when we give them some care, but they show no signs of assuming characters like those of our vegetation. What we do mean is, that where the change is not too great nor too sudden, some of the plants become slightly modified. This would indicate that the process of organization and change of form is a very slow one, and is therefore a question of time—ages it may be—in which change in environment and adaptation in form and structure have gone on slowly hand in hand.

Members of the Plant Body

The different parts into which the plant body has become differentiated are from one point of view, spoken of as members. It is evident that the simplest forms of life spoken of above do not have members. It is only when differentiation has reached the stage in which certain more or less prominent parts perform certain functions for the plant that members are recognized. In the algæ and fungi there is no differentiation into stem and leaf, though there is an approach to it in some of the higher forms. Where this simple plant body is flattened, as in the sea-wrack, or ulva, it is a *frond*. The Latin word for frond is *thallus*, and this name is applied to the plant body of all the lower plants, the algæ and fungi. The algæ and fungi together are sometimes called *thallophytes*, or *thallus plants*. The word thallus is also sometimes applied to the flattened body of the liverworts. In the foliose liverworts and mosses there is an axis with leaf-like expansions. These are believed by some to represent true stems and leaves; by others to represent a flattened thallus in which the margins are deeply and regularly divided, or in which the expansion has only taken place at regular intervals.

In the higher plants there is usually great differentiation of the plant body, though in many forms, as in the duckweeds, it is in the form of a frond. While there is a great variety in the form and function of the members of the plant body, they are all reducible to a few fundamental members. Some reduce these forms to three, the *root*, *stem*, *leaf*; while others to two, the *root*, and *shoot*, which is perhaps the best primary subdivision, and the shoot is then divided into stem and leaf, the leaf being a lateral outgrowth of the stem, and can be indicated by the following diagram:

Plant body····
- Shoot····
  - Stem.
  - Leaf.
- Root.

Kinds of Shoots

Since it is desirable to consider the shoot in its relation to environment, for convenience in discussion we may group shoots into four prominent kinds: (1) *Foliage shoots*; (2) *Shoots without foliage leaves*; (3) *Floral shoots*; (4) *Winter conditions of shoots and buds.* Topic (4) will be treated in Chapter 39, section IV.

Figure 413. Lupinus perennis. Foliage shoot and floral shoot.

(1st) Foliage Shoots

Foliage shoots are either aerial, when their relation is to both light and air; or they are aquatic, when their relation is to both light and water. They bear green leaves, and whether in the air or water we see that light is one of the necessary relations for all. Naturally there are several ways in which a shoot may display its leaves to the light and air or water. Because of the great variety of conditions on the face of the earth and the multitudinous kinds of plants, there is the greatest diversity presented in the method of meeting these conditions. There is to be considered the problem of support to the shoot in the air, or in the water. The methods for solving this problem are fundamentally different in each case, because of the difference in the density of air and water, the latter being able to buoy up the plant to a great degree, particularly when the shoot is provided with air in its intercellular spaces or air cavities. In the solution of the problem in the relation of the shoot to aerial environment, stem and leaf have in most cases coöperated;[38] but in view of the great variety of stems and their modifications, as well as of leaves, it will be convenient to discuss them in separate chapters.

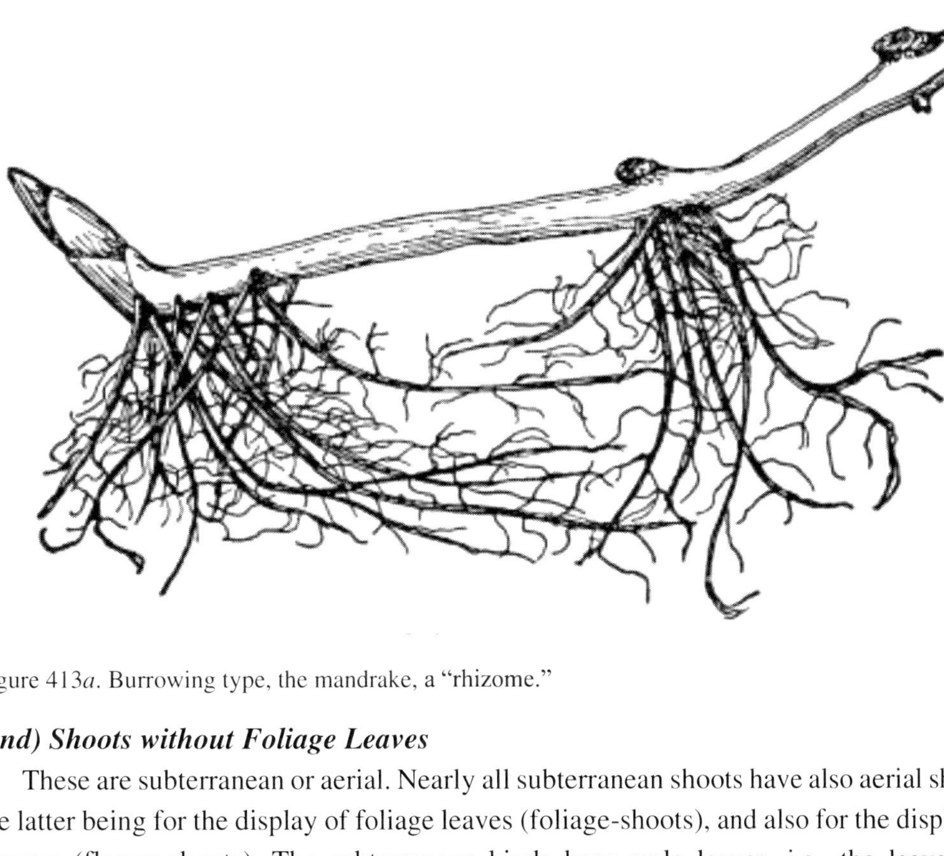

Figure 413*a*. Burrowing type, the mandrake, a "rhizome."

(2nd) Shoots without Foliage Leaves

These are subterranean or aerial. Nearly all subterranean shoots have also aerial shoots, the latter being for the display of foliage leaves (foliage-shoots), and also for the display of flowers (flower-shoots). The subterranean kinds bear scale leaves, i.e., the leaves not having a light relation are reduced in size, being small, and they lack chlorophyll. Examples are found in Solomon's seal, mandrake (Figure 413*a*), etc. Here the scale leaves are on the bud at the end of the underground stem from which the foliage shoot arises. Aerial shoots which lack foliage leaves are the dodder, Indian-pipe-plant, beech drops, etc. These plants are saprophytes or parasites (see Chapter 9). Deriving their carbohydrate food from other

[38] It is interesting to note that in some foliage shoots the stem is entirely subterranean. See discussion of the bracken fern and sensitive fern in Chapter XXXIX.

living plants, or from humus, they do not need green leaves. The leaves have, therefore, probably been reduced in size to mere scales, and accompanying this there has been a loss of the chlorophyll. Other interesting examples of aerial shoots without foliage leaves are the cacti where the stem has assumed the leaf function and the leaves have become reduced to mere spines. The various modifications which shoots have undergone accompanying a change in their leaf relation will be discussed under stems in Chapter 39.

(3d) Floral Shoots

The floral shoot is the part of the plant bearing the flower. As interpreted here it may consist of but a single flower with its stalk, as in Trillium, mandrake, etc., or of the clusters of flowers on special parts of the stem, termed flower clusters, as the *catkin*, *raceme*, *spike*, *umbel*, *head*, etc. In the floral shoot as thus interpreted there are several peculiarities to observe which distinguish it from the foliage shoot and adapt it to its life relations.

The floral shoot in many respects is comparable to the foliage shoot, as seen from the following peculiarities:

1) It usually possesses, beside the flowers, small green leaves which are in fact foliage though they are very much reduced in size, because the function of the shoot as a foliage shoot is subordinated to the function of the floral shoot. These small leaves on the floral shoot are termed *bracts*.
2) It may be (*a*) unbranched, when it would consist of a single flower, or (*b*) branched, when there would be several to many flowers in the flower cluster.
3) The flower bud has the same origin on the shoot as the leaf bud; it is either terminal or axillary, or both.
4) The members of thc flower belong to the leaf series, i.e., they are leaves, but usually different in color from foliage leaves, because of the different life relation which they have to perform. Evidence of this is seen in the transition of sepals, petals, stamens, or pistils, to foliage leaves in many flowers, as in the pond lily, the abnormal forms of trillium, and many monstrosities in other flowers (see Chapter 34).
5) The position of the members of the flower on its axis, though usually more crowded, in many cases follows the same plan as the leaves on the stem.

The various kinds of floral shoots or flower clusters will be discussed in Chapter 42, on the Floral Shoot.

Organization of Plant Tissues

A tissue is a group of cells of the same kind having a similar position and function. In large and bulky plants different kinds of tissue are necessary, not only because the work of the plant can be more economically performed by a division of labor, but also cells in the interior of the mass or at a distance from the source of the food could not be supplied with food and air unless there were specialized channels for conducting food and specialized tissue for support of the large plant body. In these two ways most of the higher plants differ from the simple ones. The tissues for conduction are sometimes called collectively the *mestome*, while tissues for mechanical support are called *stereome*. Division of labor has gone further also so that there are special tissues for absorption, assimilation, perception, reproduction, and the like. The tissues of plants are usually grouped into three systems: (1) The Fundamental System, (2) The Fibrovascular System, (3) The Epidermal System. Some of the principal tissues are as follows:

The Fundamental System

Parenchyma

Tissue composed of thin-walled cells which in the normal state are living. Parenchyma forms the loose and spongy tissue in leaves, as well as the palisade tissue (see Chapter 4); the soft tissue in the cortex of root and stem (Figure 414); as well as that of the pith, of the pith-rays or medullary rays of the stem; and is mixed in with the other elements of the vascular bundle where it is spoken of as wood parenchyma and bast parenchyma; and it also includes the undifferentiated tissue (meristem) in the growing tips of roots and shoots; also the "intrafascicular" cambium (i.e., between the bundles, some also include the cambium within the bundle).

Collenchyma

This is a strengthening tissue often found in the cortex of certain shoots. It also is composed of living cells. The cells are thickened at the angles, as in the tomato and many other herbs (Figure 414).

Sclerenchyma, or Stone-Tissue

This is also a strengthening tissue and consists of cells which do not taper at the ends and the walls are evenly thickened, sometimes so thick that the inside (lumen) of the cell has nearly disappeared. Usually such cells contain no living contents at maturity. Sclerenchyma is very common in the hard parts of nuts, and underneath the epidermis of stems and leaves of many plants, as in the underground stems of the bracken fern, the leaves of pines (Figure 415), etc.

Cork

In many cases there is a development of "cork" tissue underneath the epidermis. Cork tissue is developed by repeated division of parenchyma cells in such a way that rows of parallel cells are formed toward the outside. These are in distinct layers, soon lose their protoplasm and die; there are no intercellular spaces and the cells are usually of regular shape and fit close to each other. In some plants the cell walls are thin (cork oak), while in others they are thickened (beech). The tissue giving rise to cork is called "cork cambium," or phellogen, and may occur in other parts of the plant. For example, where plants are wounded the living exposed parenchyma cells often change to cork cambium and develop a protective layer of cork. The walls of cork cells contain a substance termed *suberin*, which renders them nearly waterproof.

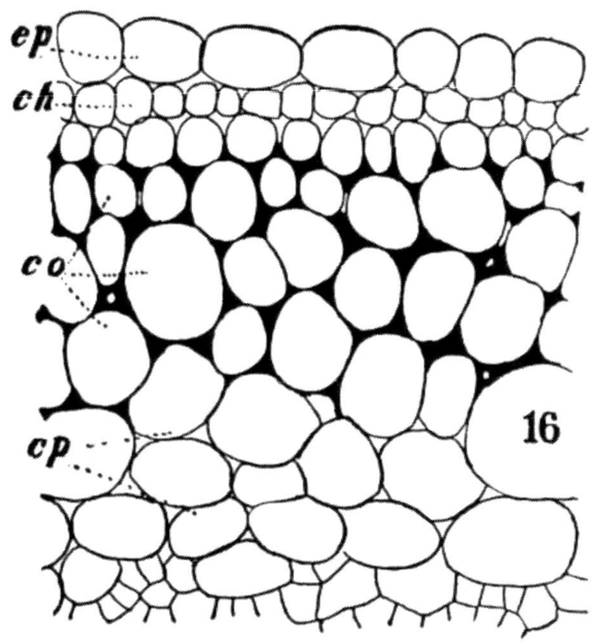

Figure 414. Transverse section of portion of tomato stem. *ep*, epidermis; *ch* chlorophyll-bearing cells; *co*, collenchyma; *cp*, parenchyma.

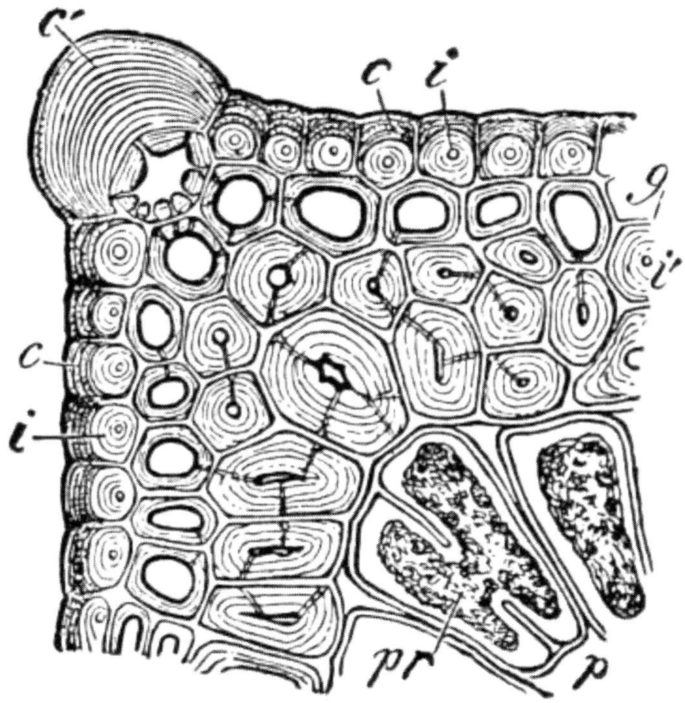

Figure 415. Margin of leaf of Pinus pinaster, transverse section, *c*, cuticularized layer of outer wall of epidermis; *i*, inner non-cuticularized layer; *c´*, thickened outer wall of marginal cell; *g*, *i´*, hypoderma of elongated sclerenchyma; *p*, chlorophyll-bearing parenchyma; *pr*, contracted protoplasmic contents. ×800. (After Sachs.)

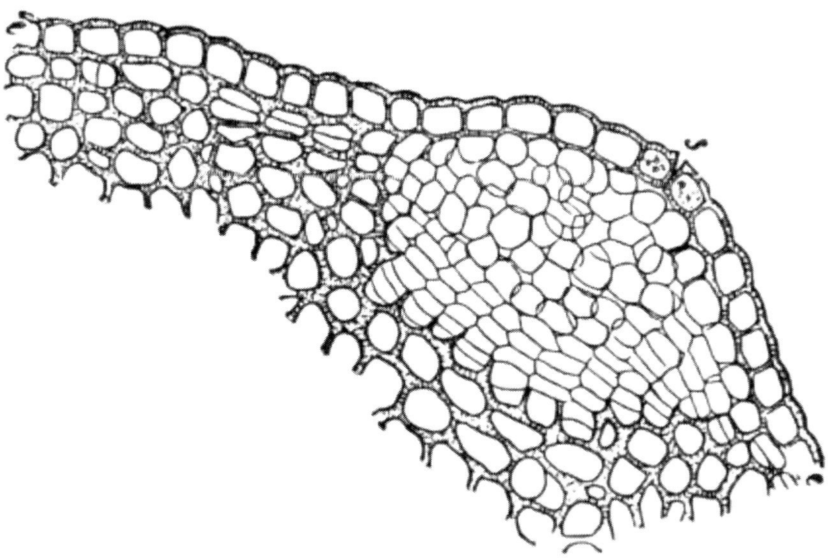

Figure 416. Section through a lenticel of Betula alba showing stoma at top, phellogen below producing rows of flattened cells, the cork. (After De Bary.)

Lenticels

These are developed quite abundantly underneath stomates on the twigs of birch, cherry, beech, elder, etc. The phellogen underneath the stoma develops a cushion of cork which presses outward in the form of an elevation at the summit of which is the stoma (Figure 416). The lenticels can easily be seen.

The Fibrovascular System

Fibrous Tissue[39]

This consists of thick-walled cells, usually without living contents which are elongated and taper at the ends so that the cells, or fibers, overlap. It is common as one of the elements of the vascular bundles, as wood fibers and bast fibers.

Vascular Tissue, or Tracheary Tissue

This consists of the vessels or ducts, and tracheides, which are so characteristic of the vascular bundle (see Chapter 5) and forms a conducting tissue for the flow of water. The vascular tissue contains spiral, annular, pitted, and scalariform vessels and tracheides according to the marking on the walls (Figures 58, 59). These are all without protoplasmic contents when mature. There are also thin-walled living cells intermingled called wood parenchyma. In the conifers (pines, etc.) the tracheary tissue is devoid of true vessels except a few spiral vessels in the young stage, while it is characterized by tracheides with peculiar markings. These marks on the tracheides are due to the "bordered" pits appearing as two concentric rings one within the other. These can be easily seen in a longitudinal section of wood of conifers.

Sieve Tissue

This consists of elongated tubular cells connected at the ends, the cross walls being perforated at the ends. These are in the phloem part of the bundle, and serve to conduct downwards the dissolved substances elaborated in the leaves.

Fascicular Cambium

This is the living, cell-producing tissue in the vascular bundle, which in the open bundle adds to the phloem on one side and the xylem on the other.

[39] Some fibers occur also very frequently in the Fundamental System, forming bundle-sheaths, or strands of mechanical tissue in the cortex.

The Epidermal System

To the epidermal system belong the epidermis and the various outgrowths of its cells in the form of hairs, or *trichomes*, as well as the guard cells of the stomates, and probably some of the reproductive organs.

The Epidermis

The epidermis proper consists of a single layer of external cells originating from the outer layer of parenchyma cells at the growing apex of the stem or root. These cells undergo various modifications of form. In many cases they lose their protoplasmic contents. In many cases the outer wall becomes thickened, especially in plants growing in dry situations or where they are exposed to drying conditions. The epidermal cells generally become considerably flattened, and are usually covered with a more or less well developed waterproof cuticle, a continuous layer over the epidermis. In many plants the cuticle is covered with a waxy exudation in the form of a thin layer, or of rounded grains, or slender rods, or grains and needles in several layers. These waxy coverings are sometimes spoken of as "bloom" on leaves and fruit.

Trichomes

Trichome is a general term including various hair-like outgrowths from the epidermis, as well as scales, prickles, etc. These include root hairs, rhizoids, simple or branched hairs, glandular hairs, glandular scales, etc. Glandular hairs are found on many plants, as tomato, verbena, primula, etc.; glandular scales on the hop; simple-celled hairs on the evening primrose, cabbage, etc.; many-celled hairs on the primrose, pumpkin; branched hairs on the shepherd's-purse, mullein, etc., stellate hairs on some oak leaves.

For stomates see Chapter 4.

Origin of the Tissues

Meristem Tissue

The various tissues consisting of cells of dissimilar form are derived from young growing tissue known as *meristem*. Meristem tissue consists of cells nearly alike in form, with thin cell walls and rich in protoplasm. It is situated at the growing regions of the plants. In the higher plants these regions in general are three in number, the stem and root apex, and the cambium cylinder beneath the cortex. Tissues produced from the stem and root apex are called *primary*, those from the cambium *secondary*. In most cases the main bulk of thc plant is secondary tissue, while in the corn plant it is all primary.

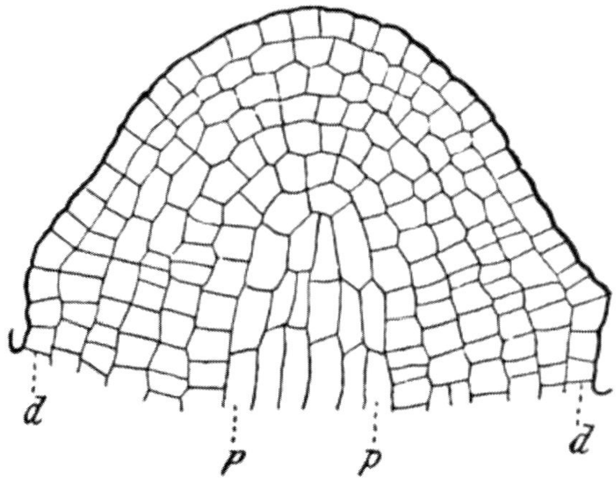

Figure 417. Section through growing point of stem, *d*, dermatogen; *p*, plerome; periblem between. (After De Bary.)

Origin of Stem Tissues

Just back of the apical meristem in a longitudinal section of a growing point it can be seen that the cells are undergoing a change in form, and here are organized three formative regions. The outer layer of cells is called *dermatogen* (skin producer), because later it becomes the epidermis. The central group of elongating cells is the *plerome* (to fill). This later develops the *central cylinder*, or *stele*, as it is called (Figure 417). Surrounding the plerome and filling the space between it and the dermatogen is the third formative tissue called the *periblem*, which later forms the cortex (bark or rind), and consists of parenchyma, collenchyma, sclerenchyma, or cork, etc., as the case may be. It should be understood that all these different forms and kinds of cells have been derived from meristem by gradual change. In the mature stems, therefore, there are three distinct regions, the central cylinder or stele, the cortex, and the epidermis.

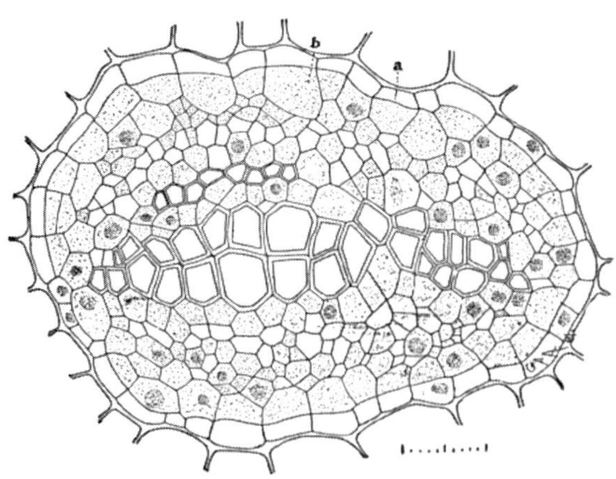

Figure 418. Concentric bundle from stem of Polypodium vulgare. Xylem in the center, surrounded by phloem, and this by the endodermis. (From the author's Biology of Ferns.)

Central Cylinder or Stele

As the central cylinder is organized from the plerome it becomes differentiated into the vascular bundles, the pith, the pith-rays (medullary rays) which radiate from the pith in the center between the bundles out to the cortex, and the pericycle, a layer of cells lying between the central cylinder and the cortex. The bundles then are farther organized into the xylem and phloem portions with their different elements, and the fascicular cambium (meristem) separating the xylem and phloem, as described in Chapter 5. Such a bundle, where the xylem and phloem portions are separated by the cambium is called an open bundle (as in Figure 58). Where the phloem and xylem lie side by side in the same radius the bundle is a *collateral* one. Dicotyledons and conifers are characterized by open collateral bundles. This is why trees and many other perennial plants continue to grow in diameter each year. The cambium in the open bundle forms new tissue each spring and summer, thus adding to the phloem on the outside and the xylem on the inside. In the spring and early summer the large vessels in the xylem predominate, while in late summer wood fibers and small vessels predominate and this part of the wood is firmer. Since the vascular bundles in the stem form a circle in the cylinder, this difference in the size of the spring and late summer wood produces the "annual" rings, so evident in the cross-section of a tree trunk. Branches originate at the surface involving epidermis, cortex, and the bundles.

In monocotyledonous plants (corn, palm, etc.) the bundles are not regularly arranged to form a hollow cylinder, but are irregularly situated through the stele. There is no meristem, or cambium, left between the xylem and phloem portions of the bundle and the bundle is thus *closed* (as in Figure 60), since it all passes over into permanent tissue. In most monocotyledons there is, therefore, practically no annual increase in diameter of the stem.

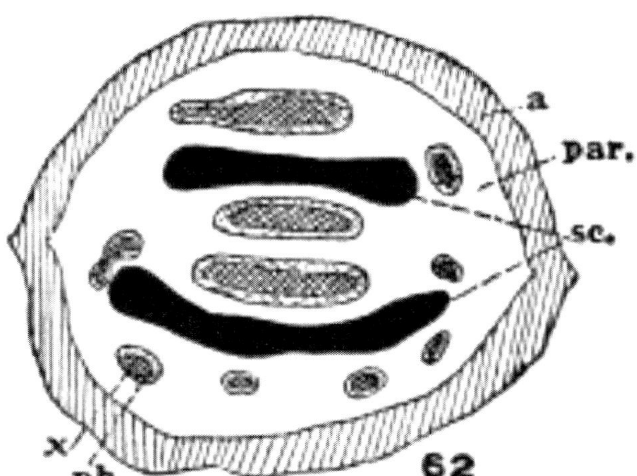

Figure 419. Section of stem (rhizome) of Pteris aquilina. *sc*, thick-walled sclerenchyma; *a*, thin-walled sclerenchyma; *par*, parenchyma.

Ferns

In the ferns and most of the Pteridophytes an apical meristem tissue is wanting, its place being taken by a single apical cell from the several sides of which cells are successively cut off, though Isoetes and many species of Lycopodium have an apical meristem group. In most of the Pteridophytes also the bundles are *concentric* instead of collateral. Figure 418 represents one of the bundles from the stem of the polypody fern. The xylem is in the center, this surrounded by the phloem, the phloem by the phloem sheath, and this in turn by the endodermis, giving a concentric arrangement of the component tissues. A cross-section of the stem (Figure 419) shows two large areas of sclerenchyma, which gives the chief mechanical support, the bundles being comparatively weak.

Origin of Root Tissues

A similar apical meristem exists in roots, but there is in addition a fourth region of formative tissue in front of the meristem called *calyptrogen* (Figure 420). This gives rise to the "root cap" which serves to protect the meristem. The vascular cylinder in roots is very different from that of the stem. There is a solid central cylinder in which the groups of xylem radiate from the center and groups of phloem alternate with them but do not extend so near the center (Figure 421).

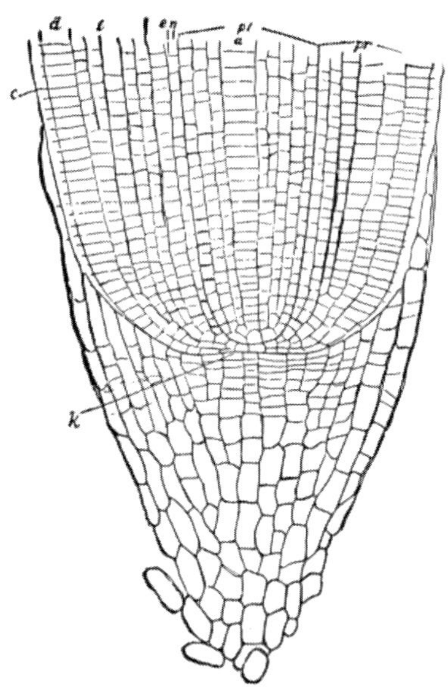

Figure 420. Median longitudinal section of the apex of a root of the barley, Hordeum vulgare. *k*, calyptrogen; *d*, dermatogen; *c*, its thickened wall; *pr*, periblem; *pl*, plerome; *en*, endodermis; *i*, intercellular air-space in process of formation; *a*, cell row destined to form a vessel; *r*, exfoliated cells of the root cap. (After Strasburger.)

As the root ages, changes take place which obscure this arrangement more or less. Branches of the roots arise from the central cylinder. In fern roots the apical meristem is replaced by a single four-sided (tetrahedral) apical cell, the root cap being cut off by successive divisions of the outer face, while the primary root tissues are derived from the three lateral faces.

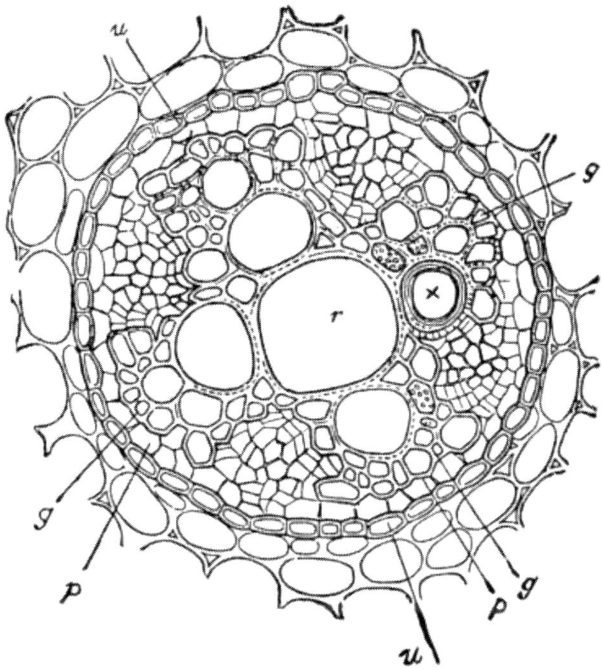

Figure 421. Cross-section of fibrovascular bundle in adventitious root of Ranunculus repens. *w*, pericycle; *g*, four radial plates of xylem; alternating with them are groups of phloem. This is a radial bundle. (After De Bary.)

Function of the Root Cap

The root cap serves an important function in protecting the delicate meristem or cambium at the tip of the root. These cells are, of course, very thin-walled, and while there is not so much danger that they would be injured from dryness, since the soil is usually moist enough to prevent evaporation, they would be liable to injury from friction with the rough particles of soil. No similar cap is developed on the end of the stem, but the meristem here is protected by the overlapping bud-scales. One of the most striking illustrations of a root cap may be seen in the case of the Pandanus, or screw-pine, often grown in conservatories (see Figure 447). On the prop roots which have not yet reached the ground the root caps can readily be seen, since they are so large that they fit over the end of the root like a thimble on the finger.

Descriptive Classification of Tissues

Epidermal System.····	Epidermis.		
	Trichomes.	Simple hairs.	
		Many-celled hairs.	
		Branched hairs, often stellate.	
		Clustered, tufted hairs.	
		Glandular hairs.	
		Root hairs.	
		Prickles.	
	Guard cells of stomates.		
Fibrovascular System.····	Xylem (wood).	Spiral vessels.	
		Pitted vessels.	
		Scalariform vessels.	
		Annular vessels.	
		Tracheides.	
		Wood fibers.	
		Wood parenchyma.	
	Cambium (fascicular).		
	Phloem (bast).	Sieve tubes.	
		Bast fibers.	
		Companion cells.	
		Bast parenchyma.	
Fundamental System.····	Stem and root.	Cortex.····	Cork.
			Collenchyma.
			Parenchyma.
			Fibers.
			Milk tissue.
		Pith-ray.··	Parenchyma.
			Intrafascicular cambium.
		Pith.······	Parenchyma.
			Sclerenchyma.
		Bundle-sheath.	
		Endodermis.	
	Leaves.	Palisade tissue.	
		Spongy parenchyma.	
	Reproductive Organs (mainly fundamental).		

PHYSIOLOGICAL CLASSIFICATION OF TISSUES

Formative Tissue

Thin-walled cells composing the meristem, capable of division and from which other tissues are formed.

Protective Tissue

- *Tegumentary System.*—Epidermis, periderm, bark protecting the plant from external contact.
- *Mechanical System.*—Bast tissue, bast-like tissue, collenchyma, sclerenchyma, afford protection against harmful bending, pulling, etc.
- *Nutritive Tissues.*
- *Absorptive System.*—Root hairs and cells, rhizoids, aerial root tissue, absorptive leaf glands, absorptive organs in seeds, haustoria of parasites, etc.
- *Assimilatory System.*—Assimilating cells in leaf and stem.
- *Conductive System.*—Sieve tissue, tracheary tissue, milk tissue, conducting parenchyma, etc.
- *Food-storing System.*—Water reservoir, water tissue, slime tissue, fleshy roots and stems, endosperm and cotyledons, etc.
- *Aerating System.*—Air spaces and tubes, special air tissue, air-seeking roots, stomates, lenticels, etc.
- *Secretory and Excretory System.*—Water glands, digestive glands, resin glands, nectaries, tannin, pitch and oil receptacles, etc.

Apparatus and Tissues for Special Duties

- Holdfasts.
- Tissues of movement, parachute hairs, floating tissue, hygroscopic tissue, living tissue.
- For perceiving stimuli.
- For conducting stimuli, etc.

Chapter 39

THE DIFFERENT TYPES OF STEMS. WINTER SHOOTS AND BUDS

ERECT STEMS

Columnar Type

The columnar type of stem may be simple or branched. When branching occurs the branches are usually small and in general subordinate to the main axis. The sunflower (Helianthus annuus) is an example. The foliage part is mainly simple. The main axis remains unbranched during the larger part of the growth-period. The principal flowerhead terminates the stem. Short branches bearing small heads then arise in the axils of a few of the upper leaves. In dry, poor soil, or where other conditions are unfavorable, there may be only the single terminal flowerhead, when the stem is unbranched. The mullein is another columnar stem. The foliage part is rarely branched, though branches sometimes occur where the main axis has become injured or broken. The flower stem is terminal. The corn plant and the Easter lily are good illustrations also of the columnar stem.

Among trees the Lombardy poplar (Populus fastigiata) is an excellent example of the columnar type. Though this is profusely branched, the branches are quite slender and small in contrast with the main axis, unless by some injury or other cause two large axes may be developed. As the technical name indicates, the branching is fastigiate, i.e., the branches are crowded close together and closely surround the central axis. The royal palm and some of the tree ferns have columnar, simple stems, but the large, wide-spreading leaves at the top of the stem give the plant anything but a cylindrical habit. Some cedars and arbor-vitæ are also columnar.

The advantages of the columnar habit of stem are three: (1) That the plant stands above other neighboring ones of equal foliage area and thus is enabled to obtain a more favorable light relation; (2) where large numbers of plants of the same species are growing close together, they can maintain practically the same habit as where growing alone; (3) the advantage gained by other types in their neighborhood in less shading than if the type were spreading. The cylindrical type can, therefore, grow between other types with less competition for existence.

Figure 422. Cylindrical stem of mullein.

Figure 423. Conical type of larch.

The Cone Type

This is well exampled in the larches, spruces, the gingko tree, some of the pines, cedars, and other gymnosperms. In the cone type, the main axis extends through the system of branches like a tall shaft, i.e., the trunk is excurrent. The lower branches are wide-spreading, and the branches become successively shorter, usually uniformly, as one ascends the stem. The branching is of two types: (1) the branches are in false whorls; (2) the branches are distributed along the stem. To the first type belong the pines, Norway spruce, Douglas spruce, etc. The white pine is an exquisite example, and in young and middle-aged trees shows the style of branching to very good advantage. The branches are nearly horizontal, with a slight sigmoid graceful curve, while towards the top the branches are ascending. This direction of the branches is due to the light relation. The few whorls at the top are ascending because of the strong light from above. They soon become extended in a horizontal direction as the main source of light is shifting to the side by the shading of the top. The ascending direction first taken by the upper branches and their subsequent turning downward, while the ends often still have a slight ascending direction gives to the older branches their sigmoid curve.

The young vernal shoots of the pines show some very interesting growth movements. There are two growth periods: (1) the elongation of the shoot, and (2) the elongation of the leaves. The elongation of the shoot takes place first and is completed in about six weeks or two months' time. The direction of the shoot in the first period seems to be entirely influenced by geotropism. It grows directly upward and stands up as a very conspicuous object in strong contrast with the dark green foliage of the more or less horizontal shoots. When the second period of growth takes place, and the leaves elongate, the shoot bends downward and outward in a lateral direction.

The rate of growth of the pines can be very easily observed since each whorl of branches (between the whorls of long shoots there are short shoots bearing the needle leaves), whether on the main axis or on the lateral branches, marks a year, the new branches arising each year at the end of the shoot of the previous year. The rate of growth is sometimes as high as twelve to twenty-four inches or more per year.

The spruces form a more perfect cone than the pines. The long branches are mostly in whorls, but often there are intermediate ones, though the rate of growth per year can usually be easily determined. In the hemlock-spruce, the branching is distributed. The larch has a similar mode of branching, but it is deciduous, shedding its leaves in the autumn, and it has a tall, conical form.

It would seem that trees of the cone type possessed certain advantages in some latitudes or elevations over other trees. (1) A conical tree, like the spruces and larches and the pines, and hemlocks also, before they get very old, meets with less injury during high winds than trees of an oval or spreading type. The slender top of the tree where the force of the wind is greatest presents a small area to the wind, while the trunk and short slender branches yield without breaking. Perhaps this is one reason why trees of this type exist in more northern latitudes and at higher elevations in mountainous regions, and why the spruce type reaches a higher latitude and altitude even than the pines. (2) The form of the tree is such as to admit light to a large foliage area, even where the trees are growing near each other. The evergreen foliage, persistent for several years, on the wide-spreading lower branches, probably affords some protection to the trees since this cover would aid in maintaining a more equable temperature in the forest cover than if the trees were bare during the winter. (3) There is less danger of injury from the weight of snow since the greater load of snow would lie on the lower branches. The form of the branches also, especially in the spruces, permits them to bend downward without injury, and if necessary unload the snow if the load becomes too heavy.

The Oval Type

This type is illustrated by the oak, chestnut, apple, etc. The trees are usually deciduous, i.e., cast their leaves with the approach of winter. The main axis is sometimes maintained, but more often disappears (trunk is deliquescent), because of the large branches which maintain an ascending direction, and thus lessen the importance of the central axis which is so marked in the cone type. Trees of this type, and in fact all deciduous trees, exhibit their character or habit to better advantage during the winter season when they are bare. Trees of this type are not so well adapted to conditions in the higher altitudes and latitudes as the cone type, for the reason given in the discussion of that type. The deciduous habit of the oaks, etc., enables them to withstand heavy winds far better than if they retained their foliage through the winter, even were the foliage of the needle kind and adapted to endure cold.

The Deliquescent Type

The elm is a good illustration of this type. The main axes and the branches fork by a false dichotomy, so that a trunk is not developed except in the forest. The branches rise at a narrow angle, and high above diverge in the form of an arch. The chief foliage development is lofty and spreading.

Trees possess several advantages over vegetation less lofty. They may start their growth later, but in the end they outgrow the other kinds, shade the ground and drive out the sun-loving kinds.

Creeping, Climbing, and Floating Stems

Prostrate Type

This type is illustrated by creeping or procumbent stems, as the strawberry, certain roses, of which a good type is one of the Japanese roses (Rosa wichuriana), which creeps very close to the ground, some of the raspberries, the cucurbits like the squash, pumpkin, melons, etc. These often cover extensive areas by branching and reaching out radially on the ground or climbing over low objects. The cucurbits should perhaps be classed with the climbers, since they are capable of climbing where there are objects for support, but they are prostrate when grown in the field or where there are no objects high enough to climb upon. In the prostrate type, there is economy in stem building. The plants depend on the ground for support, and it is not necessary to build strong, woody trunks for the display of the foliage which would be necessary in the case of an erect plant with a foliage area as great as some of the prostrate stems. This gain is offset, at least to a great extent, by the loss in ability to display a great amount of foliage, which can be done only on the upper side of the stem.

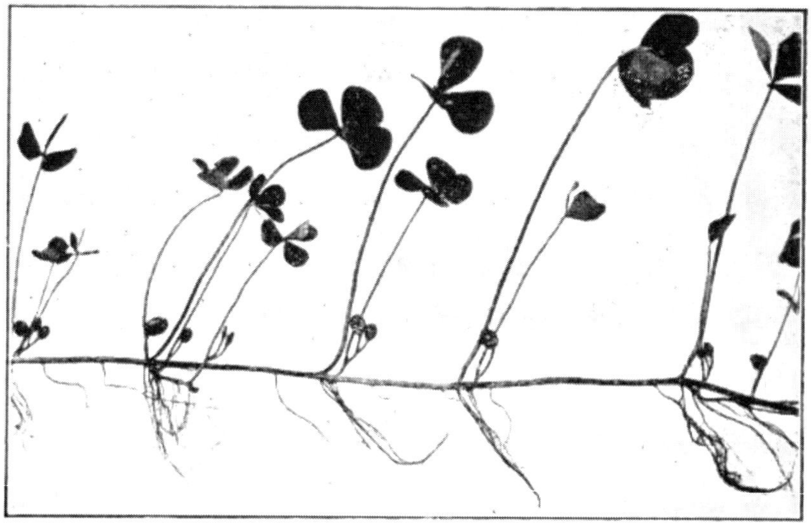

Figure 424. Prostrate type of the water fern (*marsilia*).

Other advantages gained by the prostrate stems are protection from wind, from cold in the more rigorous climates, and some propagate themselves by taking root here and there, as in certain roses, the strawberry plant, etc. Some plants have erect stems, and then send out runners below which take root and aid the plant in spreading and multiplying its numbers.

The Decumbent Type

In this type the stem is first erect, but later bends down in the form of an arch, and strikes root where the tip touches the ground. Some of the raspberries and blackberries are of this type.

The Climbing Type

The grapes, clematis, some roses, the ivies, trumpet-creeper, the climbing bittersweet, etc., are climbing stems. Like the prostrate type, the climbers economize in the material for stem building. They climb over shrubs, up the trunks of trees and often reach to a great height and acquire the power of displaying a great amount of foliage by sending branches out on the limbs of the trees, sometimes developing an amount of foliage sufficient to cover and nearly smother the foliage of large trees; while the main stem of the vine may be not over two inches in diameter and the trunk of the supporting tree may be three feet in diameter.

Floating Stems

These are necessarily found in aquatic plants. The stems may be ascending or horizontal. The stems are usually not very large, nor very strong, since the water bears them up. The plants may grow in shallow water, or in water 10-12 feet or more deep, but the leaves are usually formed at or near the surface of the water in order to bring them near the light. Various species of Potamogeton, Myriophyllum, and other plants common along the shores of lakes, in ponds, sluggish streams, etc., are examples. Among the algæ are examples like Chara, Nitella, etc., in fresh water; Sargassum, Macrocystis, etc., in the ocean. In these plants, however, the plant body is a thallus, which is divided into stem-like (*caulidium*) and leaf-like (*phyllidium*) structures.

The Burrowing Type, or Rhizomes

These are horizontal, subterranean stems. The bracken fern, sensitive fern, the mandrake (see Figure 413*a*), Solomon's seal, Trillium, Dentaria, and the like, are examples. The subterranean habit affords them protection from the cold, the wind, and from injury by certain animals. Many of these stems act as reservoirs for the storage of food material to be used in the rapid growth of the short-lived aerial shoot. In the ferns mentioned, the subterranean is the only shoot, and this bears scale leaves which are devoid of chlorophyll, and foliage leaves which are larger, and the only member of the plant body which is aerial. The foliage leaf has assumed the function of the aerial shoot. The latter is not necessary since flowers are not formed. The mandrake, Solomon's seal, Trillium, etc., have scale leaves on the fleshy underground stems, while foliage leaves are formed on the aerial stems, the latter also bearing the flowers. Some of the advantages of the rhizomes are protection from injury, food storage for the rapid development of the aerial shoot, and propagation.

Many of the grasses have subterranean stems which ramify for great distances and form a dense turf. For the display of foliage and for flower and seed production, aerial shoots are developed from these lateral upright branches.

Specialized Shoots and Shoots for Storage of Food[40]

The Bulb

The bulb is in the form of a bud, but the scale leaves are large, thick, and fleshy, and contain stored in them food products manufactured in the green aerial leaves and transported to the underground bases of the leaves. Or when the bulb is aerial in its formation, it is developed as a short branch of the aerial stem from which the reserve food material is transported. Examples are found in many lilies, as Easter lily, Chinese lilies, onion, tulip, etc. The thick scale leaves are closely overlapped and surround the short stem within (also called a *tunicated* stem). In many lilies there is a sufficient amount of food to supply the aerial stem for the development of flower and seed. There are roots, however, from the bulb and these acquire water for the aerial shoot, and when planted in soil additional food is obtained by them.

Figure 425. Bulb of hyacinth.

[40] Besides these specialized shoots for the storage of food, food substances are stored in ordinary shoots. For example, in the trunks of many trees starch is stored. With the approach of cold weather the starch is converted into oil, in the spring it is converted into starch again, and later as the buds begin to grow the starch is converted into glucose to be used for food. In many other trees, on the other hand, the starch changes to sugar on the approach of winter.

Figure 426. Corm of Jack-in-the-pulpit.

Corm

A corm is a thick, short, fleshy, underground stem. A good example is found in the jack-in-the-pulpit (Arisæma).

Tubers

These are thickened portions of the subterranean stems. The most generally known example is the potato tuber ("Irish" potato, not the sweet potato, which is a root). The "eyes" of the potato are buds on the stem from which the aerial shoots arise when the potato sprouts. The potato tuber is largely composed of starch which is used for food by the young sprouts.

Phylloclades

These are trees, shrubs, or herbs in which the leaves are reduced to mere bracts and stems, are not only green and function as leaves, but some or all of the branches are flattened and resemble leaves in form as in Phyllanthus, Ruscus, Semele, Asparagus, etc. The flowers are borne directly on these flattened axes. The prickly-pear cactus (Opuntia) is also a phylloclade. Examples of phylloclades are often to be found in greenhouses.

Undifferentiated stems are found in such plants as the duckweed, or duckmeat (Lemna, Wolffia, etc. See Chapter 3).

Annual Growth and Winter Protection of Shoots and Buds[41]

Winter Conditions[42]

While herbs are subjected only to the damp warm atmosphere of summer, woody plants are also exposed during the cold dry winter, and must protect themselves against such conditions. The air is dryer in winter than in summer; while at the same time root absorption is much retarded by the cold soil. Then, too, the osmotic activity of the dormant twig-cells being much reduced, the water-raising forces are at a minimum. It is easy to see, therefore, that a tree in winter is practically under desert conditions. Moreover, it has been found by various investigators, contrary to the general belief, that cold in freezing is only indirectly the cause of death. The real cause is the abstraction of water from the cell by the ice crystals forming in the intercellular spaces. Death ensues because the water content is reduced below the danger-point for that particular cell. It was formerly thought that on freezing, the cells in the tissue were ruptured. This is not so. Ice almost never forms within the cell, but in the spaces between. Freezing then is really a drying process, and dryness, not cold, causes death in winter. To protect themselves in winter, trees provide various waterproof coverings for the exposed surfaces and reduce the activity of the protoplasm so that it will be less easily harmed by the loss of water abstracted by the freezing process.

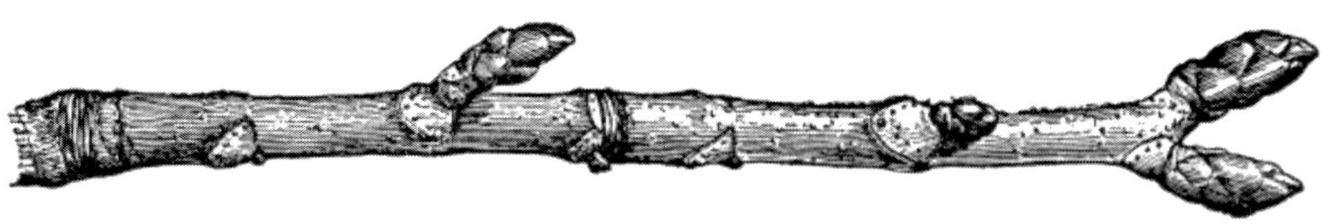

Figure 427. Two-year-old twig of horse-chestnut, showing buds and leaf-scars. (A twig with a terminal bud should have been selected for this Figure.)

[41] This topic was prepared by Dr. K. M. Wiegand.

[42] See discussion of Tropophytes in Chapter 46.

Protection of the Twig

Woody plants protect the living cells within the twigs by the production of a dull or rough corky bark, or by a thick glossy epidermis over the entire surface. At intervals occur small whitish specks called lenticels, which here perform nearly the same function as do stomates in the leaf.

Bark of Trunk

A similar service is performed by the bark for the main trunk and branches of the tree. To admit of growth in diameter the old bark is constantly being thrown off in strips, flakes, etc., and replaced by a new but larger cylinder of young bark. The external appearance thus produced enables experienced persons to recognize many kinds of trees by the trunk alone.

Leaf-Scars and Bundle-Scars

The presence of foliage leaves during the winter would greatly increase the transpiring surface without being of use to the plant; hence they are usually thrown off on the approach of winter. The scars left by the fallen leaves are termed leaf-scars. The small dots on the leaf-scars left by the vascular bundles which extended through the petiole into the twig are termed bundle-scars. Sometimes stipule-scars are left on each side of the leaf-scar by the fallen stipules.

Nodes and Internodes

The region upon a stem where a leaf is borne is termed a node. The space between two nodes is an internode.

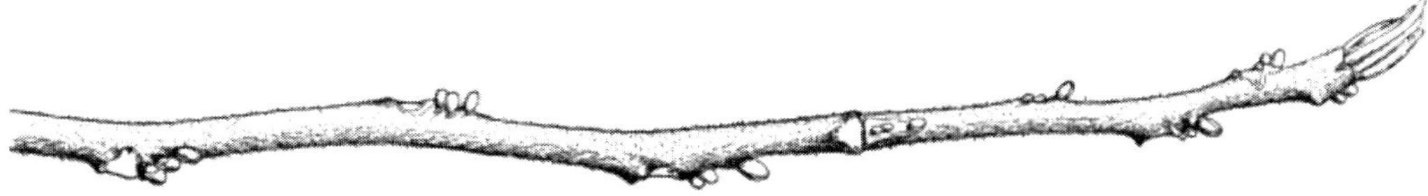

Figure 428.—Shoot of butternut showing leaf-scars, axillary buds, and adventitious buds (buds coming from above the axils).

Figure 429.—Shoot and bud of white oak.

Phyllotaxy

Investigation of a horse-chestnut or willow twig will show that the leaf-scars occupy definite positions which are constant for each plant but different for the two species. The arrangement of the leaves on the stem in any plant is termed phyllotaxy. In the horse-chestnut we find two scars placed at the same node, but on opposite sides of the stem. Somewhat higher up we find two more similarly placed, but in a position perpendicular to that of the first pair. Such phyllotaxy is termed opposite. If in any plant several leaves occur at a node, the phyllotaxy is whorled. If but one at each node, as in the willow, the phyllotaxy is alternate. The opposite and alternate types are very commonly met with. Closer observation will show that in the willow, if a line be drawn connecting the successive leaf-scars, it will pass spirally up the twig until at length a scar is reached directly over the one taken as a starting-point. Such spiral arrangement always accompanies alternate phyllotaxy. The section of the spiral thus delineated is termed a cycle. We express the nature of the cycle by the fractions $\frac{1}{2}$, $\frac{1}{3}$, $\frac{2}{5}$, $\frac{3}{8}$, $\frac{5}{13}$, etc., in which the numerator denotes the number of turns around the stem in each cycle, and the denominator the number of leaf-scars in the same distance. In a general way we find in plants only such arrangements as are represented by the fractions given above.

These fractions show the curious condition that the numerator and denominator of each is equal to the sum of the numerator or denominator of the two preceding fractions. Much speculation has been indulged in regarding the significance of these definite laws of leaf arrangement. In part they may be due to the desire that each leaf receive the maximum amount of light. Only certain definite geometrical conditions will insure this. More likely it is due to the economy of space allotted to the leaf-fundaments in the bud. Here, again, geometrical laws govern this economy. The phyllotaxy is nearly constant for a given species.

Buds

The growing point of the stem or branch together with its leaf or flower fundaments and protective structures is termed a bud. Winter buds on woody plants are terminal when inclosing the growing point of the main axis of the twig; lateral when the growing point is that of a branch of the main axis. Lateral buds are always axillary, i.e., situated on the upper angle between a leaf and the main axis.

Buds Occupying Special Positions

Several species of trees and shrubs produce more than one bud in each leaf-axil. The additional ones are termed accessory or supernumerary buds. These may be lateral to one another or they may be superposed as in the walnut or butternut. In such cases some of the buds usually contain simply floral shoots and are termed flower buds. In some species buds are frequently produced on the side of the branches and trunk at some distance from the leaf-axils, and entirely without regard for the latter; or more rarely may occur upon the root. Such buds are termed adventitious, and are the source of the feathery branchlets upon the trunks of the American elm.

Branching Follows the Phyllotaxy

Since the lateral or branch-producing buds are always located in the axil of a leaf, the branches necessarily follow the same arrangement upon the main axis as do the leaves. Since, however, many of the axillary buds fail to develop, this arrangement may be more or less obscured.

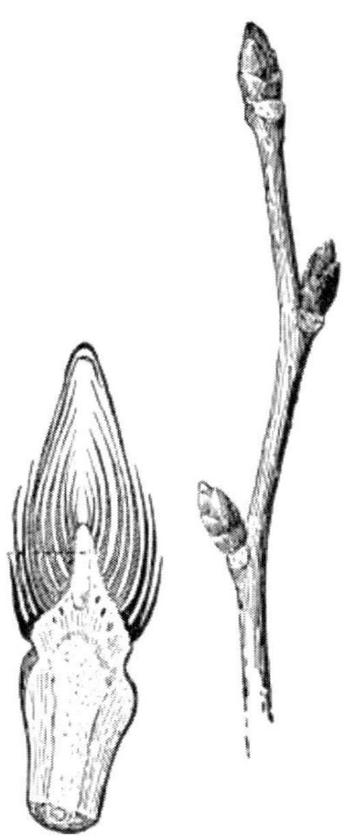

Figure 430. Bud of European elm in section, showing overlapping of scales.

Coverings of Winter Buds

These are of two sorts, hair and cork, or scales. Buds protected simply by dense hair or sunk in the cork of the twig are termed naked buds, and are comparatively rare. Most species protect their buds by the addition of an imbricated covering of closely appressed scales, the whole frequently being rendered still more waterproof by the excretion of resin between the scales or over the whole surface. The scales when studied carefully are found to be much reduced leaves or parts of leaves. In some cases they represent a modified whole leaf, when they are said to be laminar, or a leaf-petiole, when they are petiolar, or stipular, when they are much-specialized stipules of a leaf which itself is usually absent. The latter type is much the less common. The form of the bud, the nature and form of the scales, when combined with characters furnished by the leaf- and bundle-scars, enable one to recognize and classify the winter twigs of the various woody species.

Phyllotaxy of the Bud-Scales

Since the bud-scales are leaves, they follow a definite phyllotaxy. This may or may not be the same as that of the foliage leaves. Twigs with opposite leaves have opposite bud-scales, or if with alternate leaves, then alternate bud-scales, but the fractions vary. If the scales are stipular, then there are of course two at each node.

Function of the Bud Coverings

It is popularly believed that the scales and hairy coverings serve to keep the bud warm. Research, however, shows this to be almost entirely erroneous, and that the thin bud coverings are entirely inadequate to keep out the cold of winter. They cannot keep the bud even a degree or two warmer than the outside air, except when the changes are very rapid. Experiment also shows that the modifying effect of the covering when the bud thaws out is so slight as to be almost negligible. Indeed, a thermometer bulb covered with scales taken from a horse-chestnut bud warmed up more rapidly than a naked one when exposed to sunshine. The wool in the horse-chestnut bud is not for the purpose of keeping it warm, but to protect the young shoot from too great transpiration after the bud opens the following spring. Research has also shown that such tempering of the heat conditions is not especially beneficial to the plant, as was once thought. Neither can we find the main function in the prevention of water from entering the bud. This might be accomplished in much simpler ways, even if we could demonstrate the desirability of keeping the water out at all.

The true functions of the bud-scales are two in number: Firstly, the prevention of too great loss of water from the young and delicate parts within; and secondly, the protection of these same parts from mechanical injury. Without some such protection the delicate young structures would be beaten off by the wind, or become the food for hungry birds during the long winter months.

Opening of the Buds

When the young shoot begins to grow in the spring, the bud-scales are forced apart or open of their own accord. During the young condition the shoot is very soft and brittle, and also possesses a very thin, little cutinized epidermis. In this condition it is especially liable to mechanical injury and to injury from drying out. We find, therefore, a tendency for the inner bud-scales to elongate during vernation, thus forming a tube around the delicate tissue much like the opening out of a telescope. The young leaves and internodes themselves are often provided with a woody or hairy covering to retard transpiration. When the epidermis becomes more efficient the hairy covering often falls away.

In the case of naked buds protection is afforded in other ways: by the protection of hairy covering, by physiological adaptation of the tissue, or in many cases by the late appearance of the shoot in spring after the very dry April and May winds have ceased.

Bud-Scars, and How to Tell the Age of the Plant

In general the bud-scales when they fall away in the spring leave scars termed scale-scars, and the whole aggregate of scale-scars makes up the bud-scar. The position of the buds of previous winters is, therefore, marked. It becomes an easy matter to determine the age of a branch, since all that is necessary is to follow back from one bud-scar to another, the portion of the stem between representing, except in rare cases, one year's growth.

A woody plant grows in height only by the formation of new sections of stem added to the apex or side of similar sections produced the previous season, never, as is commonly supposed, by the further elongation of the previous year's growth. Hence a branch once formed upon a tree is fixed as regards its distance from the ground. The apparent rise of the branches away from the ground in forest trees is an illusion caused by the dying away of the lower branches.

Figure 431. Opening buds of hickory.

Definite and Indefinite Growth

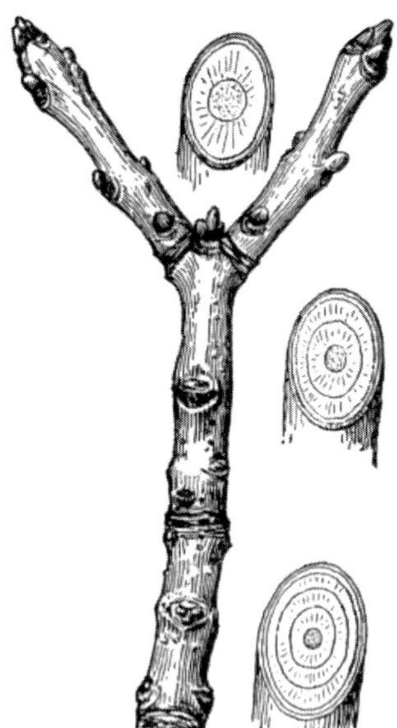

Figure 432. Three-year-old twig of the American ash, with sections of each year's growth showing annual rings.

With the opening of the buds in spring, growth begins. In some cases, when all the members for the season were formed, but still minute, within the bud, such growth consists solely in the expansion of parts already formed; in others only a few members are thus present to expand, while new ones are produced by the growing point as the season progresses. In most cases growth is completed by the middle of July, soon after which buds are formed for next year's growth. Such a method of growth is termed definite.

In a few woody plants, as, for example, sumach, locust, and raspberry, growth continues until late in the autumn. In such cases the most recently formed nodes and internodes are unable to become sufficiently "hardened" before winter sets in, and are killed back more or less. Next season's shoot is a branch from some axillary bud. Such growth is termed indefinite.

Structure of Woody Stems

If we make a cross-section of a woody twig three general regions are presented to view. On the outside is the rather soft, often greenish "bark," so called, made up of sieve tubes, ordinary parenchyma cells, and in many cases long fibrous cells composing the "fibrous bark." From a growing layer in this region, termed the phellogen, the true corky bark of the older trunk is formed.

Next within the bark we find the so-called "woody" portion of the twig. This is strong and resistant to both breaking and cutting. The microscope shows it to be composed of the ordinary already known woody elements,[43] wood fibers, for strengthening purposes, pitted and spiral vessels as conducting tissue; and intermixed with these some living parenchyma cells. A cross-section of the stem also shows narrow radial lines through the wood. These are pith-rays, composed of vertical plates of living parenchyma cells. These cells, unlike the others in the wood, are elongated radially, not vertically. The height of the pith-rays as well as their thickness varies with the species studied. In the older trunk only the outer portion, a few inches in thickness, remains light-colored and fresh, and is called sap-wood. The inner wood is usually darker and harder, and is termed heart-wood. Living parenchyma cells, in general, are present only in the sap-wood, and in this almost solely the ascent of sap occurs. Dyestuffs and other substances are frequently deposited in the walls of the heart-wood.

The third region occupying the center of the twig is the pith. This is composed ordinarily of angular, little elongated, parenchyma cells, when mature mostly without cell-contents and filled with air. The pith region in different trees is quite diversified. It may be hollow, chambered, contain scattered thick-walled cells, have woody partitions, or rarely be entirely thick-walled.

The nature of the woody ring is rather perplexing at first; but its origin is simple. We may conceive that it has developed from a stem-type like the sunflower, in which the

[43] Chapter 5, and Organization of Tissues in Chapter 38.

bundles, though separate, are connected by a continuous cambium ring. In the woody twigs the numerous bundles are closely packed together, and only separated by the primary pith-rays extending from the pith to the cortex. Other secondary pith-rays are produced within each bundle, but they usually extend only part way from the cortex to the pith. The wood represents the xylem of the bundle, and the sieve tubes of the bark, the phloem.

Growth in Thickness

Although the year's growth does not increase in length after the first season has passed, it does increase in diameter very much. From the size of an ordinary little twig it may at length become a large tree trunk several feet in thickness. Only a portion of the first year's growth is produced by the growing point. All the rest is a product of the cambium, a cylinder of wood being added to the exterior of the old wood each season. The cambium, here, as in the sunflower, lies between the phloem and the xylem, forming a cylinder entirely around the stem. In spring, when active, it becomes soft and delicate, thus enabling one to easily strip off the bark from some trees, such as willow, etc., at that season.

Annual Rings in Woody Stems

The wood produced by the cambium each season is not homogeneous throughout, but is usually much denser toward the outer part of the yearly cylinder, wood fibers here predominating. In the inner portion vessels predominate, giving a much more porous effect. The transition from one year's growth to another is very abrupt, giving rise to the appearance of rings in cross-section. Since ordinarily in temperate climates but one cylinder of wood is added each year, the number of rings will indicate the age of the trunk or branch. This is not absolutely accurate, since in some trees under certain conditions more than one ring may be produced in a summer. The porous part of the ring is often termed "spring wood," and the denser portion "fall wood," but since growth from the cambium ceases in most trees by the middle of July, "summer wood" would be more appropriate for the latter. It is mainly the alternation of the cylinders of the spring and summer wood that gives the characteristic grain to lumber. Pith-rays play an important part in wood graining only in a few woods, as, for instance, in quartered oak. The reason for the production of porous spring wood and dense summer wood is still one of the unsolved problems of botany.

Chapter 40

FOLIAGE LEAVES

GENERAL FORM AND ARRANGEMENT OF LEAVES

Influence of Foliage Leaves on the Form of the Stem

The marked effect which foliage has upon the aspect of the plant or upon the landscape is evident to all observers. Perhaps it is usual to look upon the stem as having been developed for the display of the foliage without taking into account the possibility that the foliage may have a great influence upon the form or habit of the stem. It is very evident, however, that the foliage exercises a great influence on the form of the stem. For example, as trees increase in age and size, the development of branches on the interior ceases and some of those already formed die, since the dense foliage on the periphery of the trees cuts off the necessary light stimulus. The tree, therefore, possesses fewer branches and a more open interior. In the forest also, the dense foliage above makes possible the shapely, clean timber trunks. Note certain trees where by accident, or by design, the terminal foliage-bearing branches have been removed that foliage-bearing branches may arise in the interior of the tree system.

Without foliage leaves the stems of green plants would develop a very different habit from what they do. This development could take place in three different directions under the influence of light: (1) The light stimulus would induce profuse branching, so that there would be many small branches. (2) The stem would develop fewer branches, but they would be flattened. (3) Massive trunks with but few or no branches. In fact, all these forms are found in certain green stems which do not bear leaves. An example of the first is found in asparagus with its numerous crowded slender branches. But such forms in our climate are rare, since foliage leaves are more efficient. The second and third forms are found among cacti, which usually grow in dry regions under conditions which would be fatal to ordinary thin foliage leaves.

Relation of Foliage Leaves to the Stem

In the study of the position of the leaves on the stem we observe two important modes of distribution: (1) the distribution along the individual stem or branch which bears them, usually classed under the head of *Phyllotaxy*; (2) the distribution of the leaves with reference to the plant as a whole.

Phyllotaxy, or Arrangement of Leaves

In examining buds on the winter shoots of woody plants, we cannot fail to be impressed with some peculiarities in the arrangement of these members on the stem of the plant.

In the horse-chestnut, as we have already observed, the leaves are in pairs, each one of the pair standing opposite its partner, while the pair just below or above stand across the stem at right angles to the position of the former pair. In other cases (the common bed-straw) the leaves are in whorls, that is, several stand at the same level on the axis, distributed around the stem. By far the larger number of plants have their leaves arranged alternately. A simple example of alternate leaves is presented by the elm, where the leaves stand successively on alternate sides of the stem, so that the distance from one leaf to the next, as one would measure around the stem, is exactly one half the distance around the stem. This arrangement is one half, or the angle of divergence of one leaf from the next is one half. In the case of the sedges the angle of divergence is less, that is one-third.

By far the larger number of those plants which have the alternate arrangement have the leaves set at an angle of divergence represented by the fraction two fifths. Other angles of divergence have been discovered, and much stress has been laid on what is termed a law in the growth of the stem with reference to the position which the leaves occupy. Singularly by adding together the numerators and denominators of the last two fractions gives the next higher angle of divergence. Example:

$$\frac{1 + 2}{3 + 5} = \frac{3}{8}; \quad \frac{2 + 3}{5 + 8} = \frac{5}{13};$$

and so on. There are, however, numerous exceptions to this regular arrangement, which have caused some to question the importance of any theory like that of the "spiral theory" of growth propounded by Goethe and others of his time.

Adaptation in Leaf Arrangement

As a result, however, of one arrangement or another we see a beautiful adaptation of the plant parts to environment, or the influence which environment, especially light, has had on the arrangement of the leaves and branches of the plant. Access to light and air are of the greatest importance to green plants, and one cannot fail to be profoundly impressed with the workings of the natural laws in obedience to which the great variety of plants have worked out this adaptation in manifold ways.

Distribution of Leaves With Reference to the Entire Plant

In this case, as in the former, we recognize that it is primarily a light relation. As the plant becomes larger and more branched the lower and inner leaves disappear. The trees and shrubs have by far the larger number of leaves on the periphery of the branch system. A comparison of different kinds of trees in this respect shows, however, that there is great variation. Trees with dense foliage (elm, Norway maple, etc.) present numerous leaves on the periphery which admit but little light to the interior where leaves are very few or wanting. The sugar maple and red maple do not cast such a dense shade and there are more leaves in the interior. This is more marked in the silver maple, and still more so in the locust (Gleditschia tricanthos).

Color of Foliage Leaves

The great majority of foliage leaves are green in color. This we have learned (Chapter 7) is due to the presence of a green pigment, chlorophyll, in the chloroplastids thickly scattered in the cells of the leaf. We have also learned that in the great majority of cases, the light stimulus is necessary for the production of chlorophyll green. There are many foliage leaves which possess other colors, as red (Rosa rubrifolia), purple (the purple barberry, hazel, beech, birch, etc.), yellow (the golden oak, elder, etc.); while many others have more or less deep tints of pink, red, purple, yellow, when young. All of these leaves, however, possess chlorophyll in addition to red, yellow, purple or other pigment. These other pigments are sometimes developed in great quantity in the cell-sap. They obscure the chlorophyll from view, but do not interfere seriously with the action of light and the function of chlorophyll, and perhaps in some cases serve as a screen to protect the protoplast.

Autumn Colors

Foliage leaves of many trees display in the autumn gorgeous colors. These colors are principally shades of red or yellow, and sometimes purple. The autumn color is more marked in some trees than in others. In the red maple, the red and scarlet oak, the sourwood, etc., red predominates, though sometimes yellow may be present with the red in a single leaf. Sugar maples, poplars, hickories, etc., are principally yellow in autumn. The sweet gum has a rich variety of color-red, purple, maroon, yellow; sometimes all these colors are present on the same tree.

The red and purple colors are found suffused in the cell-sap of certain cells in the leaf much as we have found it in the cells of the red beet. The yellow color is chiefly due to the disappearance and degeneration of the chlorophyll while the leaf is in a moribund state. A similar phenomenon is seen in the yellowing of crops when the soil becomes too wet, or in the blanching of grass when covered with a board, or of celery as the earth is ridged up over the leaves in late summer and autumn. A number of different theories have been advanced to explain autumn coloring, i.e., the appearance of the red coloring matter. It has been attributed to the approach of cold weather, and this has likely led to the erroneous belief on the part of some that it is caused by frost. It very often precedes frost. Some have attributed it to the action of the more oblique light rays during autumn, and still others to the diminishing water-supply with the approach of cool weather. The question is one which has not met as yet with a satisfactory solution, and is certainly a very obscure one. It is likely that the low temperature or the declining activities of the leaf affect certain organic substances in the leaf and give rise to the red color, and it is quite certain that in some years the display is more brilliant than in others. The color is more striking in some regions than in others and the different soil, as well as climate, has been supposed to have some influence. The North American forests are noted for the brilliant display of autumnal color. This is perhaps due to some extent to the great variety or number of species which display color. It would seem that there is some specific as well as individual peculiarities in certain trees. Some individuals, for example, exhibit brilliant colors every autumn, while others near of the same species are more subdued.

It has been shown by experiment that when sunlight passes through red colors the temperature is slightly increased, and it has been suggested that this may be of protection to the living substance which has ceased working and is in danger of injury from cold. There does not seem to be much ground for this suggestion, however. It certainly could not protect the protoplasm of the leaf at night when the cold is more intense, and during the day would only aggravate matters by supplying an increased amount of heat, since extremes of heat and cold in alternation are more harmful to plant life than uniform cold. Especially would this be the case in alpine climates where the alternation of heat and cold between day and night is extreme, and brilliancy of the colors of alpine plants is well known. It seems more reasonable to suppose that the red color acts as a screen, as the chlorophyll is disappearing, to protect from the injurious action of light, certain organic substances which are to be transferred back from the leaf to the stem for winter storage. So in the case of many stems in the spring or early summer when the young leaves often have a reddish color, it is likely that it acts as a screen to protect the living substance from the strong light at that season of the year until the chlorophyll screen, which is weak in young leaves, becomes darker in color and more effective, when the red color often disappears.

Function of Foliage Leaves

In general the function of the foliage leaf as an organ of the plant is fivefold (see Chapters 4, 7, 8, 11), (1) that of carbon dioxide assimilation or *photosynthesis*, (2) that of transpiration, (3) that of the synthesis of other organic compounds, (4) that of respiration, and (5) that of assimilation proper, or the making of new living substance. While none of these functions are solely carried on in the leaf, it is the chief seat of the first three of these processes, its form, position, and structure being especially adapted to the purpose. Assimilation proper, as well as respiration, probably take place equally in all growing or active parts.

Parts of the Leaf

All foliage leaves possess a *blade* or *lamina*, so called because of its *expanded* and *thin* character. The blade is the essential part. Many leaves, however, are provided with a stalk or *petiole* by which the blade is held out at a greater or lesser distance from the stem. Leaves with no petiole are *sessile*, the blade is attached by one end directly on the stem. In some cases the base of the blade is wrapped partly around the stem, or in others it extends entirely around the stem and is *perfoliate*. Besides, many leaves have short appendages, termed *stipules*, attached usually on opposite sides of the petiole at its junction with the stem. In some species of magnolia the stipules are so large that each one envelops the entire portion of the bud which has not yet opened. Many leaves possess outgrowths in the form of hairs, scales, etc. (See leaf protection.)

Simple Leaves

Simple leaves are those in which the blade is plane along the edge, not divided. The edge may be entire or indented (serrate) to a slight extent as in the elm. The form of the simple leaf varies greatly but is usually constant for a given species, or it may vary in shape in the same species on different parts of the plant. Some of the terms applied to the outline of the leaf are ovate, oval, elliptical, lanceolate, linear, needle-like, etc., but it is idle for one to waste time on matters of minute detail in form until it becomes necessary for those in the future who pursue taxonomic work. It is evident that a simple leaf, except those of minute size, possesses advantages over a divided leaf in the amount of surface it exposes to the light. But in other respects it is at a disadvantage, especially as it increases in size, since it casts a deeper shade and does not admit of such a free circulation of air. It will be found, however, in our study of the relation of leaves to light and air that the balance between the leaf and its environment is obtained in the relation of the leaves to each other.

Venation of leaves.—A very prominent character of the leaf is its "venation." This is indicated by the presence of numerous "veins," indicated usually by narrow depressed lines on the upper surface, and by more or less distinct elevated lines on the under surface. There are two general types: (1) In the corn, Smilacina, Solomon's seal, etc., the veins extend lengthwise of the leaf and are nearly parallel. Such leaves are said to be *parallel-veined*. It is generally, though not always, a character of monocotyledenous plants. (2) In the elm, rose, hawthorn, maple, oak, etc., the veins are not all parallel. The larger ones either diverge from the base of the blade (palmate leaf, maple), or the midvein extends through the middle line of the leaf, while other prominent ones branch off from this and extend, nearly parallel, toward the edge of the leaf (pinnate venation). The smaller intermediate veins which are also very distinct extend irregularly and branch and anastomose in such a fashion as to give the Figure of a net with very fine meshes. These are *netted-veined* leaves. These are characteristic of most of the dicotyledenous plants. It is evident from what has been said of the examples cited that there are two types of netted-veined leaves, the *palmate* and *pinnate*.

Note. As we have already learned in Chapter 5 the veins contain the vascular bundles of the leaf. Through them the water and food solutions are distributed to all parts of the leaf, and the return current of food material elaborated in the leaf moves back through the bast portion into the shoot. The veins also possess a small amount of mechanical tissue. This forms the framework of the leaf and aids in giving rigidity to the leaf and in holding it in the expanded position. The mechanical tissue in the framework alone could not support the leaf. Turgescence of the mesophyll is needed in addition.

Cut or Lobed Leaves

In many leaves, the indentations on the margin are few and deep. Such leaves present several lobes the proportionate size of which is dependent upon the depth of the indentation or "incision." Several of the maples, oaks, birches, the poison-ivy, thistles, the dandelion, etc., have lobed leaves. Where the indentation reaches to or very near the midrib the leaf is said to be cut. A study of various leaves will show all gradations from simple leaves with plane edges to those which are cut or divided, as in compound leaves, and the lobes are often variously indented.

Figure 433. Lobed leaves of oak forming a mosaic.

Figure 434. Twice compound leaf. Leaflets arranged in one plane, but open spaces permit free circulation of air through the large leaf.

Divided, or Compound Leaves

The rose, sumac, elder, hickory, walnut, locust, pea, clover, American creeper, etc., are examples of divided or compound leaves. The former are pinnately compound, and the latter are palmately compound. The leaf of the honey-locust is twice pinnately compound or bipinnate, and some are three times pinnately compound.[44] It is evident that compound leaves are only extreme forms of lobed or cut leaves and that the form of all bears a definite relation to the primary venation. There has been a reduction of mesophyll and of the area of smaller venation.

These forms of leaves probably have some definite significance. It is not quite clear why they should have developed as they have; though it is possible to explain several important relations of these forms to their environment. (1) The reduction of the surface of the leaf, with the retention of the firmer portions, allows freer movement of the air and affords the leaf greater protection from injury during violent winds, just as the finely dissected leaves of some water plants are less liable to injury from movement of the more dense medium in which they live. It is possible that here we may have an explanation of one of the factors involved in this reduction of leaf surface. (2) In trees with compound leaves, like the hickory, walnut, locust, ailanthus, etc., the midvein, and in the case of the Kentucky coffee-tree (Gymnocladus) the primary lateral veins also, serve in place of terminal branches of the stem. By the increase in the outline of the leaf and the reduction of its surface between the larger veins, the tree has attained the same leaf development that it would were the larger veins replaced by stems bearing simple leaves. The tree as it is, however, has the advantage of being able to cast off for the winter period a layer of what otherwise would have been a portion of the stem system, to retain which through the winter would use more energy than with the present reduced stem system, and the stouter stem is less liable to dry out. In the case of herbaceous plants, in the case of plants like most of the ferns where the stem is on the underground rootstock (Pteris), or a very short erect stem, as in the Christmas fern, the leaf replaces the aerial stem, and the division (or branching, as it is sometimes styled) of the leaf corresponds to the branching of the stem. This is more marked in the gigantic exotics like Cibotium regale, and in the tree ferns which have quite tall trunks, the massive compound leaves replace branches. In the palms and cycads are similar examples. Those who choose to observe can doubtless find many examples close

---

[44] Some of the different terms used to express the kinds of compound leaves are as follows:

Unifoliate (for a single leaflet, as in orange and lemon where the compound leaf is greatly reduced and consists of one pinna attached to the petiole by a joint). *Bifoliate* for one with two leaflets; *trifoliate* for one with three leaflets, as in the clover; *plurifoliate* for many leaflets. *Odd pinnate* for a pinnate leaf with one or more pairs of leaflets and one odd leaflet at the end.

So leaves are *palmately bifoliate*, etc., *pinnately bifoliate*, etc. *Decompound* leaves are those where they are more than twice compound, as *ternately decompound* in the common meadow rue (Thalictrum).

Perfoliate leaves are seen in the bellwort (Uvularia), *connate perfoliate*, as in some of the honeysuckles where the bases of opposite leaves are joined together around the stem. *Equitant* leaves are found in the iris, where the leaves fit over one another at the base like a saddle.

at hand. (3) While divided leaves have probably not been evolved in response to the light relation, still their relation in this respect is an important one, since if the leaf with its present size were entire, it would cast too dense a shade on other leaves below.

General Structure of the Leaf

The general structure of the leaf has been already studied (see Chapters 4, 5, 7). It is only necessary to recall the main points. The upper and lower surfaces of the leaf are provided with a layer of cells usually devoid of chlorophyll. The mesophyll of the leaf consists usually of a layer of palisade cells beneath the epidermis, and the remainder consists of loose parenchyma with large intercellular spaces. Through the mesophyll course the "veins," or fibrovascular strands, consisting of the xylem and phloem portions and serving as conduits for water, salts, and foodstuffs. In the epidermis are the stomata, each one protected by the two guard cells. The guard cells as well as the mesophyll contain chlorophyll. The stomata and the communicating intercellular spaces furnish the avenues for the ingress and egress of gases, and for the escape of water vapor.

Protection of Leaves

There are many modifications of the general plan of structure in different leaves, many of them being adaptations for the protection of the leaf under adverse or trying conditions. Many leaves are also capable of assuming certain positions which afford them protection. The discussion of this subject may be presented under two general heads: Protective modifications; protective positions.

PROTECTIVE MODIFICATION OF LEAVES

General Directions in Which These Modifications Have Taken Place

The usual type of foliage leaf selected is that of deciduous trees or shrubs or of our common herbs. Such a leaf is usually greatly expanded and thin in order to present as great a surface as possible in comparison with its mass, since the kind of work which the leaf has to do can be more effectually carried on when it possesses this form. This form of leaf is best adapted for work in regions where there is a medium amount of moisture such as exists in the temperate zones. But since there are very great variations in the climatic and soil conditions of these regions, and even greater changes in desert and arctic regions, the type of leaf described is unsuited for all. Its own life would be endangered, and it would also endanger the life of the plant. Modifications have therefore taken place to meet these conditions, or at least those plants whose leaves have become modified in those directions which are suited to the surrounding conditions have been able to persist. Excessive cold or heat, drought, winds, intense light, rain, etc., are some of the conditions which endanger leaves. The protective modifications of leaves may be grouped under four general heads: (1) Structural adaptations; (2) Protective covering; (3) Reduction of surface; (4) Elimination of the leaf through the complete assumption of the leaf function by the stem.

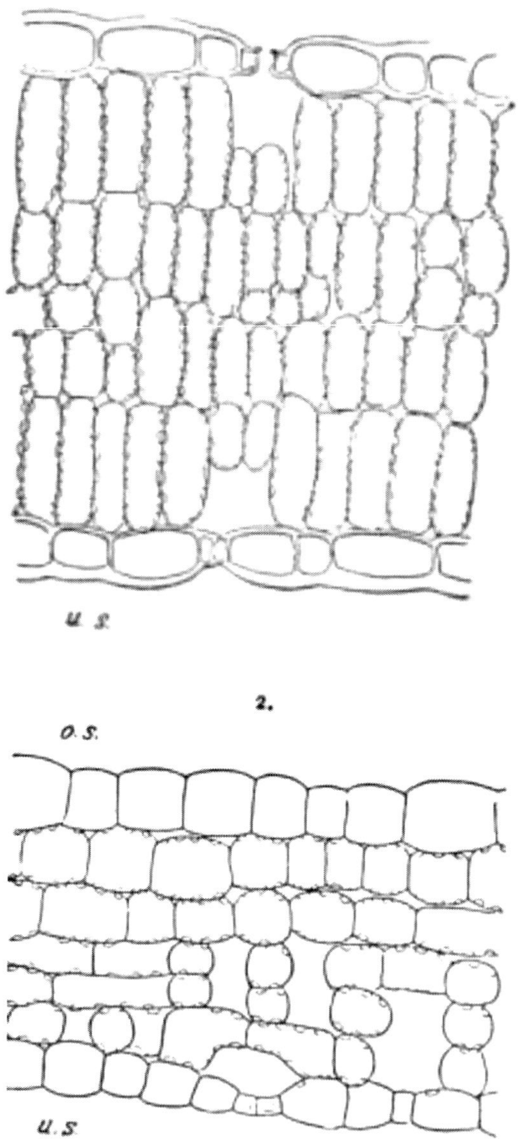

Figure 435. Structure of leaf of Lactuca scariola. Upper one grown in sunlight, palisade cells on both sides. Lower one grown in shade, no palisade tissue.

Structural Adaptations

The general structure of the leaf presents certain features which are protective. The palisade layer of cells found usually beneath the upper epidermis forms a compact layer of long cells which not only acts as a light screen cutting off a certain amount of the light, since too intense light would be harmful; it also aids in lessening the loss of water from the upper surface, where radiation is greater. The stomata are usually on the under side of aerial leaves, and the mechanism which closes them when the leaf is losing too much water is protective. As a protection against intense light the number of palisade layers is sometimes increased or the cells of this layer are narrow and long. This is often beautifully shown when comparing leaves of the same plant grown in strong light with those grown in the shade. The compass plant (Lactuca scariola) affords an interesting example. The leaves grown in the light are usually vertical, so that the light reaches both sides. Such leaves often have all of the mesophyll organized into palisade cells (Figure 435), while leaves grown in the deep shade may have no palisade cells.

Protective Covering—Epidermis and Cuticle

The walls of the epidermal cells are much thickened in some plants. Sometimes this thickening occurs in the outer wall, or both walls may be thickened. Variation in this respect as well as the extent of the thickening occur in different plants and are often correlated with the extremes of conditions which they serve to meet. The cuticle, a waxy exudation from the thick wall of the epidermis of many leaves, also serves as a protection against too great loss of water, or against the leaf becoming saturated with water during rains. The cabbage, carnation, etc., have a well-developed cuticle. The effect of the cuticle in shedding water can be nicely shown by spraying water on a cabbage leaf or by immersing it in water. Sunken stomata also retard the loss of water vapor.

Covers of Hair or Scales

In many leaves certain of the cells of the epidermis grow out into the form of hairs or scales of various forms, and they serve a variety of purposes. When the hairs form a felt-like covering as in the common mullein, some antennarias, etc., they lessen the loss of water vapor because the air-currents close to the surface of the leaf are retarded. Spines (see the thistles, etc.) also afford a protection against certain animals.

Reduction of Surface

Reduction of leaf surface is brought about in a variety of ways. There are two general modes: (1st) Reduction of surface along with reduction of mass; (2d) Reduction of surface inversely as the mass. Examples of the first mode are seen in the dissected leaves of many aquatic plants. In this finely dissected condition the mass of the leaf substance is much reduced as well as the leaf surface, but the leaf is less liable to be injured by movement of the water. In addition it has already been pointed out that lobed and divided aerial leaves are much less liable to injury from violent movements of the air, than if a leaf with the same general outline were entire. The needle leaves of the conifers are also examples, and they show as well structural provisions for protection in the thick, hard cell-walls of the epidermis. To offset the reduced surface there are numerous crowded leaves. Reduction of surface inversely as the mass, i.e., the mass of the leaf may not be reduced at all, or it may be more or less increased. In other words, there is less leaf surface in proportion to the mass of leaf substance. It is probable in many cases, example: the crowded, overlapping small scale leaves of the juniper, arbor-vitæ, cypress, cassiope, pyxidanthera, etc., that there has been a reduction in the size of the leaf, and at the same time an increase in thickness. This with the crowding together of the leaves and their thick cell-walls greatly lessens the radiation of moisture and heat, thus protecting the leaves both in dry and cold weather. The succulents, like "live-forever," have a small amount of surface in proportion to the mass of the leaf. In the yucca, though the leaves are often large, they are very thick and expose a comparatively small amount of surface to the dry air and intense sunlight of the desert regions. The epidermal covering is also hard and thick. In addition, such leaves, as well as those of many succulents, are so thick they provide water storage sufficient for the plants, which radiate so slowly from their surface.

Figure 436. A "Phylloclade," leaves absent, stems broadened to function as leaves, on the edges numerous flowers are borne.

Elimination of the Leaf

Perhaps the most striking illustration of the reduction of leaf surface is in those cases where the leaf is either completely eliminated as in certain euphorbias, or in certain of the cacti where the leaves are thought to be reduced to spines. Whether the cactus spine belongs to the leaf series or not, the leaf as an organ for assimilation and transpiration has been completely eliminated and the same is true in the phylloclades. The leaf function has been assumed by the stem. The stem in this case contains all the chlorophyll; is bulky, and provides water storage.

PROTECTIVE POSITIONS

In many cases the leaves are arranged either in relation to the stem, or to each other, or to the ground, in such a way as to give protection from too great radiation of heat or moisture. In the examples already cited the imbricated leaves of cassiope, pyxidanthera, juniper, etc., come also under this head. In the junipers the leaves spread out in the summer, while in the winter they are closely overlapped. An interesting example of protective position is to be seen in the case of the leaves of the white pine. During quite cold winter weather the needles are appressed to the stem, and sometimes the trees present a striking appearance in contrast with the spreading position of the needles in summer. On windy days in winter, the needles turn with the wind and become rigid in that position so that they remain in a horizontal position for some time, often until the wind dies down, or until milder weather. The following day, should there be a cold strong wind from the opposite direction, the needles again assume a leeward direction. In quiet weather appressed to the stem and in the form of a brush there is less radiation of heat than if they diverged. In strong winds by turning in the leeward direction the wind is not driven between the needle bases and scales. Some plants, especially many of those in arctic and alpine regions, have very short stems and the leaves are developed near the ground, or the rock. Lying close on the ground they do not feel the full force of the drying winds, there is less radiation from them, and the radiation of heat from the ground protects them. Many plants exhibit movement in response to certain stimuli which place them in a position for protection. Some of these examples have been discussed under the head of irritability (see Chapter 8). The night position of leaves and cotyledons presented by many plants, but especially by many of the Leguminosæ, is brought about by the removal of the light stimulus at evening. In many leaves, when the light influence is removed, the influence of growth turns the leaves downward, or the cotyledons of some plants upward. In this vertical position of the leaf-blade there is less radiation of heat during the cool night. The most striking cases of protection movements are seen in the sensitive plant. As we have seen, the leaves of mimosa close in a vertical position at midday if the light and heat are too strong. Excessive transpiration is thus prevented. At night the vertical position prevents excessive radiation of heat. The vertical or profile position of the leaves of the compass plant already referred to not only lessens transpiration, but the intense heat and light of the midday sun is avoided. This profile position is characteristic of certain plants in the dry regions of Australia, and the topmost leaves of tropical forests.

RELATION OF LEAVES TO LIGHT

It is very obvious from our study of the function of the foliage leaf that its most important relation to environment is that which brings it in touch with light and air. It is necessary that light penetrate the leaf tissue that the gases of the air and plant may readily diffuse and that water vapor may pass out of the leaf. The thin expanded leaf-blade is the most economical and efficient organ for leaf work. We have seen that leaves respond to fight stimulus in such a way as to bring their upper sides usually to face the source of fight, at right angles to it or nearly so (*heliotropism*, see Chapter 8). How fully this is brought about depends on the kind of plant, as well as on other elements of the environment, for as we have seen in our study of leaf protection there is danger to some plants in any region, and to other plants in certain regions that the intense light and heat may harm the protoplast, or the chlorophyll, or both.

The statement that leaves usually face the light at right angles is to be taken as a generalized one. The source of the strongest illumination varies on different days and again at different times of the day. On cloudy days the zenith is the source of strongest illumination. The horizontal position of a leaf, where there are no intercepting lateral or superior objects would receive its strongest light rays perpendicular to its surface. The fact is, however, that leaves on the same stem, because of taller or shorter adjacent stems, are so situated that the rays of greatest illuminating power are directed at some angle between the zenith and horizon. Many leaves, then, which may have their upper sides facing the general source of strongest illumination, do not necessarily face the sun, and they are thus protected from possible injury from intense light and heat because the direct rays of sunlight are for the most part oblique. This does not apply, of course, to those leaves which "follow the sun" during the day. Their specific constitution is such that intense illumination is beneficial.

The leaf is adjusted as well as may be in different species of varying constitution, and under different conditions, to a certain balance in its relation to the factors concerned. The problem then is to interpret from this point of view the positions and grouping of leaves. Because of the specific constitution of different plants, and because of a great variety of conditions in the environment, we see that it is a more or less complex question.

Figure 437. Mosaic form by trailing shoots of Panicum variegatum, "ribbon-grass."

Figure 438. Sunflower with young head turned toward morning sun.

Day and Night Positions Contrasted

In many plants the day and night positions of the leaves are different. At night the leaves assume a position more or less vertical, known as the *profile* position. This is generally regarded as a protective position, since during the cool of the night the radiation of heat is less than if the leaf were in a vertical position. In many of these plants, however, the leaves in assuming the night position become closely appressed which would also lessen the radiation. This peculiarity of leaves is largely possessed by the members of the family Leguminoseæ (clovers, peas, beans, etc.), and by the sensitive plants.[45] But it is also shared by some other plants as well (oxalis, for example). The leaves of these plants are usually provided with a mechanism which enables them to execute these movements with ease. There is a cushion (*pulvinus*) of tissue at the base of the petiole, and in the case of compound leaves, at the base of the pinnæ and pinnules which undergoes changes in turgor in its cells. The collapsing of the cells by loss of water into the intercellular spaces causes the leaf to droop. When the cells regain their turgor by the absorption of the water from the intercellular spaces the leaf is raised to the horizontal, or day position. The light stimulus induces turgor of the pulvinus, the disappearance of the stimulus is accompanied by a loss of turgor. It is a remarkable fact that in some sensitive plants, intense light stimuli are alarm signals which result in the same movement as if the light stimulus were entirely removed. As we know also contact or pressure stimulus, or jarring produces the same result in "sensitive" plants like mimosa, some species of rubus, etc. In many plants there is no well-developed pulvinus, and yet the leaves show similar movements in assuming the day and night positions. Examples are seen in the sunflower, and in the cotyledons of many plants. A little observation will enable any one interested to discover some of these plants.[46] In these cases the night position is due to epinastic growth, and while this influence is not removed during the day the light stimulus overcomes it and the leaf is raised to the day position.

Leaves Which Rotate with the Sun

During the growth period the leaves of the sunflower as well as the growing end of the stem respond readily to the direct sunlight. The response is so complete that during sunny days the leaves toward the growing end of the stem are drawn close together in the form of a rosette and the entire rosette as well as the end of the stem are turned so that they face the sun directly. In the morning under the stimulus of the rising sun the rosette is formed and

---

[45] The most remarkable case is that of the "telegraph" plant (Desmodium gyrans). Aside from the day and night positions which the leaves assume, there is a pair of small lateral leaflets to each leaf which constantly execute a jerky motion, and swing around in a circle like the second hand of a watch.

[46] Seedlings are usually very sensitive to light and are good objects to study.

faces the east. All through the day, if the sun continues to shine, the leaves follow it, and at sundown the rosette faces squarely the western horizon. For a week or more the young sunflower head will also face the sun directly and follow it all day as surely as the rosette of leaves. At length, a little while before the flowers in the head blossom, the head ceases to turn, but the rosette of leaves and the stem also, to some extent, continue to turn with the sun. When the leaves become mature they also cease to turn. This is well shown in all three photographs (Figures 438-439). The lower leaves on the stem being older have assumed the fixed horizontal position usually characteristic of the plant with cylindrical habit.

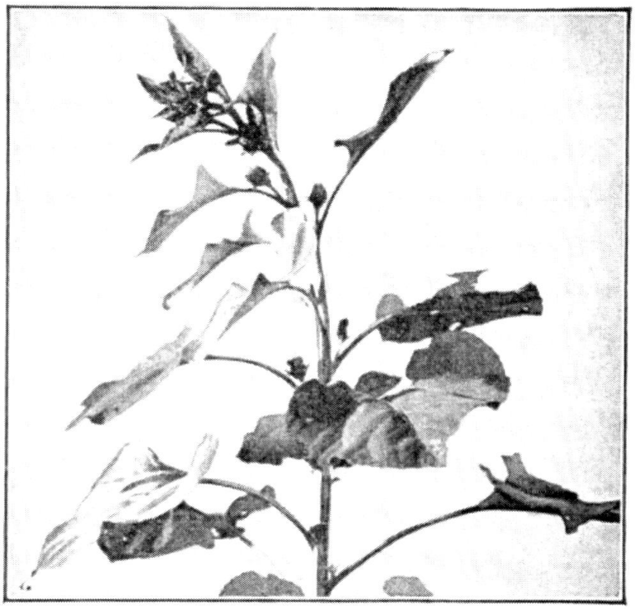

Figure 439. Same sunflower plant photographed just at sundown.

It is not true, as is commonly supposed, that the fully opened sunflower head turns with the sun. But I have observed young heads four or five inches in diameter rotate with the sun all day. This is because the growing end of the stem as well as the young head responds to the light stimulus. So there is some truth as well as a great deal of fiction in the popular belief that the sunflower head follows the sun. The young head will follow the sun all day even if all the leaves are cut off, and the growing stem will also if all the leaves as well as the flower head are cut away. Young seedlings will also turn even if the cotyledons and plumule are cut off.

This phenomenon of the rotation of leaves with the sun is much more general than one would infer, as may be seen from a little careful observation of rapidly growing plants on bright sunny days. In Alabama I have observed beautiful rosettes of *Cassia marilandica* rotate with the sun all day. The peculiarity is very striking in the cotton plant, especially when the rows extend north and south. In the forenoon or afternoon it is most striking as the entire row shows the leaves tilted up facing the sun. There are many of our weeds and common flowers of field and garden which show this rotation of the leaves. Some of these form rotating rosettes; while in others the leaves rotate independently as in the sweet clover.

Figure 440. Same plant a little older when the head does not turn, but the stem and leaves do.

Fixed Position of Old Leaves

In many of the cases cited in the preceding paragraph, the rotation of the leaf only occurs on sunny days. During cloudy days the leaves of the sunflower, for example, are in a nearly horizontal position, or the lower ones may be somewhat oblique, since the stronger illumination on such a plant would be the oblique rays rather than the zenith rays. As the leaves reach maturity also the epinastic growth is equalized by hyponastic growth so that the growth movements bring the leaf to stand in a nearly horizontal position, or that position in which it receives the best illumination. In age, then, many leaves have a fixed position and this corresponds with the position assumed on cloudy days.

Position on Horizontal Stems

On horizontal stems the leaves have a horizontal position, and if such a stem is stood in an erect position the appearance is very odd. If the leaf arises directly from the horizontal stem, its petiole will be twisted part way around in order to bring the face of the leaf uppermost.

It is interesting to observe the different relation of stem, petiole and blade and the amount of twisting as the horizontal stem or vine trails over irregularities in the surface, or climbs over and through other vegetation.

Position of Leaflets on Divided Leaves

An interesting comparison can be made with entire, lobed, divided and dissected leaves. The entire leaf usually lies in one plane, since usually the problem of adjustment is the same for the entire surface. So the lobes of a leaf usually lie all in the same plane as they would if the leaf were entire. We find the same is true usually of the compound leaf. It forms an incomplete mosaic. Some of the pieces having been removed allow much of the light to pass through to leaves beneath. Leaves, especially those of some size rarely lie in a flat plane. Some are more or less depressed. Some curve downward. Compound leaves often curve more or less and the leaflets often droop more or less in a graceful fashion. It is interesting, however, that these far separated leaflets all lie in the same gencral plane. This is because the area of the leaf, if not too large, makes the problem of position with reference to light much the same as if the leaf were entire. The leaflets or divisions, though separated, are laminate, and they can work more efficiently facing the light. But suppose we extend our observation to the finely dissected capillary leaves of some of the parsley family (Umbelliferæ), or to the upper leaves of the fennel-leaved thoroughwort (Eupatorium fœniculaceum) among the aerial plants, and to Myriophyllum among the aquatic plants. The divisions are thread-like or cylindrical. One side of the leaflet is just as efficient when presented to the light as another. As a result the leaflets are not arranged in the same plane, but stand out in many directions.

Occasionally one finds a divided or compound leaf in such a position that one portion, because of being shaded above, receives the stronger light stimulus from the side, while the other portion is lighted from above. If this relation continues throughout the growth-period of the leaf the leaflets of one portion may lie in a different plane from those of the other portion. In such cases, some of the leaflets are permanently twisted to bring them into their proper light relation.

Leaf Patterns

Mosaics, or Close Patterns

Where the leaves of a plant, or a portion of a plant, are approximate and arranged in the form of a pattern, the leaves fitting together to form a more or less even and continuous surface, such patterns are sometimes termed "mosaics," since the relation of leaves to one another is roughly like the relation of the pieces of a mosaic. A good illustration of a mosaic is presented by a greenhouse plant Fittonia (Figure 441). The stems are prostrate and the erect branches quite short, but it may have quite a wide system by the spreading of the runners; the branches of such a length that the leaves borne near the tips all fit together forming a broad surface of leaves so closely fitted together often that the stems cannot be seen. The advantage of a mosaic over a separate disposition of leaves at somewhat different levels is that the leaves do not shade one another.

Figure 441. Fittonia showing leaves arranged to form compact mosaic. The netted venation of the leaf is very distinctly shown in this plant. (Photo by the Author.)

Were all the light rays coming down at right angles to the leaves, there would not be any shading of the lower ones, but the oblique rays of light would be cut off from many of the leaves. In the case of a mosaic all the rays of light play upon all the leaves. Some of the mosaics which can be observed are as follows:

Figure 442. Rosette pattern of leaves.

Rosette Pattern

The rosette pattern is presented by many plants with "radial" leaves, or leaves which arise in a cluster near the surface of the ground, and are thus more or less crowded in their arrangement on the stem. The pretty gloxinia often presents fine examples of a loose rosette. In the rosette pattern the petioles of the lower leaves are longer than the upper ones, and the blade is thus carried out beyond the inner leaves. The leaves being so crowded in their attachment to the stem lie very nearly in the same plane.

Vines and Climbers

Some of the most extensive mosaic patterns are shown in creeping and climbing vines. A very common example is that of the ivies trained on the walls of buildings, covering in some instances many square yards of surface. Where the vines trail over the ground or clamber over other vegetation, it is interesting to observe the various patterns, and the distortion of petioles brought about by turning of the leaves. Of examples found in greenhouses, the Pellonia is excellent, and the trailing ribbon-grass often forms loose mosaics.

Branch Patterns

These patterns are very common. They are often formed in the woods on the ends of branches by the leaves adjusting themselves so as to largely avoid shading each other. Figure 443 illustrates one of them from a maple branch. It is interesting to note the way in which the leaves fit themselves in the pattern, how in some the petioles have elongated, while others have remained short. Of course, it should be understood that the pattern is made during the growth of the leaves.

Figure 443. Spray of leaves of striped maple, showing different lengths of leafstalks.

Figure 444. Cedar of Lebanon, strong light only from one side of tree (Syria).

The Tree Pattern

Mosaics are often formed by the exterior foliage on a tree, though they are rarely so regular as some of those mentioned above. Still it is common to see in some trees with drooping limbs like the elm, beautiful and large mosaics. The weeping elm sometimes forms a very close and quite even pattern over the entire outer surface. In most trees the leaf arrangement is not such as to form large patterns, but is more or less open. While the conifers do not form mosaics there are many interesting examples of grouping of foliage on branch systems into broadly expanded areas, as seen in the branches of white pine trees, especially in the edge of a wood, or as seen in the arbor-vitæ.

Other Patterns

Imbricate Pattern of Short Stems

This pattern is quite common, and differs from the rosette in that the leaves are distributed further apart on the stem so that the central ones are considerably higher up than in the mosaic. The lower petioles are longer, as in the rosette, so that the outer lower leaves extend further out. Some begonias show fine imbricate patterns.

Figure 445. Imbricate pattern of leaves; Begonia.

Spiral Patterns

They are very common on stems of the cylindrical type, which are unbranched, or but little branched. The sunflower, mullein, chrysanthemum, as it is grown in greenhouses, the Easter lily, etc., are examples. The spiral arrangement of the leaves provides that each successive leaf on the stem, as one ascends the stem, is a little to one side so that it does not cast shade on the leaf just below. In some stems, according to the leaf arrangement (or phyllotaxy), one would pass several times around in ascending the stem before a leaf would be found directly above another, which would be such a distance below that it would not be shaded to an appreciable extent. Interesting observations can be made on different plants to work out the relation of distance of leaves on the stem to length of the upper and lower leaves; the number of vertical rows on the stem compared to the width of the leaves; and the relation of these facts to the problem of light supply. Related to the spiral pattern is that of erect stems with opposite leaves. Here each pair is set at right angles to the direction of the pair above or below.

Figure 446. Palm showing radiate arrangement of leaves and the petiole of the leaf functions as stem in lifting leaf to the light.

Radiate Pattern

This pattern is present in many grasses and related plants with narrow leaves and short stems. The leaves are often very crowded at the base, but by radiating in all directions from the horizontal to the vertical, abundant exposure to light is gained with little shading. The dragon tree screw-pine, and plants grown in greenhouses also illustrate this type. It is also shown in cycads, palms, and many ferns, although these have divided leaves.

Figure 447. Screw-pine (Pandanus) showing prop roots and radiate pattern of leaves.

Compass Plants

These plants with vertical leaf arrangement, and exposure of both surfaces to the lateral rays of light have been mentioned in other sections (Lactuca scariola).

Open Patterns

Open patterns are presented by divided or "branched" leaves. Where the leaves are very finely dissected, they may be clustered in great profusion and yet admit sufficient light for some depth below. Where the leaflets are broader, the leaves are likely to be fewer in number and so arranged as to admit light to a great depth so that successive leaves below on the same or adjacent stems may not be too much shaded. On such plants, often the leaves lying next the ground are entire or less divided.

Chapter 41

The Root

Function of Roots

The most obvious function of the roots of ordinary plants are two: 1st, To furnish anchorage and partial support, and 2d, absorption of liquid nutriment from the soil. The environmental relation of such roots, then, in broad terms, is with the soil. It is very clear that in some plants the root serves both functions, while in other plants the root may fulfil only one of these requirements.

The problems which the plant has to solve in working out these relations are:

- Permeation of the soil or substratum.
- Grappling the substratum.
- A congenial moisture or water relation.
- Distribution of roots for the purpose of reaching food-laden soil.
- Exposure of surface for absorption.
- The renewal of the delicate structures for absorption.
- Aid in preparation of food from raw material.
- The maintenance of the required balance between the environment
- as a whole and the increasing or changing requirements of the plant.

Permeation of the Soil or Substratum

The fundamental divergence of character in the environmental relations of root and stem are manifest as soon as they emerge from the germinating seed. Under the influence of the same stimulus (*gravity*) the root shows its geotropic character by growing downward, while the geotropic character of the stem is shown in its upward growth.

The medium which the root has to penetrate offers considerable resistance, and the form of the root as well as its manner of growth is adapted to overcome this difficulty. The slender, conical, penetrating root-tip wedges its way between the minute particles of soil or into the minute crevices of the rock, while the nutation of the root enables it to search for the points of least resistance. The root-tips having penetrated the soil, the older portions of the root continue this wedge action by growth in diameter, though, of course, elongation of the old parts of the root does not take place. It is the widening growth of the tapering root that produces the wedge-like action. The crevices of the rock are sometimes broadened, but the resistance here is so great, the root is often greatly flattened out.

Grappling the Substratum

The mere penetration of a single root into the soil gives it some hold on the soil and it offers some resistance to a "pull" since it has wedged its way in and the contact of soil particles offers resistance. The root hairs formed on the first entering root growing laterally in great numbers and applying themselves very closely to the soil particles, increase greatly the hold of the plant on the soil, as one can readily see by pulling up a young seedling. Lateral roots are soon formed, and as these continue to extend and ramify in all directions, the hold is increased until in the case of some of the larger plants the resistance their hold would offer would equal many tons. Even in some of the smaller shrubs and herbs the resistance is considerable, as one can easily test by pulling with the hand. To obtain some idea of the amount of resistance the roots of these smaller plants offer, they can be tested by pulling with the ordinary spring scales.

A Congenial Moisture, or Water Relation

In general, the roots seek those portions of the soil provided with a modicum of moisture. Usually a suitable moisture condition is present in those portions of the soil containing the plant food. But if portions of the soil are too dry and very nearby other portions containing moisture, the roots grow mainly into the moist substratum (*hydrotropism*). If the soil is too wet, the roots grow away from it to soil with less water, or in some cases will grow to and upon the surface of the soil.

The roots need *aeration*, and where the supply of water is too great, the air is shut out, and we know that corn, wheat, and many other plants become "sickly" in low and undrained soil in wet seasons. This can only be said in the case of our ordinary dry land plants, i.e., those that occupy an intermediate position between *water-loving* plants and *dry-conditioned* plants. This phase of the subject must be reserved for special treatment. (See Chapter 46.)

Distribution of Roots for the Purpose of Reaching Food-Laden Soil

This is one of the essential relations of the root in the case of the land plant, and probably accounts for the very extensive ramification of the roots. To some extent it also explains the different root systems in some plants. The pines, spruces, etc., usually grow in regions where the soil is very shallow. The root system does not extend deeply into the soil. It spreads laterally and extends widely through the shallow surface soil and presents a very different aspect from the stem system in the air. The root system of the broad-leaved trees usually extends more deeply into the soil, while of course, extending laterally to great distances. The hickory, walnut, etc., especially have strong tap-roots which extend deeply into the soil, and the root system of such a tree is more comparable in aspect, if it were entirely uncovered, to the stem system in the air. The tap-root is more pronounced in some trees than in others. It may be that in the hickory and walnut the deep tap-root is important in supplying the tree with water in dry seasons, especially when growing on dry, gravelly soil which does not retain moisture on the surface nor hold it within two or three feet of the surface. Experiment has demonstrated, by pot culture of plants, that where soil rich in plant food lies adjacent to poor soil, no matter in what part of the pot the rich soil is, the greatest growth and branching of roots is in the rich soil.

Exposure of Root Surface for Absorption

The principal part of root absorption takes place in the young root and the root hairs growing near the root-tip. The root-tips and root hairs in their relation to the root systems on which they are borne are not to be compared morphologically with the leaves and stem system. But the root-tips and hairs are absorbing organs of the roots while the main root system supports them, brings them into relation with the soil and moisture, and conducts food and other substances to and from them. One of the important relations of the leaf is that of light, and since the source of light is restricted, i.e., it is not equally strong from all sides, an expanded and thin leaf-blade is more effective than an equal expenditure of plant material in the form of thread-like outgrowths. It is different, however, with the plant food dissolved in the soil water. It is equally accessible on all sides. A greater surface for absorption is exposed with the same expenditure of material by multiplication of the organs and a reduction in their size. Numerous delicate root hairs present a greater absorbing surface than if the same amount of material were massed into leaf-like expansions. There is another important advantage also. Its slender roots and thread-like root hairs allow greater freedom of circulation of water, food solutions, and air than if the absorbing organs of the roots were broadly expanded.

The Renewal of the Delicate Structures for Absorption

The delicate root hairs are easily injured. The thin cell-walls through which food solutions flow become more or less choked by the gradual deposit of substances in solution in the water, and continued growth of the root in diameter forms a firmer epidermis and cortex through which the solutions taken up by the root hairs would pass with difficulty. For this reason new root hairs are constantly being formed on the growing root-tip throughout the growing season, and in the case of perennial plants, through each season of their growth.

Aid in Preparation of Food from Raw Materials

For most plants the food obtained from the soil is already in solution in the soil water. But there are certain substances (examples, some of the chemical compounds of potash, phosphoric acid, etc.) which are insoluble in water. Certain acids excreted by the roots aid in making these substances soluble (see Chapter 3). In a number of plants the roots have become associated with fungus or bacterial organisms which assist in the manufacture of nitrogenous food substances, or even in the absorption of ordinary food solution from the soil, or in making use of the decaying humus of the forest (see Chapter 9).

The Maintenance of the Required Balance between the Environment and the Increasing or Changing Requirements of the Plant

In this matter the entire plant participates. Mention is made here only of the general relation which the root sustains to its own environment and the increased burden placed upon it by the shoot. The increase in the root system keeps pace with the increasing size of the stem system. The roots become stronger, their ramifications wider, and the number of absorbing rootlets more numerous. The observation is sometimes offered that the correlation between the root system of a plant, and the form of the stem system and position of the leaves, is of such a nature that plants with a tap-root system have their leaves so arranged as to shed the water to the center of the system, while plants with a fibrous-root system have their leaves so arranged as to shed the water outward. In support of this attention is called to the radiate type of the leaf system of the dandelion, beet, etc. In the second place the imbricate type as manifested in broad-leaved trees, and in the overlapping branch systems of many pines, etc. One should note, however, that in the former class the leaves are often arranged to shed as much water outward as inward. As to the latter class, there is need of experiment to determine whether these empirical observations are correct, for the following reasons: 1st, Root and leaf distribution are governed by other and more important laws, the root being influenced by the location of food in the soil which usually forms a very thin stratum while the shoot and leaf is mainly influenced by light, and root distribution is much wider in a lateral direction than that of the branches. 2d, In light rains the leaf surface holds back practically all the rain which is then evaporated into the air and lost to the root systems. 3d, In heavy and long-continued rains the water breaks through the leaf system to such an extent that roots under the tree would be as well supplied as those outside, and the ground outside being saturated anyway, the roots do not need the small additional water which may have been shed outward. 4th, It is the habit of plants where left undisturbed (except in rare cases), to grow in more or less dense formations or societies. Here there is no opportunity for any appreciable centrifugal distribution of rainfall and yet the root distribution is practically the same, except that the root systems of adjacent plants are interlaced.

Kinds of Roots

The Root System

From the foregoing, it will be understood that the roots of a plant taken together form the *root system* of that plant. In soil-roots in general we usually recognize two kinds of root systems.

The Fibrous-Root System

Roots which are composed of numerous slender branching roots resembling "fibers," are termed *fibrous*, or the plant is said to have a *fibrous-root system*. The bean, corn, most grasses, and many other plants have fibrous-root systems.

The Tap-Root System

Plants with a recognizable central shaft-like root, more or less thickened and considerably stouter than the lateral roots, are said to have *tap* roots, or they have a *tap-root system*. The dandelion, beet, carrot (see crown tuber) are examples. The hickory, walnut, and some other trees have very prominent tap-roots when young. The tap-root is maintained in old age, but the lateral roots often become finally as large as the tap-root. Besides tap-roots and fibrous-roots, which include the larger number, several other kinds of roots are to be enumerated.

Aerial Roots

Aerial roots are most abundantly developed in certain tropical plants, especially in the orchids and aroids. Many examples of these plants are grown in conservatories. The amount of moisture is so great in these tropical regions that the roots are abundantly supplied without the soil relation. Certain of the roots hang free in the air and are provided with a special sheath of spongy tissue called the *velamen*, through which moisture is absorbed from the air. Other roots attach themselves to the trunk or branches of the tree on which the orchid is growing, and furnish the support to the *epiphyte*, as such plants are often called. Among the tangle of these clinging roots falling leaves are caught. Here they decay and nourishing roots grow from the clinging roots into this mass of decaying leaves and supply some of the plant food. Aerial roots sometimes possess chlorophyll.

There are a number of plants, however, in temperate regions which have aerial roots. These are chiefly used to give the stem support as it climbs on trees or on walls. They are sometimes called clinging roots. A common example is the climbing poison-ivy (Rhus radicans), the trumpet creeper, etc. Such aerial roots are called *adventitious* roots.

Bracing Roots, or Prop Roots

These are developed in a great variety of plants and serve to brace or prop the plant where the fibrous-root system is insufficient to support the heavy shoot system, or the shoot system branches so widely props are needed to hold up the branches. In the common Indian corn several whorls of bracing roots arise from the nodes near the ground and extend outward and downward to the ground, though the upper whorls do not always succeed in reaching the ground. The screw-pine so common in greenhouses affords an excellent example of prop roots. The roots are quite large, and long before the root reaches the soil the large root cap is evident. The banyan tree of India is a classic example of prop roots for supporting the wide-reaching branches. The mangrove in our own subtropical forests of Florida is a nearer example.

Buttresses are formed at the junction of the root and trunk, and therefore are part root and part stem. Splendid examples of buttresses are formed on the silk-cotton tree. They are sometimes formed on the elm and other trees in low swampy ground.

Fleshy Roots, or Root Tubers

These are enlargements of the root in the form of tubers, as in the sweet potato, the dahlia, etc. They are storage reservoirs for food. Portions of the roots become thick and fleshy and contain large quantities of sugar, as in the sweet potato, or of *inulin* (a carbohydrate) in the root tubers of the dahlia and other composites.

Water-Roots and Roots of Water Plants

These are roots which are developed in the water, or in the soil. Water-roots are sometimes formed on land plants where the root comes in contact with a body of water, or a stream. Water-roots usually possess no root hairs, or but a few, as can be seen by comparing water-roots with soil-roots, or by comparing roots of plants grown in water cultures. The greater body of water in contact with the root and the more delicate epidermis of the root render less necessary the root hairs. The duck-meats (Lemna) are good examples of plants having only water-roots. Other aquatic plants like the potamogetons, etc., have true roots which grow into the soil and serve to anchor the plant, but they are not developed as special organs of absorption, since the stem and leaves largely perform this function.

Figure 448. Bracing roots of Indian corn.

Figure 449. Buttresses of silk-cotton tree, Nassau.

Holdfasts

These are organs for anchorage which are not true roots. These are especially well developed in some of the algæ (Fucus, Laminaria, etc.). They are usually called *holdfasts*. The holdfasts of the larger algæ are mainly for anchoring the plant. They do not function as absorbing organs, and the structure is different from that of true roots.

Haustoria or suckers is a name applied to another kind of holdfast employed by parasitic plants. In the dodder the haustorium penetrates the tissue of the *host* (the plant on which the parasite grows), and besides furnishing a means of attachment, it serves as an absorbing organ by means of which the parasite absorbs food from its host. The parasitic fungi like the powdery mildews which grow on the surface of their hosts have simple haustoria which serve both as organs of attachment and absorption, while in the rusts which grow in the interior of their hosts the haustoria are merely absorbing organs.

Rootlets, or Rhizoid

Many of the algæ, liverworts and mosses have slender, hair-like organs of attachment and absorption. These plants do not have true roots. Because of the slender form and small size of these organs, they are called *rhizoids*, or *rootlets*. In form many of them resemble the root hairs of higher plants.

Chapter 42

THE FLORAL SHOOT

THE PARTS OF THE FLOWER

The portion of the stem on which the flowers are borne is the *flower* shoot or axis, or taken together with the flowers, it is known as the *Flower Cluster*.

The Flower

The flower is best understood by an examination, first of one of the types known as a "complete" flower, as in the buttercup, the spring-beauty, the blood-root, the apple, the rose, etc.

There are two sets of organs or members in the complete flower—(1) the floral envelope; (2) the essential or necessary members or organs.

The floral envelope when complete consists of—1st, an outer envelope, the *calyx*, made up of several leaf-like structures (*sepals*), very often possessing chlorophyll, which envelop all the other parts of the flower when in bud; 2d, an inner envelope, the *corolla*, also made up of several leaf-like parts (*petals*), usually bright colored and larger than the sepals. The outer and inner floral envelopes are usually in whorls (though in close spirals in many of the buttercup family, etc.), and for reasons discussed elsewhere (Chapter 34) represent leaves. The essential or necessary members of the flower are also usually in whorls and likewise represent leaves, but only in rare cases is there any suggestion, either in their form or color, of a leaf relationship. These members are in two sets: (1) The outer, or *andrœcium*, consisting of a few or many parts (*stamens*); (2) the inner set, the *gynœcium*, consisting of a few or many parts (*carpels*).

Purpose of the Flower

While the ultimate purpose of all plants is the production of seed or its equivalent through which the plant gains distribution and perpetuation, the flower is the specialized part of the seed plant which utilizes the food and energies contributed by other members of the plant organization for the production of seed. In addition to this there are definite functions performed by the members of the flower, which come under the general head of plant work, or flower work.

The Calyx, or the Sepals

These are chiefly protective, affording protection to the young stamens and carpels in the flower bud. Where the corolla is absent, sepals are usually present and then assume the function of the petals. In a few instances the calyx may possibly ultimately join in the formation of the fruit (examples: the butternut, walnut, hickory).

The Corolla, or Petals

The petals are partly protective in the bud, but their chief function where well developed seems to be that of attracting insects, which through their visits to the flower aid in "*pollination*," especially "*cross pollination*."

The Stamens

The stamens (= microsporophylls) are flower organs for the production of *pollen*, or *pollen-spores* (= microspores). The *stalk* (not always present) is the *filament*, the *anther* is borne on the filament when the latter is present. The anther consists of the *anther sacs* or *pollen sacs* (microsporangium) containing the pollen-spores, and the *connective*, the sterile tissue lying between and supporting the anther sac. The stamens are usually separate, but sometimes they are united by their filaments, or by their anthers. When the pollen is ripe they open by slits or pores and the pollen is scattered; or in rarer cases the pollen mass (*pollinium*) is removed through the agency of insects (see Insect pollination, Chapter 43).

The Pistil

The pistil consists of the "*ovary*," the *style* (not always present), and the *stigma*. These are well shown in a *simple pistil*, common examples of which are found in the buttercup, marsh marigold, the pea, bean, etc. The simple pistil is equivalent to a *carpel* (= macrosporophyll), while the *compound pistil* consists of two or several carpels joined, as in the toothwort, trillium, lily, etc. The *ovary* is the enlarged part which below is attached to the receptacle of the flower, and contains within the *ovules*. The *style*, when present, is a slender elongation of the upper end of the ovary. The *stigma* is supported on the end of the style when the latter is present. It is often on a capitate enlargement of the style or extends down one side, or when the style is absent it is usually seated directly on the upper end of the ovary. The stigmatic surface is glutinous or "sticky," and serves to hold the pollen-spores when they come in contact with it.

The *ovules* are within the ovary and are arranged in different ways in different plants. The pollen grain (or better pollen-spore = microspore), after it has been transferred to the stigma, "germinates," and the pollen tube grows down through the tissue of the stigma and style, or courses down the stylar canal until it reaches the ovule. Here it usually enters the ovule (macrosporangium) at the *micropyle* (in some of the ament-bearing plants it enters at the *chalaza*), and the sperm cells are emptied into the embryo sac in the interior of the ovule.

Fertilization

One of the sperms unites with the egg in the embryo sac. This is *fertilization*, and from the fertilized egg the young embryo is formed still within the ovule. *Double fertilization*,— the other sperm cell sometimes unites with one or both of the "polar" nuclei which have united to form the "definitive" or "endosperm" nucleus. As a result of fertilization, the embryo plant is formed within the ovule, the coats of which enlarge by growth forming the seed coats, and altogether forming the seed. (See Chapters 34, 35, 36.)

KINDS OF FLOWERS

Absence of Certain Flower Parts

The *complete* flower contains all the four series of parts. When any one of the series of parts is lacking, the flower is said to be *incomplete*. Where only one series of the floral envelopes is present the flowers are said to be *apetalous* (the petals are absent), examples:

elm, buckwheat, etc. Flowers which lack both floral envelopes are *naked*. When pistils are absent but stamens are present the flowers are *staminate*, whether floral envelopes are present or not; and so when stamens are absent and pistils present the flower is *pistillate*. If both stamens and pistils are absent the flower is said to be *sterile* or *neutral* (snowball, marginal or showy flowers in hydrangea). Flowers with both stamens and pistils, whether or not they have floral envelopes, are *perfect* (or hermaphrodite), so if only one of these sets of *essential organs* of the flower is present the flower is *imperfect*, or *diclinous*. Sometimes the imperfect, or diclinous, flowers are on the same plant, and the plant is said to be *monœcious* (of one household). When staminate flowers are on certain individual plants, and the pistillate flowers of the same species are on other individuals, the plant is *diœcious* (or of two households). When some of the flowers of a plant are diclinous and others are perfect, they are said to be *polygamous*.

Many of these variations relating to the presence or absence of flower parts in one way or another contribute to the well-being of the plant. Some indicate a division of labor; thus in the neutral flowers of certain species of hydrangea or viburnum, the showy petals serve to attract insects which aid in the pollination of the fertile flowers. It must not be understood, however, that all variations in plants which results in new or different forms of flowers is for the good of the species. For example, under cultivation the flowers of viburnum and hydrangea sometimes are all neutral and showy. While such variations sometimes contribute to the happiness of man, the plant has lost the power of developing seed. In diclinous flowers cross pollination is necessitated.

Form of the Flower

The flower as a whole has *form*. This is so characteristic that in general all flowers of the different individuals of a species are of the same shape, though they may vary in size. In general, flowers of closely related plants of different species are of the same type as to form, so that often in the shape of the flower alone we can see the relationship of kind, though the form of the flower is not the most important nor always the sure index of kinship. Since many flowers resemble certain familiar objects, names are often used which relate to these objects.

Flowers are said to be *regular*, or *irregular*. In a regular flower all of the parts of a set or series are of the same shape and size, while in irregular flowers the parts are of a different shape or size in some of the sets. The flowers of the pea family (*Papilionaceæ*), of the mint family (*Labiatæ*), of the morning glory, larkspur, monkshood, etc., are irregular (Figure 450). The corolla usually gives the characteristic form to the flower, and the name is usually applied to the form of the corolla.

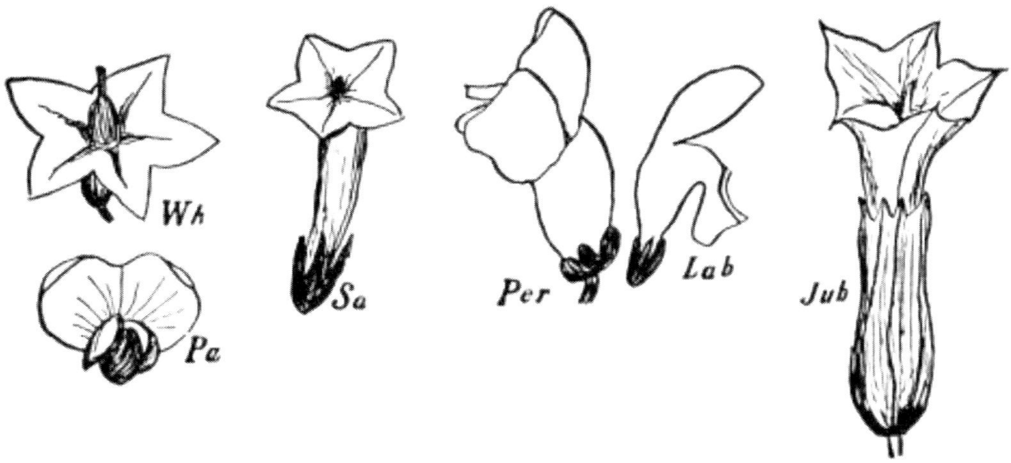

Figure 450. Several forms of flowers. Regular flowers. *wh*, wheel-shaped corolla; *sa*, salver-shaped; *tub*, tubular-shaped. Irregular flowers. *pa*, butterfly or papilionaceous; *per*, personate or masked flower; *lab*, gaping or ringent corolla. The two latter are called bilabiate flowers.

Some of the different forms are wheel-shaped or *rotate* corolla when the petals spread out at once like the spokes of a wheel, as in the potato, tomato, or bittersweet; *salver-shaped* when the petals spread out at right angles from the end of a corolla tube, as in the phlox; *bell-shaped*, or *campanulate*, as in the harebell or campanula; *funnel-shaped*, as in the morning glory; *tubular*, when the ends of the petals spread but little or none from the end of the corolla tube, as in the turnip flower or in the disk florets of the composites. The *butterfly*, or *papilionaceous* corolla is peculiar as in the pea or bean. The upper petal is the "banner," the two lateral ones the "wings," and the two lower the "keel."

The *labiate* corolla is characteristic of the mint family where the gamosepalous corolla is unequally divided, so that the two upper lobes are sharply separated from the three lower forming two "lips." The labiate corolla of the toad-flax, or snapdragon is *personate*, or *masked*, because the lower lip arches upward like a palate and closes the entrance to the corolla tube; that of the dead nettle (*Lamium*) is *ringent* or *gaping*, because the lips are spread wide apart. In some plants the labiate corolla is not very marked and differs but slightly from a regular form.

The *ligulate* or *strap-shaped* corolla is characteristic of the flowers of the dandelion or chicory, or of the ray flowers of other composites (Figure 451). The lower part of the gamosepalous corolla is tubular, and the upper part is strap-shaped, as if that part of the tube were split on one side and spread out flat.

These forms of the flower should be studied in appropriate examples.

Union of Flower Parts

In the buttercup flower all the parts of each series are separate from one another and from other series of parts. Each one is attached to the *receptacle* of the flower, which is a very much shortened portion of the flower axis. The calyx being composed of separate and distinct parts is said to be *polysepalous*, and the corolla is likewise *polypetalous*. The stamens are *distinct*, and the pistils are *simple*. In many flowers, however, there is a greater or lesser *union* of parts.

Union of Parts of the Same Series or Cycle

The parts *coalesce*, either slightly or to a great extent. Usually they are not so completely coalesced but what the number of parts of the series can be determined. Where the sepals are united the calyx is *gamosepalous*, when the petals are united the corolla is *gamopetalous*.

Union of the sepals or of the corolla is quite common, but union of the stamens is rare except in a few families where it is quite characteristic. When the stamens are united by their anthers, they are *syngenœsious*. This is the case in most flowers of the composite family. When all the stamens are united into one group by their filaments, they are *monadelphous* (one brotherhood), as in hollyhock, hibiscus, cotton, marsh-mallow, etc. When they are united by their filaments in two groups, they are *diadelphous* (two brotherhoods), as in the pea and most members of the pea family. In most species of St. John's wort (Hypericum), the stamens are united in threes (*triadelphous*).

The Carpels Are Often United

The pistil is then said to be *compound*. Where the pistils are consolidated, usually the adjacent walls coalesce and thus separate the cavity of each ovary. Each cavity in the compound pistil is a *locule*. In some cases the adjacent walls disappear so that there is one common cavity for the compound pistil (examples: purslane, chickweeds, pinks, etc.). In a few cases there is a false partition (example, in the toothwort and other crucifers). The compound pistil is very often lobed slightly, so that the different pistils can be discerned. More often the styles or stigmas are distinct, and thus indicate the number of pistils united.

Union of the Parts of Different Series

While in the buttercup and many other flowers, all the different parts are inserted on the torus or receptacle, in other flowers one series of parts may be joined to another. This is *adnation* of parts, or the two or more series are *adnate*. In the morning glory the stamens are inserted on the inner face of the corolla tube; the same is true in the mint family, and there are many other examples. The insertion of parts, whether free or adnate, is usually spoken of in reference to their relation to the pistil. Thus, in the buttercup the floral envelopes and stamens are all free and *hypogynous*, they are *below* the pistil. The pistil in this case is *superior*. In the cherry, pear, etc., the petals and stamens are borne on the edge of the more or less elevated tube of the calyx, and are said to be *perigynous*, i.e., around the pistil. In the cranberry, huckleberry, etc., the calyx is for the most part united with the wall of the ovary with the short calyx limbs projecting from the upper surface. The petals and stamens are inserted on the edge of the calyx above the ovary; they are, therefore, *epigynous*, and the ovary being under the calyx, as it were, is *inferior*.

Arrangement of Flowers, or Mode of Inflorescence

Flowers Are Solitary or Clustered

Solitary flowers are more simple in their arrangement, i.e., it is easier for us to determine and name their relation to each other and to other parts of the plant. They are either *axillary*, i.e., on short lateral shoots in the axils of ordinary foliage leaves, or they are *terminal*, i.e., they are borne on the end of the main axis of an ordinary foliage shoot. In either case they are so far separated, and the foliage leaves are so prominent, they do not form recognizable groups or clusters. The manner of arrangement of flowers on the shoot is called *inflorescence*, while the group of flowers so arranged is the *flower cluster*.

Two different modes of inflorescence are usually recognized in the arrangement of flowers on the stem. (1) The *corymbose*, or *indeterminate inflorescence* (also indefinite inflorescence), in which the flowers arise from axillary buds, and the terminal bud may continue to grow. (2) The *cymose* or *determinate inflorescence* (also *definite inflorescence*) in which the flowers arise from terminal buds. This arrests the growth of the shoot in length.

There are several advantages to the plant in the different modes of inflorescence, chief among which is the massing of the flowers, thus increasing the chances for effective pollination.

Flower Clusters with Indeterminate Inflorescence

The simplest mode of indeterminate inflorescence is where the flowers arise in the axils of normal foliage leaves, while the terminal bud, as in the florist's smilax, the bellwort, moneywort, apricot, etc., continues to grow. The flowers are *solitary* and *axillary*. In other cases which are far more numerous, the flowers are associated into more or less definite clusters in which are a number of recognizable types. The word type used in this sense, it should be understood, does not refer to an original structure which is the source of others. It merely refers to a mode of inflorescence which we attempt to recognize, and about which we group those forms which have a resemblance to one another. There are many forms of flower clusters which do not conform to any one of our recognized types, and are very puzzling. The evolution of the flower clusters has been *natural*, and we cannot make them all conform to an *artificial* classification. These *types* are named merely as a matter of convenience in the expression of our ideas. The types usually recognized are as follows:

- *The raceme.*—The flower-shoot is more or less elongated, and the leaves are reduced to a minute size termed *bracts*, while the flowers on lateral axes are solitary in the axils of the bracts. The reduction in the size of the leaves and the somewhat limited growth of the shoot in length, makes the flowers more prominent, and brings them into closer relation than if they were formed in the axils of the leaves on the ordinary foliage shoot. The choke cherry, currant, pokeweed, sourwood, etc., are examples of a raceme (Figure 569). In most plants with the raceme type, while the inflorescence is indeterminate, and the uppermost flowers (those toward the end of the main shoot) are younger, still the period of flowering is somewhat restricted and the raceme stops growing. In a few plants, however, as in the common "shepherd's-purse," the raceme continues to grow throughout the summer, so that the lower flowers may have ripened their seed while the terminal portion of the raceme is still growing and producing new flowers. Compound racemes are formed when by branching of the flower-shoot there are several racemes in a cluster, as in the false Solomon's seal (Smilacina racemosa).
- *The panicle.*—The panicle is developed from the raceme type by the branching of the lateral flower-axes forming a loose open flower cluster, as in the *oat*.
- *The thyrsus* is a compact panicle of pyramidal form, as in the lilac, horse-chestnut, etc.
- *The corymb.*—The corymb shows likewise an easy transition from the raceme type, by the shortening of the main axis of inflorescence, and the lengthening of the lower, lateral flower peduncles so that the flower cluster is more or less flattened on top. This represents the *simple corymb*. A *compound corymb* is one in which some of the flower peduncles branch again forming secondary corymbs, as in the mountain-ash. It is like a panicle with the lower flower stalks elongated.

- *The umbel.*—The umbel is developed from the raceme, or corymb. The main flower-shoot remains very short or undeveloped with several flowers on long peduncles arising close together around this shortened axis, in the form of a whorl or cluster. Examples are found in the milkweed, water pennywort (Hydrocotyle), the oxheart cherry, etc. A *compound* umbel is one in which the peduncles are branched, forming secondary umbels, as in the caraway, parsnip, carrot, etc.
- *The spike.*—In the spike the main axis is long, and the solitary flowers in the axils of the bracts are usually sessile, and often very much crowded. The plaintain, mullein (Figure 422), etc., are examples. The spike is a raceme, only the flowers are sessile and crowded. In the grasses the flower cluster is branched, and the branchlets bearing a few flowers are spikelets.
- *The head.*—When the flower axis is very much shortened and the flowers crowded and sessile or nearly so, forming a globose or compressed cluster, it is a *head* or *capitulum*. The transition is from a spike by the shortening of the main axis, as in the clover, button bush (*Cephalanthus*), etc., or in the shortening of the peduncles in an umbel, as in the daisy, dandelion, and other composite flowers. In these the head is surrounded by an involucre, which in the young head often envelopes the mass of flowers, thus affording them protection. In some other composites (Lactuca, for example) the involucre affords protection for a longer period, even while the seeds are ripening.
- *The spadix.*—When the main axis of the flower cluster is fleshy, the spike or head forms a *spadix*, as in the Indian turnip, the skunk-cabbage, the calla, etc. The spadix is usually more or less enclosed in a *spathe*, a somewhat strap-shaped leaf.
- *The catkin.*—A spike which is usually caducous, i.e., falls away after the maturity of the flower or fruit, is called a catkin, or an *ament*. The flower clusters of the alder, willow, (Figure 555), poplar, and the staminate flower clusters of the oak, hickory, hazel, birch, etc., are *aments*. So characteristic is this mode of inflorescence that the plants are called *amentiferous*, or *amentaceous*.
- *Anthesis of flowers with indeterminate inflorescence.*—In the anthesis of the raceme as well as in other corymbose forms the lower (or outer) flowers being older, open first. The opening of the flowers then takes place from below, upward; or from the outside, inward toward the center of inflorescence. The *anthesis*, i.e., the opening of the flowers of corymbose forms is said to be *centripetal*, i.e., it progresses from outside, inward. The anthesis of the fuller's teazel is peculiar, since it shows both types. There are several distinct advantages to the plant where anthesis extends over a period of time, as it favors cross pollination, favors the formation of seed in case conditions should be unfavorable at one period of anthesis, distributes the drain on the plant for food, etc.

Figure 451. Head of sunflower showing centripetal inflorescence of tubular flowers. (Photo by the Author.)

Figure 452. Heads of fuller's teazel in different stages of flowering.

Flower Clusters with Determinate Inflorescence

The simplest mode of determinate inflorescence is a plant with a solitary terminal flower, as in the hepatica, the tulip, etc. The leaves in these two plants are clustered in the form of a rosette, and the aerial shoot is naked and bears the single flower at its summit. Such a flower-shoot is a *scape*. As in the case of the indeterminate inflorescence, so here the larger number of flower-shoots are more complex and specialized, resulting in the evolution of flower clusters or masses. Accompanying the association of flowers into clusters there has been a reduction in leaf surface on the flower-shoot so that the flowers predominate in mass and are more conspicuous. Among the recognized modes of determinate inflorescence, the following are the chief ones:

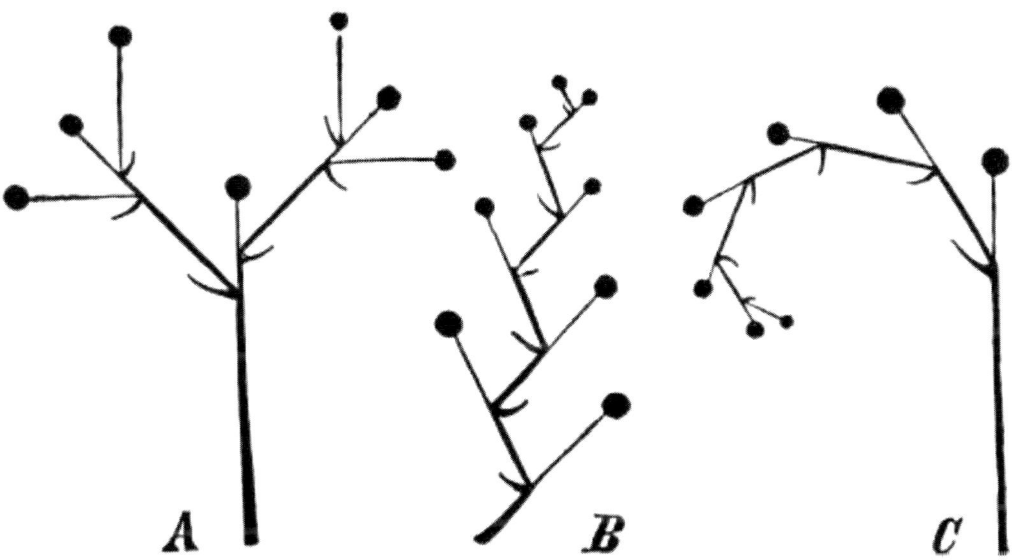

Figure 453. Diagrams of cymose inflorescence. A, dichasium; B, scorpioid cyme; C, helicoid cyme. (After Strasburger.)

- *The cyme.*—In the cyme the terminal flower on the main axis opens first and the remaining flowers are borne on lateral shoots, which arise from the axils of leaves or bracts, below. These lateral shoots usually branch and elongate so that the terminal flowers on all the branches reach nearly the same height as the terminal flower on the main shoot, forming a somewhat flattened or convex top of the flower cluster. This is illustrated in the basswood flower. The anthesis of the cyme is *centrifugal*, i.e., from the inside outward to the margin. But it is often more or less mixed, since the lateral shoots if they bear more than one flower are diminutive cymes and the terminal flower opens before the lateral ones. Where the flower cluster is quite large and the branching quite extensive, *compound cymes* are formed, as in the dogwood, hydrangea, etc.

- *The helicoid cyme.*—Where successive lateral branching takes place, and always continues on the same side a curved flower cluster is formed, as in the forget-me-not and most members of the borage family. This is known as a *helicoid cyme* (Figure 453, *C*). Each new branch becomes in turn the "false" axis bearing a new branch on the same side.
- *The scorpioid cyme.—A scorpioid cyme* (Figure 453, *B*) is formed where each new branch arises on alternate sides of the "false" axis.
- *The forking cyme* is where each "false" axis produces two branches opposite, so that it represents a false dichotomy (example, the flower cluster of chickweed).

Some of these flower clusters are peculiar and it is difficult to see how the helicoid, or scorpioid, cymes are of any advantage to the plant over a true cyme. The inflorescence of the plant being determinate, if the flowering is to be extended over a considerable period a peculiar form would necessarily result. In the *helicoid cyme* continued branching takes place on one side, and the result in the forget-me-not is a continued inflorescence in its effect like that of a continued raceme (compare shepherd's-purse). But we should not expect that all of the complex and specialized structures from simple and generalized ones are beneficial to the plant. In many plants we recognize evolution in the direction of advantageous structures. But since the plant cannot consciously evolve these structures, we must also recognize that there may be phases of retrogression in which the structures evolved are not so beneficial to the plant as the more simple and generalized ones of its ancestors. Variation and change do not result in advancing the plant or plant structures merely along the lines which will be beneficial. The tendency is in all directions. The result in general may be diagramed by a tree with divergent and wide-reaching branches. Some die out; others remain subordinate or dormant; while still others droop downward, showing a retrogression. But in this backward evolution they do not return to the condition of their ancestors, nor is the same course retraced. A new downward course is followed just as the downward-growing branch follows a course of its own, and does not return in the trunk.

Chapter 43

POLLINATION

ORIGIN OF HETEROSPORY, AND THE NECESSITY FOR POLLINATION

Both Kinds of Sexual Organs on the Same Prothallium

In the ferns, as we have seen, the sexual organs are borne on the prothallium, a small, leaf-like, heart-shaped body growing in moist situations. In a great many cases both kinds of sexual organs are borne on the same prothallium. While it is perhaps not uncommon, in some species, that the egg-cell in an archegonium may be fertilized by a spermatozoid from an antheridium on the same prothallium, it happens many times that it is fertilized by a spermatozoid from another prothallium. This may be accomplished in several ways. In the first place antheridia are usually found much earlier on the prothallium than are the archegonia. When these antheridia are ripe, the spermatozoids escape before the archegonia on the same prothallium arc mature.

Cross Fertilization in Monœcious Prothallia

By swimming about in the water or drops of moisture which are at times present in these moist situations, these spermatozoids may reach and fertilize an egg which is ripe in an archegonium borne on another and older prothallium. In this way what is termed cross fertilization is brought about nearly as effectually as if the prothallia were diœcious, i.e. if the antheridia and archegonia were all borne on separate prothallia.

Tendency Toward Diœcious Prothallia

In other cases some fern prothallia bear chiefly archegonia, while others bear only antheridia. In these cases cross fertilization is enforced because of this separation of the sexual organs on different prothallia. These different prothallia, the male and female, are largely due to a difference in food supply, as has been clearly proven by experiment.

The Two Kinds of Sexual Organs on Different Prothallia

In the horsetails (equisetum) the separation of the sexual organs on different prothallia has become quite constant. Although all the spores are alike, so far as we can determine, some produce small male plants exclusively, while others produce large female plants, though in some cases the latter bear also antheridia. It has been found that when the spores are given but little nutriment they form male prothallia, and the spores supplied with abundant nutriment form female prothallia.

Permanent Separation of Sexes By Different Amounts of Nutriment Supplied the Spores

This separation of the sexual organs of different prothallia, which in most of the ferns, and in equisetum, is dependent on the chance supply of nutriment to the germinating spores, is made certain when we come to such plants as isoetes and selaginella. Here certain of the spores receive more nutriment while they are forming than others. In the large sporangia (macrosporangia) only a few of the cells of the spore-producing tissue form spores, the remaining cells being dissolved to nourish the growing macrospores, which are few in number. In the small sporangia (microsporangia) all the cells of the spore-producing tissue form spores. Consequently each one has a less amount of nutriment, and it is very much smaller, a microspore. The sexual nature of the prothallium in selaginella and isoetes, then, is predetermined in the spores while they are forming on the sporophyte. The microspores are to produce male prothallia, while the macrospores are to produce female prothallia.

Heterospory

This production of two kinds of spores by isoetes, selaginella, and some of the other fern plants is *heterospory*, or such plants are said to be *heterosporous*. Heterospory, then, so far as we know from living forms, has originated in the fern group. In all the higher plants, in the gymnosperms and angiosperms, it has been perpetuated, the microspores being represented by the pollen, while the macrospores are represented by the embryo sac; the male organ of the gymnosperms and angiosperms being the antherid cell in the pollen or pollen tube, or in some cases perhaps the pollen grain itself, and the female organ in the angiosperms perhaps reduced to the egg-cell of the embryo sac.

In the Pteridophytes Water Serves as the Medium for Conveying the Sperm Cell to the Female Organ

In the ferns and their allies, as well as in the liverworts and mosses, surface water is a necessary medium through which the generative or sperm cell of the male organ, the spermatozoid, may reach the germ cell of the female organ. The sperm cell is here motile. This is true in a large number of cases in the algæ, which are mostly aquatic plants, while in other cases currents of water float the sperm cell to the female organ.

In the Higher Plants a Modification of the Prothallium is Necessary

As we pass to the gymnosperms and angiosperms, however, where the primitive phase (the gametophyte) of the plants has become dependent solely on the modern phase (the sporophyte) of the plant, surface water no longer serves as the medium through which a motile sperm cell reaches the egg-cell to fertilize it. The female prothallium, or macrospore, is, in nearly all cases, permanently enclosed within the sporangium, so that if there were motile sperm cells on the outside of the ovary, they could never reach the egg to fertilize it. But a modification of the microspore, the pollen tube, enables the sperm cell to reach the egg-cell. The tube grows through the nucellus, or first through the tissues of the ovary, deriving nutriment therefrom.

But here an important consideration should not escape us. The pollen grains (microspores) must in nearly all cases first reach the pistil, in order that in the growth of this tube a channel may be formed through which the generative cell can make its way to the egg cell. The pollen passes from the anther locule, then, to the stigma of the ovary. This process is termed *pollination*.

Pollination

Self Pollination, or Close Pollination

Perhaps very few of the admirers of the pretty blue violet have ever noticed that there are other flowers than those which appeal to us through the beautiful colors of the petals.

How many have observed that the brightly colored flowers of the blue violet rarely "set fruit"? Underneath the soil or débris at the foot of the plant are smaller flowers on shorter, curved stalks, which do not open. When the anthers dehisce, they are lying close upon the stigma of the ovary, and the pollen is deposited directly upon the stigma of the same flower. This method of pollination is *self pollination*, or *close pollination*. These small, closed flowers of the violet have been termed "*cleistogamous*," because they are pollinated while the flower is closed, and fertilization takes place as a result.

But self pollination takes place in the case of some open flowers. In some cases it takes place by chance, and in other cases by such movements of the stamens, or of the flower at the time of the dehiscence of the pollen, that it is quite certainly deposited upon the stigma of the same flower.

Wind Pollination

The pine is an example of wind-pollinated flowers. Since the pollen floats in the air or is carried by the "wind," such flowers are *anemophilous*. Other anemophilous flowers are found in other conifers, in grasses, sedges, many of the ament-bearing trees, and other dicotyledons. Such plants produce an abundance of pollen and always in the form of "dust," so that the particles readily separate and are borne on the wind.

Pollination by Insects

A large number of the plants which we have noted as being anemophilous are monœcious or diœcious, i.e. the stamens and pistils are borne in separate flowers. The two kinds of flowers thus formed, the male and the female, are borne either on the same individual (monœcious) or on different individuals (diœcious). In such cases cross pollination, i.e. the pollination of the pistil of one flower by pollen from another, is sure to take place, if it is pollinated at all. Even in monœcious plants cross pollination often takes place between flowers of different individuals, so that more widely different stocks are united in the fertilized egg, and the strain is kept more vigorous than if very close or identical strains were united.

But there are many flowers in which both stamens and pistils are present, and yet in which cross pollination is accomplished through the agency of insects.

Figure 454. Viola cucullata; blue flowers above, cleistogamous flowers smaller and curved below. Section of pistil at right.

Pollination of the Bluet

In the pretty bluet the stamens and styles of the flowers are of different length as shown in Figures 455, 456. The stamens of the long-styled flower are at about the same level as the stigma of the short-styled flower, while the stamens of the latter are on about the same level as the stigma of the former. What does this interesting relation of the stamens and pistils in the two different flowers mean? As the butterfly thrusts its "tongue" down into the tube of the long-styled flower for the nectar, some of the pollen will be rubbed off and adhere to it. When now the butterfly visits a short-styled flower this pollen will be in the right position to be rubbed off onto the stigma of the short style. The positions of the long stamens and long style are such that a similar cross pollination will be effected.

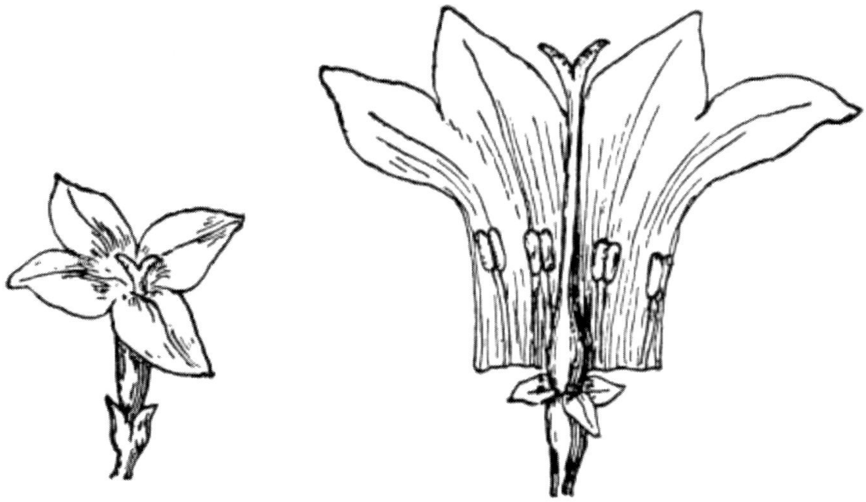

Figure 455. Dichogamous flower of the bluet (Houstonia cœrulea), the long-styled form.

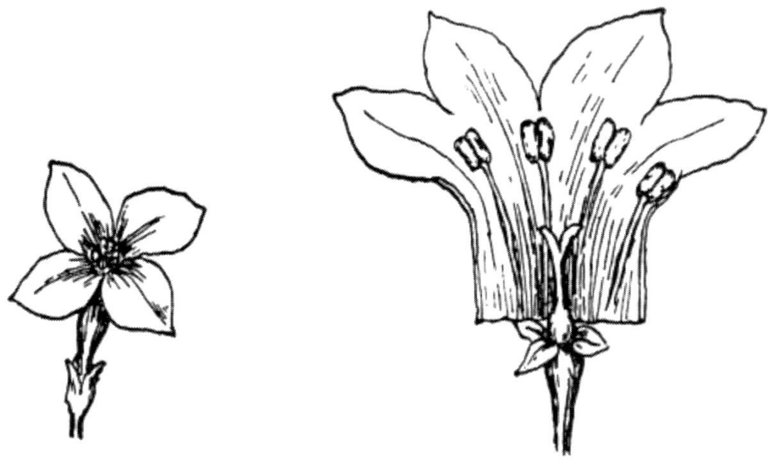

Figure 456. Dichogamous flower of bluet (Houstonia cœrulea), the short-styled form.

Pollination of the Primrose

In the primroses, of which we have examples growing in conservatories, that blossom during the winter, we have almost identical examples of the beautiful adaptations for cross pollination by insects found in the bluet. The general shape of the corolla is the same, but the parts of the flower are in fives, instead of in fours as in the bluet. While the pollen of the short-styled primulas sometimes must fall on the stigma of the same flower, Darwin has found that such pollen is not so potent on the stigma of its own flower as on that of another, an additional provision which tends to necessitate cross pollination.

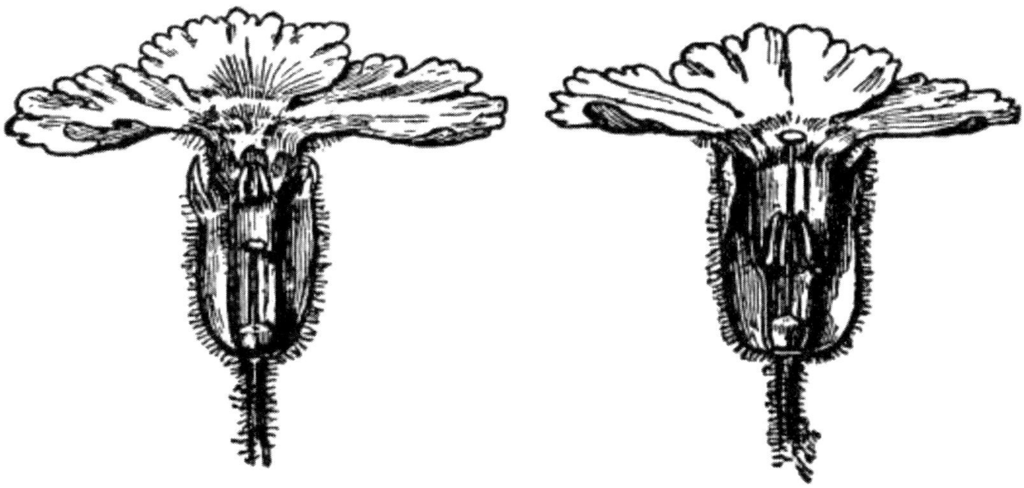

Figure 457. Dichogamous flowers of primula.

In the case of some varieties of pear trees, as the Bartlett, it has been found that the flowers remain largely sterile not only to their own pollen, or pollen of the flowers on the same tree, but to all flowers of that variety. However, they become fertile if cross pollinated from a different variety of pear.

Pollination of the Skunk's Cabbage

In many other flowers cross pollination is brought about through the agency of insects, where there is a difference in time of the maturing of the stamens and pistils of the same flower. The skunk's cabbage (Spathyema fœtida), though repulsive on account of its fetid odor, is nevertheless a very interesting plant to study for several reasons. Early in the spring, before the leaves appear, and in many cases as soon as the frost is out of the hard ground, the hooked beak of the large fleshy spathe of this plant pushes its way through the soil.

If we cut away one side of the spathe as shown in Figure 459 we shall have the flowering spadix brought closely to view. In this spadix the pistil of each crowded flower has pushed its style through between the plates of armor formed by the converging ends of the sepals, and stands out alone with the brush-like stigma ready for pollination, while the stamens of all the flowers of this spadix are yet hidden beneath. The insects which pass from the spadix of one plant to another will, in crawling over the projecting stigmas, rub off some of the pollen which has been caught while visiting a plant where the stamens are scattering their pollen. In this way cross pollination is brought about. Such flowers, in which the stigma is prepared for pollination before the anthers of the same flower are ripe, are *proterogynous*.

Figure 458. Skunk's cabbage.

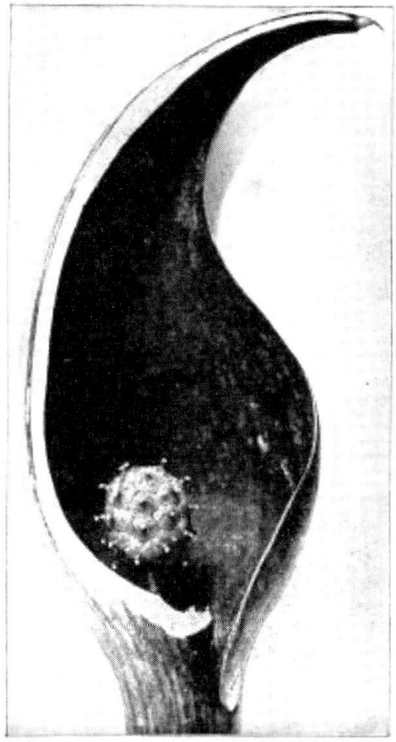

Figure 459. Proterogyny in skunk's cabbage. (Photograph by the author.)

Figure 460. Skunk's cabbage; upper flowers proterandrous, lower ones proterogynous.

Now if we observe the spadix of another plant we may see a condition of things similar to that shown in Figure 460. In the flowers in the upper part of the spadix here the anthers are wedging their way through between the armor-like plates formed by the sepals, while the styles of the same flowers are still beneath, and the stigmas are not ready for pollination. Such flowers are *proterandrous*, that is, the anthers are ripe before the stigmas of the same flowers are ready for pollination. In this spadix the upper flowers are proterandrous, while the lower ones are proterogynous, so that it might happen here that the lower flowers would be pollinated by the pollen falling on them from the stamens of the upper flowers. This would be cross pollination so far as the flowers are concerned, but not so far as the plants are concerned. In some individuals, however, we find all the flowers proterandrous.

Spiders Have Discovered This Curious Relation of the Flowers and Insects

On several different occasions, while studying the adaptations of the flowers of the skunk's cabbage for cross pollination, I was interested to find that the spiders long ago had discovered something of the kind, for they spread their nets here to catch the unwary but useful insects. I have not seen the net spread over the opening in the spathe, but it is spread over the spadix within, reaching from tip to tip of either the stigmas, or stamens, or both. Behind the spadix crouches the spider-trapper. The insect crawls over the edge of the spadix, and plunges unsuspectingly into the dimly lighted chamber below, where it becomes entangled in the meshes of the net.

Flowers in which the ripening of the anthers and maturing of the stigmas occur at different times are also said to be *dichogamous*.

Pollination of Jack-In-The-Pulpit

The jack-in-the-pulpit (Arisæma triphyllum) has made greater advance in the art of enforcing cross pollination. The larger number of plants here are, as we have found, diœcious, the staminate flowers being on the spadix of one plant, while the pistillate flowers are on the spadix of another. In a few plants, however, we find both female and male flowers on the same spadix.

The pretty bell-flower (Campanula rotundifolia) is dichogamous and proterandrous (Figure 462). Many of the composites are also dichogamous.

Pollination of Orchids

But some of the most marvellous adaptations for cross pollination by insects are found in the orchids, or members of the orchis family. The larger number of the members of this family grow in the tropics. Many of these in the forests are supported in lofty trees where they are brought near the sunlight, and such are called "epiphytes." A number of species of orchids are distributed in temperate regions.

Figure 461. A group of jacks.

Cypripedium, or Lady-Slipper

One species of the lady-slipper is shown in Figure 468. The labellum in this genus is shaped like a shoe, as one can see by the section of the flower in Figure 468. The stigma is situated at *st*, while the anther is situated at *a*, upon the style. The insect enters about the middle of the boat-shaped labellum. In going out it passes up and out at the end near the flower stalk. In doing this it passes the stigma first and the anther last, rubbing against both. The pollen caught on the head of the insect, will not touch the stigma of the same flower, but will be in position to come in contact with the stigma of the next flower visited.

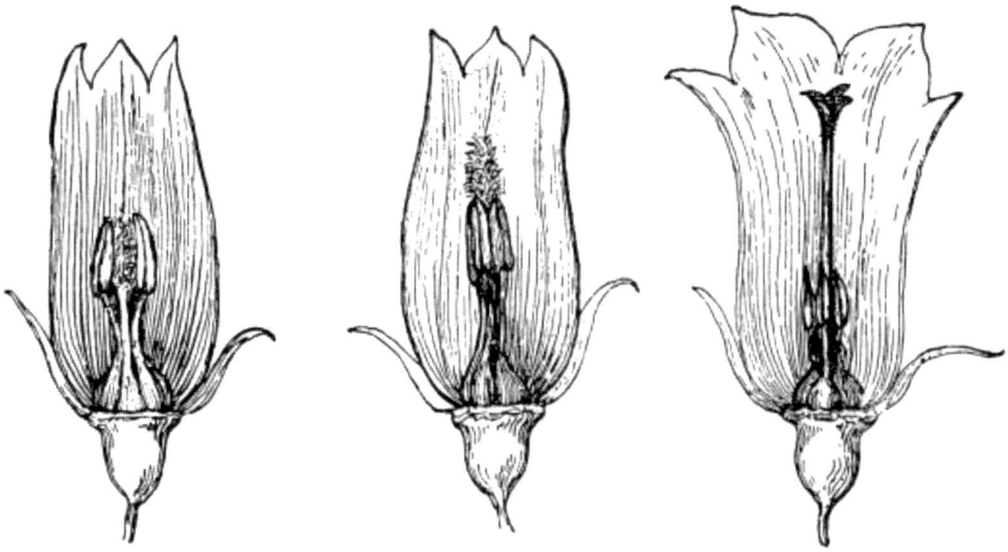

Figure 462. Proterandry in the bell-flower (campanula). Left Figure shows the syngenœcious stamens surrounding the immature style and stigma. Middle Figure shows the immature stigma being pushed through the tube and brushing out the pollen; while in the right-hand Figure, after the pollen has disappeared, the lobes of the stigma open out to receive pollen from another flower.

Epipactis

In epipactis, shown in Figure 469, the action is similar to that of the blue iris.

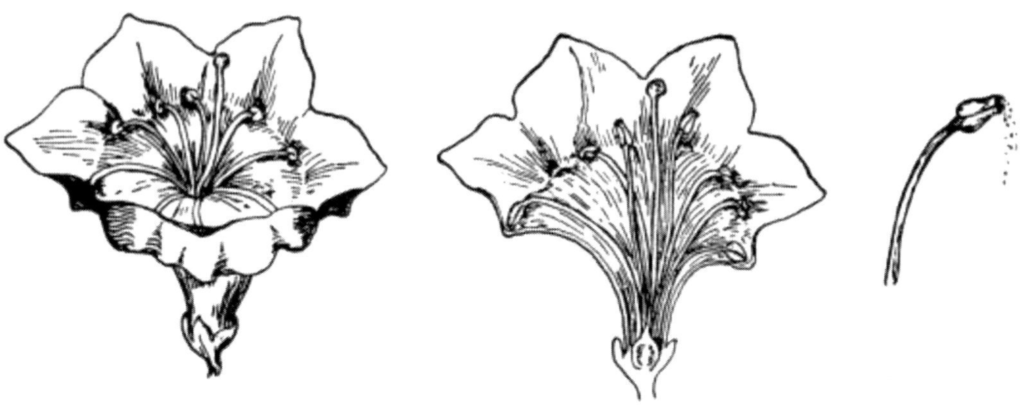

Figure 463. Kalmia latifolia, showing position of anthers before insect visits, and at the right the scattering of the pollen when disturbed by insects. Middle Figure section of flower.

In some of the tropical orchids the pollinia are set free when the insect touches a certain part of the flower, and are thrown in such a way that the disk of the pollinium strikes the insect's head and stands upright. By the time the insect reaches another flower the pollinium has bent downward sufficiently to strike against the stigma when the insect alights on the labellum. In the mountains of North Carolina I have seen a beautiful little orchid, in which, if one touches a certain part of the flower with a lead-pencil or other suitable object, the pollinium is set free suddenly, turns a complete somersault in the air, and lands with the disk sticking to the pencil. Many of the orchids grown in conservatories can be used to demonstrate some of these peculiar mechanisms.

Figure 464. Spray of leaves and flowers of cytisus.

Figure 465. Flower of cytisus grown in conservatory. Same flower scattering pollen.

Pollination of the Canna

In the study of some of the marvellous adaptations of flowers for cross pollination one is led to inquire if, after all, plants are not intelligent beings, instead of mere automatons which respond to various sorts of stimuli. No plant has puzzled me so much in this respect as the canna, and any one will be well repaid for a study of recently opened flowers, even though it may be necessary to rise early in the morning to unravel the mystery, before bees or the wind have irritated the labellum. The canna flower is a bewildering maze of petals and petal-like members. The calyx is green, adherent to the ovary, and the limb divides into three, lanceolate lobes. The petals are obovate and spreading, while the stamens have all changed to petal-like members, called *staminodia*. Only one still shows its stamen origin, since the anther is seen at one side, while the filament is expanded laterally and upwards to form the *staminodium*.

Figure 466.Spartium, showing the dusting of the pollen through the opening keels on the under side of an insect. (From Kerner and Oliver.)

The ovary has three locules, and the three styles are usually united into a long, thin, strap-shaped style, as seen in the Figure, though in some cases three, nearly distinct, filamentous styles are present. The end of this strap-shaped style has a peculiar curve on one side, the outline being sometimes like a long narrow letter S. It is on the end of this style, and along the crest of this curve, that the stigmatic surface lies, so that the pollen must be deposited on the stigmatic end or margin in order that fertilization may take place.

If we open carefully canna flower buds which are nearly ready to open naturally, by unwrapping the folded petals and staminodia, we shall see the anther-bearing staminodium is so wrapped around the flattened style that the anther lies closely pressed against the face of the style, near the margin *opposite that on which the stigma lies*.

Figure 467. Cypripedium.

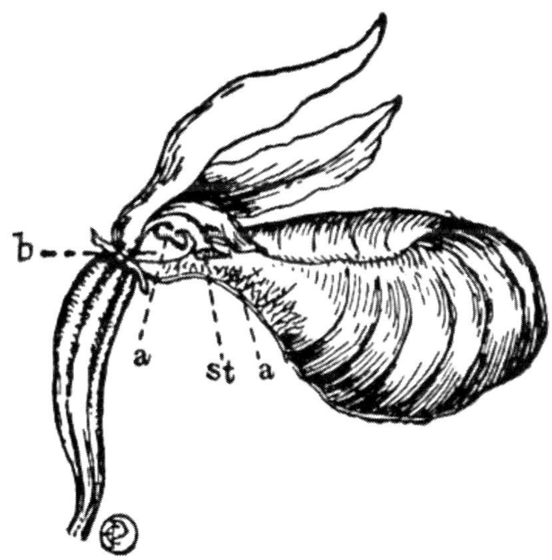

Figure 468. Section of flower of cypripedium. *st*, stigma; *a*, at the left stamen. The insect enters the labellum at the center, passes under and against the stigma, and out through the opening *b*, where it rubs against the pollen. In passing through another flower this pollen is rubbed off on the stigma.

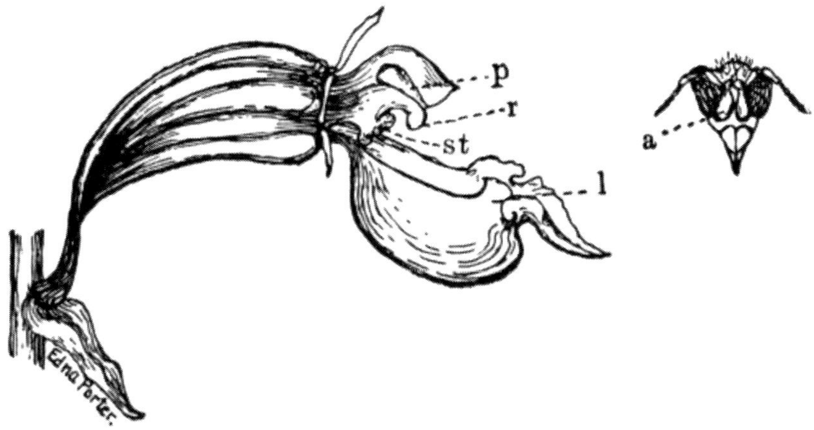

Figure 469. Epipactis with portion of perianth removed to show details. *l*, labellum; *st*, stigma; *r*, rostellum; *p*, pollinium. When the insect approaches the flower its head strikes the disk of the pollinium and pulls the pollinium out. At this time the pollinium stands up out of the way of the stigma. By the time the insect moves to another flower the pollinia have moved downward so that they are in position to strike the stigma and leave the pollen. At the right is the head of a bee, with two pollinia (*a*) attached.

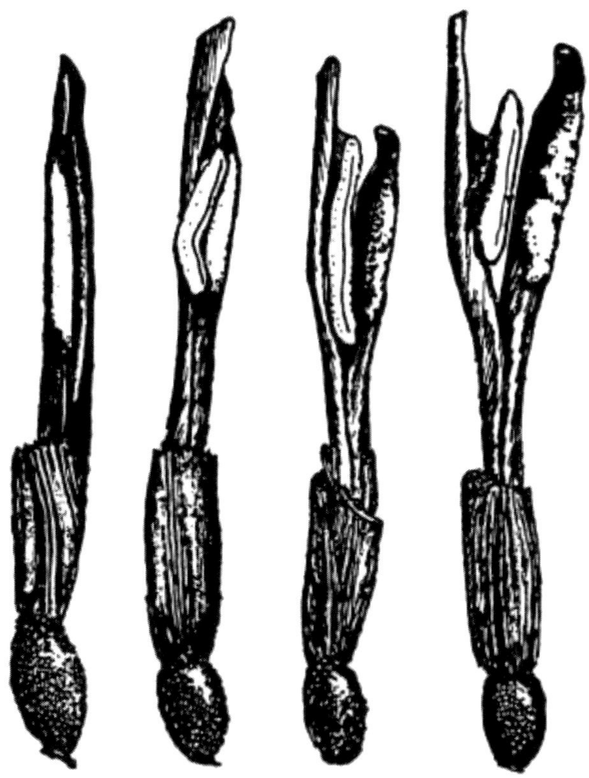

Figure 470. Canna flowers with the perianth removed to show the depositing of the pollen on the style by the stamen.

The walls of the anther locules which lie against the style become changed to a sticky substance for their entire length, so that they cling firmly to the surface of the style and also to the mass of pollen within the locules. The result is that when the flower opens, and this staminodium unwraps itself from the embrace of the style, the mass of pollen is left there deposited, while the empty anther is turned around to one side.

Why does the flower deposit its own pollen on the style? Some have regarded this as the act of pollination, and have concluded, therefore, that cannas are necessarily self pollinated, and that cross pollination does not take place. But why is there such evident care to deposit the pollen on the side of the style away from the stigmatic margin? If we visit the cannas some morning, when a number of the flowers have just opened, and the bumblebees are humming around seeking for nectar, we may be able to unlock the secret.

Pollination of the canna flower by bumblebee.

We see that in a recently opened canna flower, the petal which directly faces the style in front stands upward quite close to it, so that the flower now is somewhat funnel-shaped. This front petal is the *labellum*, and is the landing place for the bumblebee as he alights on the flower. Here he comes humming along and alights on the labellum with his head so close to the style that it touches it. But just the instant that the bee attempts to crowd down in the flower the labellum suddenly bends downward, as shown in Figure 468. In so doing the head of the bumblebee scrapes against the pollen, bearing some of it off. Now while the bee is sipping the nectar it is too far below the stigma to deposit any pollen on the latter. When the bumblebee flies to another newly opened flower, as it alights, some of the pollen of the former flower is brushed on the stigma.

One can easily demonstrate the sensitiveness of the labellum of recently opened canna flowers, if the labellum has not already moved down in response to some stimulus. Take a lead-pencil, or a knife blade, or even the finger, and touch the upper surface of the labellum by thrusting it between the latter and the style. The labellum curves quickly downward.

Sometimes the bumblebees, after sipping the nectar, will crawl up over the style in a blundering manner. In this way the flower may be pollinated with its own pollen, which is equivalent to self pollination. Undoubtedly self pollination does take place often in flowers which are adapted, to a greater or less degree, for cross pollination by insects.

Canna flower. Pollen on style, stamen at left.

Chapter 44

THE FRUIT

PARTS OF THE FRUIT

After the Flower Comes The Fruit

With the perfection of the fruit the seed is usually formed. This is the end towards which the energies of the plant have been directed. While the seed consists only of the ripened ovule and the contained embryo, the fruit consists of the ripened ovary in addition, and in many cases with other accessory parts, as calyx, receptacle, etc., combined with it. The wall of the ripened ovary is called a *pericarp*, and the walls of the ovary form the walls of the fruit.

Pericarp, Endocarp, Exocarp, Etc

This is the part of the fruit which envelops the seed and may consist of the carpels alone, or of the carpels and the adherent part of the receptacle, or calyx. In many fruits the pericarp shows a differentiation into layers, or zones of tissue, as in the cherry, peach, plum, etc. The outer, which is here soft and fleshy, is *exocarp*, while the inner, which is hard, is the *endocarp*. An intermediate layer is sometimes recognized and is called *mesocarp*. In such cases the skin of the fruit is recognized as the *epicarp*. Epicarp and mesocarp are more often taken together as exocarp.

In general fruits are *dry* or *fleshy*. Dry fruits may be grouped under two heads. Those which open at maturity and scatter the seed are *dehiscent*. Those which do not open are *indehiscent*.

INDEHISCENT FRUITS

- *The akene.*—The thin dry wall of the ovary encloses the single seed. It usually does not open and free the seed within. Such a fruit is an *akene*. An *akene* is a dry, *indehiscent* fruit. All of the crowded but separate pistils in the buttercup flower when ripe make a head of akenes, which form the fruit of the buttercup. Other examples of akenes are found in other members of the buttercup family, also in the composites, etc. The sunflower seed is a good example of an akene.

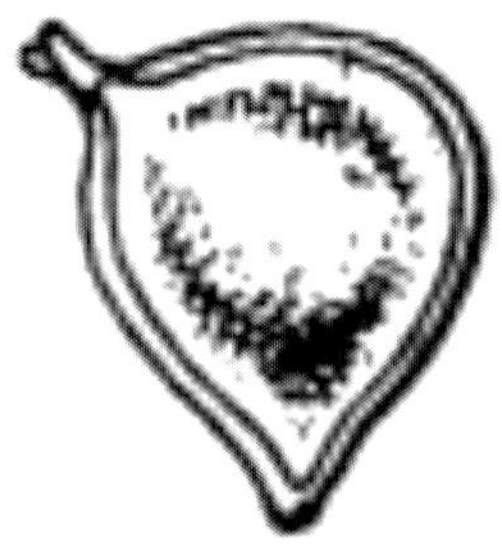

Figure 472. Seed, or akene, of buttercup.

Figure 473. Fruit of red oak. An acorn.

- *The samara.*—The winged fruits of the maple, elm, etc., are indehiscent fruits. They are sometimes called key fruits.
- *The caryopsis* is a dry fruit in which the seed is consolidated with the wall of the ovary, as in the wheat, corn, and other grasses.
- *The schizocarp* is a dry fruit consisting of several locules (from a *syncarpous gynœcium*). At maturity the carpels separate from each other, but do not themselves dehisce and free the seed, as in the carrot family, mallow family.

- *The acorn.*—The acorn fruit consists of the acorn and the "cup" at the base in which the acorn sits. The cup is a curious structure, and is supposed to be composed of an involucre of numerous small leaves at the base of the pistillate flower, which become consolidated into a hard cup-shaped body. When the acorn is ripe it easily separates from the cup, but the hard pericarp forming the "shell" of the acorn remains closed. Frost may cause it to crack, but very often the pericarp is split open at the smaller end by wedge-like pressure exerted by the emerging radicle during germination.

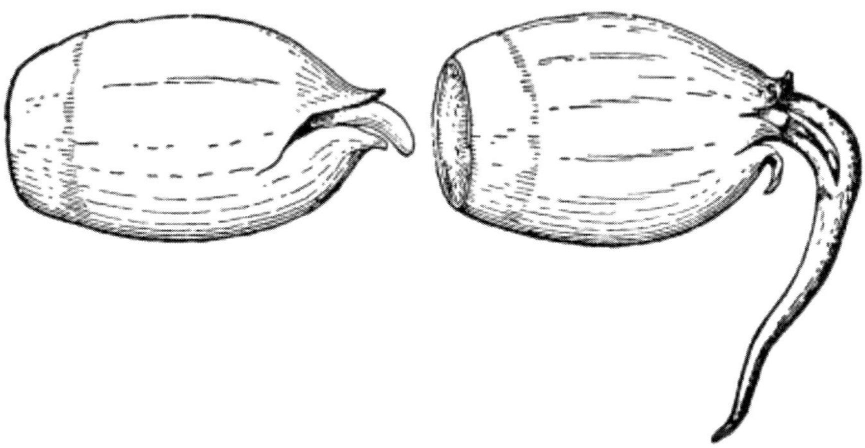

Figure 474. Germinating acorn of white oak.

- *The hazelnut, chestnut, and beechnut.*—In these fruits a crown of leaves (involucre) at the base of the flower grows around the nut and completely envelops it, forming the husk or burr. When the fruit is ripe the nut is easily shelled out from the husk. In the beechnut and chestnut the burr dehisces as it dries and allows the nut to drop out. But the fruit is not dehiscent, since the pericarp is still intact and encloses the seed.
- *The hickory-nut, walnut, and butternut.*—In these fruits the "shuck" of the hickory-nut and the "hull" of the walnut and butternut are different from the involucre of the acorn or hazelnut, etc. In the hickory-nut the "shuck" probably consists partly of calyx and partly of involucral bracts consolidated, probably the calyx part predominating. This part of the fruit splits open as it dries and frees the "nut," the pericarp being very hard and indehiscent. In the walnut and butternut the "hull" is probably of like origin as the "shuck" of the hickory nut, but it does not split open as it ripens. It remains fleshy. The walnut and butternut are often called *drupes* or *stone-fruits*, but the fleshy part of the fruit is not of the same origin as the fleshy part of the true drupes, like the cherry, peach, plum, etc.

Dehiscent Fruits

Of the dehiscent fruits several prominent types are recognized, and in general they are sometimes called pods. There is a single carpel (simple pistil), and the pericarp is dry (gynœcium apocarpous); or where there are several carpels united the pistil is compound (gynœcium syncarpous).

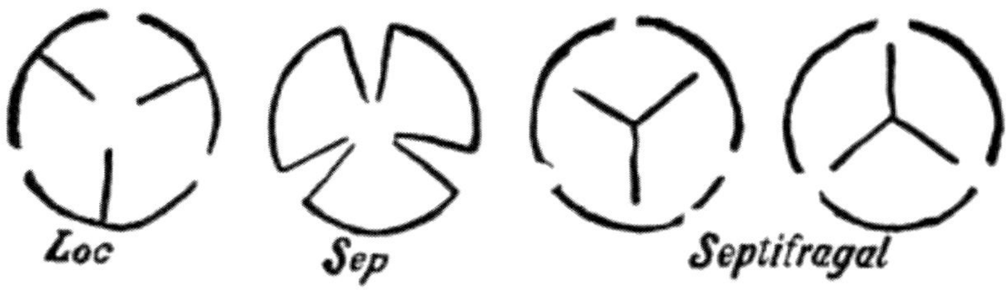

Figure 475. Diagrams illustrating three types (in cross-section) of the dehiscence of dry fruits. *Loc*, loculicidal; *Sep*, Septicidal, Septifragal.

Figure 476. Fruit of sweet pea; a pod.

- *The capsule.*—When the capsule is *syncarpous* it may dehisce in three different ways: 1st. When the carpels split along the line of their union with each other longitudinally (*septicidal dehiscence*), as in the azalea or rhododendron. 2d. When the carpels *split down the middle line* (*loculicidal dehiscence*), as in the fruit of the iris, lily, etc. 3d. When the carpels open by pores (*poricidal dehiscence*), as in the poppy. Some syncarpous capsules have but one locule, the partitions between the different locules when young having disappeared. The "bouncing-bet" is an example, and the seeds are attached to a central column in four rows corresponding to the four locules present in the young stage.

- *A follicle* is a capsule with a single carpel which splits open along the ventral or upper suture, as in the larkspur, peony.
- *The legume, or true pod*, is a capsule with a single carpel which splits along both sutures, as the pea, bean, etc. As the pod ripens and dries, a strong twisting tension is often produced, which splits the pod suddenly, scattering the seeds.
- *The silique.*—In the toothwort, shepherd's-purse, and nearly all of the plants in the mustard family the fruit consists of two united carpels, which separate at maturity, leaving the partition wall persistent. Such a fruit is a *silique*; when short it is a *silicle*, or *pouch*.

A pyxidium, or pyxis, is a capsule which opens with a lid, as in the plantain.

Fleshy and Juicy Fruits

- *The drupe, or stone-fruit.*—In the plum, cherry, peach, apricot, etc., the outer portion (exocarp) of the pericarp (ovary) becomes fleshy, while the inner portion (endocarp) becomes hard and stony, and encloses the seed, or "pit." Such a fruit is known as a drupe, or as a stone-fruit. In the almond the fleshy part of the fruit is removed.

Figure 477. Drupe, or stone-fruit, of plum.

- *The raspberry and blackberry.*—While these fruits are known popularly as "berries," they are not berries in the technical sense. Each ovary, or pericarp, in the flower forms a single small fruit, the outer portion being fleshy and the inner stony, just as in the cherry or plum. It is a *drupelet* (little drupe). All of the drupelets together make the "berry," and as they ripen the separate drupelets cohere more or less. It is a collection, or aggregation, of fruits, and consequently they are sometimes called *collective fruits*, or *aggregate fruits*. In the raspberry the fruit separates from the receptacle, leaving the latter on the stem, while the drupelets of the blackberry and dewberry adhere to the receptacle and the latter separates from the stem.
- *The berry.*—In the true berry both exocarp (including mesocarp) and endocarp are fleshy or juicy. Good examples are found in cranberries, huckleberries, gooseberries, currants, snowberries, tomatoes, etc. The calyx and wall of the pistil are adnate, and in fruit become fleshy so that the seeds are imbedded in the pulpy juice. The seeds themselves are more or less stony. In the case of berries, as well as in strawberries, raspberries, and blackberries, the fruits are eagerly sought by birds and other animals for food. The seeds being hard are not digested, but are passed with the other animal excrement and thus gain dispersal.

REINFORCED, OR ACCESSORY, FRUITS

When the torus (receptacle) is grown to the pericarp in fruit, the fruit is said to be *reinforced*. The torus may enclose the pericarps, or the latter may be seated upon the torus.

Figure 478. Fruit of raspberry.

- *In the strawberry* the receptacle of the flower becomes larger and fleshy, while the "seeds," which are akenes, are sunk in the surface and are hard and stony. The strawberry thus differs from the raspberry and blackberry, but like them it is not a true berry.
- *The apple, pear, quince, etc.*—In the flower the calyx, corolla, and stamens are perigynous, i.e., they are seated on the margin of the receptacle, or torus, which is elevated around the pistils. In fruit the receptacle becomes consolidated with the wall of the ovary (with the pericarp). The torus thus *reinforces* the pericarp. The torus and outer portion of the pericarp become fleshy, while the inner portion of the pericarp becomes papery and forms the "core." The calyx persists on the free end of the fruit. Such a fruit is called a *pome*. The receptacle, or torus, of the rose-flower, closely related to the apple, is instructive when used in comparison. The rose-fruit is called a "hip."
- *The pepo.*—The fruit of the squash, pumpkin, cucumber, etc., is called a *pepo*. The outer part of the fruit is the receptacle (or torus), which is consolidated with the outer part of the three-loculed ovary. The calyx, which, with the corolla and stamens, was epigynous, falls off from the young fruit.

FRUITS OF GYMNOSPERMS

The fruits of the gymnosperms differ from nearly all of the angiosperms in that the seed formed from the ripened ovule is naked from the first, i.e., the ovary, or carpel, does not enclose the seed.

- *The cone-fruit* is the most prominent fruit of the gymnosperms, as can be seen in the cones of various species of pine, spruce, balsam, etc.
- *Fleshy fruits of the gymnosperms.*—Some of the fleshy fruits resemble the stone-fruits and berries of the angiosperms. The *cedar "berries*," for example, are fleshy and contain several seeds. But the fleshy part of the fruit is formed, not from pericarp, since there is no pericarp, but from the outer portion of the ovules, while the inner walls of the ovules form the hard stone surrounding the endosperm and embryo. An examination of the pistillate flower of the cedar (juniper) shows usually three flask-shaped ovules on the end of a fertile shoot subtended by as many bracts (carpels?). The young ovules are free, but as they grow they coalesce, and the outer walls become fleshy, forming a berry-like fruit with a three-rayed crevice at the apex marking the number of ovules. The red fleshy fruit of the yew (taxus) resembles a drupe which is open at the apex. The stony seed is formed from the single ovule on the fertile shoot, while the red cup-shaped fleshy part is formed

from the outer integument of the ovule. The so-called "aril" of the young ovule is a rudimentary outer integument.

- The fruit of the maidenhair tree (ginkgo) is about the size of a plum and resembles very closely a stone-fruit. But it is merely a ripened ovule, the outer layer becoming fleshy while the inner layer becomes stony and forms the pit which encloses the embryo and endosperm. The so-called "aril," or "collar," at the base of the fruit is the rudimentary carpel, which sometimes is more or less completely expanded into a true leaf. The fruit of cycas is similar to that of ginkgo, but there is no collar at the base. In zamia the fruit is more like a cone, the seeds being formed, however, on the under sides of the scales.

THE "FRUIT" OF FERNS, MOSSES, ETC.

The term "fruit" is often applied in a general or popular sense to the groups of spore-producing bodies of ferns (*fruit dots*, or *sori*), the spore-capsules of mosses and liverworts, and also to the fruit-bodies, or spore-bearing parts, of the fungi and algæ.

Chapter 45

Seed Dispersal

Means for Dissemination of Seeds

During late summer or autumn a walk in the woods or afield often convinces us of the perfection and variety of means with which plants are provided for the dissemination of their seeds, especially when we discover that several hundred seeds or fruits of different plants are stealing a ride at our expense and annoyance. The hooks and barbs on various seed-pods catch into the hairs of passing animals and the seeds may thus be transported considerable distances. Among the plants familiar to us, which have such contrivances for unlawfully gaining transportation, are the beggar-ticks or stick tights, or sometimes called bur-marigold (bidens), the tick-treefoil (desmodium), or cockle-bur (xanthium), and burdock (arctium).

Figure 479. Bur of bidens or bur-marigold, showing barbed seeds.

Figure 480. Seed pod of tick-treefoil (desmodium); at the right some of the hooks greatly magnified.

Other plants like some of the sedges, etc., living on the margins of streams and of lakes, have seeds which are provided with floats. The wind or the flowing of the water transports them often to distant points.

Many plants possess attractive devices, and offer a substantial reward, as a price for the distribution of their seeds. Fruits and berries are devoured by birds and other animals; the seeds within, often passing unharmed, may be carried long distances. Starchy and albuminous seeds and grains are also devoured, and while many such seeds are destroyed, others are not injured, and finally are lodged in suitable places for growth, often remote from the original locality. Thus animals willingly or unwillingly become agents in the dissemination of plants over the earth. Man in the development of commerce is often responsible for the wide distribution of harmful as well as beneficial species.

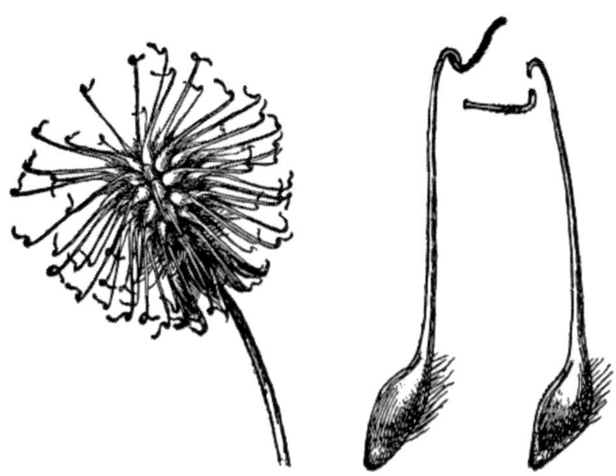

Figure 481. Seeds of geum showing the hooklets where the end of the style is kneed.

Other plants are more independent, and mechanisms are employed for violently ejecting seeds from the pod or fruit. The unequal tension of the pods of the common vetch (Vicia sativa) when drying causes the valves to contract unequally, and on a dry summer day the valves twist and pull in opposite directions until they suddenly snap apart, and the seeds are thrown forcibly for some distance. In the impatiens, or touch-me-not as it is better known, when the pods are ripe, often the least touch, or a pinch, or jar, sets the five valves free, they coil up suddenly, and the small seeds are thrown for several yards in all directions. During autumn, on dry days, the pods of the witch hazel contract unequally, and the valves are suddenly spread apart, and the seeds are hurled away.

Other plants have seeds provided with tufts of pappus, or hair-like masses, or wing-like outgrowths which serve to buoy them up as they are whirled along, often miles away. In late spring or early summer the pods of the willow burst open, exposing the seeds, each with a tuft of white hairs making a mass of soft down. As the delicate hairs dry, they straighten out in a loose spreading tuft, which frees the individual seeds from the compact mass. Here they are caught by currents of air and float off singly or in small clouds.

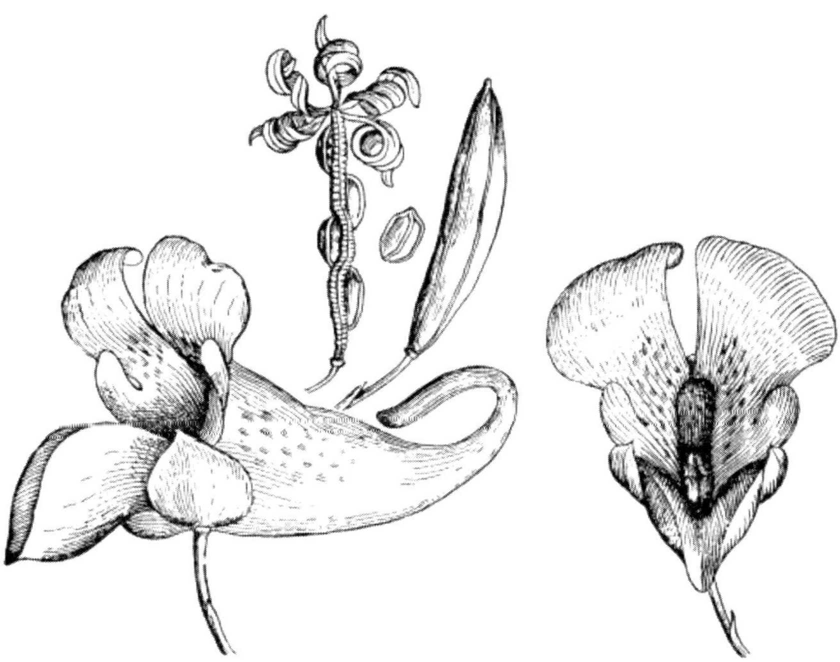

Figure 482. Touch-me-not (Impatiens fulva); side and front view of flower below; above unopened pod, and opening to scatter the seed.

- *The prickly lettuce.*—In late summer or early autumn the seeds of the prickly lettuce (Lactuca scariola) are caught up from the roadsides by the winds, and carried to fields where they are unbidden as well as unwelcome guests. This plant is shown in Figure 483.

- *The wild lettuce.*—A related species, the wild lettuce (Lactuca canadensis) occurs on roadsides and in the borders of fields, and is about one meter in height. The heads of small yellow or purple flowers are arranged in a loose or branching panicle. The flowers are rather inconspicuous, the rays projecting but little above the apex of the enveloping involucral bracts, which closely press together, forming a flowerhead more or less flask-shaped.

Figure 483. Lactuca scariola.

At the time of flowering the involucral bracts spread somewhat at the apex, and the tips of the flowers are a little more prominent. As the flowers then wither, the bracts press closely together again and the head is closed. As the seeds ripen the bracts die, and in drying bend outward and downward, around the flower stem below, or they fall away. The seeds are thus exposed. The dark brown achenes stand over the surface of the receptacle, each one tipped with the long slender beak of the ovary. The "pappus," which is so abundant in many of the plants belonging to the composite family, forms here a pencil-like tuft at the tip of this long beak. As the involucral bracts dry and curve downward, the pappus also dries, and in doing so bends downward and stands outward, bristling like the spokes of a small wheel. It is an interesting coincidence that this takes place simultaneously with the pappus of all the seeds of a head, so that the ends of the pappus bristles of adjoining seeds meet, forming a many-sided dome of a delicate and beautiful texture. This causes the beaks of the achenes to be crowded apart, and with the leverage thus brought to bear upon the achenes they are pried off the receptacle. They are thus in a position to be wafted away by the gentlest zephyr, and they go sailing away on the wind like a miniature parachute. As they come slowly to the ground the seed is thus carefully lowered first, so that it touches the ground in a position for the end which contains the root of the embryo to come in contact with the soil.

- *The milkweed, or silkweed.*—The common milkweed, or silkweed (Asclepias cornuti), so abundant in rich grounds, is attractive not only because of the peculiar pendent flower clusters, but also for the beautiful floats with which it sends its seeds skyward, during a puff of wind, to finally lodge on the earth.

Figure 484. Milkweed (Asclepias cornuti); dissemination of seed.

The large boat-shaped, tapering pods, in late autumn, are packed with oval, flattened, brownish seeds, which overlap each other in rows like shingles on a roof. These make a pretty picture as the pod in drying splits along the suture on the convex side, and exposes them to view. The silky tufts of numerous long, delicate white hairs on the inner end of each seed, in drying, bristle out, and thus lift the seeds out of their enclosure, where they are caught by the breeze and borne away often to a great distance, where they will germinate if conditions become favorable, and take their places as contestants in the battle for existence.

- *The virgin's bower.*—The virgin's bower (Clematis virginiana), too, clambering over fence and shrub, makes a show of having transformed its exquisite white flower clusters into grayish-white tufts, which scatter in the autumn gusts into hundreds of arrow-headed, spiral plumes. The achenes have plumose styles, and the spiral form of the plume gives a curious twist to the falling seed (Figure 485).

Figure 485. Seed distribution of virgin's bower (clematis).

Chapter 46

Vegetation in Relation to Environment[47]

Factors Influencing Vegetation Types

All plants are subject to the influence of environment from the time the seed begins to germinate until the seed is formed again, or until the plant ceases to live. A suitable amount of warmth and moisture is necessary that the seed may germinate. Moisture may be present, but if it is too cold, germination will not take place. So in all the processes of life there are several conditions of the environment, or the "outside" of plants, which must be favorable for successful growth and reproduction. Not only is this true, but the surroundings of plants to a large extent determine the kind of plants which can grow in particular localities. It is also evident that the reaction of environment on plants has in a large measure caused them to take on certain forms and structures which fit them better to exist under local conditions. In other cases where plants have varied by mutation (p. 338) some of the new forms may be more suited to the conditions of environment than others and they are more apt to survive. These conditions of environment acting on the plant are *factors* which have an important determining influence on the existence, habitat, habit, and form of the plant. These factors are sometimes spoken of as *ecological factors*, and the study of plants in this relation is sometimes spoken of as ecology,[48] which means a study of plants in their home or a study of the household relations of plants. These factors are of three sorts: 1st, physical factors; 2d, climatic factors; 3d, biotic factors.

Physical Factors

Some of these factors are water, light, heat, wind, chemical or physical condition of the soil, etc. *Water* is a very important factor for all plants. Even those growing on land

---

[47] For a fuller discussion of this subject by the author see Chapters XLVI-LVII of his "College Text-book of Botany" (Henry Holt & Co.).

[48] οἶκος = house, and λόγος = discourse.

contain a large percentage of water, which we have seen is rapidly lost by transpiration, and unless water is available for root absorption the plant soon suffers, and aquatic plants are injured very quickly by drying when taken from the water. Excess of soil water is injurious to some plants. *Light* is important in photosynthesis, in determining direction of growth as well as in determining the formation of suitable leaves in most plants, and has an influence in the structure of the leaf according as the light may be strong, weak, etc. *Heat* has great influence on plant growth and on the distribution of plants. The growth period for most vegetation begins at 6 °C. (= 43 °F.), or in the tropics at 10°-12 °C., but a much higher temperature is usually necessary for reproduction. Some arctic algæ, however, fruit at 1.8 °C. The upper limit favorable for plants in general is 45°-50 °C., while the optimum temperature is below this. Very high temperatures are injurious, and fatal to most plants, but some algæ grow in hot springs where the temperature reaches 80°-90 °C. Some desert plants are able to endure a temperature of 70 °C., while some flowering plants of other regions are killed at 45 °C. Some plants are specifically susceptible to cold, but most plants which are injured by freezing suffer because the freezing is a drying process of the protoplasm (see p. 374). *Wind* may serve useful purposes in pollination and in aeration, but severe winds injure plants by causing too rapid transpiration, by felling trees, by breaking plant parts, by deforming trees and shrubs, and by mechanical injuries from "sand-blast." *Ground covers* protect plants in several ways. Snow during the winter checks radiation of heat from the ground so that it does not freeze to so great a depth, and this is very important for many trees and shrubs. It also prevents alternate freezing and thawing of the ground, which "heaves" some plants from the soil. Leaves and other plant remains mulch the soil and check evaporation of water. The influence of the *chemical condition* of the soil is very marked in alkaline areas where the concentration of salt in the soil permits a very limited range of species. So the physical and mechanical conditions of the soil influence plants because the moisture content of the ground is so closely dependent on its physical condition. Rocky and gravelly soil, other things being equal, is dry. Clay is more retentive of moisture than sand, and moisture also varies according to the per cent of humus mixed with it, the humus increasing the percentage of moisture retained.

Climatic Factors

These factors are operative over very wide areas. There are two climatic factors: rainfall or atmospheric moisture, and temperature. A very low annual rainfall in warm or tropical countries causes a desert; an abundance of rain permits the growth of forests; extreme cold prevents the growth of forests and gives us the low vegetation of arctic and alpine regions.

Biotic Factors

These are animals which act favorably in pollination, seed distribution, or unfavorably in destroying or injuring plants, and man himself is one of the great agencies in checking the growth of some plants while favoring the growth of others. Plants also react on themselves in a multitude of ways for good or evil. Some are parasites on others; some in symbiosis (see p. 85) aid in providing food; shade plants are protected by those which overtop them; mushrooms and other fungi disintegrate dead plants to make humus and finally plant food; certain bacteria by nitrification prepare nitrates for the higher plants (see p. 83).

VEGETATION TYPES AND STRUCTURES

Responsive Type of Vegetation

In studying vegetation in relation to environment we are more concerned with the form of the plants which fits them to exist under the local conditions than we are with the classification of plants according to natural relationships. Plants may have the same vegetation type, grow side by side, and still belong to very different floristic types. For example, the cactus, yucca, three-leaved sumac, the sage-brush, etc., have all the same general vegetation type and thrive in desert regions. The red oaks, the elms, many goldenrods, trillium, etc., have the same general vegetation type, but represent very different floristic types. The latter plants grow in regions with abundant rainfall throughout the year, where the growing season is not very short and temperature conditions are moderate. Some goldenrods grow in very sandy soil which dries out quickly. These have fleshy or succulent leaves for storing water, and while they are of the same floristic type as goldenrods growing in other places, the vegetation type is very different. The types of vegetation which fit plants for growing in special regions or under special conditions, they have taken on in response to the influence of the conditions of their environment. While we find all gradations between the different types of vegetation, looking at the vegetation in a broad way, several types are recognized which were proposed by Warming as follows:

Mesophytes

These are represented by land plants under temperate or moderate climatic and soil conditions. The normal land vegetation of our temperate region is composed of mesophytes, that is, the plants have mesophytic structures during the growing season. The deciduous forests or thickets of trees and shrubs with their undergrowth, the meadows, pastures, prairies, weeds, etc., are examples. In those portions of the tropics where rainfall is great the vegetation is mesophytic the year around.

Xerophytes

These are plants which are provided with structures which enable them to live under severe conditions of dryness, where the air and soil are very dry, as in deserts or semideserts, or where the soil is very dry or not retentive of moisture, as in very sandy soil which is above ground water, or in rocky areas. Since the plants cannot obtain much water from the soil they must be provided with structures which will enable them to retain the small amount they can absorb from the soil and give it off slowly. Otherwise they would dry out by evaporation and die. Some of the structures which enable xerophytic plants to withstand the conditions of dry climate and soil are lessened leaf surface, increase in thickness of leaf, increase in thickness of cuticle, deeply sunken stomates, compact growth, also succulent leaves and stems, and in some cases loss of the leaf. Evergreens of the north temperate and the arctic regions are xerophytes.

Hydrophytes

These are plants which grow in fresh water or in very damp situations. The leaves of aerial hydrophytes are very thin, have a thin cuticle, and lose water easily, so that if the air becomes quite dry they are in danger of drying up even though the roots may be supplied with an abundance of water. The aquatic plants which are entirely submerged have often thin leaves, or very finely divided or slender leaves, since these are less liable to be torn by currents of water. The stems are slender and especially lack strengthening tissue, since the water buoys them up. Removed from the water they droop of their own weight, and soon dry up. The stems and leaves have large intercellular spaces filled with air which aids in aeration and in the diffusion of gases. Some use the term *hygrophytes*.

Halophytes

These are salt-loving plants. They grow in salt water, or in salt marshes where the water is brackish, or in soil which contains a high per cent of certain salts, for example the alkaline soils of the West, especially in the so-called "Bad Lands" of Dakota and Nebraska, and in alkaline soils of the Southwest and California. These plants are able to withstand a stronger concentration of salts in the water than other plants. They are also found in soil about salt springs.

Tropophytes[49]

Tropophytes are plants which can live as mesophytes during the growing season, and then turn to a xerophytic habit in the resting season. Deciduous trees and shrubs, and perennial herbs of our temperate regions, are in this sense tropophytes, while many are at the same time mesophytes if they exist in the portions of the temperate region where rainfall is abundant. In the spring and summer they have broad and comparatively thin leaves, transpiration goes on rapidly, but there is an abundance of moisture in the soil, so that root absorption quickly replaces the loss and the plant does not suffer. In the autumn the trees shed their leaves, and in this condition with the bare twigs they are able to stand the drying effect of the cold and winds of the winter because transpiration is now at a minimum, while root absorption is also at a minimum because of the cold condition of the soil. Perennial herbs like trillium, dentaria, the goldenrods, etc., turn to xerophytic habit by the death of their aerial shoots, while the thick underground shoot which is also protected by its subterranean habit carries the plant through the winter.

While these different vegetation types are generally dominant in certain climatic regions or under certain soil conditions, they are not the exclusive vegetation types of the regions. For example, in desert or semidesert regions the dominant vegetation type is made up of xerophytes. But there is a mesophytic flora even in deserts, which appears during the rainy season where temperature conditions are favorable for growth. This is sometimes spoken of as the rainy-season flora. The plants are annuals and by formation of seed can tide over the dry season. So in the region where mesophytes grow there are xerophytes, examples being the evergreens like the pines, spruces, rhododendrons; or succulent plants like the stonecrop, the purslane, etc. Then among hydrophytes the semiaquatics are really xerophytes. The roots are in water, and absorption is slow because there are no root hairs, or but few, and the aerial parts of the plant are xerophytic.

Plant Formations

The term plant formation is applied to associations of plants of the same kind, though there is a great difference in the use of the word by different writers which leads to some confusion.[50] It is sometimes applied to an association of individuals of a species, or of

---

[49] Term used by Schimper.

[50] See the author's "College Text-book of Botany." Chapter 49.

several species occupying a rather definite area of ground where the soil conditions are not greatly different (individual formation); by others it is applied to the plants of a definite physiographic area, as a swamp, moor, strand, or beach, bank, rock hill, clay hill, ravine, bluff, etc. (principal formation); and in a broad sense it is applied to the plants of climatic regions, of those in bodies of water, etc. (general formations). Space here is too limited to discuss all these kinds of formations, but the nature of the general formations will be pointed out. The general formations may be grouped into four divisions:

- 1st. Climatic formations.
- 2d. Edaphic formations.
- 3d. Aquatic formations.
- 4th. Culture formations.

Climatic Formations

Climatic influences extend over wide regions, so that climate controls the general type of vegetation of a region. In the sense of control there are two climatic factors, temperature and moisture, especially soil moisture. Temperature exerts a controlling influence over the vegetation type only where the total heat during the period of growth and reproduction is very low. This occurs in polar lands and at high elevations where the climate is alpine. In the temperate and tropical regions of the globe moisture, not heat, controls the general vegetation type. These vegetation types in general are coincident with rainfall distribution, and Schimper recognizes here three types, which with the arctic-alpine type would make four climatic formations as follows:

1st. The Woodland Formation

This formation is characterized by trees and shrubs, and it is what is called a *close* formation. By this it is meant that so far as the climate is concerned the conditions are favorable for the development of trees and shrubs in such abundance that they become the dominant vegetation type of the region and grow close together. Other plants, as herbs, grasses, etc., occur, but they grow as subordinate elements of the general vegetation type, and as undergrowth. The land portion of the globe, therefore, outside of arctic and alpine regions, where the annual precipitation is 40 to 60 or more inches, is the area for woodland formation. In some places, the eastern part of England, for example, the annual precipitation is 25 to 30 inches, but the cool temperature permits a forest growth. It is true there are places where forests do not grow,—where man cuts them down, for example. But if cultivated lands in this region were allowed to go to waste, they would in time grow up to forest again. So there are swamps where the soil is too wet for trees, or sandy or rocky areas where there is not a sufficient amount of soil or water to support forest trees. But here it is the soil conditions, not climatic conditions, which prevent the development of the forest. But we know that swamps are being filled in and the ground gradually becoming higher and drier, and that soil is slowly accumulating in rocky areas, so that in time if left to natural forces these places would become forested. So this area of heavy annual rainfall is a *potential* forest area. These areas are determined by warm currents of moisture-laden air from the ocean moving over cooler land areas where the moisture is precipitated. In general these areas are along the coasts of great continents and on mountains. Therefore the interior of a continent is apt to be dry because most of the moisture has been precipitated before it reaches the interior. Deserts or steppes are therefore usually near the interior of continents. Some exceptions to this general rule are found: central South America, which is a region of exceptional rainfall because the moisture-laden winds here come from the warmest part of the ocean; the desert region west of the Andes mountains, where the winds are not favorable; southern California, where the winds come chiefly from a cooler portion of the Pacific ocean and move over an area of high temperature, etc.

Figure 486. Typical prairie scene, a few miles west of Lincoln, Nebraska. (Bot. Dept., Univ. Nebraska.)

2nd. Grassland Formation

Grasses form the dominant vegetation type where the annual rainfall is approximately 15 to 25 inches. In true grasslands the formation is a close one since there is still a sufficient amount of moisture to provide for all the plants which can stand on the ground. Yet there is not enough moisture to permit the growth of forest as the dominant type without aid and protection by man. The so-called prairie regions are examples. Trees and shrubs do occur, but they cannot compete successfully with the grasses because the climatic conditions are favorable for the latter and unfavorable for the former. On the border line between forest and prairie the line of division is not a clear-cut one because conditions grade from one to the other. The two formations are somewhat mixed, like the outposts of contending armies, arms of the forest or prairie extending out here and there. In the United States the prairies extend from Illinois to about the 100th meridian, and beyond this to the foothills of the Rockies and southwest to the Sonora Nevada desert the region is drier, the rainfall varying from 10 to 20 inches. This is the area of the Great Plains, and while grasses of the bunch type are dominant, they make a more or less open formation because the moisture is not sufficient to supply all the plants which could be crowded on the ground, each individual tuft needing an area of ground surrounding it on which it can draw for moisture. Such a formation is an open one, and in this respect is similar to desert formations.

Figure 487. Winter range in northwestern Nevada, showing open formation; white sage (*Eurotia lanata*) in foreground, salt-bush (*Atriplex confertifolia*) and bud-sage (*Artemisia spinescens*) at base of hill, red sage (*Kochia americana*) on the higher slope. (After Griffiths, Bull. 38, Bureau Plant Ind., U. S. Dept. Agr.)

3rd. Desert Formations

These occur where the annual rainfall is still lower, 10 to 4 inches or even less, 2 to 3 inches, while in one place in Chili it is as low as ½ inch. In the great Sahara desert it is about 8 inches, while in the Sonora Nevada desert in the southwestern United States it is 4 to 8 inches. Here the formation is an open one. In the forest and prairie formations the plants compete with each other for occupancy of the ground, since climatic conditions are favorable, so that the struggle against climate is not severe. But in the desert plants do not compete with each other; since the climate is so austere, the struggle is against the climate. Hence plants stand at some distance from each other because the roots need the moisture from the ground for some distance around them. There is not enough moisture for all the plants that begin, and those which get the start take the moisture away from the intervening ones, which then die. Since the struggle is against the adverse conditions of climate and not a competition between plants to occupy the ground, no one floristic type dominates as in the case of the grasses and forests of the grassland and woodland formations, but grassland and woodland types grow together. So we find grasses, trees, and shrubs growing without competition in the desert. The dominant vegetation type is xerophytic.

Figure 488. Northern limit of tree growth, Alaska. (Copyright, 1899, by E. H. Harriman.)

4th. Arctic-Alpine Formation

This formation extends from the limit of tree growth to the region of perpetual ice and snow. The forest here comes in competition with climate, with the severe cold of the long winter night, so that tree growth is limited, and on the border line with the woodland formation the trees are stunted, bent to one side by the heavy snows, or the tops are killed by the cold wind. The arctic zone of plant growth is sometimes spoken of as the "cold waste," since conditions here are somewhat similar to those in the desert, the extreme cold exercising a drying effect on vegetation, and the vegetation type then is largely xerophytic.

Edaphic[51] Formations

Edaphic formations may occur in any of the climatic-formation areas. They are controlled by the condition of soil or ground. The condition of the soil is unfavorable for the growth of the general vegetation type of that region, or is more favorable for another vegetation type, so that soil conditions overcome the climatic conditions. These areas include swamps, moors, the strand or beach, rocky areas, etc., as well as oases in the desert, warm oases in the arctic zone, river bottoms in the prairie and plains region, alkaline areas, etc. The edaphic formations may be close or open according to the nature of the soil. The edaphic formations then are infiltrated in the climatic formations, the different vegetation types fitting together like pieces of mosaic, which can be seen in some places from a mountain top, or if one could take a bird's-eye view of the landscape or from a balloon.

Aquatic Formations

These are made up of water plants and are of two general kinds: fresh-water plant formations in ponds, lakes, streams; and salt-water plant formations in the ocean and inland salt seas.

Culture Formations

Culture formations are largely controlled by man, who destroys the climatic or edaphic formation and by cultivation protects cultivated types, or by allowing land to go to "waste" permits the growth of weeds, though weeds are often abundant in the culture areas. In general the culture formations may be grouped into two subdivisions: 1st, the vegetation of cultivated places; and 2d, the vegetation of waste places, as abandoned fields, roadsides, etc.

Plant Societies

Plant societies are somewhat definite associations of the vegetation of an area marked by physiographic conditions. A single plant society is nearly if not altogether identical with a "*principal formation*," but is a more popular expression, and besides includes all the plants growing on the area, while in the use of the term "principal formation" we have reference mainly to the dominant plants and the most conspicuous subordinate species.

---

[51] ἔδαφος = ground.

Complex Character of Plant Societies

In their broadest analysis all plant societies are complex. Every plant society has one or several dominant species, the individuals of which, because of their number and size, give it its peculiar character. The society may be so nearly pure that it appears to consist of the individuals of a single species. But even in those cases there are small and conspicuous plants of other species which occupy spaces between the dominant ones. Usually there are several or more kinds in the same society. The larger individuals come into competition for first place in regard to ground and light, the smaller ones come into competition for the intervening spaces for shade, and so on down in the scale of size and shade tolerance. Then climbing plants (lianas) and epiphytes (lichens, algæ, mosses, ferns, tree orchids, etc.) gain access to light and support by growing on other larger and stouter members of the society.

Parasites (dodder, mistletoes, rusts, smuts, mildews, bacteria, etc.) are present, either actually or potentially, in all societies, and in their methods of obtaining food sap the life and health of their hosts. Then come the scavenger members, whose work it is to clean house, as it were, the great army of saprophytic fungi (molds, mushrooms, etc.), and bacteria ready to lay hold on dead and dying leaves, branches, trunks, roots, etc., disintegrate them, and reduce them to humus, where other fungi change them into a form in which the larger members of the plant society can utilize them as plant food and thus continue the cycle of matter through life, death, decay, and into life again. Mycorhiza (see Chapter 9) or other forms of mutualistic symbiosis occur which make atmospheric nitrogen available for food, or shorten the path from humus to available food, or the humus plants feed on the humus directly. Nor should we leave out of account the myriads of nitrate and nitrite bacteria (see Chapter 9) which make certain substances in the soil available to the higher members of the society. Most plant societies are also benefited or profoundly influenced in other ways by animals, as the flower-visiting insects, birds which feed on injurious insects, the worms which mellow up the soil and cover dead organic matter so that it may more thoroughly decay. In short, every plant society is a great cosmos like the universe itself of which it is a part, where multitudinous forms, processes, influences, evolutions, degenerations, and regenerations are at work.

Forest Societies[52]

Each different climatic belt or region has its characteristic forest. For example, the forests of the Hudsonian zone in North America are different from those of the Canadian zone, and these in turn different from those in the transition zone (mainly in northern United States). The forests of the Rocky mountains and of the Pacific coast differ from those of

---

[52] For a full discussion of forest societies see Chapter L in the author's "College Text-book of Botany."

the Alleghanian, Carolinian (mainly middle United States) or Austroriparian (southern United States) areas. Finally, tropical forests are strikingly different from those of other regions. Similar variations occur in the forests of other regions of the globe. The character of these forests depends largely on climatic factors. The character of the forest varies, however, even in the same climatic area, dependent on soil conditions, or success in seeding and ground-gaining of the different species in competition, etc.

General Structure of the Forest

Structurally the forest possesses three subdivisions: the floor, the canopy, and the interior. The floor is the surface soil, which holds the rootage of the trees, with its covering of leaf-mold and carpet of leaves, mosses, or other low, more or less compact vegetation. The canopy is formed by the spreading foliage of the tree crowns, which, in a forest of an even and regular stand, meet and form a continuous mass of foliage through which some light filters down into the interior.

Figure 489. Mature forest of redwood (Sequoia sempervirens). (Bureau of Forestry, U. S. Dept. Agr., Bull. 38.)

Where the stand is irregular, i.e., the trees of different heights, the canopy is said to be "compound" or "storied." Where it is uneven, there are open places in the canopy which admit more light, in which case the undergrowth may be different. The interior of the forest lies between the canopy and the floor. It provides for aeration of the floor and interior occupants, and also room for the boles or tree trunks (called by foresters the wood mass of the forest) which support the canopy and provide the channels for communication and food exchange between the floor and canopy. The canopy manufactures the carbohydrate food and assimilates the mineral and proteid substances absorbed by the roots in the soil; and also gets rid of the surplus water needed for conveying food materials from the floor to the place where they are elaborated. It is the seat where energy is created for work, and also the place for seed production.

Longevity of the Forest

The forest is capable of self-perpetuation, and, except in case of unusual disaster or the action of man, it should live indefinitely. As the old trees die they are gradually replaced by younger ones. So while trees may come and trees may go, the forest goes on forever.

Autumn Colors

One of the striking effects produced by the deciduous forests is that of the autumn coloring of the leaves. It is more pronounced in the forests of the United States than in corresponding life zones in the eastern hemisphere because of the greater number of species. With the disintegration of the chlorophyll bodies, other colors, which in some cases were masked by the green, appear. In other cases decomposition products result in the formation of other colors, as red, scarlet, yellow, brown, purple, maroon, etc., in different species. These coloring substances to some extent are believed to protect the nitrogenous substances in the leaf from injury. The colors absorb the sun's rays, which otherwise might destroy these nitrogenous substances before they have passed back through the petiole of the leaf into the stem, where they may be stored for food. The gorgeous display of color, then, which the leaves of many trees and shrubs put on is one of the many useful adaptations of the plants.

Importance of the Forest in the Disposal of Rainfall

The importance of the forest in disposing of the rainfall is very great. The great accumulation of humus on the forest floor holds back the water both by absorption and by checking its flow, so that it does not immediately flow quickly off the slopes into the drainage system of the valley. It percolates into the soil. Much of it is held in the humus and soil. What is not retained thus filters slowly through the soil and is doled out more gradually into the valley streams and mountain tributaries, so that the flood period is extended, and its injury lessened or entirely prevented, because the body of water moving at any one time is not dangerously high. The winter snow is shaded and in the spring melts slowly, and the spring freshets are thus lessened. The action of thc leaves and humus in retarding the flow of the water prevents the washing away of the soil; the roots of trees bind the soil also and assist in holding it.

Absence of Forest Encourages Serious Floods

The great floods of the Mississippi and its tributaries are due to the rapidity with which heavy rainfall flows from the rolling prairies of the west, and from the deforested areas west of the Alleghany system. The serious floods in recent years in some of the South Atlantic States are in part due to the increasing area of deforestation in the Blue Ridge and southern Alleghany system.

The Prairie and Plains Societies

These are to be found in the grassland formation. In the prairies "meadows" are formed in the lower ground near river courses where there is greater moisture in soil. The grasses here are principally "sod-formers" which have creeping underground stems which mat together, forming a dense sod. On the higher and drier ground the "bunch" grasses, like buffalo-grass, beard-grass, or broom-sedge, etc., are dominant, and in the drier regions as one approaches desert conditions the vegetation gradually takes on more the character of the desert, so that in the plains sage-brush, the prickly-pear cactus, etc., occur. Besides the dominant vegetation of the society there are subordinate species, and the societies are especially marked by a spring and autumn flora of conspicuous flowering plants which are mixed with the grasses.

Desert Societies

These are composed of plants which possess a form or structure which enables them to exist in a very dry climate where the air is very dry and the soil contains but little moisture. The true desert plants are perennial.

Figure 490. Desert vegetation, Arizona, showing large succulent trunks of cactus with shrubs and stunted trees. Open formation. (Photograph by Tuomey.)

Figure 491. Polar tundra with scattered flowers, Alaska. (Copyright by E. H. Harriman.)

Figure 492. Perennial rosette plant from alpine flora of the Andes, showing short stem, rosette of leaves, and large flower. (After Schimper.)

The growth and flowering period occurs during the rainy season, or those portions of the rainy season when the temperature is favorable, and they rest during the very dry season and cold. Characteristic desert plants are the cacti with thick succulent green stems or massive trunks, the leaves being absent or reduced to mere spines which no longer function in photosynthesis; yuccas with thick, narrow and long leaves with a firm and thick cuticle; small shrubs or herbs with compact rounded habit and small thick gray leaves. All of these structures conserve moisture. The mesquite tree is one of the common trees in portions of the Sonora Nevada desert. Besides the true desert plants, desert societies have a rainy-season flora consisting of annuals, which can germinate, vegetate, flower, and seed during the period of rain and before the ground moisture has largely disappeared, and these pass the resting period in seed.

Arctic-Alpine Societies

The most striking of the arctic plant societies are the "polar tundra," extensive mats of vegetation largely made up of mosses, lichens, etc., only partially decayed because of the great cold of the subsoil, and perhaps also because of humus acid in the partially decayed vegetation. These tundras are brightened by numerous flowering plants which are characterized by short stems, a rosette of leaves near the ground, and by large bright-colored flowers. Heaths, saxifrages, and dwarf willow abound. Alpine plant societies are similar to the arctic, although some of the conditions are more severe than in the arctic region. This is principally due to the fact that during the summer while the plants are growing they are subject to a high temperature during the day and a very low temperature at night, whereas during the summer in arctic regions while the plants are growing there is continuous warmth for growth and continuous light for photosynthesis. Five types of alpine plants are recognized by some. 1st. *Elfin tree.* This type has short, gnarled, often horizontal stems, as seen in pines, birches, and other trees growing in alpine heights. 2d. *The alpine shrubs.* In the highest alpine belts they are dwarfed and creeping, richly branched and spreading close to the ground, while at lower belts they are more like lowland shrubs. 3d. *The cushion type.* The branching is very profuse and the branches are short and touch each other on all sides, forming compact masses (examples saxifrages, androsace, mosses, etc.). 4th. *Rosette plants.* These are perennial, short stems and very strong roots, and play an important part in the alpine meadows. 5th. *Alpine grasses.* These usually have much shorter leaves than grasses of the lowlands and consequently form a low sward.

Edaphic Plant Societies

These are equivalent to edaphic plant formations, and the vegetation is of course controlled by the peculiar conditions of the soil. There are a number of different kinds of edaphic plant societies determined by the character of the physiographic areas. 1st. *Sphagnum moors.* These are formed in shallow basins originally with more or less water. The growth of the sphagnum moss along with other vegetation and its partial decay in the water builds up ground rapidly so that in course of time the pond may be completely filled in. This filling in proceeds from the shore toward the center, and in the early stages of course there would be a pond in the center. The partial decay of vegetation creates an excess of humus acid which retards absorption by the roots. The conditions are such, then, as require aerial structures for retarding the loss of water, and plants growing in such moors are usually xerophytes. Some of the plants are identical with those growing in the arctic tundra. 2d. *Sand*[53] *strand of beach.* The quantity of sand with very little or no admixture of humus or plant food makes it difficult for plants to obtain a sufficient amount of water even where rainfall is abundant. The same may be said of the sand dunes farther back from the shore. The plants of these areas are then usually xerophytes. Some of the plants accustomed to growing in such localities are American sea-rocket, seaside spurge, bugseed, sea-blite, sea-purslane, the sandcherry, dwarf willow, marram-grass, certain species of beard-grass, etc. 3d. *Rocky shores or areas.* Here lichens and mosses first grow, later to be followed by herbs, grasses, shrubs, and trees, as decayed plant remains accumulate in the rock crevices. 4th. *Shores of ponds, or swamp moors.* Here the vegetation often takes on a zonal arrangement if the ground gradually slopes to the shore and out into the pond. In Figure

[53] See Chapter LIV of the author's "College Text-book of Botany."

493 is shown zonal distribution of plants. The different kinds of plants are drawn into these zones by the varying amount of ground water in the soil, or the varying depth of the water on the margin of the pond as one proceeds from the land towards the deeper water. On the border lines or tension lines between the different zones the plants are struggling to occupy here ground which is suitable for each adjacent individual formation. Other edaphic societies are those of marl ponds, alkaline areas, oases in deserts, warm oases in arctic lands, the forested areas along river bottoms in prairie or plains regions, etc.

Aquatic Plant Societies

In general we might distinguish three kinds, 1st. *Fresh-water plant societies*, with floating algæ like spirogyra, œdogonium, etc., the floating duck-meats, riccias; the plants of the lily type with roots and stems attached to the bottom and leaves floating on the surface, like the water-lily and certain pondweeds, and finally the completely submerged ones like certain pondweeds, the bassweed (Chara), etc. 2d. *Marine plant societies*, which are made up mostly of the red and brown algæ or "seaweeds," though some green algæ and flowering plants also occur. 3d. *The salt marshes* where the water is brackish and there is usually a luxuriant growth of marsh-grasses.

Figure 493. Macrophytes in the upper zone of the photic region. Ascophyllum and Fucus at low tide, Hunter's Island, New York City. (Photograph by M. A. Howe.)

Figure 494. Zonal distribution of plants, South Shore, Cayuga Lake.

Chapter 47

Classification of the Angiosperms

Relation of Species, Genus, Family, Order, etc.

Species

It is not necessary for one to be a botanist in order to recognize, during a stroll in the woods where the trillium is flowering, that there are many individual plants very like each other. They may vary in size, and the parts may differ a little in form. When the flowers first open they are usually white, and in age they generally become pinkish. In some individuals they are pinkish when they first open. Even with these variations, which are trifling in comparison with the points of close agreement, we recognize the individuals to be of the *same kind*, just as we recognize the corn plants, grown from the seed of an ear of corn, as of the same kind. Individuals of the same kind, in this sense, form a *species*. The white wake-robin, then, is a species.

But there are other trilliums which differ greatly from this one. The purple trillium (T. erectum) shown in Figure 495 is very different from it. So are a number of others. But the purple trillium is a species. It is made up of individuals variable, yet very like one another, more so than any one of them is like the white wake-robin.

Genus

Yet if we study all parts of the plant, the perennial rootstock, the annual shoot, and the parts of the flower, we find a great resemblance. In this respect we find that there are several species which possess the same general characters. In other words, there is a relationship between these different species, a relationship which includes more than the individuals of one kind. It includes several kinds. Obviously, then, this is a relationship with broader limits, and of a higher grade, than that of the individuals of a species. The grade next higher than species we call *genus*. Trillium, then, is a genus. Briefly the characters of the genus trillium are as follows:

Figure 495. Trillium erectum (purple form), two plants from one rootstock.

Genus Trillium

Perianth of six parts: sepals 3, herbaceous, persistent; petals colored. Stamens 6 (in two whorls), anthers opening inward. Ovary 3-loculed, 3-6-angled; stigmas 3, slender, spreading. Herbs with a stout perennial rootstock, with fleshy, scale-like leaves, from which the low annual shoot arises, bearing a terminal flower and 3 large netted-veined leaves in a whorl.

Note.—In speaking of the genus the present usage is to say trillium, but two words are usually employed in speaking of the species, as *Trillium grandiflorum*, *T. erectum*, etc.

Genus Erythronium

The yellow adder-tongue, or dogtooth violet (*Erythronium americanum*), shown in Figure 496, is quite different from any species of trillium. It differs more from any of the species of trillium than they do from each other. The perianth is of six parts, light yellow, often spotted near the base. Stamens are 6. The ovary is obovate, tapering at the base, 3-valved, seeds rather numerous, and the style is elongated. The flower stem, or scape, arises from a scaly bulb deep in the soil, and is sheathed by two elliptical-lanceolate, mottled leaves. The smaller plants have no flower and but one leaf, while the bulb is nearer the surface. Each year new bulbs are formed at the end of runners from a parent bulb. These runners penetrate each year deeper into the soil. The deeper bulbs bear the flower stems.

Genus Lilium

While the lily differs from either the trillium or erythronium, yet we recognize a relationship when we compare the perianth of six colored parts, the 6 stamens, and the 3-sided and long 3-loculed ovary.

Figure 496. Adder-tongue (erythronium). At left below pistil, and three stamens opposite three parts of the perianth. Bulb at the right.

Family Liliaceæ

The relationship between genera, as between trillium, erythronium, and lilium, brings us to a still higher order of relationship, where the limits are broader than in the genus. Genera which are thus related make up the *family*. In the case of these genera the family has been named after the lily, and is the lily family, or *Liliaceæ*.

Order, Class, Group

In like manner the lily family, the iris family, the amaryllis family, and others which show characters of close relationship are united into an order which has broader limits than the family. This order is the lily order, or order *Liliales*. The various orders unite to make up the *class*, and the classes unite to form a *group*.

Variations in Usage of the Terms Class, Order, Etc

Thus, according to the system of classification adopted by some, the angiosperms form a *group*. The group angiosperms is then divided into two *classes*, the *monocotyledones* and *dicotyledones*. (It should be remembered that all systematists do not agree in assigning the same grade and limits to the classes, subclasses, etc. For example, some treat of the angiosperms as a class, and the monocotyledons and dicotyledons as subclasses; while others would divide the monocotyledons and dicotyledons into classes, instead of treating each one as a class or as a subclass. Systematists differ also in usage as to the termination of the ordinal name; for example, some use the word *Liliales* for *Liliifloræ*, in writing of the order.)

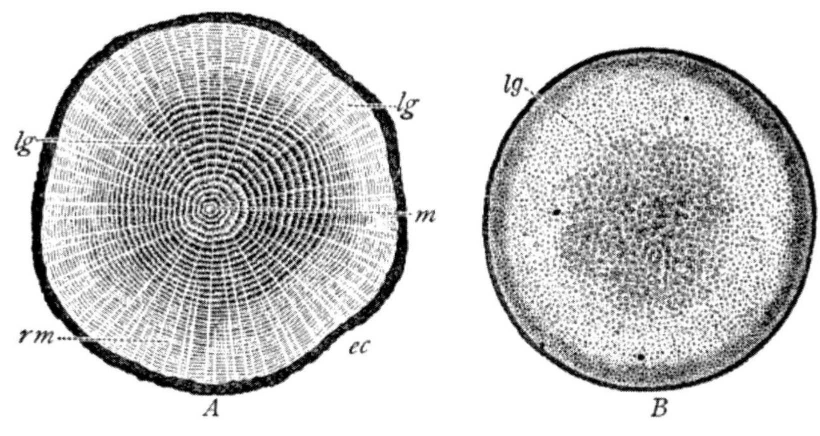

Figure 497. A. Cross-section of the stem of an oak tree thirty-seven years old, showing the annual rings. rm, the medullary rays; m, the pith (medulla). B. Cross-section of the stem of a palm tree, showing the scattered bundles.

Monocotyledones

In the monocotyledons there is a single cotyledon on the embryo; the leaves are parallel-veined; the parts of the flower are usually in threes; endosperm is usually present in the seed; the vascular bundles are usually closed, and are scattered irregularly through the stem as shown by a cross-section of the stem of a palm (Figure 497), or by the arrangement of the bundles in the corn stem (Figure 57). Thus a single character is not sufficient to show relationship in the class (nor is it in orders, nor in many of the lower grades), but one must use the sum of several important characters.

Dicotyledones

In the dicotyledons there are two cotyledons on the embryo; the venation of the leaves is reticulate; the endosperm is usually absent in the seed; the parts of the flower are frequently in fives; the vascular bundles of the stem are generally open and arranged in rings around the stem, as shown in the cross-section of the oak (Figure 497). There are exceptions to all the above characters, and the sum of the characters must be considered, just as in the case of the monocotyledons.

Taxonomy

This grouping of plants into species, genera, families, etc., according to characters and relationships is *classification*, or *taxonomy*.

To take Trillium grandiflorum for example, its position in the system, if all the principal subdivisions should be included in the outline, would be indicated as follows:

- Group, Angiosperms.
- Class, Monocotyledones.
- Order, Liliales.
- Family, Liliaceæ.
- Genus, Trillium.
- Species, grandiflorum.

In the same way the position of the toothwort would be indicated as follows:

- Group, Angiosperms.
- Class, Dicotyledones.
- Order, Papaverales.
- Family, Cruciferæ.
- Genus, Dentaria.
- Species, diphylla.

But in giving the technical name of the plant only two of these names are used, the genus and species, so that for the toothwort we say *Dentaria diphylla*, and for the white wake-robin we say *Trillium grandiflorum*.

Kingdom and Subkingdom

Organic beings form altogether two kingdoms, the Animal Kingdom and the Plant Kingdom. The Plant Kingdom is then divided into a number of subkingdoms as follows: 1st, Subkingdom Thallophyta, the thallus plants, including the Algæ and Fungi; 2d, Subkingdom Bryophyta, the moss-like plants, including the Liverworts and Mosses; 3d, Subkingdom Pteridophyta, the fern-like plants, including Ferns, Lycopods, Equisetum, Isoetes, etc.; 4th, Subkingdom Spermatophyta, the seed plants, including Gymnosperms and Angiosperms. Subkingdoms are divided into groups of lower order down to the classes. So there are subclasses, subfamilies or tribes, subgenera, and even subspecies. But taking the principal taxonomic divisions from the greater to the lesser rank, the order would be as follows:

- Plant Kingdom.
- Subkingdom, Spermatophyta.
- Group (not used in a definite sense).
- Class, Gymnospermæ.
- Order, Pinales.
- Family, Pinaceæ.
- Genus, Pinus.
- Species, strobus, or, in full,
- Pinus strobus, the white pine.

GROUP ANGIOSPERMÆ

Class Monocotyledones

- *Order Pandanales.*—Aquatic or marsh plants. The cattail flags (Typha) and the bur-reeds (Sparganium), each representing a family. The name of the order is taken from the tropical genus Pandanus (the screw-pine often grown in greenhouses).
- *Order Naiadales.*—Aquatic or marsh herbs. Three families are mentioned here.
  - The pondweed family (Naiadaceæ), named after one genus, Naias. The largest genus is Potamogeton, the species of which are known as pondweeds. Ruppia

occidentalis occurs in saline ponds in Nebraska, and R. maritima along the seacoast and in saline districts in the interior.
  - The water-plantain family (Alismaceæ) includes the water-plantain (Alisma) and the arrow-leaves (Sagittaria).
  - The tape-grass family (Vallisneriaceæ) includes the tape-grass, or eel-grass (the curious *Vallisneria spiralis*).
- *Order Graminales.*—Two families.
  - The grass family (Gramineæ), the grasses and grains.
  - The sedge family (Cyperaceæ), the sedges.
- *Order Palmales*, with one family, Palmaceæ, includes the palms, abundant in the tropics and extending into Florida. Cultivated in greenhouses.
- *Order Arales.*
  - The arum family (Araceæ). Flowers in a fleshy spadix. Examples: Indian turnip (Arisæma), sweet-flag (Acorus), skunk-cabbage (Spathyema).
  - The duckweed family (Lemnaceæ). (Examples: Lemna, Spirodela, Wolffia. See paragraphs 51-53.)
- *Order Xyridales*, from the genus Xyris, the yellow-eyed grass family (Xyridaceæ). Species mostly tropical, but a few in North America. Other examples are the pipewort family (Eriocaulaceæ, example, Eriocaulon septangulare), the pineapple family (Bromeliaceæ, example, the pineapple cultivated in Florida); the Florida moss or hanging moss (Tillandsia usneoides); the spiderwort family (Commelinaceæ), including the spiderwort (Tradescantia, several species in North America); the pickerel-weed family (Pontederiaceæ), including the genus Pontederia in borders of ponds and streams.
- *Order Liliales.*—Some of the families are as follows:
  - The rush family (Juncaceæ, example, Juncus), with many species, plants of usually swamp habit.
  - The lily family (Liliaceæ, examples: Lilium, Allium = Onion, Erythronium, Yucca).
  - The iris family (Iridaceæ, examples: Iris, the blue-flag, fleur-de-lis, etc.).
  - The lily-of-the-valley family (Convallariaceæ, examples: lily-of-the-valley, Trillium, etc.)
  - The amaryllis family (Amaryllidaceæ, examples: Narcissus, the daffodil; Cooperia, in southwestern United States).
- *Order Scitaminales.*—This order includes the large showy cultivated Canna of the canna family.
- *Order Orchidales.* Example, the orchid family (Orchidaceæ) with Cypripedium, Orchis, etc.

Class Dicotyledones

Series 1. CHORIPETALÆ. Petals wanting (Apetalæ, or Archichlamydæ of some authors), or present and distinct from one another (Polypetalæ, or Metachlamydæ).

- *Order Casuarinales*, confined to tropical seacoasts (example, Casuarina).
- *Order Piperales* includes the lizard's-tail family (Saururaceæ), Saururus cernuus, lizard's-tail, in the eastern United States.
- *Order Salicales.*—Shrubs or trees, flowers in aments. Includes the willows and poplars (Salix and Populus of the willow family, Salicaceæ).
- *Order Myricales.*—Shrubs or small trees. Includes the sweet-gale (Myrica gale) in wet places in northern United States and British North America, Myrica cerifera forming thickets on sand dunes along the Atlantic coast, and the sweet-fern (Comptonia peregrina = C. asplenifolia) in the eastern United States in dry soil of hillsides.
- *Order Leitneriales.*—Shrubs or trees. Includes the cork-wood, Leitneria floridana (Leitneriaceæ).
- *Order Juglandales.*—Trees, staminate flowers in aments. The walnut family (Juglandaceæ, examples: walnut, butternut, etc. Juglans; hickory, Hicoria = Carya).
- *Order Fagales.*—Trees and shrubs. Flowers in aments, or the pistillate ones with an involucre which forms a cup in fruit, as in the acorn of the oak.
  - The birch family (Betulaceæ, examples: Betula, birch; Corylus, hazelnut; Alnus, alder, etc.).
  - The beech family (Fagaceæ = Cupuliferæ, examples: Fagus, beech; Castanea, chestnut; Quercus, oak).
- *Order Urticales.*—Trees, shrubs, or herbs. Examples: the elm family (Ulmaceæ), the mulberry family (Moraceæ), and the nettle family (Urticaceæ).
- *Order Santalales*, herbs or shrubs, mostly parasitic.
  - The mistletoe family (Loranthaceæ), with the American mistletoe (Phoradendron flavescens), parasitic on deciduous trees in the South Atlantic, Central, and Gulf States (N. J. to Ind. Ter.).
  - The sandalwood family (Santalaceæ, example, the bastard toad-flax, Comandra umbellata), widely distributed in North America.
- *Order Aristolochiales.*—Herbs or vines with heart-shaped or kidney-shaped leaves. The birthwort family (Aristolochiaceæ, example, Aristolochia serpentaria, the Virginia snake-root, eastern United States; wild ginger, or heart-leaf, Asarum canadense, eastern North America.)

- *Order Polygonales.*—Examples: the buckwheat family (Polygonaceæ), including buckwheat (Fagopyrum), and numerous species of Polygonum, known as smartweed, water-pepper, tear-thumb, bindweed, knotweed, prince's-feather, etc.
- *Order Chenopodiales.*—Herbs. There are several families; one of the largest is the goosefoot family (Chenopodiaceæ). The genus Chenopodium includes many species, known as goosefoot, lamb's-quarters, etc. Here belong also the Russian thistle (Salsola tragus) and the saltwort (S. kali). The former is sometimes a troublesome weed in the central and western United States, naturalized from Europe. The latter occurs along the Atlantic coast on seabeaches. Atriplex occurs in salty or alkaline soil, also the glasswort (Salicornia herbacea), the bugseed (Corispermum). The pokeweed family (Phytolaccaceæ), the Amaranth family (Amaranthaceæ), the purslane family (Portulacaceæ, including the purslane or "pursley," Portulaca oleracea, and the spring-beauty, Claytonia virginica), and the pink family (Caryophyllaceæ), belong here.
- *Order Ranales.*—Herbs, shrubs, or trees. Examples are:
  - The water-lily family (Nymphæaceæ), with the yellow water-lily (Nymphæa advena = Nuphar advena) and the white water-lily (Castalia odorata = Nymphæa odorata).
  - The magnolia family (Magnoliaceæ), including the magnolias (Magnolia) and the tulip-tree (Liriodendron). The crowfoot family (Ranunculaceæ), with the buttercups, hepatica, clematis, etc.
- *Order Papaverales.*—Mostly herbs. Examples are:
  - The poppy family (Papaveraceæ), including the opium or garden poppy (Papaver somniferum), the blood-root (Sanguinaria canadensis), the Dutchman's-breeches (*Bicuculla cucullaria* = *Dicentra cucullaria*), squirrel's-corn (Bicuculla canadensis = D. canadensis).
  - The mustard family (Cruciferæ), including the toothwort (Dentaria), shepherd's-purse (*Bursa bursa-pastoris* = *Capsella bursa-pastoris*), the cabbage, turnip, etc.
- *Order Sarraceniales.*—Insectivorous plants.
  - The pitcher-plant family (Sarraceniaceæ). Examples: *Sarracenia purpurea*, the pitcher-plant, in peat-bogs, northern and eastern North America.
  - The sundew family (Droseraceæ). Examples: *Drosera rotundifolia*, and other sundews.
- *Order Rosales.*—Herbs, shrubs or trees. Seventeen families are given in the eastern United States. Examples:
  - The riverweed family (Podostemaceæ), containing the riverweed (Podostemon).
  - The saxifrage family (Saxifragaceæ), containing a number of species. Example, *Saxifraga virginiensis*.

- The gooseberry family (Grossulariaceæ), including the wild and the cultivated gooseberry.
- The witch-hazel family (Hamamelidaceæ), including the witch-hazel (Hamamelis), in eastern North America, and the sweet gum (*Liquidambar styraciflua*).
- The plane-tree family (Platanaceæ), with the plane-tree, or buttonwood (*Platanus occidentalis*), eastern North America. (Other species occur in western United States.)
- The rose family (Rosaceæ), including roses, spiræas, raspberries, strawberries, the shrubby cinquefoil (*Dasiphora fruticosa*), etc.
- The apple family (Pomaceæ), including the apple, mountain-ash, pear, June-berry (or shadbush, also service-berry), the hawthorns (Cratægus).
- The plum family (Drupaceæ), including the cherries, plums, peaches, etc.
- The pea family (Papilionaceæ), including the pea, bean, clover, vetch, lupine, etc., a very large family.

- *Order Geraniales.*—Herbs, shrubs, or trees. Nine families in the eastern United States. Examples:
  - The geranium family (Geraniaceæ), with the cranesbill (Geranium maculatum) and others.
  - The wood-sorrel family (Oxalidaceæ), with the wood-sorrel (Oxalis acetosella) and others.
  - The flax family (Linaceæ). Example, flax (Linum vulgaris).
  - The spurge family (Euphorbiaceæ). Plants with a milky juice, and curious, degenerate flowers. Examples: the castor-oil plant (Ricinus), the spurges (many species of Euphorbia).
- *Order Sapindales.*—Mostly trees or shrubs. Twelve families in the eastern United States. Example:
  - The sumac family (Anacardiaceæ), containing the sumacs in the genus Rhus. Examples: the poison-ivy (*R. radicans*), a climbing vine, in thickets and along fences, in eastern United States. Sometimes trained over porches. The poison-oak (*R. toxicodendron*), a low shrub. Poison-sumac or poison-alder (*R. vernix* = *R. venenata*), sometimes called "thunderwood," or dogwood, is a large shrub or small tree, very poisonous. The smoke-tree (Cotinus cotinoides) belongs to the same family, and is often planted as an ornamental tree. The maple family (Aceraceæ), including the maples (Acer).
  - The buckeye family (Hippocastanaceæ), including the horse-chestnut (Æsculus hippocastanum), much planted as a shade tree along streets. Also there are several species of buckeye in the same genus.

  - The jewelweed family (Balsaminaceæ), including the touch-me-not (Impatiens biflora and aurea) in moist places. The garden balsam (*Imp. balsamea*) also belongs here.
- *Order Rhamnales.*—Shrubs, vines, or small trees. There are two families, the buckthorn (Rhamnaceæ), the grape family (Vitaceæ), including the grapes (Vitis), the American ivy (Parthenocissus quinquefolia = Ampelopsis quinquefolia), in woods and thickets, eastern North America, and much planted as a trailer over porches. The Japanese ivy (P. tricuspidata = A. veitchii) used as a trailer on the sides of buildings belongs here.
- *Order Malvales.*—Herbs, shrubs, or trees.
  - The linden family (Tiliaceæ). Example, the basswood or American linden (*Tilia americana.*)
  - The mallow family (Malvaceæ), including the hollyhock, the mallows, rose of Sharon (Hibiscus), etc.
- *Order Parietales,* with seven families in the eastern United States. The St. John's wort (Hypericum) and the violets each represent a family. The violets (Violaceæ) are well-known flowers.
- *Order Opuntiales.*—These include the cacti (Cactaceæ), chiefly growing in the dry or desert regions of America.
- *Order Thymeleales,* with two families and few species.
- *Order Myrtales.*—Land, marsh, or aquatic plants. The most conspicuous are in the evening primrose family (Onagraceæ), including the fireweeds, or willow herbs (Epilobium), and the evening primrose (Onagra biennis = Œnothera biennis).
- *Order Umbellales.*—Herbs, shrubs, or trees, flowers in umbels.
  - The ginseng family (Araliaceæ). This includes the spikenards and sarsaparillas in the genus Aralia, and the ginseng (or "sang"), Panax quinquefolium.
  - The carrot family (Umbelliferæ). This family includes the wild carrot (*Daucus carota*), the poison-hemlock (Cicuta), the cultivated carrot and parsnip, and a large number of other genera and species.
  - The dogwood family (Cornaceæ). The flowering dogwood (*Cornus florida*), abundant in eastern North America, is an example.

Series 2. GAMOPETALÆ (= Sympetalæ or Metachlamydæ). Petals partly or wholly united, rarely separate or wanting.

- *Order Ericales.*—There are six families in eastern United States. Examples:
  - The wintergreen family (Pyrolaceæ), including the shin-leaf (*Pyrola elliptica*).
  - The Indian-pipe family (Monotropaceæ), with the Indian-pipe (*Monotropa uniflora*) and other humus saprophytes.

- The heath family (Ericaceæ). Examples: Labrador tea (Ledum), in bogs and swamps in northern North America. The azaleas, with several species widely distributed, are beautiful flowering shrubs, and many varieties are cultivated. The rhododendrons are larger with larger flower clusters, also beautiful flowering shrubs. R. maximum in the Alleghany Mountains and vicinity, from Nova Scotia to Ohio and Georgia. R. catawbiense, usually at somewhat higher elevations, Virginia to Georgia. The mountain laurel (Kalmia latifolia) and other species rival the rhododendrons and azaleas in beauty. The trailing arbutus (*Epigæa repens*) in sandy or rocky woods is a well-known small trailing shrub in eastern North America. The sourwood (*Oxydendrum arboreum*) is a tree with white racemes of flowers in August, and scarlet leaves in autumn. The spring or creeping wintergreen (*Gaultheria procumbens*) is a small shrub with aromatic leaves, and bright red spicy berries.
- The huckleberry family (Vaccinaceæ) includes the huckleberries (example, *Gaylussacia resinosa*, the black or high-bush huckleberry, eastern United States), the mountain cranberry (*Vitis-Idæa vitisidæa* = *Vaccinium vitisidæa*) in the northern hemisphere; the bilberries and blueberries (of genus Vaccinium); the cranberries (examples: the large American cranberry, *Oxycoccus macrocarpus* and the European cranberry, *Oxycoccus oxycoccus*, in cold bogs of northern North America, the latter also in Europe and Asia).

- *Order Primulales.*—Two families here. The primrose family (Primulaceæ) contains the loosestrifes (Steironema), star-flower (Trientalis), etc.
- *Order Ebenales.*—Of the four families, the ebony family (Ebenaceæ) contains the well-known persimmon (Diospyros virginiana) and the storax family (Styracaceæ) with the silverbell, or snowdrop tree (Mohrodendron carolinum).
- *Order Gentianales.*—Herbs, shrubs, vines, or trees. Six families in the United States.
  - The olive family (Oleaceæ) includes the common lilac (Syringa), the ash trees (Fraxinus), the privet (Ligustrum).
  - The gentian family (Gentianaceæ) among other genera includes the gentians (Gentiana).
  - The milkweed family (Asclepiadaceæ) contains plants mostly with a milky juice. Asclepias with many species is one of the most prominent genera.
- *Order Polemoniales.*—Mostly herbs, rarely shrubs and trees. Fifteen families in the eastern United States.
  - The morning glory family (Convolvulaceæ) includes the bindweeds (Convolvulus), the morning glory (Ipomæa), etc.
  - The dodder family (Cuscutaceæ) includes the dodders, or "love-vines." There are nearly thirty species in the United States. The stems are slender and twine around other plants upon which they are parasitic.

- The phlox family (Polemoniaceæ). The most prominent genus is Phlox. Over forty species occur in North America.
- The borage family (Boraginaceæ) includes the heliotrope (Heliotropium), the hound's-tongue (Cynoglossum), the forget-me-not (Myosotis), and others.
- The vervain family (verbenaceæ) contains the verbenas.
- The mint family (Labiatæ) contains the mints (Mentha), skull-cap (Scutellaria), dead-nettles (Lamium).
- The potato family (Solanaceæ) includes the ground-cherry (Physalis), the nightshades (Solanum), the tomato (Lycopersicon), tobacco (Nicotiana).
- The figwort family (Scrophulariaceæ) includes the common mullein (Verbascum), the monkey-flower (Mimulus), the toad-flax (Linaria), turtle's-head (Chelone), and many other genera and species.
- The bladderwort family (Lentibulariaceæ) includes the curious bog or aquatic plants with finely dissected leaves, and with bladders in which insects are caught (Utricularia).
- The trumpet-creeper family (Bignoniaceæ) includes the trumpet-creeper (Bignonia), the catalpa tree, and others.

- *Order Plantaginales* with one family (Plantaginaceæ) includes the plantains (Plantago).
- *Order Rubiales* with three families is represented by the madder family (Rubiaceæ) with the bluets (Houstonia), the button-bush (Cephalanthus), the partridge-berry (Mitchella), the bedstraws (Galium), etc.
  - The honeysuckle family (Caprifoliaceæ) with the elder (Sambucus), the arrowwoods and cranberry trees (Viburnum), the honeysuckles (Lonicera), etc.
- *Order Valerianales* with two families includes the teasel family (Dipsacaceæ). Example, Fuller's teasel (Dipsacus).
- *Order Campanulales* with five families, the corolla usually gamopetalous.
  - The gourd family (Cucurbitaceæ) includes the pumpkin, squash, melon, and a few feral species. Example, the star-cucumber (*Sicyos angulatus*), in moist places in eastern and middle United States.
  - The bell-flower family (Campanulaceæ) includes the hare-bells or bell-flowers (Campanula), the lobelias (example, Lobelia cardinalis, the cardinal-flower), etc.
  - The chicory family (Cichoriaceæ) includes the chicory or succory (*Cichorium intybus*, known also as blue-sailors), the oyster-plant or salsify (*Tragopogon porrifolius*), the dandelion (*Taraxacum taraxacum* = *T. densleonis*), the lettuce (Lactuca), the hawkweed (Hieraceum), and others.

- The ragweed family (Ambrosiaceæ) includes the ragweeds (Ambrosia), the cockle-bur (Xanthium), and others.
- The thistle family (Compositæ) includes the thistle (Carduus), asters (Aster), goldenrods (Solidago), sunflowers (Helianthus), eupatoriums or joepye-weeds, thoroughworts (Eupatorium), cone-flowers or black-eyed Susans (Rudbeckia), tickseed (Coreopsis), bur-marigold or beggar-ticks or devil's-bootjack (Bidens), chrysanthemums, etc.

INDEX

A

B

C

D

E

F

G

H

I

J

K

L

M

N

O

P

T

U

V

W

X

Y

Z